This book covers the interplay between the behaviour of a class of stochastic processes (random walks) and structure theory. Written by leading researchers, this collection of invited papers presents links with spectral theory and discrete potential theory, besides probabilistic and structure theoretic aspects. Its interdisciplinary approach spans several areas of mathematics including geometric group theory, discrete geometry and harmonic analysis, and will be of interest to researchers and post-graduate students, both in mathematics and statistical physics.

ISTITUTO NAZIONALE DI ALTA MATEMATICA
FRANCESCO SEVERI

SYMPOSIA MATHEMATICA

Volume XXXIX

Random Walks and Discrete Potential Theory

Cortona 1997

Edited by

Massimo Picardello
Università di Roma - Tor Vergata

Wolfgang Woess
Università di Milano

CAMBRIDGE
UNIVERSITY PRESS

PUBLISHED BY THE PRESS SYNDICATE OF THE UNIVERSITY OF CAMBRIDGE
The Pitt Building, Trumpington Street, Cambridge, United Kingdom

CAMBRIDGE UNIVERSITY PRESS
The Edinburgh Building, Cambridge, CB2 2RU, UK www.cup.cam.ac.uk
40 West 20th Street, New York, NY 10011-4211, USA www.cup.org
10 Stamford Road, Oakleigh, Melbourne 3166, Australia
Ruiz de Alarcón 13, 28014 Madrid, Spain

First published 1999

Printed in the United Kingdom at the University Press, Cambridge

A catalogue record for this book is available from the British Library

ISBN 0 521 77312 1 hardback

Contents

Preface

This book contains the proceedings of the conference on "Random Walks and Discrete Potential Theory" that was held in Cortona, a town in southern Tuscany, Italy, in June 1997.

The scientific committee of the meeting consisted of V. A. Kaimanovich (Rennes), L. Saloff-Coste (Toulouse, now Ithaca), M. A. Picardello (Rome) and W. Woess (Milan), the latter two being the local organizers and editors of this volume. The papers presented here were written upon invitation of the editors, and underwent a process of (mild) refereeing. "Mild" because we are very happy to be able to unite a representative cross-section of recent work on Random Walks, Discrete Potential Theory, and several intimately related topics authored by such prominent contributors.

We are grateful to all organizations that directly or indirectly supported this conference. It was held at the "Palazzone", a 16th century castle close to Cortona. It is owned by the Scuola Nationale Superiore di Pisa, who rents it for several months every year to the INdAM - the Italian National Institute of Higher Mathematics. The meeting was patronized by the INdAM, which took care of the logistical infrastructure and the most substantial part of the funding. Significant financial support also came from the Italian National Research Council (CNR - GNAFA), the MURST project "Problemi nonlineari" coordinated by Prof. A. Fasano, and the Departments of Mathematics of the Universities of Milan and Rome II.

In some sense, we understand this conference as the continuation of a "tradition" that began in 1988 with a meeting, also held in Cortona, on "Harmonic Analysis on Trees and Graphs" (organized by M. A. Picardello and A. Figà-Talamanca), and was continued in the 1991 conference in Frascati on "Harmonic Analysis and Discrete Potential Theory" (organized by M. A. Picardello, editor of the proceedings that appeared in Plenum Press, NY, 1992) and the workshop on "Harmonic Functions on Graphs" (organized by A. Korányi, editor of the proceedings that appeared in Contemporary Mathematics, Vol. 206, 1997). The next event in this row can be expected in 2001: a special semester on "Random Walks" at the Schrödinger Institute in Vienna is in its preparatory phase.

May 1999 Massimo A. Picardello and Wolfgang Woess

Green's functions, generalized first eigenvalues and perturbation of diffusions or Markov chains

Alano Ancona

Abstract. We consider some questions related to the main results in [A4] about perturbations of diffusions in manifolds which preserve roughly the size of Green's functions. In the first section, a discrete analogue (as well as a stronger version for hyperbolic graphs) of Theorem 1 in [A4] is given. The proof follows the method there but it is much simpler here than in the diffusion case (when the generator $\mathcal{L}$ is in divergence form) because some special local features of the Potential theory with respect to such $\mathcal{L}$'s vanish in our case. In section 2, an extended form of the main results in [A4] for divergence form operators in manifolds is presented (we weaken the bounded geometry assumptions). This is applied in section 3 to perturbations in manifolds which have some (well-known) homogeneity properties (e.g. M has nonnegative Ricci curvature. See §3). In section 4, we first describe a simple procedure which makes it possible to derive some results in discrete settings from their continuous and local counterparts. This is then used to get a discrete analogue of the result in section 3.

1 Perturbation and comparability of Green's functions on graphs

Let X be a connected graph with finite valence, i.e. X is a set equipped with a symmetric relation $x \sim y$ such that $x \not\sim x$ and $V_x = \{y\,;\, y \sim x\}$ is finite with cardinality $\leqslant N_0$ for all $x \in X$ and some integer N_0; moreover the distance $d(x, y) = \inf\{m \geqslant 0\,;\, \exists\{z_j\}_{0\leqslant j\leqslant m}$ with $z_0 = x$, $z_m = y$, and $z_j \in X, z_j \sim z_{j+1}$ for $0 \leqslant j < m\}$ is finite in $X \times X$.

For $\theta \geqslant 1$, let $\mathcal{K}_\theta$ denotes the class of (adapted) kernels $P : X \times X \to \mathbb{R}_+$ such that : $P(x, y) = 0$ if $d(x, y) > \theta$, $\sum_{0\leqslant j\leqslant\theta} P^{(j)}(x, y) \geqslant \frac{1}{\theta}$ if $d(x, y) \leqslant 1$ and $P(x, y) \leqslant \theta$ for all $(x, y) \in X^2$. For each such kernel, there is a corresponding (nonlocal) potential theory in X, a function $s : X \to \overline{\mathbb{R}}_+$ being P-superharmonic iff $P(s) \leqslant s$ in X; when P is submarkovian, these functions are the excessive functions with respect to the Markov chain defined by P. A finite function s in X is called P-harmonic at $a \in X$ if $P(s)(a) = s(a)$. (ref. [Soa], [A2], [A3].)

Harnack inequalities in this setting take the following simple form. Let $c = \frac{1}{\theta(\theta+1)}$. For every $P \in \mathcal{K}_\theta$ and every P-excessive function s in X, $s(x) \geqslant c\, s(y)$ when $d(x, y) \leqslant 1$.

The Green function G_P associated to P is $G_P(x, y) = \sum_{0\leqslant j<\infty} P^{(j)}(x, y)$ for $x \in X$, $y \in X$. The pair (X, P) is transient -that is there are finite nonharmonic P-excessive functions- if and only if $G(x,y) < \infty$ for one (or for all) $(x,y) \in X \times X$. If P is submarkovian, this means that for the P random walk every state $x \in X$ is transient. If P is transient and if $a \in X$, then $s(x) = G(x,a)$ is a P- potential (i.e. s is excessive, finite and admits no positive harmonic minorant) and $(I-P)s = \delta_a$ in X (where $\delta_a = 1_{\{a\}}$).

For $P \in \mathcal{K}_\theta$, define the critical value (or generalized first eigenvalue) of P as $\lambda_1(P) := \sup\{t < 1; (1-t)^{-1}P \text{ is transient}\} = \sup\{t \leqslant 1; \exists s : X \to \mathbb{R}_+^*, Ps + ts \leqslant s\}$. From $P(1) \leqslant \theta(1+N_0)^\theta$, it follows that $\underline{\lambda}(N_0,\theta) = 1 - \theta(1+N_0)^\theta \leqslant \lambda_1(P) < 1$. If P is submarkovian or if P is transient, $\lambda_1(P) = \sup\{t \geqslant 0; P + tI \text{ is transient in } X\}$. For $\varepsilon > 0$, let $\mathcal{K}(\theta,\varepsilon) = \{P \in \mathcal{K}_\theta; \lambda_1(P) \geqslant \varepsilon\}$. This is the class of all ε-weakly coercive members P of $\mathcal{K}_\theta$, [A2].

Finally, for P_1 and P_2 in $\mathcal{K}_\theta$, set $\mathrm{dist}(P_1,P_2)(x) = \sum_{\{y;d(x,y)\leqslant\theta\}} |P_1(x,y) - P_2(x,y)|$, $x \in X$. Fix also a reference point $O \in X$ and let $d(x) = d(O,x)$ for $x \in X$.

Theorem 1.1. *Let P and Q be two members of $\mathcal{K}(\theta,\varepsilon_0)$, $\theta \geqslant 1$, $\varepsilon_0 > 0$. Assume that $\mathrm{dist}(P,Q)(x) \leqslant \Phi(d(x))$, $x \in X$, for some positive decreasing function Φ in $\mathbb{R}_+$ such that $I_\Phi = \Phi(0) + \int_0^{+\infty} \Phi(t)\,dt < \infty$. Then, there is a constant $C = C(N_0,\theta,\varepsilon_0,I_\Phi) \geqslant 1$ such that*

$$C^{-1}G_P(x,y) \leqslant G_Q(x,y) \leqslant C\,G_P(x,y) \quad \text{for } x \in X,\ y \in X.$$

Moreover, C may be chosen such that $\lim_{I\to 0} C(N_0,\theta,\varepsilon_0,I) = 1$ for fixed θ, ε_0, N_0.

Compare [A4], Theorem 1, and note the slight improvement regarding the dependence of C on I_Φ (see also theorem 2.6 below). We need an elementary continuity property of $\lambda_1(P)$. The proof in [A4] of the similar property for second order elliptic operators is more involved.

Proposition 1.2. *The function $\lambda_1 : \mathcal{K}_\theta \to \mathbb{R}$ is continuous, and in fact Lipschitz continuous with respect to the metric $\Delta(P,Q) = \sup_{x\in X} \mathrm{dist}(P,Q)(x)$.*

Proof. Denote by P_1, P_2 two elements of $\mathcal{K}_\theta$ and assume that $\lambda_1(P_1) > t \geqslant \underline{\lambda}(N_0,\theta)$. We show that there is a constant $c = c(X,\theta) > 0$ such that $\lambda_1(P_2) \geqslant t - c\,\Delta(P_1,P_2)$.

By the assumption on P_1 and t, there is a finite > 0 function s in X such that $(I-P_1)s \geqslant t\,s$. By Harnack inequalities (for the kernel $(1-t)^{-1}P_1$ if $t \leqslant 0$ and for P_1 otherwise), $s(y) \leqslant C(N_0,\theta)\,s(x)$ whenever $d(x,y) \leqslant \theta$. Hence,

$$s(x) - P_2 s(x) \geqslant t\,s(x) + (P_1 - P_2)s(x) \geqslant t\,s(x) - C(N_0,\theta)\,\Delta(P_1,P_2)\,s(x)$$

in X and $\lambda_1(P_2) \geqslant t - C\,\Delta(P_1,P_2)$.

Lemma 1.3. *Let $P \in \mathcal{K}_\theta$, let u and h be two positive functions in X, and let $f : (0,\infty) \to (0,\infty)$ be of class C^2. Set $v = \frac{u}{h}$, $w = h\, f(v)$ and $\Delta_P = P - I$. Then, the following identity holds for all $x \in X$:*

$$\Delta_P w(x) = (f(v(x)) - v(x)\, f'(v(x))\,)\,\Delta_P h(x) + f'(v(x))\Delta_P u(x)$$
$$+ \sum_{y \in X} \Big\{ P(x,y)h(y)(v(y) - v(x))^2 \big(\int_0^1 f''((1-t)v(x) + tv(y))(1-t)\,dt \big) \Big\}.$$

Proof. By Taylor's formula, $f(v(y)) = f(v(x)) + f'(v(x))[v(y) - v(x)] + R(x,y)$ where $R(x,y) = [v(y) - v(x)]^2 \int_0^1 f''(v(x) + t[v(y) - v(x)])\,(1-t)\,dt$, for x and y in X.

Thus, $w(y) = h(y)f(v(x)) + f'(v(x))[u(y) - v(x)h(y)] + h(y)\,R(x,y)$ and

$$Pw(x) = f(v(x))Ph(x) + f'(v(x))[Pu(x) - v(x)Ph(x)] + \sum_{y \in X} P(x,y)h(y)\,R(x,y).$$

Since $w(x) = f(v(x))\,h(x)$ and $u(x) - v(x)h(x) = 0$, the formula follows.

Lemma 1.4. *Let P_1 and P_2 be two members of $\mathcal{K}(\theta, \varepsilon_0)$, $\theta \geqslant 1$, $\varepsilon_0 > 0$, such that $P_1(x,y) = P_2(x,y)$ whenever $(x,y) \notin B_0 \times B_0$ where B_0 is a ball $B_0 = \overline{B}(a,R)$ in X. Then, the related Green's functions G^1 and G^2 are comparable, i.e. $G^1(x,y) \leqslant C\, G^2(x,y)$ for some constant $C \geqslant 1$ depending only on R, N_0, θ and ε_0.*

Proof. Recall that by the weak coercivity there is a constant C_1 such that $C_1^{-1} \leqslant G^j(x,x) \leqslant C_1$, $x \in X$, $j = 1,2$ ([A2]). By Harnack inequalities it follows that if $d(x,y) \leqslant 2R + 4\theta$, then $C_2^{-1}\, G^2(y,x) \leqslant G^1(y,x) \leqslant C_2\, G^2(y,x)$ for some $C_2 = C_2(N_0, \theta, \varepsilon_0, R) \geqslant 1$.

Consider then for $b \in \overline{B}(a, R + \theta)$, $u = \inf\{G^1(.,b), C_2^{-1}\, G^2(.,b)\}$; the function u is P_2 excessive (by the assumption on the P_j and the choice of C_2) and $(I - P_2)u \geqslant C_2^{-1}\delta_b$. It follows that $u \geqslant C_2^{-1}\, G^2(.,b)$, i.e. $G^1(x,b) \geqslant C_2^{-1} G^2(x,b)$ in X. Considering also the adjoint kernels, it is seen that $C_2^{-1} G^2(x,y) \leqslant G^1(x,y) \leqslant C_2\, G^2(x,y)$ if x or y is in $\overline{B}(a, R + \theta)$.

The above argument may be repeated now for any $b \in X$. The function $v = \inf\{G^1(.,b),\ C_2^{-1} G_2(.,b)\}\}$ is again P_2-excessive in X and coincides with $C_2^{-1}\, G_2(.,b)$ in $\overline{B}(b,\theta)$. Whence $G^1(.,b) \geqslant C_2^{-1}\, G_2(.,b)$. □

Proof of Theorem 1.1. A. We first prove that for every $\delta > 0$ there is an $\eta = \eta(\varepsilon_0, \theta, X) > 0$ such that whenever P_1 and P_2 are in $\mathcal{K}_\theta$ and such that dist $(P_1, P_2)(x) \leqslant \Phi(d(x))$, $x \in X$, for some continuous nonincreasing

$\Phi : \mathbb{R}_+ \to \mathbb{R}_+$ with $I_\Phi := \Phi(0) + \int_0^\infty \Phi(t)\,dt \leqslant \eta$ then the Green's functions G^1, G^2 of P_1, P_2, satisfy $G^2(x,a) \leqslant (1+\delta)\,G^1(x,a)$ for all $(a,x) \in X \times X$.

Since $\lambda_1(P_1) \geqslant \varepsilon_0$ we may choose and fix a (positive) $(P_1 + \varepsilon_0\, I)$-excessive function h in X with $h(O) = 1$. Let $a \in X$ and let $u = G^1(.,a)/G^1(O,a)$ be the P_1-Green kernel with pole at a and normalized at O.

Let $\tilde{\Phi}$ be continuous nonincreasing in $[0,\infty)$ such that $\tilde{\Phi}(t) = \Phi(0)$ for $t \leqslant \frac{1}{2}$ and $\tilde{\Phi}(t) = \Phi(t)$ for $t \geqslant 1$. Clearly $\int_0^\infty \tilde{\Phi}(t)dt \leqslant I_\Phi$. Set then $Q(t) = \frac{1}{t\sqrt{I_\Phi}}\tilde{\Phi}(\kappa^{-1}\,|\log(t)|)$ for $t > 0$ where κ is a constant $\geqslant 1$ to be fixed later depending only on X and θ (not on Φ), and consider the function $f(t) = \int_0^t \exp(-\int_0^s Q(\tau)\,d\tau)\,ds$, $t \geqslant 0$. Note that f is concave, nondecreasing and C^1 on $[0,\infty[$ with $f(0) = 0$, $f'(0) = 1$. Moreover f is a C^2 solution in $(0,\infty)$ of the differential equation $y''(t) + Q(t)y'(t) = 0$ and $f'(\infty) = \exp(-\int_0^\infty Q(s)\,ds) \geqslant \exp(-2\kappa\sqrt{I_\Phi}\,)$.

Consider now the positive function $w = hf(\frac{u}{h}) = hf(v)$, where $v = u/h$, and compute $(I - P_2)w$ using lemma 1.3. Since $(I - P_1)h \geqslant \varepsilon_0\, h$ and $(I - P_1)u(x) = |G^1(O,a)|^{-1}\delta_a(x)$,

$$\begin{aligned}(I-P_2)w(x) \geqslant \frac{f'(v(x))}{G^1(O,a)}\,\delta_a(x) + \varepsilon_0\, h(x)[f(v(x)) - v(x)f'(v(x))] \\ - \sum_{y\in X}\{P_2(x,y)h(y)(v(y)-v(x))^2\int_0^1 f''((1-t)v(x)+tv(y))(1-t)dt\} \\ + \big(f(v(x)) - v(x)f'(v(x))\big)(P_1 - P_2)h(x) + f'(v(x))(P_1 - P_2)u(x).\end{aligned} \tag{1.1}$$

Clearly the first three terms are positive, the first one being larger than $G^1(O,a)^{-1}\,f'(\infty)\,\delta_a$. To estimate from below the second term, we first observe that by Harnack inequalities and $v(O) = 1$ it follows that for some $A = A(\theta, X) > 0$ and all $x \in M$,

$$\exp(-A\,d(O,x)) \leqslant v(x) \leqslant \exp(A\,d(O,x)).$$

Since $f(s) - sf'(s) = -s^2\int_0^1 \tau\, f''(\tau s)\,d\tau = s^2\int_0^1 \tau\, Q(\tau s)\, f'(\tau s)\,d\tau$ is larger than $\frac{s^2}{4}f'(\infty)\ \times \min\{Q(t); s/2 \leqslant t \leqslant s\}$, it follows that for the second term T_2 in (1.1)

$$\begin{aligned}T_2 &:= \varepsilon_0\, h(x)[f(v(x)) - v(x)f'(v(x))] \\ &\geqslant \varepsilon_0\, h(x) \times \frac{v(x)^2}{4} \times \frac{f'(\infty)}{v(x)\sqrt{I_\Phi}} \times \tilde{\Phi}(\frac{A\,d(x) + \log(2)}{\kappa}) \\ &= \frac{\varepsilon_0}{4\sqrt{I_\Phi}} \times f'(\infty) \times u(x) \times \tilde{\Phi}(\frac{A\,d(x)+\log(2)}{\kappa}) \\ &\geqslant \frac{\varepsilon_0}{4\sqrt{I_\Phi}} \times \exp(-2\kappa\sqrt{I_\Phi}) \times u(x) \times \tilde{\Phi}(d(x)),\end{aligned}$$

for all $x \in X$ (even $x = O$), once we choose and fix $\kappa = 4(A+1)$.

On the other hand for the fourth term T_4 in the r.h.s of (1.1) we have, for another Harnack constant $c = c(\theta, \varepsilon_0, N_0) \geqslant 1$, (since $\tilde{\Phi} \leqslant \Phi(0) \leqslant I_\Phi$)

$$|T_4| \leqslant c\,|f(v(x)) - v(x)f'(v(x))|\,\tilde{\Phi}(d(x))\,h(x) = \frac{c}{\varepsilon_0}\,\tilde{\Phi}(d(x))\,T_2 \leqslant \frac{c}{\varepsilon_0}\,I_\Phi\,T_2.$$

Finally for the fifth term, using $f' \leqslant 1$ and Harnack inequalities,

$$|f'(v(x))\,(P_1 - P_2)u(x)| \leqslant c\,\tilde{\Phi}(d(x))\,u(x).$$

It follows from the above that if I_Φ is sufficiently small, (depending only on c_0, θ and X) the function w is P_2-superharmonic and such that $(I - P_2)w(a) \geqslant \exp(-2\kappa\sqrt{I_\Phi}\,)\,G^1(O,a)^{-1}\,\delta_a$. Thus, (using $f(t) \leqslant t$ and maximum principle) we have,

$$u(x) = \frac{G^1(x,a)}{G^1(O,a)} \geqslant w(x) \geqslant \exp(-2\kappa\sqrt{I_\Phi}\,)\,G^1(O,a)^{-1}\,G^2(x,a),$$

or $G^1(x,a) \geqslant \exp(-2\kappa\sqrt{I_\Phi}\,)G^2(x,a)$, $x \in X$, provided I_Φ is sufficiently small. The claim stated in the beginning of A above is proved.

B. To prove the first claim in Theorem 1.1, we use the following construction. Let R be a large positive number, let $\varphi(t) = \inf\{1, 2(1 - \frac{t}{2R})^+\}$, $t \geqslant 0$, and let

$$Q(x,y) = \varphi(d(x))\,P_1(x,y) + (1 - \varphi(d(x)))\,P_2(x,y),\ x \in X,\ y \in X.$$

If R is sufficiently large, $R \geqslant R_0(\theta)$, it is not hard to check that $Q \in \mathcal{K}(\theta')$ for some $\theta' = \theta'(N_0,\theta) > 1$ independent of R. This is because, for every ball $B = B(a,\theta^\theta)$, $a \in X$, either $Q(x,y) \geqslant \frac{1}{4}P_1(x,y)$ for all $(x,y) \in B \times X$ or $Q(x,y) \geqslant \frac{1}{4}P_2(x,y)$ for all $(x,y) \in B \times X$.

Since $\operatorname{dist}(P_1,Q)(x) \leqslant \psi_R(d(x)) := \inf\{\Phi(R/2), \Phi(d(x))\} \leqslant \frac{2}{R}I_\Phi$ (if $R \geqslant 2\theta$), it follows from proposition 1.2 that $Q \in \mathcal{K}(\theta', \varepsilon_0/2)$ if $R \geqslant R_1(N_0,\theta,\varepsilon_0,I_\Phi)$. By lemma 1.4 the Green's functions G and G_2 corresponding to the kernels Q and P_2 are then equivalent in size.

On the other hand, if $R \geqslant R_1$ is large, the first part of the proof shows that the Green's functions G and G_1 for Q and P_1 satisfy $\frac{1}{2}G_1 \leqslant G \leqslant 2\,G_1$. However, this last argument gives a comparison constant C for G_1 and G_2 depending also on the speed at which I_{ψ_R} vanishes as $R \to \infty$ (not only on I_Φ). To take care of this point, set $R = R_1(N_0,\theta,\varepsilon_0,I_\Phi)$ and introduce the kernels $S_j = (1 - \frac{j}{m})Q + \frac{j}{m}P_1$, $j = 0,\ldots,m$. Clearly, $\operatorname{dist}(P_1,S_j)(x) \leqslant \operatorname{dist}(P_1,Q)(x) \leqslant \Phi(R/2)$ and $S_j \in \mathcal{K}(\theta',\varepsilon_0/2)$ if θ' and R_1 are chosen sufficiently large (depending only on N_0, θ, ε_0 and I_Φ). Also, $\operatorname{dist}(S_j,S_{j+1})(x) \leqslant \frac{1}{m}\operatorname{dist}(Q,P_1)(x) \leqslant \frac{1}{m}\Phi(d(x))$. Thus, we may choose m depending only on N_0, θ , ε_0, I_Φ and such that, by part A above (and using obvious notations), $\frac{1}{2}G_{S_j} \leqslant G_{S_{j+1}} \leqslant 2G_{S_j}$, $0 \leqslant j < m$. The first part of Theorem 1.1. follows. □

Let us add that when the graph X is hyperbolic there is stronger version of Theorem 1.1 where the assumptions and the conclusion are localized in some regions in the graph.

Theorem 1.5. *Assume that X is Gromov-hyperbolic. Let Γ be a subset of X, let P_1, P_2 be members of $\mathcal{K}(\theta, \varepsilon_0)$, $\varepsilon_0 > 0$, such that dist$(P_1, P_2)(x) \leqslant \Phi(d(x))$ for $x \in \Gamma$ with Φ as in theorem 1.1 and let G^j denote the Green's function for P_j. Let ψ be some increasing function on $[0, \infty[$ such that $\psi(0) \geqslant 1$ and $\lim_{t\to\infty} \frac{\psi(t)}{\log(t)} = +\infty$ and let $\Gamma^0 = \{x \in \Gamma\,;\, \overline{B}(x, \psi(d(x))) \subset \Gamma\}$. Then, there is a constant $C \geqslant 1$ such that $C^{-1}G^1(x,y) \leqslant G^2(x,y) \leqslant C\,G^1(x,y)$ whenever there is a geodesic path in X connecting x to y and contained in Γ^0.*

See [A2], [A3] for definitions, and Potential theoretic results about hyperbolic graphs. Theorem 1.5 is similar to Theorem 3 in [A4] §8, but we have tried to improve the statement (the similar improvement holds in the continuous case). The proof is similar to [A4] and rests on the following : there is a constant $A = A(\theta, \varepsilon_0, X) \geqslant 1$ such that if $P \in \mathcal{K}(\theta, \varepsilon_0)$, $\varepsilon_0 > 0$, if $x_1, \ldots, x_m$ is a "geodesic" path in X, and $U = \bigcup_{1 \leqslant j \leqslant m} B(x_j, A\log(j+2))$, then $R^{X\setminus U}_{G(.,x_1)}(x_m) \leqslant \frac{1}{2}G(x_m, x_1)$, where G is the Green's function with respect to P and $R^{X\setminus U}_{G(.,x_1)}$ is the réduite ([B], [A2]) of $G(., x_1)$ over $X \setminus U$.

Sketch of proof of theorem 1.5. Observe that by hyperbolicity, Harnack inequalities and Harnack principle at infinity for P_j, P_j^* ([A2]), we may assume that x belongs to a "geodesic" $(a_0, a_1, \ldots, a_m)$ connecting $O = a_0$ to $y = a_m$ and such that $\gamma = (a_k, \ldots, a_m)$, $x = a_k$, $1 < k < m$. We may also assume that k is large. Let $U = \bigcup_{k \leqslant j \leqslant m} \overline{B}(a_j, \psi(j))$, let Q be the kernel in X such that $Q(x,y) = P_2(x,y)$ if $d(x, X\setminus U) \leqslant 2\,\theta^\theta$, $Q(x,y) = \frac{1}{2}(P_1(x,y) + P_2(x,y))$ if $2\,\theta^\theta < d(x, X\setminus U) \leqslant 4\,\theta^\theta$ and $Q(x,y) = P_1(x,y)$ otherwise. Then, Q is in some class $\mathcal{K}(\theta', \varepsilon_0/2)$ (if k is large, thanks to proposition 1.2), and $G_Q \approx G_{P_2}$ by theorem 1.1. On the other hand, the function $u = G_{P_1}(., a_k) - [G_Q(., a_k) - R^A_{G_Q(.,a_k)}]$, -where $A = \{x\,;\, d(x, X\setminus U) < 8\,\theta^\theta\}$ and the réduite is performed with respect to Q- is positive outside U and P_1-superharmonic in X. Thus, $u \geqslant 0$ by a standard maximum principle and $G_{P_1}(a_m, a_k) \geqslant \frac{1}{2}G_Q(a_m, a_k)$ if k is large. Exchanging P_1 and Q, the same argument gives $G_{P_1}(a_m, a_k) \leqslant 2\,G_Q(a_m, a_k)$. Theorem 1.5 follows.

Example 1.6. Let X be a subset of the unit ball U of $\mathbb{R}^N$ such that $d_h(x, X) \leqslant c_0$ for $x \in U$ and $d_h(x, x') \geqslant c_1$ for x, x' in X, $x \neq x'$, where d_h is the standard hyperbolic distance in U and c_j, $j = 1, 2$ are positive constants. Define a graph structure in X on setting $x \sim y$ iff $0 < d_h(x,y) \leqslant$

$5c_0$. Let $\zeta \in \partial U$ and let $\Gamma = \{x \in X\,;\, d_h(x, \overline{O\zeta}) \leqslant |\log^+(d_h(O,x))|^2\}$, $\Gamma_0 = \{x \in \Gamma\,;\, d_h(x, X\setminus\Gamma) \geqslant |\log^+(d_h(O,x))|^{\frac{3}{2}}\}$. Then, if P and Q belong to $\mathcal{K}_X(\theta,\varepsilon_0)$, $\varepsilon_0 > 0$, and if $|P(x,y) - Q(x,y)| \leqslant \Phi(d_h(O,x))$ when $x \in \Gamma$, $y \in X$, with Φ nonincreasing and $\int_0^\infty \Phi(t)\,dt < \infty$, then the corresponding Green's kernels G_P, G_Q satisfy for some constant $c \geqslant 1$: $c^{-1}G_P(x,y) \leqslant G_Q(x,y) \leqslant c\,G_P(x,y)$, x, y in Γ_0.

2 The case of elliptic operators in divergence form in Riemannian manifolds

In this section the continuous versions given in [A4] of Theorem 1.1 and proposition 1.2 are put in a more general form. This will turn out to be useful in section 3.

2.1. Let M be a complete noncompact Riemannian manifold and let σ denote its riemannian volume. It is assumed that M has bounded geometry in the following sense (see [SC1]-[SC3], [MSC]): there are positive reals r_0, c_V and c_P such that for all $a \in M$ and all $r \leqslant r_0$:

(DV) (doubling volume condition) $|B(a,2r)| \leqslant c_V\,|B(a,r)|$,

(PI) (scaled L^2 Poincaré inequalities) For every $u \in C_0^1(M)$,
$$\int_{B(a,r)} |u(y) - \overline{u}_{B(a,r)}|^2\,d\sigma(y) \leqslant c_P\,r^2 \int_{B(a,2r)} |\nabla u(y)|^2\,d\sigma(y),$$

where $|A| = \sigma(A)$ and $\overline{u}_A = |A|^{-1}\int_A u\,d\sigma$ for $A \subset M$. Denote by (PI_1) the (stronger) property that for all $u \in C_0^1(M)$, $r \leqslant r_0$, $a \in M$ and some constant c_p^1 : $\int_{B(a,r)} |u(y) - \overline{u}_{B(a,r)}|\,d\sigma(y) \leqslant c_P^1\,r \int_{B(a,2r)} |\nabla u(y)|\,d\sigma(y)$.

We fix a real $N > 2$ such that, for some $c > 0$, $|B(a,r)| \leqslant c(\frac{r}{r'})^N|B(a,r')|$ if $a \in M$, $0 < r' < r < r_0$ (thus $N \geqslant \dim(M)$). Any $N \geqslant \log c_V/\log 2$ with $N > 2$ would do. It is well-known (ref. [MSC]) that M has the following basic properties :

(WTC) (Whitney type coverings) For each $r \in (0,r_0]$, there is a cover of M by balls $B(a_j,r)$, $j \geqslant 1$ such that $\sum_{j\geqslant 1} 1_{B(a_j,5r)}(x) \leqslant C$ in M, for some constant $C = C_M(r_0) \geqslant 1$.

(SB) (Sobolev inequality) If $f \in C_0^1(M)$ is supported in $B(a,r)$, $a \in M$, $r \leqslant r_0$, then
$$\left(\int |f|^{\frac{2N}{N-2}}\,d\sigma\right)^{\frac{N-2}{2N}} \leqslant C_S\,r\,|B(a,r)|^{-\frac{1}{N}}\sqrt{\int |\nabla f|^2\,d\sigma}$$

where $C_S = C_M(r_0,N) > 0$; in particular, $\int |f|^2\,d\sigma \leqslant C_S^2\,r^2 \int |\nabla f|^2\,d\sigma$. (See [SC1] Theorem 2.1 and [MSC] Lemma 7.4.) The "usual" Poincaré-Sobolev inequalities (with the same exponents as above) are also valid, [MSC], Corollaire 5.2.

Recall also that if the Ricci curvature of M is bounded from below then (DV), (PI) (and (PI_1)) hold with $r_0 = 1$ and we may let $N = \sup\{\dim(M), 2+\varepsilon\}$, $\varepsilon > 0$.

2.2. Let θ and p be real numbers such that $N < p \leqslant \infty$, $1 \leqslant \theta < \infty$. Denote by $\mathcal{D}_M(\theta, p)$ the class of all elliptic operators $\mathcal{L}$ on M with a *given* representation in the form

$$\mathcal{L}u = \operatorname{div}(\mathcal{A}(\nabla u)) + D.\nabla u + \operatorname{div}(u\, D') + \gamma u, \tag{2.1}$$

where $\mathcal{A} : x \mapsto \mathcal{A}_x$ is a Borel section of the bundle $\operatorname{End}(T(M))$, D and D' are Borel vector fields on M, and γ is a real valued Borel function in M. Moreover,

$$\theta^{-1}\, |\xi|^2 \leqslant \langle \mathcal{A}_a(\xi), \xi\rangle \leqslant \theta\, |\xi|^2\,, \tag{2.2}$$

(2.3)
$$\|\mathcal{A}_a\|_{End(T_a(M))} + |B_a|^{-\frac{1}{p}}[\|D\|_{L^p(B_a)} + \|D'\|_{L^p(B_a)}] + |B_a|^{-\frac{2}{p}}\, \|\gamma\|_{L^{\frac{p}{2}}(B_a)} \leqslant \theta,$$

for $a \in M$ and $\xi \in T_a(M)$, $B_a = B(a, r_0)$. This differs from the definition of $\mathcal{D}_M(\theta, p)$ in [A4] (but is equivalent to it, under the stronger bounded geometry assumptions made therein).

2.3. If U is a region in M, define the Sobolev space $H^1(U)$ as the space of all $f \in L^2(U)$ with a weak gradient in $L^2(U)$, equipped with the obvious norm. $H_0^1(U)$ is the closure of $C_0^1(U)$ in $H^1(U)$. To $\mathcal{L}$ as above corresponds the local Dirichlet form $a_{\mathcal{L}}(u,v) = \int\{\mathcal{A}(\nabla u).\nabla v - v\,D.\nabla u + u\,D'.\nabla v - \gamma uv\}\, d\sigma$, for u, v in $H^1(M)$ and say compactly supported.

As well-known, each $\mathcal{L}$-solution u (of class H^1_{loc}) in a region $U \subset M$ "is" continuous in U and when u is nonnegative the following Harnack inequalities hold :

$$c^{-1}\, u(a) \leqslant u(x) \leqslant c\, u(a) \tag{2.4}$$

if $B(a,r) \subset U$, $r \leqslant r_0$ and $d(x,a) \leqslant r/2$, where $c = c_M(r_0, \theta, p) \geqslant 1$. Moreover, there are positive constants c' and β depending on r_0, θ, p and M such that

$$(1 + c'(\rho/r)^\beta)^{-1}\, u(a) \leqslant u(x) \leqslant (1 + c'(\rho/r)^\beta)\, u(a) \tag{2.5}$$

if $d(x,a) \leqslant \rho \leqslant r/2$. This follows from known extensions of Moser's method (e.g. [SE2], [GT] chap. 8) which are easily adapted to our case; see [ST] for the needed version of the John-Nirenberg theorem. See also [SC3]. There is also a useful form of Harnack inequalities for supersolutions. Namely, if u is $\mathcal{L}$-superharmonic and $\geqslant 0$ in $B(a,r)$, $a \in M$, $0 < r < r_0$,

$$u(a) \geqslant c\,|B(a,\frac{r}{2})|^{-1} \int_{B(a,\frac{r}{2})} u(x)\,d\sigma(x)\,, \quad c = c(M, r_0, \theta, p) > 0. \tag{2.6}$$

Finally, recall the available version of Caccioppoli's inequality : if $u \in H^1(B_a(r))$, $r \leqslant r_0$, is a nonnegative $\mathcal{L}$-subsolution in $B(a,r)$, then, for some $c = c_M(r_0, \theta, p) > 0$,

$$\int_{B(a,r/2)} |\nabla u|^2\,d\sigma \leqslant \frac{c}{r^2} \int_{B(a,r)} u^2\,d\sigma. \tag{2.7}$$

2.4. The *critical level* $\lambda_1(\mathcal{L})$ (or generalized first eigenvalue) of $\mathcal{L} \in \mathcal{D}_M(\theta,p)$ is the supremum of all $t \in \mathbb{R}$ such that $\mathcal{L}+tI$ has a Green's function in M. For $\varepsilon_0 > 0$, denote by $\mathcal{D}_M(\theta,p,\varepsilon_0)$ the set of all $\mathcal{L} \in \mathcal{D}_M(\theta,p)$ with $\lambda_1(\mathcal{L}) \geqslant \varepsilon_0$ (i.e. $\mathcal{L}$ is "ε_0-weakly coercive", [A1]).

If $\mathcal{L} \in \mathcal{D}_M(\theta,p,\varepsilon_0)$, there is a Green's function G for $\mathcal{L}$ and under the strong "bounded geometry" assumptions in [A4], $c^{-1} \leqslant G(x,y) \leqslant c$ for x, y in M such that $d(x,y) = r_0$ and some $c = c_M(r_0,\theta,p,\varepsilon_0) > 0$ (see [A1]). (Our convention will be that if $G = G^U$ is $\mathcal{L}$-Green's function in U, $G_y : x \mapsto G(x,y)$ satisfies $\mathcal{L}(G_y) = -\delta_y$ in the weak sense). In the present setting we have the following substitute.

Proposition 2.1. *There is a constant $C = C_M(r_0,\theta,p,\varepsilon_0) > 0$ such that the Green's function G of $\mathcal{L} \in \mathcal{D}_M(\theta,p,\varepsilon_0)$ satisfies, for x and y in M with $d(x,y) = r_0$,*

$$C^{-1}\,|B(x,r_0)|^{-1} \leqslant G(x,y) \leqslant C\,|B(x,r_0)|^{-1}. \tag{2.8}$$

We first quote the following useful fact.

Lemma 2.2. *There are numbers $\rho = \rho_M(r_0,\theta,p)$, $0 < \rho < r_0$, and $c = c_M(\theta,p,r_0) > 0$ such that for all $\mathcal{L} \in \mathcal{D}_M(\theta,p)$, $b \in M$ and $u \in H_0^1(B(b,\rho))$,*

$$a_{\mathcal{L}}(u,u) \geqslant c\,\|u\|^2_{H_0^1(B_b(\rho))}.$$

In other words the Dirichlet forms $a_{\mathcal{L}}$, $\mathcal{L} \in \mathcal{D}_M(\theta,p)$, are "uniformly" locally coercive.

Proof. If $\alpha^{-1} = \frac{Np}{p-N}$ and $V_\rho = \sigma(B_b(\rho))$, we have using (2.1), (SB) and

(DV),

$$
\begin{aligned}
|\int u\,(D.\nabla u)\,d\sigma| &\leqslant \|u\|_{\frac{2N}{N-2}}\,\|D\|_{L^N(B_b(\rho))}\,\|\nabla u\|_2\\
&\leqslant C_S\,\rho\,V_\rho^{-\frac{1}{N}}\,\|u\|^2_{H^1_0(B_b(\rho))}\,\|D\|_{L^p(B_b(\rho))}\,V_\rho^\alpha\\
&= c\,\rho\,V_\rho^{-\frac{1}{p}}\,\|u\|^2_{H^1_0(B_b(\rho))}\,\|D\|_{L^p(B_b(\rho))}\\
&\leqslant c\,\theta\,\rho\,[\frac{V_{r_0}}{V_\rho}]^{\frac{1}{p}}\,\|u\|^2_{H^1_0(B_b(\rho))} \leqslant c'\,\theta\,\rho^{1-\frac{N}{p}}\,\|u\|^2_{H^1_0(B_b(\rho))}\\
&\leqslant c\,\theta\,\rho\,[\frac{V_{r_0}}{V_\rho}]^{\frac{1}{p}}\,\|u\|^2_{H^1_0(B_b(\rho))} \leqslant c'\,\theta\,\rho^{1-\frac{N}{p}}\,\|u\|^2_{H^1_0(B_b(\rho))}.
\end{aligned}
$$

Thus, $\int u\,(D.\nabla u)\,d\sigma \leqslant \frac{1}{4\theta}\|u\|^2_{H^1_0(B_b(\rho))}$ for ρ small enough. A similar estimate holds for $\int u\,(D'.\nabla u)\,d\sigma$ and

$$
\begin{aligned}
|\int \gamma u^2\,d\sigma| &\leqslant \|\gamma\|_{L^{N/2}(B_b(\rho))}\,\|u\|^2_{L^{2(N-2)/N}(B_b(\rho))}\\
&\leqslant [V_\rho^{2(\frac{1}{N}-\frac{1}{p})}\theta\,V_{r_0}^{2/p}]\,[c\,V_\rho^{-\frac{2}{N}}\,\rho^2\,\|u\|^2_{H^1_0(B_b(\rho))}]\\
&\leqslant c'\,\theta\,\rho^{2(1-\frac{N}{p})}\,\|u\|^2_{H^1_0(B_b(\rho))}.
\end{aligned}
$$

Since $p > N$, the lemma follows. □

Remark 2.3. The proof also shows that $a_{\mathcal{L}}(\varphi,\psi) \leqslant c\,\|\varphi\|_{H^1_0(M)}\,\|\psi\|_{H^1_0(M)}$ for φ and ψ in $H^1_0(B_b(r_0))$, $\mathcal{L} \in \mathcal{D}_M(\theta,p)$ and $c = c_M(\theta,p,r_0) > 0$. This continuity estimate extends to all φ and ψ in $H^1(M)$. (See the comments in the end of this section and [A4], §2.)

Proof of Proposition 2.1. a. As in [A1], we write, using the resolvant equation $G^t = G + tG^t \circ G$ (where G^t is Green's function for $\mathcal{L} + tI$), for $0 < t < \varepsilon_0$,

$$\int G^t(\varphi)\,\varphi\,d\sigma = \int \varphi\,G(\varphi)\,d\sigma + t\int G(\varphi)\,\overset{*}{G^t}(\varphi)\,d\sigma$$

where $\varphi = 1_{B_x(r_0/2)}$ and $\overset{*}{G^t}$ is the adjoint of G. Since $\int G^t(\varphi)\,\varphi\,d\sigma < \infty$, it follows that $\int \overset{*}{G^t}(\varphi)(\varphi - tG(\varphi))\,d\sigma$ is positive so that there is an $x' \in B_x(r_0/2)$ such that $G(\varphi)(x') \leqslant t^{-1}$. In particular, by (DV) and by Harnack principle (2.4), (2.6), we have $G\varphi(y) \leqslant c\,t^{-1}$. Finally, by Harnack inequalities for the adjoint of $\mathcal{L}$, $G(y,x) \leqslant c\,t^{-1}\,|B(x,r_0)|^{-1}$.

b. To prove the lower bound, it is sufficient to show an estimate $g(a,y) \geqslant c\,|B(a,r_0)|^{-1}$ for the Green's function g of $\mathcal{L}$ in $B(a,\rho)$, with ρ as in Lemma 2.2, and $y \in \partial B(a,\rho/2)$. Let $u = g_a$, $m = \sup\{u(z);\, d(a,z) = \rho/2\}$,

$v = R^{\partial B(a,\rho/2)}_{u}$ (the réduite [B] is taken with respect to $\mathcal{L}$ in $B(a,\rho)$), and pick $\varphi \in C_0^1(B(a,\rho))$ with $0 \leqslant \varphi \leqslant 1$, $\varphi = 1$ in $\overline{B}(a,\rho/2)$, $|\nabla\varphi| \leqslant \frac{1}{3\rho}$.

By remark 2.3, and since $-\mathcal{L}v$ is (as a distribution in $B(a,\rho)$) a positive measure μ_a supported by $\partial B(a,\rho/2)$, we have $a_{\mathcal{L}}(v,v) \leqslant m\, a_{\mathcal{L}}(v,\varphi) \leqslant m\, c\,\|v\|\,\|\varphi\|$ (where $\|.\|$ is the H_0^1 norm); whence, by lemma 2.2, $a_{\mathcal{L}}(v,v) \leqslant c'\, m^2\, |B(a,\rho)|$. On the other hand, it is well known that μ_a is the $\mathcal{L}^*$ harmonic measure of a in $B(a,\rho/2)$. Thus, $a_{\mathcal{L}}(v,v) = \int v\, d\mu_a \geqslant \frac{1}{c'}m\,\|\mu_a\|$ (by Harnack). Since $\mathcal{L}^*$ is transient in $B(a,\rho)$, there is a positive $\mathcal{L}^*$-harmonic function h in $B(a,\rho)$ such that $h(a) = 1$. By Harnack, $\frac{1}{c''} \leqslant h(z) \leqslant c''$ on $\partial B_a(\frac{\rho}{2})$ and $\|\mu_a\| \geqslant \frac{1}{c''}$ since $1 = \int h\, d\mu_a$. Hence $a_{\mathcal{L}}(v,v) \geqslant C\, m$ and by the above $m \geqslant \frac{C'}{|B(a,\rho)|}$. The proposition is proven. □

The argument with μ_a and h shows in fact that $C^{-1}\, m a_{\mathcal{L}}(v,v) \leqslant C\, m$; it may also be used to prove the following (compare [Sta], §9) -where we use quasi-everywhere inequalities with quasi-continuous versions of the elements of $H_0^1(B)$, ref. [Den]. See also [Sta], definition 1.1.

Lemma 2.4. *Let $B = B(a,\rho)$ be a ball in M with $\rho = \rho_M(r_0,\theta,p)$ as in lemma 2.2. Then, for $\mathcal{L} \in \mathcal{D}_M(\theta,p)$, the Green's function g in B with respect to $\mathcal{L}$ satisfies, for $r = d(x,a) \leqslant \rho/2$ and some $c = c_M(r_0,\theta,p) \geq 1$,*

$$c^{-1} \leqslant cap(B(a,r))\, g(x,a) \leqslant c$$

where $cap(A) = \inf\{\|u\|^2_{H_0^1(B)}\,;\, u \in H_0^1(B), u \geqslant 1$ q.e. in $A\}$ for $A \subset B$.

Proof. Let now $v = R^{\partial B(a,r)}_{g(.,a)}$ and set $g(r) = \sup\{g(y,a);\, d(y,a) = r\}$. As above, we have $c^{-1}g(r) \leqslant a_{\mathcal{L}}(v,v) \leqslant c\, g(r)$, $c = c_M(\theta,p)$. On the other hand, v is the Stampacchia $a_{\mathcal{L}}$-projection of 0 on $\Gamma = \{\varphi \in H_0^1(B);\, \varphi \geqslant g(.,a)$ q.e. in $\partial B(a,r)\}$, i.e. $a_{\mathcal{L}}(v,v) \leqslant a_{\mathcal{L}}(v,\varphi)$ for $\varphi \in \Gamma$. By continuity and coercivity of $a_{\mathcal{L}}$ and since $v \approx g(r)$ in $\partial B(a,r)$, it follows that $\|v\| \leqslant c\, g(r)\sqrt{\mathrm{cap}(\partial B(a,r))}$. Clearly $\|v\| \geqslant c^{-1}\, g(r)\sqrt{\mathrm{cap}(\partial B(a,r))}$. The lemma follows. □

Remark 2.5. It easily follows from Lemma 2.4 and Proposition 2.1 that if G_j are the Green's functions of $\mathcal{L}_j$ in M, $\mathcal{L}_j \in \mathcal{D}_M(\theta,p,\varepsilon_0)$, $p > N$, $\varepsilon_0 > 0$, $j = 1,2$, then $c^{-1}\, G_1(x,y) \leqslant G_2(x,y) \leqslant c\, G_1(x,y)$ for $d(x,y) \leqslant r_0$, $c = c_M(r_0,\theta,p,\varepsilon_0)$.

2.5. Fix now a reference point $O \in M$ and set $d(x) = d(O,x)$, $x \in M$. If $q \in [1,+\infty]$ and if $\mathcal{L}_1$ and $\mathcal{L}_2$ are members of some class $\mathcal{D}_M(\theta,p)$ with representations $\mathcal{L}_j(u) = \mathrm{div}(A_j(\nabla u)) + D_j.\nabla u + \mathrm{div}(u\, D_j') + \gamma_j u$, set for

$a \in M$, $B_a = B(a, r_0)$ and

$$\mathrm{dist}_q[\mathcal{L}_1, \mathcal{L}_2](a) = \|\mathcal{A}_1 - \mathcal{A}_2\|_{L^q(B_a)} |B_a|^{-\frac{1}{q}} + |B_a|^{-\frac{1}{p}}[\|D_1 - D_2\|_{L^p(B_a)} + \|D_1' - D_2'\|_{L^p(B_a)}] + |B_a|^{-\frac{2}{p}} \|\gamma_1 - \gamma_2\|_{L^{\frac{p}{2}}(B_a)}, \tag{2.9}$$

(p is implicit in the l.h.s.). We may then extend theorems 1 and 2 in [A4] as follows.

Theorem 2.6. *Let $\mathcal{L}_1$ and $\mathcal{L}_2$ be elements of $\mathcal{D}_M(\theta, p, \varepsilon_0)$ (with $p > N$ and $\varepsilon_0 > 0$) and let G^1 and G^2 be the corresponding Green's functions. If $\mathrm{dist}_\infty(\mathcal{L}_1, \mathcal{L}_2)(x) \leqslant \Psi(d(x))$ in M for some decreasing function Ψ on $[0, \infty)$ such that $I_\psi := \int_0^{+\infty} \Psi(s)\, ds < +\infty$, then*

$$c^{-1}\, G^2(x,y) \leqslant G^1(x,y) \leqslant c\, G^2(x,y), \quad x \in M,\, y \in M, \tag{2.10}$$

for some constant $c = c_M(\theta, p, \varepsilon_0, I_\psi) > 0$. Moreover, if (PI$_1$) holds, $\mathrm{dist}_\infty(\mathcal{L}_1, \mathcal{L}_2)$ may be replaced above by $\mathrm{dist}_{q_0}(\mathcal{L}_1, \mathcal{L}_2)$, for some $q_0 \in [1, +\infty[$ depending only on c_V, c_P^1, θ and p.

Finally, for each $\delta > 0$ there exists $\eta = \eta_M(r_0, \theta, p, \varepsilon_0, \delta) > 0$ such that we may let $c = 1 + \delta$ in (2.10) if we restrict to x, y such that $d(x,y) \geqslant r_0$ and if $\int_0^{+\infty} \Psi(s)\, ds \leqslant \eta$.

Theorem 2.7. *Let $\theta > 0$, $p > N$ be fixed. For every $\delta > 0$ there is number $\eta > 0$ such that if $\mathcal{L}_1$, $\mathcal{L}_2 \in \mathcal{D}_M(\theta, p)$ and $\mathrm{dist}_1(\mathcal{L}_1, \mathcal{L}_2) \leqslant \eta$ on M, then*

$$|\lambda_1(\mathcal{L}_1) - \lambda_1(\mathcal{L}_2)| \leqslant \delta. \tag{2.11}$$

In fact, λ_1 is Lipschitz continuous in $\mathcal{D}_M(\theta, p)$ with respect to the distance $d(\mathcal{L}, \mathcal{L}') = \|\mathrm{dist}_{q_0}(\mathcal{L}, \mathcal{L}')\|_{\infty, M}$. The next criterion for $\lambda_1(\mathcal{L}) > 0$ follows from theorem 2.7.

Corollary 2.8. *Let $\mathcal{L}_1 \in \mathcal{D}_M(\theta, p, \varepsilon_0)$ and $\mathcal{L}_2 \in \mathcal{D}_M(\theta, p)$ (with $p > N$ and $\varepsilon_0 > 0$). There is a number $\delta > 0$ depending only on M, θ, p and ε_0, such that if*

(i) there is a Green's function in M for $\mathcal{L}_2$,

(ii) $\mathrm{dist}_1(\mathcal{L}_1, \mathcal{L}_2) \leqslant \delta$ outside some compact subset K of M, (e.g. $\mathrm{dist}_1(\mathcal{L}_1, \mathcal{L}_2)(x)$ tends to zero when $d(x) \to +\infty$),

then $\lambda_1(\mathcal{L}_2) > 0$. Moreover, condition (i) above can be dropped if $\mathcal{L}_j(1) = 0$ for $j = 1$, 2.

In the case $\mathcal{L}_1(1) = \mathcal{L}_2(1) = 0$, it may be seen that $\lambda_1(\mathcal{L}_2) \geqslant \varepsilon$ for some $\varepsilon > 0$ depending only on M, r_0, K, θ, p and ε_0. See [A4].

To prove Theorem 2.6 and 2.7 one may follow the method in [A4] : it turns out that (using the preliminary material above) the proofs and their different steps there extend with only very mild changes; so, we only briefly describe these in what follows. In the preliminary section 2 of [A4], Lemma 2.3 (multiplier lemma) and Lemma 2.5 extend with no change in their proofs whereas the proof of Proposition 2.1 and Corollary 2.2 require few simple changes. In the statement of Lemma 2.4 one should add in the r.h.s. of the estimate the factor $\sigma(B(a,\rho))$. The remaining parts (sections 3, 4, 5 in [A4]) of the proof extends smoothly as well (the anounced dependence on Ψ of the constant c in theorem 2.6 is obtained on using the final argument in the proof of theorem 1.1 above). For example, a factor $\sigma(B(a,r_0))^{\frac{1}{2}}$ must be put in (3.4) (statement of proposition 3.1) and similarly at the end of the proposition 3.3. Also, Lemma 5.1 is still valid as it stands, but it is best to replace in the statement A by $A/\sqrt{\sigma(B(a,r_1))}$ so that the proof extends almost word by word. Finally, the extension to our framework of a theorem of Boyarskii-Meyers (see e.g. [Me], [Gia]) is needed for the second claim in theorem 2.6: namely, $\|\nabla u\|_{L^{2+\varepsilon}(B_a)} \leqslant cu(a)$ for u positive $\mathcal{L}$-harmonic in $B(a,2r_0)$, $\varepsilon = \varepsilon_M(r_0,\theta,p) > 0$, $\mathcal{L} \in \mathcal{D}_M(\theta,p)$, $c = c_M(r_0,\theta,p)$. By a standard procedure [Gia], this may be derived (see §5.4) from the following "local" version of a theorem of Gehring [Ge]. See [Gia] for the case $M = \mathbb{R}^N$ and §5, for the simple extension to our framework.

Theorem 2.9. *Let $q > 1$, $0 < R < r_0/2$, $a \in M$, $B = B(a,R)$ and $f \in L^q(B)$ be such that*

$$\oint_{B(x,r)} |f|^q\, d\sigma \leqslant C\,[(\oint_{B(x,2r)} |f|\, d\sigma)^q + \oint_{B(x,2r)} |g|^q\, d\sigma] + \kappa \oint_{B(x,2r)} |f|^q\, d\sigma \tag{2.12}$$

whenever $B(x,2r) \subset B$, for some constants $C \geqslant 1$, $0 \leqslant \kappa < 1$, and a function g in $L^{p_0}(B)$, $q < p_0 < \infty$. Then, there exists $p = p(c_V, q, p_0, C, \kappa)$ in (q,p_0) such that

$$(\oint_{B(a,\frac{R}{2})} |f|^p\, d\sigma)^{\frac{1}{p}} \leqslant c\,[(\oint_B |f|^q\, d\sigma)^{\frac{1}{q}} + (\oint_B |g|^{p_0}\, d\sigma)^{\frac{1}{p_0}}]$$

where $c = c_M(q, p_0, \kappa, C) \in \mathbb{R}_+$.

We only need here the case $\kappa = 0$. ($\oint_A$ denotes the mean value over A with respect to σ.)

3 "Homogeneous" manifolds with volume growth of degree > 2

In this section, M denotes a noncompact complete Riemannian manifold which is homogeneous in the sense that it satisfies (DV) and (PI) in section 2 for *all* $r_0 > 0$ and some fixed constants c_V, c_P ([SC1]. See also [MSC]).

We fix $c_0 \geqslant 1$ and $N > 2$ such that $|B(x,r)| \leqslant c_0 \left(\frac{r}{r'}\right)^N |B(x,r')|$ for $x \in M$, $0 < r' < r < \infty$. It is moreover assumed that with respect to the chosen reference point $O \in M$ the following growth condition is satisfied for some $\nu > 2$:

$$(\mathrm{Gr}(\nu)) \qquad \text{if } 1 \leqslant s \leqslant r, \quad \text{then} \quad |B(O,r)| \geqslant c_\nu \left(\tfrac{r}{s}\right)^\nu |B(O,s)|,$$

where $c_\nu > 0$ is some constant. Our aim is to prove a version of Theorem 2.6 which takes into account the "global" Harnack inequalities which are available in this setting [SC1]-[SC3] and to extend known results for the case $M = \mathbb{R}^N$ ([Pi1], [Pi2], [A4] §6).

We first prove a Hardy-Sobolev type inequality. Recall that $d(x) = d(O,x)$ for $x \in M$.

Proposition 3.1. *There is a constant $c = C(c_V, c_P, \nu, c_\nu) > 0$ such that for each $f \in C_0^1(M)$*

$$\int_M \frac{|f(x)|^2}{1+d(x)^2}\, d\sigma(x) \leqslant c \int_M |\nabla f|^2 \, d\sigma(x). \tag{3.1}$$

Remark 3.2. (3.1) implies that M is transient (see e.g. [A3]); thus the assumption that $\nu > 2$ cannot be relaxed. In fact, (3.1) implies that for $0 \leqslant \varepsilon < c^{-1}$, the elliptic operator $L_\varepsilon = \Delta_M + \frac{\varepsilon}{(1+d^2)} I$ (where Δ_M is the Laplace-Beltrami operator in M) is transient, i.e. there is a non degenerate positive L_ε-supersolution for each such ε. Conversely, if there is a non trivial positive L_ε-supersolution for some $\varepsilon > 0$, then (3.1) holds with $c = \varepsilon^{-1}$.

Proof. Recall that the Green's function G of M with respect to Δ_M satisfies, for some constant $c \geqslant 1$ depending only on c_V and c_P, -see account and references in [SC1], [SC3]-

$$\frac{1}{c}\int_{d(x,y)}^\infty \frac{r\,dr}{|B(x,r)|} \leqslant G(x,y) \leqslant c \int_{d(x,y)}^\infty \frac{r\,dr}{|B(x,r)|}. \tag{3.2}$$

To show that L_ε is transient for small $\varepsilon > 0$, we will check that for $2 < \beta < \nu$ and $\mu = (1+d)^{-\beta}.\sigma$ one has $G\mu \leqslant c(1+d)^{2-\beta} \leqslant 2c(1+d^2)(1+d)^{-\beta}$ in M for some $c > 0$. Thus $G\mu$ is a finite positive supersolution for $\Delta + \frac{1}{2c(1+d^2)} I$ and (3.1) follows.

To estimate $G\mu(x) = \int_M G(x,z)\,[1+d(z)]^{-\beta}\, d\sigma(z)$ for $x \in M$, set $R = d(x)$ and let $I_1 = \int_{\{d(z) \leqslant 2R;\, d(x,z) \geqslant \frac{R}{2})\}} G(x,z)\,[1+d(z)]^{-\beta}\, d\sigma(z)$, $I_2 = \int_{B(x,\frac{R}{2})} G(x,z)\,[1+d(z)]^{-\beta}\, d\sigma(z)$, $I_3 = \int_{d(z) \geqslant 2R} G(x,z)\,[1+d(z)]^{-\beta}\, d\sigma(z)$.

We let $V(t) = |B(0,t)|$ for $t > 0$ and restrict as we may to $R \geqslant 2$.

We have $I_1 \leqslant c\,G(0,x)\, \int_{d(z)\leqslant 2R}(1+d(z))^{-\beta}\, d\sigma(z)$ by Harnack inequality (or on using (3.2)), and

$$\begin{aligned}\int_{d(z)\leqslant 2R}(1+d(z))^{-\beta}\, d\sigma(z) &= \beta \iint_{\{d(z)\leqslant 2R;\, t\geqslant d(z)\}} (1+t)^{-\beta-1}\, d\sigma(z)\, dt \\ &= \beta \int_0^\infty (1+t)^{-1-\beta}\, V(t\wedge(2R))\, dt \\ &= \frac{V(2R)}{(1+R)^\beta} + \beta \int_0^{2R} (1+t)^{-1-\beta}\, V(t)\, dt \\ &\leqslant \frac{V(2R)}{R^\beta} + c\,\frac{V(2R)}{R^\nu} \int_0^{2R} (1+t)^{\nu-1-\beta}\, dt,\end{aligned}$$

where we have used (Gr(ν)). Clearly, $\int_0^{2R}(1+t)^{\nu-1-\beta}\, dt \sim 2^{\nu-\beta}(\nu-\beta)^{-1} R^{\nu-\beta}$ and therefore $I_1 \leqslant c\,G(0,x)\,V(2R)\,R^{-\beta}$, for some constant $c>0$.

On the other hand, by (3.2) and the growth condition,

$$\begin{aligned}G(0,x) &\leqslant c \int_R^\infty \frac{s}{V(s)}\, ds \\ &\leqslant \frac{c}{V(R)} \int_R^\infty s\,(\frac{R}{s})^\nu\, ds = \frac{c}{V(R)}\, \frac{R^2}{\nu-2}.\end{aligned}$$

Using (DV), the desired estimate (i.e. $I_1 \leqslant c\,R^{2-\beta}$) for I_1 follows.

For I_2, we have, thanks again to (3.2),

$$\begin{aligned}I_2 &\leqslant c\,R^{-\beta} \int_{B(x,R/2)} G(x,z)\, d\sigma(z) \\ &\leqslant c'\,R^{-\beta} \iint_{\{s>d(x,z),\, z\subset B(x,R/2)\}} \frac{s}{|B(x,s)|}\, d\sigma(z)\, ds \\ &\leqslant c'\,R^{-\beta} \int_0^\infty \frac{s}{|B(x,s)|}\, |B(x,s\wedge\frac{R}{2})|\, ds \leqslant \frac{c'}{R^\beta}[\frac{R^2}{8} + V(2R)\,G(0,x)].\end{aligned}$$

Since we have seen that $G(0,x) \leqslant c\,\frac{R^2}{V(R)}$, the desired estimate for I_2 follows. Consider finally I_3. We have

$$\begin{aligned}I_3 &\leqslant c \iint_{\{t\geqslant d(z)\geqslant 2R\}} \frac{t}{V(t)}\, d(z)^{-\beta}\, dt\, d\sigma(z) \\ &\leqslant c \int_{2R}^\infty \frac{t}{V(t)}\, [\int_{2R}^t s^{-\beta}\, dV(s)]\, dt \\ &\leqslant c \int_{2R}^\infty \frac{t}{V(t)}\, [t^{-\beta}V(t) + \beta \int_{2R}^t s^{-\beta-1}\, V(s) ds]\, dt \\ &\leqslant \frac{c}{\beta-2}(2R)^{2-\beta} + c\,\beta \int_{2R}^\infty t\, [\int_{2R}^t s^{-\beta-1}(\frac{s}{t})^\nu\, ds]\, dt\end{aligned}$$

using again (Gr(ν)). Thus, since $2 < \beta < \nu$,

$$I_3 \leqslant \frac{c}{\beta - 2}(2R)^{2-\beta} + c \int_{2R}^{\infty} t^{1-\nu} \frac{1}{\nu - \beta} t^{\nu-\beta}\, dt \leqslant c'\, R^{2-\beta}.$$

The proof of Proposition 3.1 is complete. □

The point now is to make a change of metric in M. Fix $\ell : M \to \mathbb{R}_+$ smooth and such that $\ell(x) = 1$ if $d(x) = d(x, O) \leqslant 1$, $\frac{4}{5}\, d(x) \leqslant \ell(x) \leqslant \frac{6}{5}\, d(x)$ if $d(x) \geqslant 1$, $|\nabla \ell| \leqslant 2$ in M.

Equip M with the new metric $\tilde{g}_x(u) = \ell(x)^{-2}|u|^2$, $u \in T_x(M)$, $x \in M$, and denote $\tilde{M}$ the resulting Riemannian manifold, $\tilde{d}$ the distance in $\tilde{M}$. Clearly, in the M-ball $B(a, \frac{1}{8}\, d(a))$, the ratios of d and $\tilde{d}$ are in between $\frac{1}{2}\, d(a)$ and $2\, d(a)$, if $d(a) \geqslant 1$. Moreover,

$$\tfrac{5}{6}(\log(d(a)) + 1) \leqslant \tilde{d}(O, a) \leqslant \tfrac{5}{4}(\log(d(a)) + 1).$$

In particular, $\tilde{d}(x) = \tilde{d}(O, x) \to \infty$ as $x \to \infty_M$ and $\tilde{M}$ is complete. Moreover, it is easily checked that $\tilde{M}$ satisfies (for some positive constants r_0, c'_V, c'_P and the *same* dimension exponent $N > 2$ as M) the bounded geometry assumptions in §2. Letting a "tilde" to refer to a notion with respect to $\tilde{M}$, clearly, if $d(a) \geqslant 2$ and $r \leqslant r_0 = \frac{1}{100} \log 2$, then $\tilde{B}(a, r) \subset B(a, 2r\, d(a)) \subset \tilde{B}(a, 4\, r)$; so, if $r' < r$, by (DV) for M,

$$\begin{aligned}
\tilde{\sigma}(\tilde{B}(a,r)) &\leqslant (\frac{2}{d(a)})^m\, \sigma(\tilde{B}(a,r)) \leqslant (\frac{2}{d(a)})^m\, \sigma(B(a, 2r\, d(a))) \\
&\leqslant c\,(\frac{2}{d(a)})^m\, (\frac{r}{r'})^N \sigma(B(a, \frac{1}{2} r'\, d(a))) \\
&\leqslant 4^m\, c\, (\frac{r}{r'})^N\, \tilde{\sigma}(B(a, \frac{1}{2} r'\, d(a))) \\
&\leqslant 4^m\, c\, (\frac{r}{r'})^N\, \tilde{\sigma}(\tilde{B}(a, r')), \quad \text{where } m = \dim(M).
\end{aligned}$$

Thus (DV) holds in $\tilde{M}$ with local dimension exponent N. Similarly, if $f \in C_0^1(M)$, $a \in M$ and $r > 0$ are as before, if $\tilde{\overline{f}}_r$ is the mean of f over $\tilde{B}(a, r)$ with respect to $\tilde{\sigma}$ and $\overline{f}_{2rd(a)}$ the mean of f over $B(a, 2rd(a))$ with respect to σ,

$$\begin{aligned}
\int_{\tilde{B}(a,r)} |f - \tilde{\overline{f}}_r|^2\, d\tilde{\sigma} &\leqslant \int_{\tilde{B}(a,r)} |f - \overline{f}_{2rd(a)}|^2\, d\tilde{\sigma} \\
&\leqslant (2/d(a))^m \int_{\tilde{B}(a,r)} |f - \overline{f}_{2rd(a)}|^2\, d\sigma \\
&\leqslant (2/d(a))^m \int_{B(a,2rd(a))} |f - \overline{f}_{2rd(a)}|^2\, d\sigma \\
&\leqslant cr^2 d(a)^2 (2/d(a))^m \int_{B(a,4rd(a))} |\nabla f|^2\, d\sigma.
\end{aligned}$$

So $\int_{\tilde{B}(a,r)} |f - \tilde{\tilde{f}}_r|^2 \, d\tilde{\sigma} \leqslant c r^2 \int_{\tilde{B}(a,8r)} |\tilde{\nabla} f|^2_{\tilde{M}} \, d\tilde{\sigma}$, and (PI) is satisfied in $\tilde{M}$ for $r_0 = \frac{1}{100}\log(2)$.

Fix now a uniformly coercive section $\{\mathcal{A}_x\}_{x\in M}$ of $\mathrm{End}(T(M))$ which is a bounded measurable function of $x \in M$ (i.e. $\mathcal{A}_x > C$ and $\|\mathcal{A}_x\| \leqslant C^{-1}$ in M) and an elliptic operator

$$L = \mathrm{div}(\mathcal{A}\nabla .) + B.\nabla . + \mathrm{div}(B'.) + b$$

in M, with B, B' locally L^p, b locally $L^{p/2}$ in M for some $p > N$, and such that

$$(1+d(a))\Big(\oint_{C_a} [|B| + |B'|]^p d\sigma\Big)^{\frac{1}{p}} + (1+d(a))^2\Big(\oint_{C_a} |b|^{\frac{p}{2}} \, d\sigma\Big)^{\frac{2}{p}} \leqslant C, \quad a \in M,$$

where $C_a = \{x \in M \, ; \, d(x,a) \leqslant \frac{1}{2} d(a)\}$. Note $L_1 = \mathrm{div}(\mathcal{A}\nabla .)$.

Let $\mathcal{L}_1 = \ell^2 L_1$, $\mathcal{L} = \ell^2 L$. Simple computations show that with respect to the Riemannian structure of $\tilde{M}$, $\mathcal{L}$ may be written as

$$\mathcal{L} = \mathrm{div}_{\tilde{M}}(\mathcal{A}\nabla_{\tilde{M}} .) + D.\nabla_{\tilde{M}} . + \mathrm{div}_{\tilde{M}}(D'.) + \gamma,$$

with $D = \ell^2 B + (m-2)\,\ell\,\mathcal{A}^*(\nabla\ell)$, $D' = \ell^2 B'$, $\gamma = \ell^2 b + (m-2)\,\ell\,\nabla\ell.B'$ where in the last three equations the gradients and the scalar products are taken with respect to M. It follows that $\mathcal{L} \in \mathcal{D}_{\tilde{M}}(\theta, p)$ -and $\mathcal{L}_1 \in \mathcal{D}_{\tilde{M}}(\theta, \infty)$- for some sufficiently large θ (depending on C). Also, by proposition 3.1, $L_1 + \frac{\varepsilon}{1+r^2} I$ admits a positive supersolution in M for small $\varepsilon > 0$, which means that $\mathcal{L}_1$ is weakly coercive over $\tilde{M}$. Theorems 2.6, 2.7 and known results about L_1-Green's function [SC1]-[SC3] imply then the following result anounced in [A4] (and where all notations are related to the original Riemannian structure in M).

Theorem 3.3. *Assume that the L-Green's function G_L exists and that*

$$(1+d(a))\Big(\oint_{C_a} [|B|+|B'|]^p d\sigma\Big)^{\frac{1}{p}} + (1+d(a))^2\Big(\oint_{C_a} |b|^{\frac{p}{2}} \, d\sigma\Big)^{\frac{2}{p}} \leqslant \psi(d(a)) \quad (3.3)$$

for some nonincreasing function ψ satisfying $\int_0^\infty \psi(s)\frac{ds}{1+s} < \infty$. Then, $G_L \sim G_\Delta$, i.e., for some $c \geqslant 1$,

$$c^{-1}\int_{d(x,y)}^{\infty} \frac{s\,ds}{\sigma(B(x,s))} \leqslant G_L(x,y) \leqslant c \int_{d(x,y)}^{\infty} \frac{s\,ds}{\sigma(B(x,s))}, \; x \in M, \; y \in M.$$

Remark 3.4. **1.** Theorem 3.3 gives a sufficient condition in order that global Harnack inequalities (at all scales) hold for L. **2.** Theorems 2.6, 2.7 and

3.3 extend to more general "local" frameworks. In particular, M could be a graph with edges (of finite valence) with a riemannian structure on each of its (closed) edges, the corresponding distance d and volume (length) σ in M, and one could consider Dirichlet forms in M of the type $\alpha(\varphi, \psi) = \int_M \{a(x)\nabla\varphi.\nabla\psi - \psi D.\nabla\varphi + \varphi D'.\nabla\psi - \gamma\varphi\psi\} d\sigma$ with assumptions similar to (2.2), (2.3). Conditions (DV), (PI) (at all scales in the case of theorem 3.3) are of course also assumed.

4 A version for Markov chains on graphs of the homogeneous type

We return now to a discrete setting, slightly different from the setting in section 1 though. Let X be a graph, i.e. a set of vertices with an edge symmetric relation $x \sim y$, possibly $x \sim x$ for some $x \in X$. It is assumed that X is infinite, connected and with finite valence. Let d_X denote the natural graph distance in X and write $x \sim_s y$ when $x \neq y$ and $x \sim y$.

Let $P : X \times X \to \mathbb{R}_+$ be a kernel on X in the form

$$Pu(x) = \sum_{y \sim x} p(x,y)u(y) + \sum_{y \sim x} D(x,y)(u(y) - u(x)) + \gamma(x)u(x), \qquad (4.1)$$

for $x \in X$ and $u : X \to \mathbb{R}$. It is assumed that p is symmetric, with $p(x,y) \geqslant c_0$ when $x \sim_s y$ for some $c_0 > 0$, that $\sum_{y \sim x} p(x,y) = 1$ for all $x \in X$ (note that this condition is always satisfied after modifying γ and $\{p(x,x)\}_{x \in X}$); moreover $D(x,y) = -D(y,x)$ when $x \sim y$. Let $p(x,y) = 0$, $D(x,y) = 0$ if $x \not\sim y$.

Since P is nonnegative, $|D(x,y)| \leqslant p(x,y)$ for $x \sim y$ and $p(x,x) - \sum_{y \sim x} D(x,y) + \gamma(x) \geqslant 0$, $x \in X$, so that $\gamma(x) \geqslant -1$. We shall also assume the adaptness condition $|D(x,y)| \leqslant (1-\theta)\, p(x,y)$ in $X \times X$ for some $\theta \in (0,1)$.

P may be understood as a transition function for a Markov chain on (the vertices of) X if $\gamma(x) \leqslant 0$ for $x \in X$; $|\gamma(x)|$ is then the probability that starting from x the random walk will die before performing its first step. For sake of simplicity, the reader may assume henceforth that $\gamma \leqslant 0$ in X. Otherwise, to ensure at least the existence and the positivity of the P-harmonic measures μ_a of a in $\{a\}$ with respect to the boundary $V_a = \{b;\, b \sim_s a\}$, we require that $\gamma(a) < \sum_{y \sim_s a} [p(a,y) + D(a,y)]$, $a \in X$. Thus $|\gamma| \leqslant 2$ in X.

In order to get a discrete version of theorem 3.3 above, we shall construct a diffusion (if $\gamma \leqslant 0$ say) which is in some sense very close to the Markov chain in X driven by P.

Let M denote the one dimensional manifold (with isolated singularities) obtained by joining every two neighbors x and y in X by a smooth simple arc $[x, y]$ (a smooth loop at x if $x \sim x$) and by also adding "free" segments starting from each $x \in X$ with an end point $\tilde{x} \notin X$. Thus, as sets $[x, y]$ and $[y, x]$ are identical and any path in M connecting $\tilde{x}$ to $y \in X$ (or $\tilde{y}$), $y \neq x$, must first go through x. Set $S_M = \{x \in X\} \cup \{\tilde{x}\,;\, x \in X\}$ the set of singular point (or vertices) in M and fix a smooth Riemannian structure on $M \setminus S_M$ such that all the added arcs $[x, y]$, $[x, \tilde{x}]$, $x \in X$, $y \in X$, $x \sim y$, are of length one.

Consider now a (local, in general non symmetric) Dirichlet form in M of the type

$$\alpha(f, g) = \int_M \nabla f . \nabla g \frac{ds}{\ell} + \int_M g \nabla f . V\, ds - \int_M f\, g\, \kappa\, ds \tag{4.2}$$

for f, g compactly supported and in $H^1(M)$. Here ℓ, κ are functions in $M \setminus S_M$ which are constant on each edge of M, and V a vector field in $M \setminus S_M$ constant (in the obvious sense) in each edge of M. The value of ℓ on the edge (x, y), is denoted $\ell(x, y)$, and $V(x, y)$ is the algebraic measure of V in (x, y) oriented from x to y if $x \neq y$; thus $V(x, y) = -V(y, x)$ when $y \sim_s x$. It is assumed that $\ell \geqslant \lambda_0$ for some $\lambda_0 > 0$, that $V = 0$ on loops (x, x) and edges $(x, \tilde{x})$ and that κ vanishes except on the edges $(x, \tilde{x})$, $x \in X$, where its value is denoted $\kappa(x)$.

The Dirichlet form (4.2) is related to a local Potential theory in M in the sense of Brelot (a diffusion if $\kappa \leqslant 0$), the corresponding harmonic (resp. finite superharmonic) functions in a region $U \subset M$ being the continuous $u : U \to \mathbb{R}$ such that

$$\frac{1}{\ell} \Delta_M u - du(V) + \kappa u = 0, \tag{4.3}$$

(resp. $\leqslant 0$) weakly in $U \setminus S_M$, the derivatives $\partial_{\nu_e^x} u(x)$ exist and satisfy $\sum_{e \in E_x} \frac{1}{\ell(e)} \partial_{\nu_e^x} u(x) = 0$ (resp. $\leqslant 0$) if $x \in S_M \cap U$. Here E_x is the set of edges starting at x and ν_e^x is the unit tangent vector to $e \in E_x$ at $x \in S_M$, pointing inward in e.

For $a \in X$, define its star in M as $\mathrm{St}(a) = \{\bigcup_{y \sim a} [a, y)\} \cup [a, \tilde{a}]$ -so $\partial \mathrm{St}(a) = \{y \in X\,;\, y \sim_s x\}$- and compute the harmonic measure ω_a of a in $\mathrm{St}(a)$ with respect to α. Set $\ell(a) = \ell(a, \tilde{a})$.

Lemma 4.1. *We have* $\omega_a(y) = \dfrac{1}{\Phi_a} \dfrac{V(a, y)}{e^{V(a,y)\ell(a,y)} - 1}$ *for* $y \in \partial St(a)$, *where* $\dfrac{t}{(e^{\ell t} - 1)} = \dfrac{1}{\ell}$ *for* $t = 0$ *and* $\Phi_a = \displaystyle\sum_{y \in X, y \sim_s a} \frac{V(a, y)}{\exp(V(a, y)\ell(a, y)) - 1} +$ $\sqrt{\dfrac{|\kappa(a)|}{\ell(a)}}\, th(\sqrt{\ell(a)|\kappa(a)|})$, *if* $\kappa(a) \leqslant 0$. *If* $0 < \ell(a)\kappa(a) < (\frac{\pi}{2})^2$, *the sec-*

ond term in the r.h.s. above should be replaced by $-\sqrt{\frac{\kappa(a)}{\ell(a)}}\,\mathrm{tg}(\sqrt{\ell(a)\,\kappa(a)})$, *the measure* ω_a *being undefined as a positive measure when* $\Phi_a \leqslant 0$.

Proof. Let $u : \mathrm{St}(a) \to \mathbb{R}$ be α-harmonic at a. By (4.3), for $y \sim_s a$, $m \in [a, y]$ and $s = d(a, m)$, $u(m) = C_y\left(\exp[s\,\ell(a,y)V(a,y)] - 1\right) + C'_y$, whereas $u(m) = C\,\mathrm{ch}(\sqrt{|\kappa|\ell(a)}(s-1))$ along $[a, \tilde{a}]$ (if $\kappa(a) \leqslant 0$). Obviously $u(a) = C\,\mathrm{ch}(\sqrt{|\kappa|\ell(a)}) = C'_y$ for $y \sim_s a$ and $u(y) - u(a) = C_y(\exp(\ell(a,y)V(a,y)) - 1)$. Thus $Au(a) := \sum_{e \in E_a} \frac{1}{\ell(e)}\, \partial_{\nu_e^a} u(a)$ is such that

$$\begin{aligned} Au(a) &= \sum_{y \sim_s a} C_y V_y - C\sqrt{\frac{|\kappa|}{\ell}}\mathrm{sh}(\sqrt{|\kappa|\ell}) \\ &= \sum_{y \sim_s a} \frac{(u(y) - u(a))\,V_y}{\exp(\ell_y V_y) - 1} - C\sqrt{\frac{|\kappa|}{\ell}}\mathrm{sh}(\sqrt{|\kappa|\ell}) \end{aligned} \tag{4.4}$$

where $\ell_y = \ell(a, y)$, $V_y = V(a, y)$, $\ell = \ell(a, \tilde{a})$ and $\kappa = \kappa(a)$. The result follows. □

Going back to the kernel P in (4.1), we would like to show that the ℓ, V, and κ's can be chosen such that, for each $a \in X$, ω_a coincides with the harmonic measure μ_a of a in $\mathrm{St}'(a) = \{y \in X\,;\, y = a \text{ or } y \sim a\}$ with respect to P and the boundary $\{y\,;\, y \sim_s y\}$. The later is given by

$$\mu_a(y) = \frac{p(a, y) + D(a, y)}{-\gamma(a) + \sum_{z \in X, z \sim_s a}(p(a, z) + D(x, z))}\ ,\ y \in \partial\mathrm{St}(a).$$

Thus, it will be sufficient that for all $(x, y) \in X^2$ with $x \sim_s y$,

$$\begin{aligned} \frac{V(x,y)}{\exp(V(x,y)\ell(x,y)) - 1} &= p(x,y) + D(x,y)\ , \\ V(x,y) = -V(y,x)\ &,\ \ell(x,y) = \ell(y,x) \end{aligned} \tag{4.5}$$

$$\begin{aligned} &\kappa(x) \leqslant 0 \ \text{ and } \gamma(x) = -\sqrt{\ell(x)^{-1}|\kappa(x)|}\,\mathrm{th}(\sqrt{\ell(x)|\kappa(x)|}) \text{ if } \gamma(x) \leqslant 0 \\ &\kappa(x) \geqslant 0 \ \text{ and } \gamma(x) = \sqrt{\ell(x)^{-1}\,\kappa(x)}\,\mathrm{tg}(\sqrt{\ell(x)\,\kappa(x)}\,) \ \text{ if } \gamma(x) \geqslant 0. \end{aligned} \tag{4.6}$$

By choosing $\ell(x) = 1$, $x \in X$, $\kappa(x) > -\frac{\pi^2}{4}$ is determined by the last set of equations with $\kappa \sim \gamma$. The next lemma shows that we may determine a set of solutions for $\ell(x, y)$, $V(x, y)$.

Lemma 4.2. *The map* $\psi : (\ell, v) \to \left(\frac{v}{e^{v\ell} - 1}\,,\, \frac{-v}{e^{-v\ell} - 1}\right)$ *is a* C^∞ *diffeomorphism of* $U = \mathbb{R}_+^* \times \mathbb{R}$ *onto* $\psi(U) = (\mathbb{R}_+^*)^2$.

Proof. It is understood here that $\psi(\ell, 0) = (\frac{1}{\ell}, \frac{1}{\ell})$. Since $t \mapsto \frac{e^t - 1}{1 - e^{-t}} = e^t$ is an increasing bijection of $\mathbb{R}$ onto $\mathbb{R}_+^*$, it is clear that ψ is a bijection of

U onto $(\mathbb{R}_+^*)^2$. On the other hand, the jacobian $J_\psi(\ell, v)$ does not vanish (equivalently the jacobian of $(\ell, v) \mapsto (\frac{e^{v\ell}-1}{v}, \frac{e^{-v\ell}-1}{v})$ is $\neq 0$ in $\mathbb{R}_+^* \times \mathbb{R}$ as easily checked by explicit computation). □

It follows from the lemma that $\Phi : (\ell, v) \mapsto \frac{1}{2}(\psi_1(\ell, v)) + \psi_2(\ell, v), \psi_1(\ell, v) -\psi_2(\ell, v))$ is a smooth diffeomorphism of U onto $W = \{(p, q)\,;\, p > 0\,,\, |q| < p\}$. Thus Φ solves (4.5) and (as p varies between two positive constants and $|D(x, y| \leqslant (1-\theta)p(x, y))$ the solution is such that $|V(x, y|) = O(|D(x, y|)$ (and, as already noted, $|\kappa(x)| = O(|\gamma(x)|)$).

Fix now a form α as in (4.2) and such that the above holds. Not only the harmonic measures μ_a and ω_a coincide, but also the generators for P and α agree in the following sense.

Lemma 4.3. *Let $u : \overline{St(a)} \to \mathbb{R}$ be continuous on $\overline{St(a)}$, α-harmonic in $(a, \tilde{a})$ and in each edge (a, y), $y \sim a$. Then, $Au(a) = Pu(a) - u(a)$.*

This follows from the computations in the proof of Lemma 4.1. Then the next two statements are straightforward.

Lemma 4.4. *Let $A \subset X$, let $\partial A = \{y \in X \cap A^c\,;\, \exists x \in A\,;\, y \sim_s x\}$ and let U_A be the union of all edges $[x, y)$, $[x, \tilde{x}]$, $x \in A$. If $u : A \cup \partial A \to \mathbb{R}$ is P-harmonic in A (i.e. $u = Pu$ in A), there exists a unique extension of u to U_A which is continuous in $\overline{U_A}$ and α-harmonic in U_A.*

Lemma 4.5. *The kernel P is transient (i.e. it admits a finite Green's function) iff the Dirichlet form α is transient. If this is the case, and if one denotes G and g the corresponding Green's functions in X (resp. M) on has $g(a, b) = G(a, b)$ for a and b in X.*

Assume now that there are constants $c_V > 0$, $c_P > 0$ such that for all $a \in X$ and $r > 0$,

(DV) $|\overline{B}_X(a, 2r)| \leqslant c_V\, |\overline{B}_X(a, r)|$,where $\overline{B}_X(a, s) = \{y \in X; d_X(a, y) \leqslant s\}$,

(PI_2) $\sum_{x \in \overline{B}_X(a,r)} |f(x) - f_{a,r}|^2 \leqslant c_P\, r^2 \sum_{x \in \overline{B}_X(a,2r)} \sum_{y \sim x} |f(y) - f(x)|^2$,

where $f_{a,r} = (\sum_{x \in \overline{B}_X(a,2r)} f(x))/|\overline{B}_X(a, r)|$ and $|A|$ is the cardinality of $A \subset X$. It is easy to see that this means that M satisfies the corresponding assumptions, as in the beginning of §3. For reader's convenience, we insert a proof of (PI) in M for a ball $B(a, r)$ in M, $a \in X$, and r integer $\geqslant 1$.

Let $B = B(a, r)$, $B' = B(a, 2r)$, $B_X = \overline{B} \cap X$, $B'_X = \overline{B'} \bigcap X$ and let $f \in \mathrm{Lip}(M)$. For each $x \in B'_X$ choose an edge $I(x) = (x, y) \subset B(a, 2r)$. If f_I is the mean of f over I, if $\overline{f} = \frac{1}{|B_X|} \sum_{x \in B_X} f_{I(x)}$, we have by ($\mathrm{PI}_2$), (and using that $\int_0^1 |g(t)-g(t_0)|^2\,dt \leqslant \int_0^1 |g'(t)|^2 dt$ for $g \in \mathrm{Lip}([0,1])$, $0 \leqslant t_0 \leqslant 1$),

$$\int_B |f - \overline{f}_{B(a,r)}|^2\,d\sigma \leqslant \int_B |f - \overline{f}|^2 d\sigma \leqslant 2[\sum_{I\subset B}\int_I |f - f_I|^2 d\sigma + \sum_{I\subset B} |f_I - \overline{f}|^2]$$

$$\leqslant 2\sum_{I\subset B}\int_I |\nabla f|^2\,d\sigma + 4\sum_{x\in B_X}\sum_{I\ni x}|f_I - f_{I(x)}|^2 + 4\,c\sum_{x\in B_X}|f_{I(x)} - \overline{f}|^2$$

$$\leqslant c'\int_B |\nabla f|^2\,d\sigma + 4c'r^2 \sum_{x\in B'_X, y\sim x} |f_{I(x)} - f_{I(y)}|^2 \leqslant c'' r^2 \int_{B'} |\nabla f|^2\,d\sigma.$$

As an illustration of our construction, the following discrete Harnack inequalities are derived directly from their continuous counterparts. They are well-known (with other methods) at least in the symmetric case ([D1], references there and [D2] for the parabolic analogue).

Theorem 4.6. *Let $a \in X$, let $R > 1$ and let $C = \sup\{R \sup_{y\sim x} |D(x,y)| + R^2|\gamma(x)|\,;\, d(a,x) \leqslant R\}$. Then, if $u : B_X(a, R+1) \to \mathbb{R}_+$, is P-harmonic in the ball $B_X(a,R)$,*

$$C_H^{-1}u(a) \leqslant u(x) \leqslant C_H\, u(a)\,, \qquad x \in B_X(a, R/2),$$

with a constant C_H depending only on C, X, c_0 and θ (not on R, a or u).

Similarly, Theorem 3.3 (see remark 3.4.2) and the above construction yields the following.

Theorem 4.7. *Let $O \in X$ be such that $|B_X(O,R)| \geqslant c(\frac{R}{R'})^\nu |B_X(O,R')|$ for all R, R', $0 < R' < R$ and some reals $c > 0$, $\nu > 2$. Assume moreover that P is transient and that*

$$d_X(O,x)\sup_{y\sim x}|D(x,y)| + d_X(O,x)^2|\gamma(x)| \leqslant \Phi(d_X(O,x)) \qquad (4.7)$$

with Φ nonincreasing and such that $\int_0^\infty \Phi(r)\,\frac{dr}{1+r} < \infty$. Then, the Green's function G for P is such that

$$C^{-1}\sum_{n=d(x,y)}^{\infty}\frac{n}{|B_X(x,n)|} \leqslant G(x,y) \leqslant C \sum_{n=d(x,y)}^{\infty}\frac{n}{|B_X(x,n)|}$$

for x, y in X and some real $C \geqslant 1$. In particular, global Harnack inequalities (at all scales) hold for P.

Let $\Delta(x) = \sup_{y\sim x}|D(x,y)|$ and $U(R) = \{x \in X; R/2 \leqslant d_X(O,x) \leqslant R\}$. If $p > N$ (here $N > 2$ is a dimensional exponent for X, as in §3) we may replace (4.7) by : $R(\fint_{U(R)} |\Delta|^p \, d\mu)^{\frac{1}{p}} + R^2(\fint_{U(R)} |\gamma|^{\frac{p}{2}} \, d\mu)^{\frac{2}{p}} \leqslant \Phi(R)$, where $\fint_A d\mu$ is the mean on A w.r.t. the counting measure.

5 Appendix. About the proof of Theorem 2.9 and the $L^{2+\varepsilon}$ gradient estimate

We briefly describe how the argument in [Gia] chap. V may be adapted (and simplified) in our framework. Recall first the following elementary covering property (where $R = r_0/2$).

5.1. If $E \subset B(a,R)$ and if $\{B(x,r_x)\}_{x\in E}$ is a family of open balls in M with $0 < r_x < R$, there exists a (possibly finite) sequence $x_1,\dots,x_n,\dots$ in E such that (i) $E \subset \bigcup_{j\geqslant 1} B(x_k, r_{x_k})$, (ii) the closed balls $\overline{B}(x_k, \frac{1}{3}r_{x_k})$ are mutually disjoint.

Next, we have the following Calderon-Zygmund decomposition property (compare [Gia], p. 127). Note $\alpha(x) = |B(x, \frac{1}{2}(R - d(a,x)))|/|B(a,R)|$ for $x \in B(a,R)$.

5.2. Let $f \in L^1(B(a,R))$, $t \geqslant \fint_{B(a,R)} |f| \, d\sigma$, and $C_k = \{x \in M\,;\, R(1 - 2^{-k}) \leqslant d(a,x) \leqslant R(1 - 2^{-k-1})\}$, $k \in \mathbb{N}$. For each $k \geqslant 0$ there exists a sequence of balls $B_{k,j} = \overline{B}(x_{k,j}, r_{k_j})$ in M such that, for some large γ depending only on c_V: (i) $x_{k,j} \in C_k$, $r_{k,j} \leqslant 2^{-k-10}R$, (ii) $|f(x)| \leqslant \gamma\, t\, \alpha(x)^{-1}$ a.e. in $C_k \setminus \cup_{j\geqslant 1} B_{k,j}$, (iii) $\gamma^{-1} t \alpha(x_{k,j})^{-1} \leqslant \fint_{\overline{B}(x_{k,j},r)} |f| \, d\sigma \leqslant \gamma t \alpha(x_{k,j})^{-1}$ for $\frac{1}{6} r_{k,j} \leqslant r \leqslant 6\, r_{k,j}$, (iv) for fixed k, the balls $B(x_{k,j}, \frac{r_{k,j}}{3})$ are mutually disjoint.

Set $M'_f(x) = \sup\{\fint_B |f| \, d\sigma\,;\, B = \overline{B}(x,r),\ r < \frac{1}{4}(R - d(x,a))\}$, $F_k = \{x \in C_k\,;\, M'_f(x) \leqslant \gamma\, t\, \alpha(x)^{-1}\}$ (where γ is to be chosen). We have $|f(x)| \leqslant \gamma\, t\, \alpha(x)^{-1}$ a.e. in F_k. If $x \in C_k \setminus F_k$, $M'_f(x) > \gamma\, t\, \alpha(x)^{-1}$, whereas for $2^{-20}(R - d(a,x)) \leqslant r \leqslant \frac{1}{4}(R - d(a,x))$ we have $\fint_{\overline{B}(x,r)} |f| \, d\sigma \leqslant (\fint_{\overline{B}(a,R)} |f| \, d\sigma) \times (|B(a,R)|/|B(x,r)|) \leqslant \frac{\gamma}{4}\, t\, \alpha(x)^{-1}$ if γ is large. Select now for each $x \in C_k \setminus F_k$ the smallest radius $r_x \in (0, \frac{R-d(a,x)}{4}]$ such that $\fint_{\overline{B}(x,r)} |f| \, d\sigma \leqslant \frac{\gamma}{2} t\, \alpha(x)^{-1}$ for all $r \in [r_x, \frac{R-d(a,x)}{4}]$, and apply 5.1 to $\{B(x, 6\, r_x)\}_{x\in C_k\setminus F_k}$ to derive 5.2.

5.3. To prove Theorem 2.9, note first that after enlarging the constant C, κ may be made small (for fixed p_0 and q) in (2.12) : first, if $B(x, 2^k r) \subset B(a,R)$, on iterating (2.12) and getting $B(x, 2^k r)$ (with some large k) instead of $B(x, 2r)$ in the r.h.s.; use then this version of (2.12) with the balls $B_j = B(y_j, 2^{-k-4} r)$ of a Whitney type covering of $B = B(x,r)$ (the mean $\fint_B |F|^q \, d\sigma$ is, up to a constant factor, bounded by a convex mean of the means $\fint_{B_j} |F|^q \, d\sigma$). We may then apply 5.2 to the function $F(x) = f(x)\alpha(x)^{\frac{1}{q}}$ and derive similarly to [Ge], [Gia] the inequalities

$\int_{F>\beta t} |F|^q\,d\sigma \leqslant C\,t^{q-1}[\int_{|F|>t} |F|\,d\sigma + \int_{G>t} G\,d\sigma]$ for $t \geqslant \oint_{B(a,R)} |f|^q d\sigma$, where G^q is the maximal function of $|g|^q$, β a constant $\geqslant 1$. Then one may make $\beta = 1$, and the conclusion follows by an elementary lemma. Ref. [Gia] chap. 5 for the case $M = \mathbb{R}^N$.

5.4. *The estimate of* $\|\nabla u\|_{2+\varepsilon,B(a,R)}$. Let u be positive $\mathcal{L}$-harmonic in $B(a,3R)$, $R \leqslant r_0/3$, $\mathcal{L} \in \mathcal{D}_M(\theta,p)$. Notice that $u \leqslant c\,u(a)$ in $U = B(a,2R)$ and that $\operatorname{div}(\mathcal{A}\nabla u) = \operatorname{div}(V) + f$ with $(\oint_U |V|^p\,d\sigma)^{\frac{1}{p}} \leqslant c\,u(a)$, $(\oint_U |f|^{\frac{2p}{p+2}}\,d\sigma)^{\frac{p+2}{2p}} \leqslant c\,u(a)$, $c = c_M(\theta,p,R)$. For $B = B(x,r)$ such that $B' = B(x,2r) \subset U$, use (as [Gia], p. 139) the test function $\eta^2(u - u_{B'})$ where $\eta \in C_0^1(B')$, $\eta = 1$ in B, $|\nabla\eta| \leqslant \frac{2}{r}$. Simple calculations give that if $p' \in (N,p)$,

$$\oint_B |\nabla u|^2 d\sigma \leqslant \frac{c}{r^2}\oint_{B'} |u - u_{B'}|^2 d\sigma + c\oint_{B'} |V|^2 d\sigma + c[\oint_{B'} |f|^{\frac{2p'}{p'+2}} d\sigma]^{\frac{p'+2}{2p'}}[\oint_{B'} |u - u_{B'}|^{\frac{2p'}{p'-2}} d\sigma]^{\frac{p'-2}{2p'}}.$$

By (PI_1) and the ensuing Poincaré-Sobolev inequalities [MSC], this yields

$$\begin{aligned}\oint_B |\nabla u|^2\,d\sigma &\leqslant c\{[\oint_{B'} |\nabla u|^\rho\,d\sigma]^{\frac{2}{\rho}} + r^2(\oint_{B'} |f|^{\frac{2p'}{p'+2}} d\sigma)^{\frac{p'+2}{p'}} + \oint_{B'} |V|^2 d\sigma\}\\ &\leqslant c\{[\oint_{B'} |\nabla u|^\rho\,d\sigma]^{\frac{2}{\rho}} + K(\oint_{B'} |f|^{\frac{2p'}{p'+2}} d\sigma) + \oint_{B'} |V|^2 d\sigma\}\end{aligned}$$

where $\frac{1}{\rho} - \frac{1}{N} = \frac{p'-2}{2p'}$ and $K \leqslant c\,R^2\,(\oint_U |f|^{\frac{2p'}{p'+2}} d\sigma)^{\frac{2}{p'}}(\frac{r}{R})^{2-\frac{2N}{p'}}$ so that $K \leqslant c\,R^2\,(\oint_U |f|^{\frac{2p'}{p'+2}} d\sigma)^{\frac{2}{p'}}$. Since $\rho < 2$, the desired result follows from Theorem 2.9 (with $\kappa = 0$).

References

[A1] Ancona A., *Negatively curved manifolds, elliptic operators and the Martin boundary*, Ann. of Math. **125** (1987), 495-536.

[A2] Ancona A., *Positive harmonic functions and hyperbolicity*, Potential Theory, Surveys and Problems, Lecture Notes in Math. 1344, Springer-Verlag, 1988, pp. 1–23.

[A3] Ancona A., *Théorie du Potentiel sur les graphes et les variétés*, Ecole d'été de Probabilités de Saint-Flour XVIII - 1988, Lecture Notes in Math. 1427, Springer-Verlag, 1990, pp. 5–112.

[A4] Ancona A., *First eigenvalues and comparisons of Green's functions for elliptic operators on manifolds or domains*, J. d'Analyse Math. **72** (1997), 45-92.

[B] Brelot M., *Axiomatique des fonctions harmoniques*, Les Presses de l'Université de Montréal, 1969.

[De1] Delmotte T., *Inégalités de Harnack elliptiques sur les graphes*, Coll. Math. **72, (1)** (1997), 265-277.

[De2] Delmotte T., *Parabolic Harnack inequality and estimates of Markov chains on graphs*, Prépublication (1997).

[Den] Deny, J. (1970), *Méthodes hilbertiennes en théorie du potentiel, in Potential Theory (C.I.M.E., I Ciclo, Stresa, 1969)*, Edizioni Cremonese, Rome.

[Ge] Gehring F., *The L^p integrability of the partial derivatives of a quasiconformal mapping*, Acta Math. Série I, **130** (1973), 265-277.

[Gia] Giaquinta M., *Multiple integrals in the calculus of variations and nonlinear elliptic systems*, Annals of Math. Studies 105, Princeton University Press, Princeton N. J., 1983.

[GT] Gilbarg D., Trudinger N. S., *Elliptic partial differential equations of second order*, Springer Verlag, Berlin Heidelberg New York, 1977.

[MSC] Maheux P., Saloff-Coste L., *Analyse sur les boules d'un operateur sous-elliptique*, Math. Ann. **303 (4)** (1995), 713-740.

[Mey] Meyers N. G., *An L^p-estimate for the gradient of solutions of second order elliptic divergence equations*, Ann. Scuola Norm. Sup. Pisa **17** (1963), 189-206.

[Pi1] Pinchover Y., *Sur les solutions positives d'équations elliptiques et paraboliques*, C. R. Acad. Sci Paris **302** (1986), 447-450.

[Pi2] Pinchover Y., *On the equivalence of Green functions of second order elliptic equations*, J. Differential Integral Equations **80** (1989), 237-250.

[SC1] Saloff-Coste, L., *A note on Poincaré, Sobolev, and Harnack inequalities*, Int. Math. Res. Not. **No.2,** (1992), 27-38.

[SC2] Saloff-Coste L., *Uniformly elliptic operators on Riemannian manifolds*, J. Diff. Geo. **36** (1992), 417-450.

[SC3] Saloff-Coste L., *Parabolic Harnack inequality for divergence form second order differential operators*, Potential Analysis **4 (4)** (1995).

[SC4] Saloff-Coste L., *On global Sobolev inequalities*, Forum Math. **6 (3)** (1994), 271-280.

[Se1] Serrin J., *On the Harnack inequality for linear elliptic equations*, J. Anal. Math. **4** (1956), 292-308.

[Se2] Serrin J., *Local behavior of solutions of quasi-linear equations*, Acta Math. **111** (1964), 247-302.

[Soa] Soardi, P. M., *Potential theory on infinite networks*, Lecture Notes in Mathematics, 1590, Springer-Verlag, Berlin, 1994.

[Sta] Stampacchia G., *Le problème de Dirichlet pour les équations elliptiques du second ordre à coefficients discontinus*, Ann. Inst. Fourier **15 (1)** (1965), 189-258.

[ST] Strömberg J-O, Torchinsky A., *Weighted Hardy Spaces*, Lecture Notes in Mathematics, 1381, Springer-Verlag, Berlin Heidelberg, 1989.

Alano Ancona
Département de Mathématiques,
Campus d'Orsay, Bât. 425,
Université Paris Sud,
ORSAY, FRANCE 91405.
email:ancona@matups.math.u-psud.fr

Random walks on graphical Sierpinski carpets

Martin T. Barlow, Richard F. Bass

Abstract. We consider random walks on a class of graphs derived from Sierpinski carpets. We obtain upper and lower bounds (which are non-Gaussian) on the transition probabilities which are, up to constants, the best possible. We also extend some classical Sobolev and Poincaré inequalities to this setting.

1. Introduction.

We begin with some notation and definitions. Let $G = (V, E)$ be an infinite connected graph. We write $d(x, y)$ for the natural graph distance on V, and for $x \in V$ and $r \geqslant 0$ set $B(x, r) = \{y \in V : d(x, y) \leqslant r\}$. Let $a(x, y) = 1$ if $\{x, y\} \in E$ and $a(x, y) = 0$ otherwise, and set

$$\mu(x) = \sum_{y \in V} a(x, y).$$

We assume that G has uniformly bounded vertex degree, that is, there exists $c_1 < \infty$ such that $1 \leqslant \mu(x) \leqslant c_1$ for all $x \in V$. We define the measure μ on V by $\mu(A) = \sum_{x \in A} \mu(x)$.

The (discrete time) simple random walk (SRW) X on G is the Markov chain $X = (X_n, n \in \mathbb{Z}_+, \mathbb{P}^x, x \in V)$ with transition probabilities

$$\mathbb{P}^z(X_{n+1} = y \mid X_n = x) = \frac{a(x, y)}{a(x)};$$

thus at each time step X moves from a site x to one of the neighbors of x with equal probability for each. We define

$$p(n, x, y) = \mathbb{P}^x(X_n = y)\mu(y)^{-1}. \tag{1.1}$$

(The term $\mu(y)^{-1}$ above looks slightly unnatural, but it has the advantage of making $p(n, x, \cdot)$ the density of $\mathbb{P}^x(X_n \in \cdot)$, with respect to μ, and also makes $p(n, x, y) = p(n, y, x)$.)

Associated with X is the μ-symmetric continuous time random walk Y_t on the graph $G = (V, E)$. This is the Markov process $Y = (Y_t, t \geqslant 0, \mathbb{Q}^x, x \in V)$ with infinitesimal generator

$$\mathcal{L}f(x) = \mu(x)^{-1} \sum_y a(x, y)(f(y) - f(x)). \tag{1.2}$$

If

$$\mathcal{E}(f,f) = \tfrac{1}{2} \sum_{x \in V} \sum_{y \in V} a(x,y)(f(x) - f(y))^2, \qquad f \in L^2(V,\mu), \tag{1.3}$$

then Y is the Markov process with Dirichlet form $(\mathcal{E}, L^2(V,\mu))$ on $L^2(V,\mu)$. Note that Y waits at a vertex an independent exponential length of time (with parameter 1) and then jumps to one of the neighboring vertices, with equal probability for each. If T_n is the time of the n-th jump of Y then $\hat{X}_n = Y_{T_n}$ is a discrete time SRW on G (and so is equal in law to X). Set

$$q(t,x,y) = \mathbb{Q}^x(Y_t = y)\mu(y)^{-1}, \quad t \geqslant 0, \; x,y \in V. \tag{1.4}$$

There has been much interest in recent years in the general problem of estimating p_n and q_t from geometrical information about the graph G. For a sample of papers see [V1], [Da], [De], [P], [Co], [HSC], and for a survey see [W]. There are, of course, many similarities between this problem and the one of estimating the large time behavior of the heat kernel on a manifold M from geometrical information about M.

Given the similarity between the processes X and Y one would expect that p_n and q_t should satisfy similar bounds. This is usually the case, but the problem for p_n is generally more difficult, since X is less smooth than Y. Indeed, given bounds on p_n, simple integration gives bounds on q_t, since the number of jumps of Y in $[0,t]$ is a Poisson r.v. with mean t which is independent of $\hat{X}$. So

$$q(t,x,y) = \sum_{n=0}^{\infty} \frac{e^{-t} t^n}{n!} p(n,x,y). \tag{1.5}$$

In obtaining bounds on p_n or q_t one almost always finds that the easiest procedure is to obtain the bounds in the following order:

1. Global upper bounds,
2. Off-diagonal upper bounds,
3. On-diagonal lower bounds,
4. Off-diagonal lower bounds,

where each step builds on information obtained in the previous step.

For the global upper bounds there is a very elegant theory, based on the work in [V1], which gives equivalences between these bounds and certain isoperimetric and Sobolev inequalities.

Notation. For $f : V \to \mathbb{R}$ set

$$||f||_p^p = \sum_x |f(x)|^p \mu(x).$$

For $e = \{x, y\} \in E$ define $a(e) = a(x, y)$ $(= a(y, x))$. We regard a as a measure on E. If $f : V \to \mathbb{R}$ define $|\nabla f| : E \to \mathbb{R}$ by

$$|\nabla f|(x, y) = |f(x) - f(y)|,$$

and set

$$||\nabla f||_p^p = \sum_{e \in E} a(e)(|\nabla f|(e))^p, \quad p \geqslant 1.$$

Note that $||\nabla f||_2^2 = \mathcal{E}(f, f)$, and also that

$$c_1 \mathcal{E}(f, f) \leqslant ||f||_2^2 \leqslant ||f||_1^2. \tag{1.6}$$

For $A \subset V$ set $\partial A = \{x \in A^c : d(x, A) = 1\}$, where $d(x, A) = \min_{y \in A} d(x, y)$. The cardinality of a set A will be denoted $|A|$.

Definition 1.1. Let $p \geqslant 1$.

G satisfies a *p-isoperimetric inequality* (I_p) if there exists $c_1 < \infty$ such that

$$\mu(A)^{p-1} \leqslant c_1 \mu(\partial A)^p \text{ for all finite } A \subset V.$$

G satisfies a (α, p)*-Sobolev inequality* (S_p^α) (here $\alpha = 1$ or 2) if there exists $c_2 > 0$ such that

$$||\nabla f||_\alpha \geqslant c_2 ||f||_{\alpha p/(p-\alpha)}, \quad \text{for } f \text{ with compact support.}$$

G satisfies a *p-Nash inequality* (N_p) if for some $c_3 > 0$

$$\mathcal{E}(f, f) \geqslant c_3 ||\nabla f||_2^{2+4/p} ||f||_1^{-4/p} \quad \text{for } f \text{ with compact support.}$$

G satisfies (P_p) if for some $c_4 > 0$

$$p(n, x, y) \leqslant c_4 n^{-p/2}, \quad n > 0,\ x, y \in V.$$

G satisfies (Q_p) if for some $c_5 > 0$

$$q(t, x, y) \leqslant c_5 t^{-p/2}, \quad t > 0,\ x, y \in V.$$

Theorem 1.2. *(See [V1], [CKS]).*
(a) If $p > 1$ then $(I_p) \iff (S_p^1) \Rightarrow (N_p) \iff (Q_p) \iff (P_p)$.
(b) If $p > 2$ then $(I_p) \iff (S_p^1) \Rightarrow (S_p^2) \iff (N_p) \iff (Q_p) \iff (P_p)$.

Thus Theorem 1.2 enables one to obtain global upper bounds on $q(t,x,y)$ starting from an isoperimetric inequality on G. Bounds of this kind are called 'on-diagonal': they follow from the same bound on $q(t,x,x)$ since

$$q(t,x,y) \leqslant q(t,x,x)^{1/2} q(t,y,y)^{1/2}. \tag{1.7}$$

Off-diagonal upper bounds now follow by a variety of techniques: 'Davies method' (see [CKS]), or the wave equation approach (as in [V2], [Ca]), which yields the general bound,

$$p(n,x,y) \leqslant 2(\mu(x)/\mu(y))^{1/2} e^{-d(x,y)^2/2n}; \tag{1.8}$$

this holds for the process X on any connected graph.

Theorem 1.3. *([V2], [HSC]). Suppose G satisfies (Q_p). Then*

$$p(n,x,y) \leqslant c_1 n^{-p/2} \exp(-c_2 d(x,y)^2/n), \quad x,y \in V,\ n \geqslant 1, \tag{1.9}$$

$$q(t,x,y) \leqslant c_3 t^{-p/2} \exp(-c_4 d(x,y)^2/t), \quad x,y \in V,\ d(x,y) \leqslant t. \tag{1.10}$$

Remark. Note that $p(n,x,y) = 0$ if $d(x,y) > n$. We cannot expect the Gaussian decay in (1.10) to hold for $d(x,y) \gg t$. For example, for the SRW on $\mathbb{Z}$, if $t = 1$ and $m \gg 1$ then by (1.3)

$$q(1,0,m) \geqslant (e^{-1}/m!) p(m,0,m) \sim c m^{-1/2} (2e)^{-m} m^{-m},$$

which decays as $e^{-m \log m}$.

For regular graphs we can summarize much of the information above in terms of a number of 'dimensions' of the graph V.

Definition 1.4. We say
G has *uniform volume growth* d_f if

$$c_1 r^{d_f} \leqslant \mu(B(x,r)) \leqslant c_2 r^{d_f}, \quad r \geqslant 1, x \in V. \tag{1.11}$$

(Note that (1.11) implies that G satisfies a volume doubling condition.)
G has *isoperimetric dimension* d_i if

$$d_i = \sup\{p : G \text{ satisfies } (I_p)\}.$$

G has *spectral dimension* d_s if

$$d_s = \sup\{p : G \text{ satisfies } (Q_p)\}.$$

Theorem 1.2(a) implies that $d_i \leqslant d_s$, and it is not hard to deduce from (1.8) that $d_s \leqslant d_f$.

The estimates of Theorem 1.3, together with a bound on the volume growth of G gives easily an on-diagonal lower bound for q – see for example Theorem 6.1 below. However, the final step, obtaining off-diagonal bounds on p or q is harder – generally some more information of an analytic kind about G is needed, such as that G satisfies a Poincaré inequality.

For many graphs the upper bound techniques outlined above do provide bounds which are, up to constants, best possible. However, there are graphs which have a quite regular structure but for which the bounds in Theorem 1.3 are far from giving the true behavior of $p(n, x, y)$. These 'fractal graphs' have large holes, so that $d_i < d_f$; this acts to restrict the dispersion of the random walk, and causes p and q to decay in a non-Gaussian fashion.

We will not attempt to define 'fractal graphs' but will simply remark that given any finitely ramified fractal F in one of the families which has been studied in the diffusions on fractals literature (such as nested fractals, post-critically finite self-similar sets, or Sierpinski carpets – see [B1]), there is a natural way to construct an infinite graph G, such that the large-scale structure of G mimics the microscopic structure of F.

The first treatment of heat kernel bounds on fractal graphs was by Jones [J], on the graphical Sierpinski gasket. (See Figure 1.1). This graph has uniform volume growth $d_f = \log 3/\log 2$, but since $|\partial B(0, 2^n - 1)| = 2$ for all n it satisfies $d_i = 1$. (So the technique of obtaining bounds on q_t from an isoperimetric inequality fails completely in this case).

Figure 1: The graphical Sierpinski gasket.

In this paper we study a class of graphs derived from generalized Sierpinski carpets. We defer the full definition of these graphs to Section 2. The simplest examples of the graphs in this class are those derived from the standard d-dimensional Sierpinski carpet.

Let $H_0 = \mathbb{Z}_+^d$. For $x = (n_1, n_2, ..., n_d) \in H_0$ write n_i in ternary: so $n_i = \sum_{r=0}^{\infty} n_{ir} 3^r$, where $n_{ir} \in \{0, 1, 2\}$, and $n_{ir} = 0$ for all but finitely many r. Set

$$J_k = \{(n_1, ..., n_d) : n_{ik} = 1 \text{ for } 1 \leqslant i \leqslant d\},$$

so that J_k consists of a union of disjoint cubes of side 3^k: the cube in J_k closest to the origin is $\{3^k, \ldots, 2.3^k - 1\}^d$. Now set

$$H_n = H_0 - \bigcup_{k=1}^{n} J_k, \quad H = \bigcap_{n=0}^{\infty} H_n.$$

To match our notation below we take $V = V^{(d)} = a + H$, where $a = (\frac{1}{2}, ..., \frac{1}{2}) \in \mathbb{Z}^d$. For $x, y \in V$ let $\{x, y\} \in E$ if and only if $|x - y| = 1$.

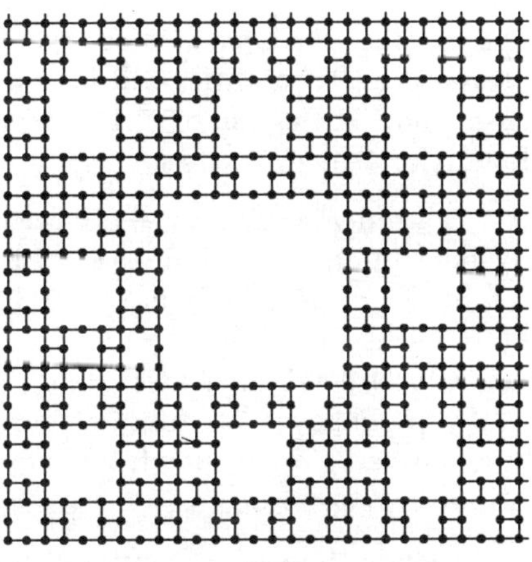

Figure 1.2: The graphical Sierpinski carpet $V^{(2)}$.

It is not hard to see that V has uniform volume growth $d_f = d_f(V^{(d)}) = \log(3^d - 1)/\log 3$. However the presence of large holes means that the process X moves less rapidly on V than on $\mathbb{Z}^d$; the holes act as obstacles which take a significant amount of time for X to find its way around. We shall see below that there exists a parameter d_w (called the *walk dimension of* V) such that writing $\sigma(x, r) = \min\{n \geqslant 0 : X_n \notin B(x, r)\}$,

$$c_1 r^{d_w} \leqslant \mathbb{E}^x \sigma(x, r) \leqslant c_2 r^{d_w}, \; r \geqslant 0. \tag{1.12}$$

(Compare this with the classical estimate $\mathbb{E}^x \sigma(x,n) \approx n^2$ for a simple random walk in $\mathbb{Z}^2$.) The estimate (1.12) shows that $p(n,x,\cdot)$ puts nearly all its mass on a ball of radius roughly n^{1/d_w}. So if $p(n,x,.)$ were reasonably evenly spread on this set, we would have (since $\mu(B(x,n^{1/d_w}) \approx n^{d_f/d_w})$ that

$$p(n,x,y) \leqslant cn^{-d_f/d_w}, \; x,y \in V, \; n \geqslant 1. \tag{1.13}$$

We prove this estimate below (Theorem 5.2) by deriving it from a Nash inequality which holds in the continuous setting. By Theorem 1.3 we deduce therefore from (1.13) that $p(n,x,y)$ satisfies (1.9) with $p = d_s/2$. However, this upper bound is still not a very good one.

We prove the following sharp (up to constants) upper and lower bounds for $p(n,x,y)$.

Theorem 1.5. *There exists $d_w > 2$, and constants c_1, c_2, c_3, c_4 such that if n and $d(x,y)$ are both odd or both even and $d(x,y) \geqslant n$, then*

$$\begin{aligned} c_1 n^{-d_f/d_w} \exp(-c_2(d(x,y)^{d_w}/n)^{1/(d_w-1)}) &\leqslant p(n,x,y) \\ &\leqslant c_3 n^{-d_f/d_w} \exp(-c_4(d(x,y)^{d_w}/n)^{1/(d_w-1)}). \end{aligned} \tag{1.14}$$

Note that since G is, like $\mathbb{Z}^d$, bipartite, the random walk has period two, and hence the transition probabilities will be 0 if one of $d(x,y)$ and n is even and the other is odd. The estimate (1.14) implies immediately that the spectal dimension d_s for G is $d_s = d_f/d_w$.

Let $C = \{x : |x_i| \leqslant \frac{1}{2}\}$ and $P = \cup_{x \in V}(x + C)$: we call P the *pre-carpet.* Our results here rely heavily on the estimates for Brownian motion on P with normal reflection on ∂P developed by us in [BB3]. We refer the reader to that paper for more discussion of the significance of d_w and d_s and for background on diffusions on fractals. See also [Ba].

Our techniques may perhaps be of some interest as well in showing how one might obtain estimates on graphs from information on a related continuous-time continuous-space process.

In Section 2 we introduce the notation we will use and recall the key inequalities we need from [BB3]. Section 3 extends the Sobolev and Poincaré inequalities to the graphical setting. In Section 4 we show how the coupling argument of [BB3] yields a Harnack inequality for the random walks on our graphs. The upper bound in (1.14) is derived in Section 5 (see Theorem 5.6) and the lower bound in Section 6 (see Theorem 6.5). Section 7 contains some further results, such as a Liouville theorem, and estimates for the process Y on the graph G.

We use the letter c with subscripts to denote constants which depend only on the graph G. We renumber the constants for each lemma, proposition, theorem, and corollary.

2. Notation and preliminaries.

Let $d \geqslant 2$, $F_0 = [0,1]^d$, and let $l_F \in \mathbb{N}$, $l_F \geqslant 3$ be fixed. For $n \in \mathbb{Z}$ let $\mathcal{S}_n$ be the collection of closed cubes of side l_F^n with vertices in $l_F^n \mathbb{Z}^d$. For $A \subseteq \mathbb{R}^d$, set

$$\mathcal{S}_n(A) = \{S: S \subset A, S \in \mathcal{S}_n\}.$$

For $S \in \mathcal{S}_n$, let Ψ_S be the orientation preserving affine map which maps F_0 onto S.

We define a decreasing sequence (F_n) of closed subsets of F_0. Let $1 \leqslant m_F \leqslant l_F^d$ be an integer, and let F_1 be the union of m_F distinct elements of $\mathcal{S}_{-1}(F_0)$. We impose the following conditions on F_1:

Hypotheses 2.1.

(H1) *(Symmetry) F_1 is preserved by all the isometries of the unit cube F_0.*

(H2) *(Connectedness)* Int(F_1) *is connected, and contains a path connecting the hyperplanes $\{x_1 = 0\}$ and $\{x_1 = 1\}$.*

(H3) *(Non-diagonality) Let B be a cube in F_0 which is the union of 2^d distinct elements of $\mathcal{S}_{-1}$. (So B has side length $2l_F^{-1}$). Then if* int $(F_1 \cap B)$ *is non-empty, it is connected.*

(H4) *(Borders included) F_1 contains the line segment $\{x : 0 \leqslant x_1 \leqslant 1, x_2 = \ldots = x_d = 0\}$.*

We may think of F_1 as being derived from F_0 by removing the interiors of $l_F^d - m_F$ squares in $\mathcal{S}_{-1}(F_0)$. Given F_1, F_2 is obtained by removing the same pattern from each of the squares in $\mathcal{S}_{-1}(F_1)$. Iterating, we obtain a sequence (F_n), where F_n is the union of m_F^n squares in $\mathcal{S}_{-n}(F_0)$. Formally, we define

$$F_{n+1} = \bigcup_{S \in \mathcal{S}_{-n}(F_n)} \Psi_S(F_1) = \bigcup_{S \in \mathcal{S}_{-1}(F_1)} \Psi_S(F_n), \quad n \geqslant 1.$$

We call the set $F = \cap_{n=0}^{\infty} F_n$ a *generalized Sierpinski carpet*. Let d_f denote the Hausdorff dimension of F; by [Hu] $d_f = \log m_F / \log l_F$.

Let

$$P = \bigcup_{r=0}^{\infty} l_F^r F_r; \tag{2.1}$$

we call P the *pre-carpet* (see [O]). Let μ_P be Lebesgue measure restricted to P.

We now define our graph $G = (V, E)$. Let V be the collection of centers of cubes in $\mathcal{S}_0(P)$, and for x, $y \in V$ let $\{x, y\} \in E$ if and only if $|x - y| = 1$.

Let a, μ, X, $p(n,x,y)$, Y, $q(t,x,y)$, and $\mathcal{E}$ be as in Section 1. If we restrict the sum in (1.3) to $x, y \in D \subset V$, we will write $\mathcal{E}_D(f,f)$. For $S \in \mathcal{S}_0(P)$ write z_S for the center of S.

Recall that $B(x,r) = \{y \in V : d(x,y) \leqslant r\}$. We will define $S(x,r) \subset P$ by

$$S(x,r) = \bigcup \{S \in \mathcal{S}_0(P) : d(z_S, x) \leqslant r\}.$$

We let

$$\sigma(x,r) = \sigma^X(x,r) = \sigma(x,r,X) = \inf\{n \geqslant 0 : X_n \notin B(x,r)\}. \tag{2.2}$$

Note the following simple result, which holds for any graph for which the vertex degree is bounded.

Lemma 2.2. *Let $G = (V,E)$, let d be the usual graph distance on G, and suppose $|B(x,1)| \leqslant M$ for all $x \in V$. Let*

$$\mathcal{E}_G^{(n)}(g,g) = \tfrac{1}{2} \sum_x \sum_y 1_{(d(x,y) \leqslant n)} (g(x) - g(y))^2.$$

Then there exists $c_1(n,M)$, depending only on n, M, such that

$$\mathcal{E}_G^{(n)}(g,g) \leqslant c_1(n,M) \mathcal{E}_G^{(1)}(g,g).$$

Finally, we need to summarize some inequalities obtained for P in [BB3]. Define

$$\mathcal{E}^0(f,f) = \tfrac{1}{2} \int_P |\nabla f(z)|^2 \, dz.$$

We let $C_0(A)$ denote the capacity of A with respect to reflecting Brownian motion (with normal reflection) on P. Associated with reflecting Brownian motion on P is a constant d_w, called the walk dimension: see [BB3] for the definition of d_w in terms of electrical resistance. Set

$$d_s = 2d_f/d_w.$$

We start with the mass-capacity inequality.

Theorem 2.3. *([BB3, Sect. 7]). Suppose $d_s > 2$. There exists c_1 such that if A is a subset of P and $\mu_P(A) \geqslant 1$, then*

$$C_0(A) \geqslant c_1 \mu_P(A)^{(d_s-2)/d_s}.$$

We have the following conditional Nash inequality for $\mathcal{E}^0$.

Theorem 2.4. *([BB3, Sect. 7]). For each $c_1 > 0$ there exists c_2 such that*

$$\|f\|_2^{2+4/d_s} \leqslant c_2 \mathcal{E}^0(f,f) \|f\|_1^{4/d_s} \qquad \text{whenever } \mathcal{E}^0(f,f) \leqslant c_1 \|f\|_1^2$$

Finally we give a Poincaré inequality for P. Write

$$\mathcal{E}^0_{S(x_0,r)}(f,f) = \int_{S(x_0,r)} |\nabla f(x)|^2 \, dx$$

and

$$f_{S(x_0,r)} = \int_{S(x_0,r)} f(z) \, dz.$$

Theorem 2.5. *([BB3, Sect. 7]). There exists c_1 such that if $x_0 \in P$ and $r \geqslant 0$, then*

$$\int_{S(x_0,r)} |f - f_{S(x_0,r)}|^2 \leqslant c_1 r^{d_w} \mathcal{E}^0_{S(x_0,r)}(f,f).$$

In [BB3] we proved the Poincaré inequality for cubical domains. The same proof works, however, for $S(n,x)$. In any case we do not use Theorem 2.5 below.

3. Analytic Inequalities for G.

In this section we will derive Sobolev and Poincaré inequalities for $\mathcal{E}$ from the corresponding inequalities for the Dirichlet form $\mathcal{E}^0$ on the pre-carpet P.

To approximate functions on G by functions on P it will be useful to introduce another graph $H = (U \cup W, E_H)$. Let D be a union of cubes in $\mathcal{S}_0(P)$. Let $U = \mathbb{Z}^d \cap D$, $W = V \cap D$, and let $\{x,y\} \in E_H$ if and only if there exists $S \in \mathcal{S}_0(D)$ such that $x, y \in S$, and exactly one of x, y is in each of U, W. (So H consists of the centers and corners of the cubes in $\mathcal{S}_0(D)$, with edges joining the center of each cube $S \in \mathcal{S}_0(D)$ to its corners). Let d_H be graph distance on H, let $b(x,y) = 1_{(d_H(x,y)=1)}$, and let

$$\begin{aligned} \mathcal{E}^H_D(h,h) &= \tfrac{1}{2} \sum_{x \in U \cup W} \sum_{y \in U \cup W} b(x,y)(h(x) - h(y))^2 \\ &= \sum_{x \in U} \sum_{y \in W} b(x,y)(h(x) - h(y))^2. \end{aligned}$$

Set $B(x) = \sum_{y \in U \cup W} b(x,y)$, and let $\alpha(x) = 2^{-d} B(x)$. We regard α as a measure on $U \cup W$; note that $\alpha(x) = 1$ for $x \in W$. For $h : U \cup W \to \mathbb{R}$ write

$$||h||_p^p = \sum_{x \in U \cap W} |h(x)|^p \alpha(x), \qquad p \geqslant 1.$$

Note that

$$||h||_p^p = \sum_{S\in\mathcal{S}_0(D)} \Big(h(z_S) + 2^{-d} \sum_{y\in U\cap S} |h(y)|^p\Big). \tag{3.1}$$

If $g \in L^2(W,\alpha)$ define Γg on $U \cup W$ by:

$$\Gamma g = \begin{cases} g(x), & x \in W, \\ B(x)^{-1} \sum_y b(x,y)g(y) & x \in U. \end{cases}$$

Lemma 3.1. *(a) If $p \geqslant 1$ and $g \in L^p(W,\alpha)$ then there exists $c_1(p)$ such that*

$$\sum_{x\in W} |g(x)|^p \leqslant \sum_{x\in U\cup W} |\Gamma g(x)|^p \alpha(x) \leqslant c_1(p) \sum_{x\in W} g(x)^p.$$

(b) If $g \in L^1(W,\alpha)$ then

$$\sum_{x\in U} \Gamma g(x)\alpha(x) = \sum_{x\in W} g(x)\alpha(x).$$

(c) There exists c_2 such that

$$\mathcal{E}_D^H(\Gamma g, \Gamma g) \leqslant c_2 \mathcal{E}_D(g,g).$$

Proof. (a) As $\Gamma g(x) = g(x)$ and $\alpha(x) = 1$ for $x \in W$ the left hand side is obvious. For $x \in U$

$$|\Gamma g(x)|^p \leqslant c_3 \Big| \sum_{y\in W} b(x,y)g(y)\Big|^p \leqslant c_4 \sum_{y\in W} b(x,y)|g(y)|^p,$$

and summing over $x \in U$ gives the right hand side of (a).
(b) This follows immediately from the calculation

$$\sum_{x\in U} \Gamma g(x)\alpha(x) = \sum_{x\in U} 2^{-d} \sum_{y\in W} b(x,y)g(y) = \sum_{y\in U} g(y)\alpha(y).$$

(c) Write $h = \Gamma g$. If $x \in U$, $y \in W$, and $\{x,y\}$ is an edge, let $z = z(x,y) \in W$ be such that $\{x,z\} \in E_H$, and $|g(z) - g(y)| \geqslant |h(x) - h(y)|$; such a choice is possible from the definition of $h(x)$. Then $d(y,z) \leqslant 2^d$, and so

$$\begin{aligned} \mathcal{E}_D^H(h,h) &= \sum_{y\in W} \sum_{x\in U} b(x,y)(h(y) - h(x))^2 \\ &\leqslant \sum_{y\in W} \sum_{x\in U} (g(y) - g(z(x,y)))^2 \\ &\leqslant \sum_{y\in W} \sum_{z\in W} 1_{(d(y,z)\leqslant 2^d)}(g(y) - g(z))^2 \\ &= \mathcal{E}_D^{(2^d)}(g,g) \leqslant c_5 \mathcal{E}_D(g,g), \end{aligned}$$

where we used Lemma 2.2 in the last line. □

We now perform a similar procedure to relate $\mathcal{E}_D^H$ and the Dirichlet form $\mathcal{E}_D^0$ for reflecting Brownian motion in D, defined by

$$\mathcal{E}_D^0(f,f) = \tfrac{1}{2} \int_D |\nabla f|^2.$$

If $h : U \cup W \to \mathbb{R}$, let

$$\tilde{h}(x) = 2^{-d} \sum_{y \in U} b(x,y) h(y), \qquad x \in W.$$

Define for $y \in U$, $z \in W$,

$$\varphi_y(x) = 0 \vee \prod_{i=1}^d (1 - |x_i - y_i|),$$

$$\psi_z(x) = 0 \vee \prod_{i=1}^d (1 - 2|x_i - z_i|).$$

Note that if $S \in \mathcal{S}_0(D)$, $y \in S \cap U$ then

$$\int_S \varphi_y(x)dx = \int_S \psi_{z_S}(x)dx = 2^{-d} \tag{3.2}$$

and

$$\int_S |\nabla \varphi_y(x)|^2 dx \leqslant 1, \qquad \int_S |\nabla \psi_y(x)|^2 dx \leqslant 2. \tag{3.3}$$

Given $h : U \cup W \to \mathbb{R}$ define $\Lambda h : D \to \mathbb{R}_+$ as follows. For $S \in \mathcal{S}_0(D)$, Λh is defined on S by

$$\Lambda h(x) = \sum_{y \in U \cap S} h(y)\varphi_y(x) + (h(z_S) - \tilde{h}(z_S))\psi_{z_S}(x), \qquad x = (x_1, \dots, x_d) \in S.$$

Note that the definition of Λh agrees on the intersection of two cubes in $\mathcal{S}_0(D)$, and that $\Lambda h(x) = h(x)$ for $x \in U \cup W$. So Λh is a continuous extension of h from $D \cap (U \cup W)$ to D.

Lemma 3.2. *(a) For $p \geqslant 1$, if $h \in L^p(U \cup W, \alpha)$, then there exist $c_1(p)$ and $c_2(p)$ such that*

$$c_1(p)\|h\|_p^p \leqslant \int_D |\Lambda h(x)|^p \, dx \leqslant c_2(p)\|h\|_p^p.$$

(b) $$\int_D \Lambda h(x)dx = 2^{-d} \sum_{x \in W} \alpha(x)h(x) + (1-2^{-d}) \sum_{y \in U} \alpha(y)h(y).$$

(c) There exists c_3 such that

$$\mathcal{E}_D^0(\Lambda h, \Lambda h) \leqslant c_3 \mathcal{E}_D^H(h,h).$$

Proof. (a) Let $S \in \mathcal{S}_0(D)$, and let $y' \in S \cap (U \cup W)$ be a point at which $|h|$ attains its maximum. Then

$$|h(z_S)|^p + 2^{-d} \sum_{y \in U \cap S} |h(y)|^p \leqslant 2|h(y')|^p. \tag{3.4}$$

On the other hand, as $\Lambda h(y') = h(y')$, and the modulus of continuity of Λh in S is bounded by a constant times $h(y')$, there exists $c_4(p) > 0$ such that

$$\int_S \Lambda h(x)dx \geqslant c_4(p)|h(y')|^p. \tag{3.5}$$

The left hand side of (a) now follows on summing over $S \in \mathcal{S}_0(D)$, and using (3.1), (3.4) and (3.5). As

$$\int_S |\varphi_y(x)|^p dx = c_5(p), \qquad \int_S |\psi_{z_S}(x)|^p dx = c_6(p),$$

the right hand side of (a) follows easily from the definition of Λh.
(b) This follows from (3.2) and the definition of Λh by an straightforward calculation.
(c) Write $f = \Lambda h$, and fix $S \in \mathcal{S}_0(D)$. Then for $x \in S$, as $\sum_{y \in U \cap S} \varphi_y = 1$ on S,

$$\begin{aligned} f(x) &= \sum_{y \in U \cap S} h(y)\varphi_y(x) + (h(z_S) - \widetilde{h}(z_S))\psi_{z_S}(x) \\ &= \sum_{y \in U \cap S} (h(y) - h(z_S))(\varphi_y(x) - 2^{-d}\psi_{z_S}(x)) + h(z_S). \end{aligned}$$

Differentiating and using Cauchy-Schwarz it follows that

$$|\nabla f(x)|^2 \leqslant 2 \sum_{y\in U\cap S} (h(y)-h(z_S))^2 \Big(\sum_{y\in U\cap S} |\nabla \varphi_y(x)|^2 + 4^{-d}|\nabla \psi_S(x)|^2 \Big).$$

Using (3.3) we therefore have

$$\int_S |\nabla f(x)|^2 \leqslant c_7 \sum_{y\in U\cap S} (h(y)-h(z_S))^2,$$

and summing over $S \in \mathcal{S}_0(D)$ gives (c). □

We can now deduce Nash and Sobolev inequalities for $\mathcal{E}$ from those for $\mathcal{E}^0$. For the next few results, we take $D = P$, so that $W = V$.

Theorem 3.3. *Let $g \in L^2(W, |\cdot|)$. Then*

$$||g||_2^{1+4/d_s} \leqslant c_1 \mathcal{E}(g,g) ||g||_1^{4/d_s}. \tag{3.6}$$

Further, if $d_s > 2$ then

$$||g||_p^2 \leqslant c_2 \mathcal{E}(g,g), \qquad p = 2d_s/(d_s-2). \tag{3.7}$$

Proof. Since $\mathcal{E}(g^+,g^+) \leqslant \mathcal{E}(g,g)$, it is sufficient to prove the theorem for non-negative g. Let $g \geqslant 0$, $h = \Gamma g$, $f = \Lambda h$. By Lemmas 3.1 and 3.2 we have, for $p = 1,2$,

$$c_3(p)||g||_p \leqslant ||f||_p \leqslant c_4(p)||g||_p, \tag{3.8}$$

and

$$\mathcal{E}^0(f,f) \leqslant c_5 \mathcal{E}_H(h,h) \leqslant c_6 \mathcal{E}(g,g). \tag{3.9}$$

Therefore, using (2.4), $\mathcal{E}^0(f,f) \leqslant c_7 ||f||_1^2$. So, by Theorem 2.4 we have

$$||f||_2^{1+4/d_s} \leqslant c_1 \mathcal{E}^0(f,f) ||f||_1^{4/d_s},$$

and (3.6) then follows from (3.8) and (3.9).

If $d_s > 2$ then (3.7) is immediate from (3.6) and Theorem 2.16 of [CKS]. □

The Sobolev inequality (3.7) implies a mass-capacity inequality. Write $C_G(A)$ for the capacity of $A \subset V$.

Proposition 3.4. *If $d_s > 2$ then for $A \subset V$*

$$C_G(A) \geqslant c_1 |A|^{(d_s-2)/d_s}.$$

Proof. We apply (3.7) to functions g that are one on A and tend to 0 as $|x| \to \infty$ to obtain

$$|A|^{2/p} \leqslant c_2 \mathcal{E}(g,g).$$

Taking the infimum over all such g and using the definition of $C_G(A)$ proves the proposition. □

We prove a Poincaré inequality for $\mathcal{E}$.

Theorem 3.5. *Let $n \geqslant 0$, $x_0 \in P$, and let $A = B(x_0, n)$. There exists c_1 such that if $g : V \cap D \to \mathbb{R}$, then*

$$\sum_{x \in D \cap V} |g(x) - g_D|^2 \leqslant c_1 n^{d_w} \mathcal{E}_D(g,g),$$

where $g_D = |D \cap V|^{-1} \sum_{x \in D \cap V} g(x)$.

Proof. Let $D = S(x_0, n)$. By subtracting a constant from g we can assume that $g_D = 0$. Let $h = \Gamma g$, $f = \Lambda h$. By part (b) of Lemmas 3.1 and 3.2, we have

$$0 = \sum_{x \in V} h(x) = \sum_{x \in U} \alpha(x) h(x) = \int_D f(x)\,dx,$$

so, by Theorem 2.5,

$$\int_D f^2 \leqslant c_2 n^{d_w} \mathcal{E}_0^D(f,f).$$

On the other hand, by Lemmas 3.1 and 3.2,

$$\sum_{x \in D \cap V} g(x)^2 \leqslant c_3 \int_D f^2, \qquad \mathcal{E}_0^D(f,f) \leqslant c_4 \mathcal{E}^D(g,g),$$

and combining these inequalities proves the theorem. □

Proposition 3.6. *Suppose $d_s > 2$. There exist c_1 and c_2 such that for all x and for all $r \geqslant 1$*

$$c_1 r^{d_f - d_w} \leqslant C_G(B(x,r)) \leqslant c_2 r^{d_f - d_w}.$$

Proof. The lower bound follows from the mass-capacity inequality for G:

$$C_G(B(x,r))^{d_s/(d_s-2)} \geqslant c_3 |B(x,r)| \geqslant c_4 r^{d_f}.$$

So
$$C_G(B(x,r)) \geqslant c_5 r^{d_f(d_s-2)/d_s} = c_5 r^{d_f-d_w}.$$

We obtain the upper bound as follows. Let π be the capacitary measure for $S(x,r)$ with respect to Brownian motion on the pre-carpet. Let $z \in \partial S(x,2r)$. Then as $u(\cdot,z)$, the Green function for reflecting Brownian motion in P, is harmonic in $S(x,r)$, by the Harnack inequality ([BB3, Sect. 4]),

$$1 \geqslant u\pi(z) = \int_{S(x,r)} u(y,z)\,\pi(dy) \geqslant c_6 u(x,z)\pi(S(x,r)).$$

Using this,
$$C_0(S(x,r)) = \pi(S(x,r)) \leqslant c_6^{-1} u(x,z)^{-1}.$$
By [BB3], $u(x,z) \geqslant c_7|x-z|^{d_f-d_w}$, so

$$C_0(S(x,r)) \leqslant c_8|x-z|^{d_f-d_w} \leqslant c_9 r^{d_f-d_w}.$$

By [FOT], $C_0(S(x,r))$ is also equal to

$$\inf\{\mathcal{E}(g,g) : g = 1 \text{ on } S(x,r), g(y) \to 0 \text{ as } |y| \to \infty\}.$$

Choose g such that $\int |\nabla g|^2 \leqslant 2c_9 r^{d_f-d_w}$, $g=1$ on $S(x,r)$, and $g(y) \to 0$ as $|y| \to \infty$.

For each $x \in V$, let S_x be the unit cube whose center is x. Define h on V by
$$h(x) = \int_{S_x} g(y)\,dy.$$

Suppose $x, y \in V$ with $|x-y| = 1$. Let $e = y - x$. Then

$$\begin{aligned}(h(y)-h(x))^2 &= \Big|\int_{S_y} g - \int_{S_x} g\Big|^2 = \Big|\int_{S_x} (g(z+e)-g(z))\,dz\Big|^2 \\ &= \Big|\int_{S_x}\int_0^1 \nabla g(z+te)\,dt\,dz\Big|^2 \leqslant \int_0^1\int_{S_x} |\nabla g(z+te)|^2\,dz\,dt \\ &\leqslant \int_{S_x\cup S_y} |\nabla g(z)|^2\,dz.\end{aligned}$$

Summing over all pairs (x,y) with $|y-x|=1$, we obtain (with $D=P$)
$$\mathcal{E}_D^H(h,h) \leqslant 2^d \mathcal{E}_0^D(g,g).$$

Our result follows as

$$C_G(B(x,r)) = \inf\{\mathcal{E}_D^G(h,h) : h = 1 \text{ on } B(x,r), h(y) \to 0 \text{ as } |y| \to \infty\},$$

$\mathcal{E}_G^D(h,h) \leqslant c_{11}\mathcal{E}_H^D(h,h)$, and $C_G(B(x,r)) \leqslant C_G(S(x,2r))$. □

4. Coupling and Harnack inequalities.

The most difficult step in [BB3] was the proof of a coupling result for reflecting Brownian motion on the pre-carpet P. We now translate this theorem into the graph setting.

Call $A \subset \mathbb{R}^d$ a *half-face* if there exists $i \in \{1,\dots,d\}$, $a = (a_1,\dots,a_d) \in \frac{1}{2}\mathbb{Z}^d$ with $a_i \in \mathbb{Z}$ such that

$$A = \{x : x_i = a_i, \quad a_j \leqslant x_j \leqslant a_j + 1/2 \quad \text{for } j \neq i\}.$$

Let $\mathcal{A}$ be the collection of half-faces, and for $A \in \mathcal{A}$ let w_A be the center of A. For $x, y \in \mathbb{R}^d$ write $[x,y]$ for the line segment with endpoints x, y. For $S \in \mathcal{S}_0$ set

$$\Gamma_S = \{[z_S, w_A] : A \subset S\}.$$

(So, if $d = 3$, Γ_S consists of the 24 line segments, each joining the center of S to the center of one of the half-faces bordering S). Set

$$\Gamma = \bigcup\{\Gamma_S : S \in \mathcal{S}(P)\}.$$

Let $Z = (Z_t, t \geqslant 0, \mathbb{Q}^x, x \in \Gamma)$ be the 'cable process' of Varopoulos [V2] on the set Γ. Z is a diffusion on Γ: on the cables $[z_S, w_A]$ Z behaves like a standard 1-dimensional Brownian motion, while at the vertices z_S (and the 'trivial' vertices w_A) Z behaves like a Walsh Brownian motion (see [BPY]), taking excursions along each of the cables meeting z_S with equal probability.

Let $\gamma(dx)$ be 1-dimensional Lebesgue measure on Γ. Z is the diffusion associated with the Dirichlet form

$$\tfrac{1}{2}\int_\Gamma |\nabla g(x)|^2\, dx,$$

on $\mathcal{D}^Z \subset L^2(\Gamma,\gamma)$. Here, $\mathcal{D}^Z$ is the set of all $g \in L^2(\Gamma,\gamma)$ such that the integral above is finite, and for $x \in \Gamma - V$, $\nabla g(x)$ is the derivative of g in the direction of the line segment containing z. (Since $\gamma(V) = 0$ the definition of $\nabla g(x)$ for $x \in V$ is not relevant).

We can now use the argument of [BB3, Sect. 3] to prove a coupling theorem for Z. While the arguments of [BB3] are written for reflecting Brownian motion W on P, all they use is that W satisfies certain symmetry conditions, and the fact that the set of boundaries between half faces

$$A^* = \bigcup\{A \cap B; A, B \in \mathcal{A}, A \neq B\}$$

is polar for W. (This last condition, while convenient, is no doubt not an essential feature of the argument). Z satisfies the same symmetry conditions, and since $A^* \cap \Gamma = \emptyset$, A^* is polar for Z. Hence we deduce the following coupling result.

Given two processes, X, Y, define

$$T_C(X,Y) = \inf\{t \geqslant 0 : X_t = Y_t\}.$$

Theorem 4.1. *There exists $r_0 \geqslant 1$, and constants $p_4 = p_4(d, l_F) > 0$, $c_1 > 0$ such that if $x_0 \in V$, $r \geqslant r_0$, $x, y \in \Gamma \cap S(x, c_1 r)$ then there exists a probability space $(\Omega, \mathcal{F}, (\mathcal{F}_t), \mathbb{P})$ carrying processes $(Z_t^x, t \geqslant 0)$, $(Z_t^y, t \geqslant 0)$ with the following properties.*
(a) *For $z = x, y$ the process $(Z_t^z, t \geqslant 0)$ is equal in law to $(Z_t, t \geqslant 0)$ under $\mathbb{Q}^z$. In particular $Z_0^x = x$, $Z_0^y = y$.*
(b)

$$\mathbb{P}\big(T_C(Z^x, Z^y) < \sigma(x_0, r, Z^x) \wedge \sigma(x_0, r, Z^y)\big) > p_4 > 0.$$

Definition 4.2. Let $\mathcal{L}_\Gamma$ be the infinitesimal generator of Z. If D is an open subset of Γ we say h is *Z-harmonic on D* if $\mathcal{L}_\Gamma h = 0$ on D.

If A is a subset of V we say h is *X-harmonic* (or just harmonic) on A if $\mathcal{L}h = 0$ on D, where $\mathcal{L}$ is given by (1.2).

Note that if D is an open subset of Γ, and $J = \{x \in V \cap D : B(x, 1) \subset D\}$ then if h is Z-harmonic on D, $h|_V$ is X-harmonic on J.

Once we have coupling, we deduce a uniform Harnack inequality for Z_t. The proof is the same word for word as Section 4 of [BB3].

Theorem 4.3. *Let $x_0 \in V$. There exists c_1 such that if u is nonnegative and Z-harmonic in $B(x_0, 2r) \subset \Gamma$ then*

$$u(y) \leqslant c_1 u(x), \qquad x, y \in S(x_0, r) \cap \Gamma.$$

Using the fact that $u|_V$ is (X)-harmonic we therefore obtain immediately a Harnack inequality for the discrete random walk X on G.

Theorem 4.4. *Let $x_0 \in V$. There exists c_1 such that if u is nonnegative and harmonic in $B(x_0, 2r)$, then*

$$u(y) \leqslant c_1 u(x), \qquad x, y \in B(x_0, r).$$

Proof. Let u be harmonic for $B(x_0, 2r)$. By linear interpolation extend u to a function v on $S(x_0, 2r) \cap \Gamma$. Since v is harmonic with respect to the cable process Z_t, the result follows by Theorem 4.3. □

Let $\operatorname{Osc}_A h$ denote the difference of the supremum and infimum of h on A. A consequence of the Harnack inequality is the following.

Proposition 4.5. *If h is bounded and harmonic on $B(x, 2r)$, then there exists $\theta < 1$ independent of x, r, and h such that*

$$\operatorname*{Osc}_{B(x,r)} h \leqslant \theta \operatorname*{Osc}_{B(x,2r)} h.$$

Proof. By looking at $Ah+B$ for suitable constants A and B, we may assume $\sup_{B(x,2r)} h = 1$ and $\inf_{B(x,2r)} h = 0$. By looking at $1-h$ if necessary, we may assume $h(x) \geqslant 1/2$. By the Harnack inequality, $h \geqslant c_1/2$ on $B(x,r)$. Therefore $\operatorname{Osc}_{B(x,r)} h \leqslant 1 - c_1/2$, and our result follows by setting $\rho = 1 - c_1/2$. □

5. Upper bounds.

Remark 5.1. It is easy to see that the random walk X_n on G has period 2, and so $p(n, x, y)$ is equal to 0 unless n and $d(x, y)$ have the same parity (i.e., both are even or both are odd).

Upper bounds for $p(n, x, x)$ are quick.

Theorem 5.2. *There exists c_1 such that if $x \in V$, then*

$$p(n, x, x) \leqslant c_1 n^{-d_s/2}.$$

Proof. This follows from the Nash inequality (3.6) and Theorem 2.1 of [CKS]. □

The main work is in obtaining good upper bounds for $p(n, x, y)$ when $|x - y|$ is relatively large. For simplicity we consider only the case $d_s > 2$ in this section, and consider the case $d_s \leqslant 2$ in Section 7. First we obtain estimates on the Green function $g(x, y)$ for the random walk on G, defined by

$$g(x, y) = \sum_{m=0}^{\infty} p(m, x, y).$$

Theorem 5.3. *There exist c_1 and c_2 such that*

$$c_1 d(x,y)^{d_w - d_f} \leqslant g(x,y) \leqslant c_2 d(x,y)^{d_w - d_f}. \tag{5.1}$$

Proof. Fix $x \in V$. Write $N(x, r)$ for $\partial(B(x,r)^c)$; note that $N(x, r)$ consists of those points of $B(x, r)$ which have a neighbor in $B(x, r)^c$. It is clear from the geometry of V that there exist r_0 and M, both independent of x, with the following property: if $r \geqslant r_0$, there exist M points $x_1, \ldots, x_M$ in $N(x, 4r)$

such that $x_{i+1} \in B(x_i, r)$, $i = 1, \ldots, M-1$, and $N(x, 4r) \subseteq \cup_{i=1}^{M} B(x_i, r)$; we emphasize that M does not depend on r.

Suppose $r \geqslant r_0$. By the Harnack inequality repeated M times on the sets $B(x_i, 2r)$, there exists c_3 such that if u is nonnegative and harmonic on $V - \{x\}$, then

$$\sup_{N(x,4r)} u \leqslant c_3 \inf_{N(x,4r)} u. \tag{5.2}$$

We apply this to $g(x, \cdot)$. Fix $x_0 \in N(x, 4r)$. Let π be the capacitary measure for $B(x, 4r)$. It is well-known that $g\pi$ is equal to 1 on $B(x, 4r)$, and moreover π is supported on $N(x, 4r)$. We then have

$$1 = g\pi(x) = \sum_{z \in N(x,4r)} g(x, z)\pi(\{z\}).$$

By the Harnack inequality (5.2) there exist c_4 and c_5 such that

$$c_4 g(x, x_0) \leqslant g(x, z) \leqslant c_5 g(x, x_0), \qquad z \in N(x, 4r).$$

Thus

$$1 \leqslant c_5 g(x, x_0)\pi(N(x, 4r)) = c_5 g(x, x_0) C_G(B(x, 4r))$$

and

$$1 \geqslant c_4 g(x, x_0)\pi(N(x, 4r)) = c_4 g(x, x_0) C_G(B(x, 4r)).$$

Using Proposition 3.6 we have the inequality (5.1) for $g(x, x_0)$. By the Harnack inequality we have our result for $g(x, y)$ when $r \leqslant d(x, y) \leqslant 8r$.

It remains to consider the case when $d(x, y) < r_0$. As

$$g(x, y) = \sum_{m=0}^{\infty} p(m, x, y),$$

and there is positive probability that X_n will hit y before $2r$ steps, the lower bound is clear.

We examine the upper bound. Since $g(x, y) \leqslant g(x, x)$ by the maximum principle, we need only show $g(x, x) \leqslant c_6$. There exists $c_7 > 0$ independent of y such that starting at y, there is probability at least c_7 that X_n will leave $B(x, r_0)$ within $2r_0$ steps. The probability that X_n remains within $B(x, r_0)$ after $2jr_0$ steps is then, by a standard renewal argument, bounded by $(1 - c_7)^j$. Thus

$$\mathbb{E}^x \sum_{m=0}^{\sigma(x,r_0)} 1_{\{x\}}(X_m) \leq \mathbb{E}^x \sigma(x, r_0) \leqslant c_8.$$

By the strong Markov property,

$$g(x,x) \leqslant \mu(x)^{-1}\mathbb{E}^x \sum_{m=0}^{\sigma(x,r_0)} 1_{\{x\}}(X_m) + \mathbb{E}^x g(X(\sigma(x,r_0)),x),$$

and this is bounded by $c_8 + c_2 r_0^{d_w - d_f} < \infty$. □

The next step is to obtain upper and lower bounds on $\mathbb{E}^y \sigma(x,r)$.

Theorem 5.4. *There exist c_1 and c_2 such that if $x \in V$,*

$$c_1 r^{d_w} \leqslant \mathbb{E}^x \sigma(x,r),$$

and

$$\mathbb{E}^y \sigma(x,2r) \leqslant c_2 r^{d_w}, \qquad y \in B(x,r).$$

Proof. The upper bound comes from Theorem 5.3 and

$$\begin{aligned}\mathbb{E}^y \sigma(x,2r) &\leqslant \mathbb{E}^y \sum_{m=0}^{\infty} 1_{B(x,2r)}(X_m)\\ &\leqslant \sum_{B(x,2r)} g(y,z)\mu(z) \leqslant c_3 \sum_{B(x,2r)} d(y,z)^{d_w - d_f} \leqslant c_4 r^{d_w}.\end{aligned}$$

For the lower bound we will choose $\theta \in (0,1)$ and r_0 in a moment and establish the lower bound first for $r \geqslant r_0$. We write

$$\begin{aligned}&\mathbb{E}^y \sum_{m \leqslant \sigma(x,r)} 1_{B(x,\theta r)}(X_m) \qquad (5.3)\\ &\qquad = \mathbb{E}^y \sum_{m=0}^{\infty} 1_{B(x,\theta r)}(X_m) - \mathbb{E}^y \mathbb{E}^{X(\sigma(x,r))} \sum_{m=0}^{\infty} 1_{B(x,\theta r)}(X_m).\end{aligned}$$

The first term on the right hand side of (5.2) is equal to

$$\sum_{z \in B(x,\theta r)} g(x,z)\mu(z) \geqslant \sum_{z \in B(x,\theta r)} (c_1 d(x,z))^{d_w - d_f} \geqslant c_5 (\theta r)^{d_w},$$

provided $\theta r \geqslant 1$. The second term on the right hand side of (5.3) is equal to

$$\mathbb{E}^x \sum_{z \in B(x,\theta r)} g(X(\sigma(x,2r)),z)\mu(z). \qquad (5.4)$$

For $z \in B(x, \theta r)$ and $y \in \partial B(x, 2r)$, we have $g(y, z) \leqslant c_6 r^{d_w - d_f}$, so (5.4) is bounded above by

$$c_7(\theta r)^{d_f} c_6 r^{d_w - d_f} = c_8 \theta^{d_f} r^{d_w}.$$

Therefore

$$\begin{aligned} \mathbb{E}^x \sigma(x,r) &\geqslant \mathbb{E}^x \sum_{m \leqslant \sigma(x,r)} 1_{B(x,\theta r)}(X_m) \\ &\geqslant c_5 \theta^{d_w} r^{d_w} - c_8 \theta^{d_f} r^{d_w}, \end{aligned}$$

provided $\theta r \geqslant 1$.

As $d_s > 2$, then $d_f > d_w$. Choose θ small so that $c_5 \theta^{d_w} \geqslant 2c_8 \theta^{d_f}$, and then r_0 large so that $\theta r_0 \geqslant 1$. For $r \geqslant r_0$ we then have

$$\mathbb{E}^x \sigma(x,r) \geqslant c_8 \theta^{d_f} r^{d_w}.$$

The case $r \leqslant r_0$ is trivial if we adjust c_1, since $\mathbb{E}^x \sigma(x,r) \geqslant 1$. □

We now follow [BB1] and [BB2] closely to get a tightness estimate.

Theorem 5.5. *There exist c_1 and c_2 such that*

$$\mathbb{P}^x\big(\sup_{m \leqslant n} |X_m - X_0| > \lambda\big) \leqslant c_1 \exp\Big(-c_2\Big(\frac{\lambda^{d_w}}{n}\Big)^{1/(d_w-1)}\Big).$$

Proof. From Theorem 5.4 we have

$$\begin{aligned} c_3 r^{d_w} &\leqslant \mathbb{E}^x \sigma(x,r) \leqslant n + \mathbb{E}^x \Big[\mathbb{E}^{X_n} \sigma(x,r); \sigma(x,r) > n\Big] \\ &\leqslant n + c_4 r^{d_w} \mathbb{P}^x(\sigma(x,r) > n) \\ &= n + c_4 r^{d_w} - c_4 r^{d_w} \mathbb{P}^x(\sigma(x,r) \leqslant n). \end{aligned}$$

We thus obtain

$$\mathbb{P}^x(\sigma(x,r) \leqslant n) \leqslant c_5 + c_6 n r^{-d_w},$$

or

$$\mathbb{P}^x(\sigma(x,r) r^{-d_w} \leqslant x) \leqslant c_5 + c_6 x, \qquad x > 0. \tag{5.5}$$

In order for $|X_m - X_0|$ to exceed λ for some $m \leqslant n$, X_m must exit disjoint balls of the form $B(x_i, r)$ at least $M = \lambda/2r$ times. Let Y_k, $k = 1, \dots, M$, be r^{-d_w} times the time to exit the kth ball. By the strong Markov property,

$$\mathbb{P}(Y_i \leqslant x \mid Y_1, \dots, Y_{i-1}) \leqslant c_5 + c_6 x.$$

So

$$\mathbb{E}\Big(e^{-uY_i} \mid Y_1, \ldots, Y_{i-1}\Big) \leqslant c_5 + \int_0^{(1-c_5)/c_6} e^{-ux} c_6 \, dx \leqslant c_5 + c_6 u^{-1}.$$

Then

$$\begin{aligned}\mathbb{P}\Big(\sum_{i=1}^{M} Y_i \leqslant x\Big) = \mathbb{P}\Big(e^{-u\sum Y_i} \geqslant e^{-ux}\Big) &\leqslant e^{ux}\mathbb{E}e^{-u\sum Y_i} \\ &\leqslant e^{ux}(c_5 + c_6 u^{-1})^M \\ &\leqslant c_5^M \exp\Big(ux + \frac{c_6 M}{c_5 u}\Big).\end{aligned}$$

Taking $u = (c_6 M / c_5 x)^{\frac{1}{2}}$,

$$\mathbb{P}(\sum Y_i \leqslant x) \leqslant \exp\Big(2(c_6 M x / c_5)^{\frac{1}{2}} - M \log(1/c_5)\Big).$$

We now choose r so that $M \in [c_7 \log(c_8/n), 2c_7 \log(c_8/n)]$ for appropriate c_7 and c_8. □

We now get, following [BB2], the off-diagonal estimate.

Theorem 5.6. *There exist c_1 and c_2 such that*

$$p(n, x, y) \leqslant c_1 n^{-d_s/2} \exp\Big(-c_2\Big(\frac{d(x,y)^{d_w}}{n}\Big)^{1/(d_w-1)}\Big). \tag{5.6}$$

Proof. Fix x, y, n, let $A_x = \{z \in V : d(x,z) \leqslant d(x,y)\}$, $A_y = \{z \in V : d(y,z) \leqslant d(x,y)\}$, and $S = \inf\{n \geqslant 0 : |X_n - X_0| \geqslant d(x,y)/3\}$. Let us first suppose n is even. Then

$$\mu(y)p(n,x,y) = \mathbb{P}^x(X_n = y, X_{n/2} \in A_x) + \mathbb{P}^x(X_n = y, X_{n/2} \in A_y). \tag{5.7}$$

For the second term,

$$\begin{aligned}\mathbb{P}^x(X_n = y, X_{n/2} \in A_y) &= \mathbb{P}^x(S < n/2, X_n = y, X_{n/2} \in A_y) \\ &\leqslant \mathbb{P}^x(S < n/2) \sup_{y' \in A_y} \mathbb{P}^{y'}(X_{n/2} = y) \\ &\leqslant \mathbb{P}^x(S < n/2) \sup_{y' \in A_y} p(n/2, y', y)\mu(y).\end{aligned}$$

For the first term in (5.7), by symmetry,

$$\mu(x)\mathbb{P}^x(X_n = y, X_{n/2} \in A_x) = \mu(y)\mathbb{P}^y(X_n = x, X_{n/2} \in A_x),$$

which is bounded the same way.

Using Theorems 5.2 and 5.5, we have our result when n is even. If n is odd, we have by the Markov property that

$$p(n,x,y) = \sum_{\{z:(y,z)\in E\}} p(n-1,x,z)p(1,z,y)\mu(z);$$

applying (5.5) to $p(n-1,x,z)$ establishes (5.6) when n is odd. □

6. Lower bounds.

Theorem 6.1. *There exists c_1 such that if n is even,*

$$p(n,x,x) \geqslant c_1 n^{-d_s/2}.$$

Proof. Let $m = n/2$. Recall

$$\mathbb{P}^x(\sigma(x,r) \leqslant n) \leqslant c_2 \exp(-c_3(r^{d_w}/n)^{1/(d_w-1)}). \tag{6.1}$$

Pick r to be the first integer such that $c_2 \exp(-c_3(r^{d_w}/n)^{1/(d_w-1)}) \leqslant 1/2$. Then

$$\mathbb{P}^x(X_m \in B(x,r)) \geqslant \mathbb{P}^x(\sigma(x,r) > n/2) \geqslant \tfrac{1}{2},$$

while

$$\mu(B(x,r)) \lessapprox c_4 r^{d_f} \lessapprox c_5 m^{d_s/2}.$$

By Cauchy-Schwarz,

$$\begin{aligned}\tfrac{1}{4} \leqslant [\mathbb{P}^x(X_{n/2} \in B(x,r))]^2 &= \Big[\sum_{B(x,r)} p(n/2,x,y)\mu(y)\Big]^2 \\ &\leqslant \mu(B(x,r)) \sum_{B(x,r)} p(n/2,x,y)^2 \mu(y) \\ &\leqslant \mu(B(x,r))p(n,x,x).\end{aligned}$$

Using the estimate for $\mu(B(x,r))$ yields the theorem. □

Fix $n \geqslant 1$, such that n is even, let A be a large constant such that $c_2 \exp(-c_3 A^{d_w/(d_w-1)}) \leqslant 1/2$, where c_2 and c_3 are as in (6.1), and let $D_n = S(x, An^{1/d_w})$. Let $\overline{p}(m,y,z)$ be the transition probabilities for the random walk X_m killed on exiting D_n and let $\overline{g}(y,z) = \sum_{m=0}^{\infty} \overline{p}(m,y,z)$ be the corresponding Green function.

The same as argument as that in Theorem 6.1 proves

Corollary 6.2. *There exists c_1 such that if n is even, $\overline{p}(n,x,x) \geqslant c_1 n^{-d_s/2}$.*

Fix x and let $F_x = \{y \in D_n : d(x,y) \text{ is even}\}$. To get a lower bound for the off-diagonal terms, we start with

Proposition 6.3. *There exist c_1 and $h : D_n \to \mathbb{R}$ such that $|h|$ is bounded by $c_1 n^{-d_s/2-1}$ and if n is even $\overline{g}h(y) = \overline{p}(n,x,y)$ if $y \in F_x$.*

Proof. By the semigroup property, $\overline{p}(2,\cdot,\cdot)$ is a symmetric nonnegative definite matrix. Thus if we consider $\overline{p}(2,\cdot,\cdot)$ restricted to $F_x \times F_x$, we still have a symmetric nonnegative definite matrix, and by the spectral theorem there exist a collection of eigenvalues $\lambda_1, \lambda_2, \cdots, \lambda_M$ and a collection of corresponding orthonormal eigenvectors $\varphi_i : F_x \to \mathbb{R}$ (orthonormal with respect $\mu|_{F_x}$) such that

$$\overline{p}(2,y,z) = \sum_{i=1}^{M} \lambda_i \varphi_i(y)\varphi_i(z).$$

We extend φ_i to all of D_n by setting it equal to zero on $D_n - F_x$.

By the semigroup property,

$$\overline{p}(2m+2,y,z) = \sum_{w \in D_n} \overline{p}(2m,y,w)\overline{p}(2,z,w)\mu(w).$$

For $y, z \in F_x$ we know $\overline{p}(2,z,w)$ is 0 unless w is also in F_x; hence we may replace the sum in the above equation by a sum over $w \in F_x$. Using the eigenvalue expansion and induction, we then obtain

$$\overline{p}(2m,y,z) = \sum_{i=1}^{M} \lambda_i^m \varphi_i(y)\varphi_i(z), \qquad y,z \in F_x.$$

Since $\overline{p}(2,y,z)$ is nonegative definite, it follows that $\lambda_i \geqslant 0$. Since $\overline{p}$ is a submarkovian kernel, $\lambda_i \leqslant 1$ for all i.

Let

$$h(z) = \sum_{i=1}^{M} (1-\lambda_i)\lambda_i^n \varphi_i(x)\varphi_i(z)$$

if $z \in F_x$ and zero otherwise. If $y \in F_x$, then

$$\overline{g}h(y) = \sum_{z \in D_n} \overline{g}(y,z)h(z) = \sum_{m=0}^{\infty} \sum_{z \in D_n} \overline{p}(m,y,z)h(z).$$

Since h is zero on $D_n - F_x$ and $y \in F_x$, then $\overline{p}(m,y,z)h(z)$ is nonzero only if m is even and $z \in F_x$. Therefore

$$\begin{aligned}\overline{g}h(y) &= \sum_{m=0}^{\infty}\sum_{z\in F_x} \overline{p}(2m,y,z)h(z)\\ &= \sum_{m=0}^{\infty}\sum_{i=1}^{M}(1-\lambda_i)\lambda_i^m\lambda_i^n\varphi_i(x)\varphi_i(y) = \sum_{i=1}^{M}\lambda_i^n\varphi_i(x)\varphi_i(y)\\ &= \overline{p}(n,x,y).\end{aligned}$$

It remains to bound $|h(z)|$. By the Cauchy-Schwarz inequality,

$$|h(z)| \leqslant \Big(\sum_{i=1}^{M}(1-\lambda_i)\lambda_i^n\varphi_i(x)^2\Big)^{1/2}\Big(\sum_{i=1}^{M}(1-\lambda_i)\lambda_i^n\varphi_i(z)^2\Big)^{1/2}$$

Note the function $\lambda \to (1-\lambda)\lambda^{n/2}$ is bounded by c_2/n for $\lambda \geqslant 0$. We then observe that

$$\begin{aligned}\sum_{i=1}^{M}(1-\lambda_i)\lambda_i^n\varphi_i(z)^2 &\leqslant \frac{c_2}{n}\sum_{i=1}^{M}\lambda_i^{n/2}\varphi_i(z)^2\\ &\leqslant \frac{c_2}{n}\overline{p}(n/2,z,z) \leqslant \frac{c_2}{n}p(n/2,z,z)\\ &\leqslant c_3 n^{-d_s/2-1}.\end{aligned}$$

This completes the proof. □

Proposition 6.4. *There exist c_1 and c_2 such that if n and $d(x,y)$ are even and $d(x,y) \leqslant c_1 n^{1/d_w}$, then $p(n,x,y) \geqslant c_2 n^{-d_s/2}$.*

Proof. We will choose a constant c_3 in a moment. Let $\rho = \min\{n \geqslant 0; n \in 2\mathbb{Z}, n \geqslant \sigma(x, c_3 n^{1/d_w})$. For any $y \in F_x$,

$$\overline{p}(n,x,y) = \overline{g}h(y) = \mathbb{E}^y\sum_{m=0}^{\rho} h(X_m) + \mathbb{E}^y\overline{g}h(X_\rho). \tag{6.2}$$

The first term on the right is bounded by

$$\|h\|_\infty \mathbb{E}^y\rho \leqslant c_4 c_3^{d_w} n\|h\|_\infty,$$

using Theorem 5.4. So by Proposition 6.3, the first term on the right is bounded by $c_4 c_3^{d_w} n^{-d_s/2}$. The second term on the right of (6.2) is harmonic in $S(x, c_3 n^{1/d_w})$. By Proposition 4.5, there exists $c_5 < 1$ such that if $y \in S(x, 2^{-k} c_3 n^{1/d_w})$, then

$$|\mathbb{E}^y \overline{g}h(X_\rho) - \mathbb{E}^x \overline{g}h(X_\rho)| \leqslant c_6 c_5^k \|\overline{g}h\|_\infty \leqslant c_7 c_5^k n^{-d_s/2}.$$

Therefore,

$$\overline{p}(n, x, x) - \overline{p}(n, x, y) \leqslant 2c_4 c_3^{d_w} n^{-d_s/2} + c_7 c_5^k n^{-d_s/2}$$

for $y \in S(x, 2^{-k} c_3 n^{1/d_w}) \cap F_x$.

Recall there exists c_8 such that $\overline{p}(n, x, x) \geqslant c_8 n^{-d_s/2}$. Now take c_3 small so that $2c_4 c_3^{d_w} < c_8/3$ and then k large so that $c_7 c_5^k < c_8/3$. Let $c_1 = 2^{-k} c_3$ and let n_0 be chosen so that $c_1 n_0^{1/d_w}$ is larger than 4. Provided $n \geqslant n_0$ and $y \in S(x, c_1 n^{1/d_w}) \cap F_x$, we have

$$p(n, x, y) \geqslant \overline{p}(n, x, y) \geqslant c_8 n^{-d_s}/3.$$

We thus have our result if $n \geqslant n_0$. It is easy to see that the proposition also holds if $n \leqslant n_0$ provided we adjust c_2. □

Finally we use chaining to obtain our result.

Theorem 6.5. *If n and $d(x, y)$ have the same parity and $d(x, y) \leqslant n$, then*

$$p(n, x, y) \geqslant c_1 n^{-d/2} \exp(-c_2 (d(x, y)^{d_w}/n)^{1/(d_w - 1)}).$$

Proof. Write D for $d(x, y)$. If z is a neighbor of y,

$$p(n, x, y) \geqslant p(n-1, x, z) p(1, z, y) \mu(z) \geqslant c_3 p(n-1, x, z),$$

with c_3 independent of x, y, z, and n. So it suffices to consider the case when n is even.

The case $n \leqslant 4$ is obvious, so we assume $n > 4$. If $D \leqslant c_4 n^{1/d_w}$, where c_4 is the constant c_2 in Theorem 6.1, the result is immediate from Theorem 6.1 or Corollary 6.2. So we need to look at the case $n \geqslant D > c_4 n^{1/d_w}$.

Let k be the smallest integer greater than or equal to $(D^{d_w}/An)^{1/(d_w - 1)}$, where A will be chosen in a moment. Let $m_1, \ldots, m_k$ be even integers, each between $[n/k]/4$ and $4[n/k] + 2$, so that $m_1 + \ldots + m_k = n$. Note

$$\frac{D}{k} \leqslant \frac{c_4}{4} \left(\frac{n}{k}\right)^{1/d_w}$$

if we take A sufficiently large.

We can find points $z_0, z_1, \dots, z_k$ such that $z_0 = x, z_k = y$, and

$$d(z_i, z_{i+1}) \leqslant c_4[n/k]^{1/d_w}, \qquad i = 0, 1, 2, \dots, k-1.$$

We now let $C_i = \{w \in G : d(w, z_i) \leqslant c_4[n/k]^{1/d_w}, d(w, z_i) \text{ even}\}$. So $|C_i| \geqslant c_5(n/k)^{d_f/d_w} = c_5[n/k]^{d_s/2}$. If $w_i \in C_i$ and $w_{i+1} \in C_{i+1}$, then by the triangle inequality $d(w_i, w_{i+1}) \leqslant 2c_4[n/k]^{1/d_w}$. We can then write

$$\begin{aligned}
&p(n, x, y) \\
&\quad \geqslant \sum_{w_1 \in C_1} \sum_{w_2 \subset C_2} \cdots \sum_{w_{k-1} \in C_{k-1}} p(m_1, z_0, w_1)\mu(w_1) \\
&\qquad\qquad \times p(m_2, w_1, w_2)\mu(w_2) \dots p(m_k, w_{k-1}, z_k)\mu(w_{k-1}) \\
&\quad \geqslant (\Pi_{i=1}^{k-1} \mu(C_i))(c_7[n/k]^{-d_s/2})^k \\
&\quad \geqslant (c_5(n/k)^{d_s/2})^{k-1}(2c_7 n/k)^{-kd_s/2} \\
&\quad \geqslant c_8^k n^{-d_s/2} k^{d_s/2} \\
&\quad = n^{-d_s/2} \exp(-k \log(1/c_8) + d_s(\log k)/2) \\
&\quad \geqslant n^{-d_s/2} \exp(-c_9 k).
\end{aligned}$$

Recalling the definition of k, this proves the theorem. □

7. Further results.

(a) $d_s \leqslant 2$. The place in the argument where $d_s > 2$ was used was in obtaining upper and lower bounds on the expectation of $\sigma(x, r)$. If $d_s \leqslant 2$, we modify the argument as follows. Instead of C_G and $g(\cdot, \cdot)$, we use the capacity and Green function for X_n killed on exiting $B(x, c_1 r)$ for suitably large c_1. With this change, the argument goes through as before.

(b) Transience and recurrence. From the estimates on $p(n, x, x)$ we see $\sum_{n=0}^{\infty} p(n, x, x)$ is finite for all x if $d_s > 2$ and identically infinite if $d_s \leqslant 2$. So X_n is transient if and only if $d_s > 2$.

(c) Bounded harmonic functions. We show that the only bounded harmonic functions on G are the constant ones.

Theorem 7.1. *If h is bounded and harmonic on V, then h is identically constant.*

Proof. Let $r > 1$. By iterating Proposition 4.5

$$\operatorname*{Osc}_{B(x,r)} h \leqslant \rho^m \operatorname*{Osc}_{B(x,2^m r)} h \leqslant 2\rho^m \|h\|_\infty.$$

Letting $m \to \infty$, $\mathrm{Osc}_{B(x,r)} h = 0$. Since r is arbitrary, h is constant. □

(d) Continuous time random walk. Much as in Sections 5 and 6 we see there exist c_1, c_2, c_3, c_4 such that for $t \geqslant d(x,y)$

$$c_1 t^{-d_s/2} \exp(-c_2(d(x,y)^{d_w}/t)^{1/(d_w-1)}) \leqslant q(t,x,y) \leq c_3 t^{-d_s/2} \exp(-c_4(d(x,y)^{d_w}/t)^{1/(d_w-1)}).$$

References

[B1] *M. T. Barlow, Diffusions on fractals.* in Lectures in Probability Theory and Statistics, École d'Été de Probabilités de Saint-Flour XXV - 1995, 1-121. Lect. Notes Math. 1690, Springer 1998.

[BB1] M. T. Barlow and R. F. Bass, *The construction of Brownian motion on the Sierpinski carpet.* Ann. Inst. H. Poincaré **25** (1989) 225–257.

[BB2] M.T. Barlow and R.F. Bass, *Transition densities for Brownian motion on the Sierpinski carpet.* Probab. Th. Rel. Fields **91** (1992) 307-330.

[BB3] M.T. Barlow and R.F. Bass, *Brownian motion and harmonic analysis on Sierpinski carpets.*To appear in Canadian J. Math. 1999.

[BPY] M.T. Barlow, J. Pitman and M. Yor: *On Walsh's Brownian Motions.* Sém. Prob. XXIII, 275-293, Lect. Notes Math. **1372** (1989).

[Ba] R.F. Bass, *Diffusions on the Sierpinski carpet.* Trends in Probability and related Analysis: Proceedings of SAP '96. World Scientific, Singapore, 1997. To appear.

[CKS] E.A. Carlen, S. Kusuoka, and D.W. Stroock, *Upper bounds for symmetric Markov transition functions.* Ann. Inst. H. Poincaré **23** (1987) 245–287.

[Ca] T.K. Carne: *A transmutation formula for Markov Chains.* Bull. Sc. Math. **109** (1985), 399-405.

[Co] T. Coulhon, *Dimension á l'infini d'un semi-groupe analytique.* Bull. Sci. Math. **114** (1990) 485-500.

[Da] E.B. Davies: *Large deviations for heat kernels on graphs.* J. London Math. Soc. (2) **47** (1993), 65-72.

[De] T. Delmotte: *Parabolic Harnack inequality and estimates for Markov chains on graphs.* Preprint 1997.

[FOT] M. Fukushima, Y. Oshima, and M. Takeda, *Dirichlet Forms and Symmetric Markov Processes.* de Gruyter, Berlin, 1994.

[HSC] W. Hebisch, L. Saloff-Coste: *Gaussian estimates for Markov Chains and random walks on groups.* Ann. Prob. **21** (1993), 673-709.

[Hu] J.E. Hutchinson, *Fractals and self-similarity.* Indiana J. Math. **30** (1981) 713-747.

[J] O.D. Jones, *Transition probabilities for the simple random walk on the Sierpinski graph.* Stoch. Proc. Appl. **61** (1996), 45-69.

[O] H. Osada, *Isoperimetric dimension and estimates of heat kernels of pre–Sierpinski carpets.* Probab. Th. Rel. Fields **86** (1990) 469–490.

[P] M.M.H. Pang, *Heat kernels of graphs.* J. London Math. Soc. (2) **47** (1993), 50-64.

[SC] L. Saloff-Coste, *A note on Poincaré, Sobolev, and Harnack inequalities.* Inter. Math. Res. Notices (1992) 27–38.

[V1] N. Th. Varopoulos, *Isoperimetric inequalities and Markov chains.* J. Funct. Anal. **63** (1985) 215-239.

[V2] N. Th. Varopoulos, *Long range estimates for Markov chains.* Bull. Sc. Math. **109** (1985), 225-252.

[W] W. Woess, *Random walks on infinite graphs and groups – a survey on selected topics.* Bull. London Math. Soc. **26** (1994), 1-60.

Martin T. Barlow
Department of Mathematics
University of British Columbia
Vancouver, V6T 1Z2 Canada

Richard F. Bass
Department of Mathematics
University of Washington
Seattle, WA 98195 U.S.A.

Percolation perturbations in potential theory and random walks

Itai Benjamini, Russell Lyons, and Oded Schramm

Abstract. We show that on a Cayley graph of a nonamenable group, a.s. the infinite clusters of Bernoulli percolation are transient for simple random walk, that simple random walk on these clusters has positive speed, and that these clusters admit bounded harmonic functions. A principal new finding on which these results are based is that such clusters admit invariant random subgraphs with positive isoperimetric constant.

We also show that percolation clusters in any amenable Cayley graph a.s. admit no nonconstant harmonic Dirichlet functions. Conversely, on a Cayley graph admitting nonconstant harmonic Dirichlet functions, a.s. the infinite clusters of p-Bernoulli percolation also have nonconstant harmonic Dirichlet functions when p is sufficiently close to 1. Many conjectures and questions are presented.

1. Introduction.

The question of whether various potential-theoretic properties of graphs and manifolds are preserved under perturbations or approximations has been studied for more than a decade. For example, invariance under quasi-isometries of transience (i.e., the existence of nonconstant positive superharmonic functions) or of existence of harmonic functions in certain classes has been studied by Kanai (1986), T. Lyons (1987), Saloff-Coste (1992), Soardi (1993), Benjamini and Schramm (1996a), Thm. 3.5, and Holopainen and Soardi (1997).

In this paper, we study perturbations of graphs that are more radical than quasi-isometries and that are random. Namely, edges are deleted at random to form a percolation subgraph ω and the behavior of **simple random walk** $\langle X(t) \rangle$ on ω is examined (where each neighbor of $X(t)$ in ω is equally likely to be $X(t+1)$).

We recall some definitions. Given a graph $G = \big(V(G), E(G)\big)$ and $p \in [0,1]$, the random subgraph ω_p formed by deleting each edge independently with probability $1-p$ is called p**-Bernoulli bond percolation**. The **critical probability** $p_c(G)$ is the infimum over all $p \in [0,1]$ such that there is positive probability for the existence of an infinite connected component in ω_p. The connected components of ω_p are also called **clusters**. In the case that G is an amenable Cayley graph, Burton and Keane (1989) show that p-Bernoulli percolation has a.s. at most one infinite cluster. For background on percolation, especially in $\mathbb{Z}^d$, see Grimmett (1989). Following earlier work of Grimmett and Newman (1990) on the direct product of a regular tree and $\mathbb{Z}$, a general study of percolation on discrete groups was initiated in Benjamini and Schramm (1996b). One phenomenon that was conjectured there to be general was a converse to the Burton and Keane result, namely, that on any nonamenable group, for some p, there are a.s. infinitely many infinite clusters in ω_p. This led to the definition

$$p_u(G) := \inf\Big\{p : \mathbf{P}[\omega_p \text{ has exactly one infinite cluster}] = 1\Big\}.$$

Thus, $p_u(G) = p_c(G)$ when G is an amenable Cayley graph. Häggström and Peres (1998) show that on a Cayley graph G, for *every* $p > p_u$, there is exactly one infinite cluster a.s. in p-Bernoulli percolation. It is known that $p_u < 1$ in many cases besides amenable groups, e.g., finitely presented groups with one end (Babson and Benjamini 1999) and Kazhdan groups (Lyons and Schramm 1998).

The **uniqueness phase** of Bernoulli percolation is the range of p where there is precisely one infinite cluster a.s.; the **nonuniqueness phase** is the range of p where there is more than one infinite cluster a.s.

The unique infinite percolation cluster of supercritical Bernoulli percolation on a graph, if there is such, can be viewed as a random perturbation of the graph. It is then natural to ask which properties of the graph are inherited by such a percolation cluster.

After presenting further definitions and reviewing some background in Section 2, we begin by studying in Section 3 purely geometric aspects of percolation clusters, namely, how the isoperimetric constant $\iota_E(G)$ (see Section 2) behaves under percolation. If ω is a random configuration of Bernoulli percolation on a Cayley graph G, then, of course, $\iota_E(\omega) = 0$ a.s. However,

Theorem 1.1. *If G is a nonamenable Cayley graph and ω is a random configuration of Bernoulli percolation on G, then a.s. every infinite cluster of ω contains a subgraph ω' with $\iota_E(\omega') > 0$.*

(See Theorem 3.9.) In fact, we show in Theorem 3.10 that one can further require ω' to be a tree. Note that ω is a random variable, representing the configuration; in the sequel, we shall often say, however, that ω is a "percolation".

The theorem raises the following question about percolation in $\mathbb{Z}^d$. Recall that the **isoperimetric dimension** of a graph G is the supremum of all s such that

$$\inf\left\{\frac{|\partial_E V_1|^s}{|V_1|^{s-1}} : V_1 \subset V(G),\ 0 < |V_1| < \infty\right\} > 0.$$

Question 1.2. Let ω be supercritical Bernoulli percolation in $\mathbb{Z}^d$. Is it true that for every $\epsilon > 0$, a.s. ω contains a subgraph with isoperimetric dimension at least $d - \epsilon$? Does it contain a subgraph with isoperimetric dimension d?

It is well known that if G is nonamenable, then the speed of simple random walk $\langle X(t) \rangle$ on G is positive, i.e.,

$$\lim_{t \to \infty} \frac{\operatorname{dist}_G(X(0), X(t))}{t} > 0 \quad \text{a.s.},$$

which results from the fact that the spectral radius is less than 1 (Kesten 1959); see Section 2. We prove the following extension in Section 4 as a consequence of (a more precise version of) Theorem 1.1:

Theorem 1.3. *Let G be a nonamenable Cayley graph and ω be Bernoulli percolation on G. Let $\langle X(t) \rangle$ be simple random walk on ω. Given that the cluster K of $X(0)$ is infinite, we have a.s. that the speed of X is positive.*

(See Theorem 4.4.)

We conjecture the following generalization:

Conjecture 1.4. *If G is a Cayley graph on which simple random walk has positive speed, then a.s., simple random walk on each infinite cluster of p-Bernoulli percolation has positive speed.*

There are (nontransitive) graphs on which simple random walk has zero speed, but for which Bernoulli percolation a.s. produces clusters where simple random walk has positive speed. An example is a binary tree with a copy of $\mathbb{Z}$ attached to every vertex.

Conjecture 1.5. *If G is a Cayley graph on which simple random walk has zero speed, then a.s., simple random walk on every cluster of Bernoulli percolation also has zero speed.*

Any possible counterexample would have to be amenable and of exponential growth by Lemma 4.6 below.

The speed of random walk is related to other probabilistic behavior through the following theorem due to the work of Avez (1974), Derriennic (1980), Kaimanovich and Vershik (1983), and Varopoulos (1985). Recall that a function $F : V(G) \to \mathbb{R}$ is called **harmonic** if $F(x) = \sum_{y \sim x} F(y) / \deg_G x$ for all $x \in V$.

Theorem 1.6. **(Speed, Entropy, and Bounded Harmonic Functions)** *The following conditions are equivalent for a given Cayley graph:*

(i) *the speed of simple random walk is zero;*
(ii) *the asymptotic entropy of simple random walk is zero;*

(iii) *there are no nonconstant bounded harmonic functions.*

Furthermore, Kaimanovich (1990) extended the equivalence of (ii) and (iii) to many random walks in a random environment (RWRE) that have a stationary measure; the extension to include (i) is easy (Lemma 4.6). Since simple random walk restricted to percolation clusters has an equivalent invariant measure (see Lemma 4.1), our conjectures and results about the speed of random walk have some alternative formulations in terms of entropy and bounded harmonic functions.

Transience holds more generally, of course, than positive speed. Transience of the infinite clusters of Bernoulli percolation in $\mathbb{Z}^d$, $d > 2$, was established in Grimmett, Kesten and Zhang (1993) (see Benjamini, Pemantle and Peres (1998) for a different proof). De Masi, Ferrari, Goldstein and Wick (1989) proved an invariance principle for simple random walk on the supercritical cluster in $\mathbb{Z}^2$.

Conjecture 1.7. *If G is a transient Cayley graph, then a.s. every infinite cluster of Bernoulli percolation on G is transient.*

The nonamenable case follows from Theorem 1.3; a slight extension is:

Theorem 1.8. *Let G be a Cayley graph such that the ball of radius n has cardinality at least ζ^n for all n, where $\zeta > 1$. Then a.s., every infinite cluster of p-Bernoulli percolation on G is transient when $p > 1/\zeta$.*

(See Theorem 4.9.)

In Section 5, we study the existence of nonconstant harmonic Dirichlet functions on the percolation clusters of Cayley graphs. A function $F : V(G) \to \mathbb{R}$ is called **Dirichlet** if $\sum_x \sum_{y \sim x} |F(x) - F(y)|^2 < \infty$. Recall that $\mathcal{O}_{\mathbf{HD}}$ denotes the class of graphs that do not admit any nonconstant harmonic Dirichlet functions. As we shall see below, the picture in the nonamenable situation is rather involved and our understanding is far from complete. In the amenable case, we have

Theorem 1.9. *If G is an amenable Cayley graph and ω is Bernoulli percolation, then a.s. every cluster of ω is in $\mathcal{O}_{\mathbf{HD}}$.*

(See Theorem 5.5.)

Medolla and Soardi (1995) proved that amenable transitive graphs are in $\mathcal{O}_{\mathbf{HD}}$ (see Remark 7.5 of Benjamini, Lyons, Peres, and Schramm (1998) for a short proof via uniform spanning forests). Soardi (1993) proved that $\mathcal{O}_{\mathbf{HD}}$ is invariant under quasi-isometries. (See Lyons and Peres (1998) for a simple proof due to Schramm.)

Our proof of Theorem 1.9 uses the uniform spanning forest measures and their connection to harmonic Dirichlet functions as presented in BLPS (1998). Such an ingredient can be motivated as follows: In order to study the influence of geometric properties on potential-theoretic behavior, it is useful to have a geometric representation of the potential-theoretic objects. The uniform spanning forest achieves this by representing the analytic space of harmonic Dirichlet functions by a random geometric object. The relevant definitions and properties are given in Section 5.

A positive answer to the following question would extend Theorem 1.9:

Question 1.10. Let G be a Cayley graph, and suppose that $G \in \mathcal{O}_{\mathbf{HD}}$. Let ω be Bernoulli percolation on G in the uniqueness phase. Does it follow that a.s. the infinite cluster of ω is in $\mathcal{O}_{\mathbf{HD}}$? We do not know the answer even in the case that G is the direct product of a tree and $\mathbb{Z}$ or is a lattice in hyperbolic space $\mathbb{H}^d$ $(d \geqslant 3)$.

In the nonuniqueness phase of Bernoulli percolation on a (nonamenable) Cayley graph G, the infinite clusters are not in $\mathcal{O}_{\mathbf{HD}}$ (Corollary 4.7).

In the other direction, for graphs admitting nonconstant harmonic Dirichlet functions, we believe:

Conjecture 1.11. *Let G be a Cayley graph, $G \notin \mathcal{O}_{\mathbf{HD}}$. Then a.s. all infinite clusters of p-Bernoulli percolation are not in $\mathcal{O}_{\mathbf{HD}}$.*

We can prove this for p sufficiently large:

Theorem 1.12. *Let G be a Cayley graph. If $G \notin \mathcal{O}_{\mathbf{HD}}$, then there is some $p_0 < 1$ such that for every $p \geqslant p_0$, almost surely no infinite cluster of p-Bernoulli percolation is in $\mathcal{O}_{\mathbf{HD}}$.*

(See Theorem 5.7.)

The last section of our paper discusses questions concerning the speed of simple random walk on a graph and a variant of the isoperimetric constant, called **anchored expansion**. The anchored expansion constant might be useful in studying the speed of simple random walk on infinite percolation clusters and other graphs as well.

Although in the above theorems, only Cayley graphs are mentioned, we work in the greater generality of transitive graphs. Similarly, we discuss percolation processes that are much more general than Bernoulli percolation.

Acknowledgments. We thank the organizers of the Cortona conference, Vadim A. Kaimanovich, Massimo A. Picardello, Laurent Saloff-Coste, and Wolfgang Woess, for a most enjoyable and productive meeting. We thank Yuval Peres for comments on an earlier version of this paper.

2. Notation and Background.

Graph terminology, isoperimetric constant and ends. We use the letter G to denote a graph and Γ to denote a closed subgroup of the automorphism group $\mathrm{Aut}(G)$ of G. The vertices and edges of G will be denoted $V(G)$ and $E(G)$, respectively. When there is an edge in G joining vertices u, v, we write $u \sim v$. The **degree** $\deg v = \deg_G v$ of a vertex $v \in V(G)$ is the number of edges incident with it. A **tree** is a connected graph with no cycles. A **forest** is a graph whose connected components are trees. The **distance** between two vertices $v, u \in V(G)$ is denoted by $\mathrm{dist}(v, u) = \mathrm{dist}_G(v, u)$, and is the least number of edges of a path in G connecting v and u.

For a set of vertices $V_1 \subset V(G)$, let $\partial_E V_1$ denote the set of edges in $E(G)$ that have one endpoint in V_1 and one endpoint in $V(G) - V_1$. The graph G is **amenable** if there is a sequence of finite vertex subsets $V_1 \subset V_2 \subset \cdots \subset V_n \subset \cdots \subset V(G)$, such that $\bigcup_n V_n = V(G)$ and $|\partial_E V_n|/|V_n| \to 0$ as $n \to \infty$. Here and in the sequel, $|A|$ denotes the cardinality of a set A. The (edge) **isoperimetric constant** of a graph G, also known as the **Cheeger constant**, is defined by

$$\iota_E(G) := \inf\{|\partial_E V_0|/|V_0| : \emptyset \neq V_0 \subset V(G),\ |V_0| < \infty\}.$$

An infinite set of vertices $V_0 \subset V(G)$ is **end-convergent** if for every finite $K \subset V(G)$, there is a component of $G - K$ that contains all but finitely many vertices of V_0. Two end-convergent sets V_0, V_1 are **equivalent** if $V_0 \cup V_1$ is end-convergent. An **end** of G is an equivalence class of end-convergent sets.

Spectral radius, speed and entropy. Given $v, u \in V(G)$, let $p_t(v, u)$ be the probability that simple random walk starting at v will be at u at time t. The **spectral radius** $\rho(G)$ is defined by

$$\rho(G) := \limsup_{t \to \infty} p_t(v, u)^{1/t}$$

and does not depend on the choice of v and u. Dodziuk's (1984) discrete version of Cheeger's inequality states that if G has bounded degrees and $\iota_E(G) > 0$, then $\rho(G) < 1$.

The **speed** of a random walk X starting at $o \in V(G)$ is

$$\Lambda = \Lambda(X) := \lim_{t \to \infty} \mathrm{dist}\big(o, X(t)\big)/t$$

when the limit exists. The lim inf speed is defined by

$$\Lambda^- = \Lambda^-(X) := \liminf_{t \to \infty} \mathrm{dist}\big(o, X(t)\big)/t.$$

If G has bounded degrees and $\rho(G) < 1$, then there are constants $\zeta > 0$ and $\beta < 1$ such that

$$\mathbf{P}\Big[\mathrm{dist}\big(o, X(t)\big) < \zeta t\Big] \leqslant \beta^t ,$$

because the probability that the random walk is inside the ball of radius ζt about o is bounded by the number of vertices in the ball times the probability that X is at the most likely vertex. Consequently, in this situation, $\Lambda^- > 0$ a.s.

The **entropy** of a probability measure μ on a finite or countable set A is defined to be

$$H(\mu) := \sum_{x \in A} -\mu(x) \log \mu(x) .$$

Let μ_t denote the distribution of the location $X(t)$ of a random walk X at time t. If $\lim_t H(\mu_t)/t$ exists, it is called the **asymptotic entropy** of the random walk. If $\langle X(t) \rangle$ is simple random walk on a Cayley graph, then the asymptotic entropy exists. In fact, the Subadditive Ergodic Theorem ensures the existence of $\lim_t -t^{-1} \log \mu_t\big(X(t)\big)$ a.s. and in L^1; see Derriennic (1980). The same is true for stationary RWRE, as observed by Kaimanovich (1990). Similar reasoning applies to graphs with a transitive unimodular automorphism group (Kaimanovich and Woess 1998).

For further information about random walks, spectral radius, harmonic functions, etc., see Kaimanovich and Vershik (1983) and Woess (1994).

Automorphism groups, unimodularity, and the Mass-Transport Principle. Let $\Gamma \subset \mathrm{Aut}(G)$ be a subgroup of automorphisms of G with the topology of pointwise convergence. We say that Γ is (vertex) **transitive** if for every $v, u \in V(G)$, there is a $\gamma \in \Gamma$ with $\gamma u = v$. The graph G is **transitive** if $\mathrm{Aut}(G)$ is transitive. Recall that every closed subgroup $\Gamma \subset \mathrm{Aut}(G)$ has a unique (up to a constant scaling factor) Borel measure that, for every $\gamma \in \Gamma$, is invariant under left multiplication by γ; this measure is called (left) **Haar measure**. The group Γ is **unimodular** if Haar measure is also invariant under right multiplication.

Most of our theorems concern percolation that is invariant under a transitive unimodular closed subgroup of $\mathrm{Aut}(G)$. For example, when G is the (right) Cayley graph of Γ and Γ acts by left multiplication, then $\Gamma \subset \mathrm{Aut}(G)$ is (obviously) closed, unimodular, and transitive. If G is an amenable graph, then every closed transitive subgroup of $\mathrm{Aut}(G)$ is unimodular (Soardi and Woess 1990).

Several illustrations of the significance of unimodularity can be found in Benjamini, Lyons, Peres, and Schramm (1999). The most important one seems to be that when $\Gamma \subset \mathrm{Aut}(G)$ is unimodular, the Mass-Transport Principle takes the following simple form:

Theorem 2.1. (Mass-Transport Principle) *Let G be a graph with a transitive unimodular closed automorphism group $\Gamma \subset \mathrm{Aut}(G)$. Let $o \in V(G)$ be an arbitrary base point. Suppose that $\phi : V(G) \times V(G) \to [0, \infty]$ is invariant under the diagonal action of Γ. Then*

$$\sum_{v \in V(G)} \phi(o, v) = \sum_{v \in V(G)} \phi(v, o). \tag{2.1}$$

See BLPS (1999) for a discussion of this principle and for a proof. In fact, Γ is unimodular iff (2.1) holds for every such ϕ. Hence, (2.1) can be taken as a definition of unimodularity.

Percolation terminology. A **bond percolation** ω on G is a random subset of $E(G)$. For a more precise definition, given a set A, let 2^A be the collection of all subsets $\eta \subset A$, equipped with the σ-field generated by the events $\{a \in \eta\}$, where $a \in A$. A bond percolation ω on G is then a random variable whose distribution is a probability measure $\mathbf{P}$ on $2^{E(G)}$. Similarly, a **site percolation** is given by a probability measure on $2^{V(G)}$, while a (mixed) **percolation** is given by a probability measure on $2^{V(G) \cup E(G)}$ that is supported on subgraphs of G. If ω is a bond percolation, then $\hat{\omega} := V(G) \cup \omega$ is the associated mixed percolation. In this case, we shall often not distinguish between ω and $\hat{\omega}$, and think of ω as a subgraph of G. Similarly, if ω is a site percolation, there is an associated mixed percolation $\hat{\omega} := \omega \cup \big(E(G) \cap (\omega \times \omega)\big)$, and we shall often not bother to distinguish between ω and $\hat{\omega}$.

Let $p \in [0, 1]$. Then the distribution of p-**Bernoulli bond percolation** ω on G is the product measure on $2^{E(G)}$ that satisfies $\mathbf{P}[e \in \omega] = p$ for all $e \in E(G)$. Similarly, one defines p-**Bernoulli site percolation** on $2^{V(G)}$.

If $v \in V(G)$ and ω is a percolation on G, the **component** (or **cluster**) $K(v)$ of v in ω is the set of vertices in $V(G)$ that can be connected to v by paths contained in ω.

Suppose that Γ is an automorphism group of a graph G. A percolation on G is **Γ-invariant** if its distribution $\mathbf{P}$ is invariant under each automorphism in Γ.

Insertion tolerance and component indistinguishability. Given a set Z, an element $z \in Z$, and a subset $\omega \in 2^Z$, let $\Pi_z \omega := \omega \cup \{z\}$. A probability measure $\mathbf{P}$ on 2^Z is **insertion tolerant** if $\mathbf{P}[\Pi_z \mathcal{A}] > 0$ for every $z \in Z$ and every measurable $\mathcal{A} \subset 2^Z$ satisfying $\mathbf{P}[\mathcal{A}] > 0$. For example, p-Bernoulli bond percolation is insertion tolerant when $p > 0$.

Let G be graph and Γ a closed transitive subgroup of $\mathrm{Aut}(G)$. Let ω be a Γ-invariant bond percolation. We say that ω has **indistinguishable components** if for every measurable $\mathcal{A} \subset 2^{V(G)} \times 2^{E(G)}$ that is invariant under the diagonal action of Γ, almost surely, for all infinite components

C of ω, we have $(C,\omega) \in \mathcal{A}$, or for all infinite components C, we have $(C,\omega) \notin \mathcal{A}$. (That is, whether $(C,\omega) \in \mathcal{A}$ does not depend on C, but may depend on ω.)

The following is from Lyons and Schramm (1998):

Theorem 2.2. (Component Indistinguishability) *Let G be a graph with a transitive unimodular closed automorphism group $\Gamma \subset \mathrm{Aut}(G)$. Every Γ-invariant, insertion-tolerant, bond percolation process on G has indistinguishable components.*

For example, this shows that in p-Bernoulli bond percolation on a Cayley graph of a nonamenable group, almost surely, either all infinite clusters are transient, or all clusters are recurrent. In fact, as indicated by Theorem 1.3, a.s. all infinite clusters are transient.

Similar statements hold for site and mixed percolations.

3. Geometry of Perturbations of Nonamenable Graphs.

Let G be an infinite graph and K a finite subgraph of G. Set

$$\alpha_K := \frac{1}{|V(K)|} \sum_{v \in V(K)} \deg_K v ,$$

where $\deg_K v$ is the degree of v in K. Define

$$\alpha(G) := \sup\{\alpha_K : K \subset G \text{ is finite}\} .$$

Note that when all vertices in G have degree d, the isoperimetric constant of G satisfies

$$\iota_E(G) = d - \alpha(G) .$$

Let T be a regular tree and $o \in V(T)$ be some base point. Häggström (1997) has shown that when ω is an automorphism-invariant percolation on T and $\mathbf{E}[\deg_\omega o] \geqslant \alpha(T)$, there are infinite clusters in ω with positive probability. In BLPS (1999), it was shown that the same result applies to transitive graphs with a unimodular automorphism group. In Theorem 3.2, we extend this result and show that with the same assumptions but with a strict inequality $\mathbf{E}[\deg_\omega o] > \alpha(G)$, with positive probability, there is a subgraph in ω with $\iota_E > 0$. This will be used to prove

Theorem 3.1. (Uniqueness Gives a Subgraph with $\iota_E > 0$) *Let G be a graph with a transitive unimodular closed automorphism group $\Gamma \subset \mathrm{Aut}(G)$, and suppose that $\iota_E(G) > 0$. Let ω be a Γ-invariant percolation in G that has a.s. exactly one infinite component. Then (on a larger probability* *space) there is a percolation $\omega' \subset \omega$ such that $\omega' \neq \emptyset$ and $\iota_E(\omega') > 0$ a.s. Moreover, the distribution of the pair (ω', ω) is Γ-invariant.*

In the following, if $K \subset G$ is a subgraph and $v \in V(G)$ is not in K, then we set $\deg_K v := 0$.

As we indicated, the proof of Theorem 3.1 is based on the following more quantitative result.

Theorem 3.2. (High Marginals Give a Subgraph with $\iota_E > 0$) *Let G be a graph with a transitive unimodular closed automorphism group $\Gamma \subset \mathrm{Aut}(G)$. Let ω be a Γ-invariant (nonempty) percolation in G. Let $h > 0$ and suppose that*

$$\mathbf{E}[\deg_\omega o \mid o \in \omega] > \big(\alpha(G) + 2h\big). \tag{3.1}$$

Then there is (on a larger probability space) a percolation $\omega' \subset \omega$ such that $\omega' \neq \emptyset$ and $\iota_E(\omega') \geqslant h$ with positive probability. Moreover, the distribution of the pair (ω', ω) is Γ-invariant.

Proof. Given any subgraph ω of G, we define percolations ω_n on ω inductively as follows. Set $\omega_0 := \omega$. Suppose that ω_n has been defined. Let β_n be a (1/2)-Bernoulli site percolation on G, independent of $\omega_0, \ldots, \omega_n$. Let γ_n be the union of the finite components K of $\beta_n \cap \omega_n$ that satisfy

$$\frac{|\partial_{E(\omega_n)} K|}{|K|} < h,$$

where $\partial_{E(\omega_n)} K$ denotes the set of edges of ω_n connecting K to its complement. Now set $\omega_{n+1} := \omega_n - \gamma_n$. Finally, define

$$\omega' := \bigcap_{n=0}^{\infty} \omega_n .$$

For future use, write $\Xi(h, \omega) := \omega'$.

Most of the proof will be devoted to showing that $\omega' \neq \emptyset$ with positive probability, but first we verify that $\iota_E(\omega') \geqslant h$. Indeed, let W be a finite nonempty set of vertices in G, let F be the set of all edges of G incident with W, and let $F_0 \subset F$. Suppose that $|F_0|/|W| < h$. To verify that $\iota_E(\omega') \geqslant h$ a.s., it is enough to show that the probability that $W \subset \omega'$ and $\omega' \cap F = F_0$ is zero. If $\omega_n \cap F = F_0$ for some n, then a.s. there is some $m > n$ such that W is a component of β_m. Now either $W \not\subset \omega_m$, in which case $W \not\subset \omega'$, or $W \subset \omega_m$, in which case W is not contained in ω_{m+1}, hence not in ω'. On the other hand, if $\omega_n \cap F \neq F_0$ for every n, then also $\omega' \cap F \neq F_0$. Consequently $\iota_E(\omega') \geqslant h$ a.s.

Now set

$$D_n := \mathbf{E} \deg_{\omega_n} o\,, \qquad D_\infty := \mathbf{E} \deg_{\omega'} o\,,$$
$$\theta_n := \mathbf{P}[o \in \omega_n]\,, \qquad \theta_\infty := \mathbf{P}[o \in \omega']\,.$$

Our goal is to prove the inequality

$$D_{n+1} \geqslant D_n - (\theta_n - \theta_{n+1})\big(\alpha(G) + 2h\big)\,. \tag{3.2}$$

This will be achieved through use of the Mass-Transport Principle. Observe that

$$\theta_n - \theta_{n+1} = \mathbf{P}[o \in \gamma_n]\,.$$

Fix n and define the random function $m : V(G) \times V(G) \to [0, \infty)$ as follows. For every vertex $v \in V(G)$, let $K(v)$ be the component of v in γ_n, which we take to be $\emptyset$ if $v \notin \gamma_n$. Let $v, u \in V$. If $u \notin \gamma_n$, set $m(v, u) := 0$. If $v \in K(u)$, let $m(v, u) := \deg_{\omega_n} v / |K(u)|$. Otherwise, let $m(v, u)$ be $|K(u)|^{-1}$ times the number of edges in ω_n that connect v to a vertex in $K(u)$. Note that v and u need not be adjacent in order that $m(v, u) \neq 0$. Clearly, $\mathbf{E}m(v, u)$ is invariant under the diagonal action of Γ on $V(G) \times V(G)$. Consequently, the Mass-Transport Principle implies that

$$\sum_{v \in V(G)} \mathbf{E}m(o, v) = \sum_{v \in V(G)} \mathbf{E}m(v, o)\,.$$

A straightforward calculation shows that

$$\sum_{v \in V(G)} m(o, v) = \deg_{\omega_n} o - \deg_{\omega_{n+1}} o\,,$$

while, if $o \in \gamma_n$, we have that $\sum_{v \in V(G)} m(v, o)$ is equal to twice the number of edges of ω_n incident with $K(o)$, divided by $|K(o)|$. The number of edges of G with both endpoints in $K(o)$ is at most $\alpha(G)|K(o)|/2$, and, by construction, $|\partial_{\omega_n} K(o)| < h|K(o)|$. Hence

$$\sum_{v \in V(G)} m(v, o) < \alpha(G) + 2h \tag{3.3}$$

when $o \in \gamma_n$ and $\sum_{v \in V(G)} m(v, o) = 0$ otherwise. Therefore,

$$\begin{aligned} D_n - D_{n+1} = \mathbf{E}[\deg_{\omega_n} o - \deg_{\omega_{n+1}} o] &= \sum_{v \in V(G)} \mathbf{E}m(v, o) \\ &\leqslant (\alpha(G) + 2h)\mathbf{P}[o \in \gamma_n] = (\alpha(G) + 2h)(\theta_n - \theta_{n+1})\,, \end{aligned}$$

which is the same as (3.2).

Induction and (3.2) give

$$D_n \geqslant D_0 - \theta_0\big(\alpha(G) + 2h\big) + \theta_n\big(\alpha(G) + 2h\big)\,;$$

taking a limit as $n \to \infty$ yields the inequality

$$D_\infty \geqslant D_0 - \theta_0\big(\alpha(G) + 2h\big) + \theta_\infty\big(\alpha(G) + 2h\big)\,. \tag{3.4}$$

This gives $D_\infty > 0$, because (3.1) is equivalent to $D_0 - \theta_0\big(\alpha(G) + 2h\big) > 0$. Consequently, $\omega' \neq \emptyset$ with positive probability. □

Remark 3.3. (The Density of ω') The following lower bound for θ_∞ is a consequence of (3.4) and the inequality $\theta_\infty \deg_G o \geqslant D_\infty$:

$$\begin{aligned} \mathbf{P}[o \in \omega'] = \theta_\infty &\geqslant \frac{D_0 - \big(\alpha(G) + 2h\big)\theta_0}{\deg_G o - \big(\alpha(G) + 2h\big)} \\ &= \mathbf{P}[o \in \omega]\left(1 - \frac{\deg_G o - \mathbf{E}[\deg_\omega o \mid o \in \omega]}{\iota_E(G) - 2h}\right). \end{aligned} \tag{3.5}$$

Remark 3.4. (A Weak Inequality Suffices) In fact, in place of (3.1), it is enough to assume the weak inequality $\mathbf{E}[\deg_\omega o \mid o \in \omega] \geqslant \big(\alpha(G)+2h\big)$. The reason is that the inequality (3.3) is strict when $o \in \gamma_n$, which implies that (3.4) is strict when $D_\infty \neq D_0$.

Theorem 3.5. (Threshold for a Forest) *If ω is a forest a.s., then Theorem 3.2 is true when $\alpha(G)$ is replaced by 2.*

Proof. In a finite tree $K \subset G$, we have $\alpha_K < 2$. Hence the proof of Theorem 3.2 applies with 2 replacing $\alpha(G)$ everywhere. □

Proof of Theorem 3.1. Fix a base point $o \in V(G)$. Let ω_* be the infinite component of ω. Conditioned on ω, for every vertex $v \in V(G)$, let $\phi(v)$ be chosen uniformly among the vertices of ω_* closest to v, with all $\phi(v)$ independent given ω, and for edges $e = [v, u] \in E(G)$, let $\phi(e)$ be chosen uniformly among shortest paths in ω_* joining $\phi(v)$ and $\phi(u)$, with all $\phi(e)$ independent given all $\phi(v)$ and ω. For integers j, let η_j be the set of edges $e \in E(G)$ such that $\phi(e)$ is contained within a ball of radius j about one of the endpoints of e. Then $\eta_1 \subset \eta_2 \subset \cdots$ are Γ-invariant bond percolations on G with $\bigcup_j \eta_j = E(G)$. Consequently, $\mathbf{E}\deg_{\eta_j} o \to \deg_G o$ as $j \to \infty$. For each j, choose independently a random sample of $\Xi(h, \eta_j)$ and denote it ξ_j. By (3.5), we have that $\mathbf{P}[\xi_j \neq \emptyset] \to 1$. Let $J := \inf\{j : \xi_j \neq \emptyset\}$. Then $J < \infty$ a.s. Set $\omega' := \phi(\xi_J)$. Since $\iota_E(\xi_J) \geqslant h$ a.s., we have also $\iota_E\big(\phi(\xi_J)\big) > 0$ a.s. □

Suppose that G is transitive, $\iota_E(G) > 0$, and ω is, say, Bernoulli percolation on G that has a.s. more than one infinite component. Then Theorem 3.1 does not apply to ω. However, as observed by Burton and Keane (1989), insertion tolerance shows that there are a.s. components of ω with at least three ends. Hence the next theorem does apply.

Theorem 3.6. (A Forest with $\iota_E > 0$ Inside Many-Ended Percolation) *Let G be a graph with a transitive unimodular closed automorphism group $\Gamma \subset \mathrm{Aut}(G)$, and let ω be a Γ-invariant percolation on G. Suppose that a.s., there are components of ω with at least three ends. Then there is (on a larger probability space) a random forest $\mathfrak{F} \subset \omega$ with $\iota_E(\mathfrak{F}) > 0$, $\mathfrak{F} \neq \emptyset$ a.s., and the distribution of the pair $(\mathfrak{F}, \omega)$ is Γ-invariant.*

We shall need the following two lemmas from BLPS (1999). Recall that $K(x)$ denotes the component of x in ω.

Lemma 3.7. (Ends, p_c and Degrees) *Let G be a graph with a transitive unimodular closed automorphism group $\Gamma \subset \mathrm{Aut}(G)$. Let ω be a Γ-invariant percolation on G that has infinite components with positive probability. If*

(i) *some component of ω has at least three ends with positive probability, then*

(ii) *some component of ω has $p_c < 1$ with positive probability and*

(iii) *for every vertex x, $\mathbf{E}\big[\deg_\omega x \mid |K(x)| = \infty\big] > 2$.*

If ω is a forest a.s., then the three conditions are equivalent.

Lemma 3.8. (Trimming to a Forest) *Let G be a graph with a transitive unimodular closed automorphism group $\Gamma \subset \mathrm{Aut}(G)$. Let ω be a Γ-invariant percolation on G such that a.s. there is a component of ω with at least three ends. Then (on a larger probability space) there is a random forest $\mathfrak{F} \subset \omega$ such that the distribution of the pair $(\mathfrak{F}, \omega)$ is Γ-invariant and a.s. whenever a component K of ω has at least three ends, there is a component of $K \cap \mathfrak{F}$ that has infinitely many ends.*

Proof of Theorem 3.6. By Lemma 3.8, there is a random forest $\mathfrak{F}' \subset \omega$ with some components having infinitely many ends a.s. and the distribution of $(\mathfrak{F}', \omega)$ is Γ-invariant. Let $\mathfrak{F}''$ be the union of the infinite components of $\mathfrak{F}'$. By Lemma 3.7, $\mathbf{E}[\deg_{\mathfrak{F}''} o \mid o \in \mathfrak{F}''] > 2$. Given ω, for each j, let ξ_j be an independent sample of $\Xi(1/j, \mathfrak{F}'')$. Put $\mathfrak{F} := \xi_J$, where $J := \inf\{j : \xi_j \neq \emptyset\} < \infty$ a.s.: Clearly, $J < \infty$ with positive probability. If the set $\mathcal{A}$ of ω where $J = \infty$ had positive probability, then we would obtain a contradiction to what has just been proved by noting that $\mathcal{A}$ is Γ-invariant and by conditioning on $\mathcal{A}$. ∎

We can now deduce the following extension of Theorem 1.1:

Theorem 3.9. *Let G be a graph with a transitive unimodular closed automorphism group $\Gamma \subset \mathrm{Aut}(G)$, and suppose that $\iota_E(G) > 0$. Let ω be a Γ-invariant percolation on G that has infinite clusters a.s. Then in each of the following cases (on a larger probability space) there is a percolation $\omega' \subset \omega$ such that $\omega' \neq \emptyset$, $\iota_E(\omega') > 0$ a.s., and the distribution of the pair (ω', ω) is Γ-invariant:*

(i) ω *is Bernoulli percolation;*
(ii) ω *has a unique infinite cluster a.s.;*
(iii) ω *has a cluster with at least three ends a.s.;*
(iv) $\mathbf{E}[\deg_\omega o \mid o \in \omega] > \alpha(G)$ *and* ω *is ergodic.*

Proof. In Bernoulli percolation, if there is more than one infinite cluster, then there is a cluster with at least three ends by insertion tolerance and ergodicity. Consequently, (i) follows from (ii) and (iii). Parts (ii)–(iv) follow from Theorems 3.1, 3.6, and 3.2. □

Although it will not be needed in the sequel, we note that Theorem 3.1 can be strengthened as follows.

Theorem 3.10. (A Forest in the Uniqueness Regime with $\iota_E > 0$) *Let G be a graph with a transitive unimodular closed automorphism group $\Gamma \subset \mathrm{Aut}(G)$, and suppose that $\iota_E(G) > 0$. Let ω be a Γ-invariant percolation on G that has a.s. exactly one infinite component. Then (on a larger probability space) there is a random forest $\mathfrak{F} \subset \omega$ with $\mathfrak{F} \neq \emptyset$ and $\iota_E(\mathfrak{F}) > 0$ a.s., and the distribution of the pair $(\mathfrak{F}, \omega)$ is Γ-invariant.*

Proof. Let $\omega' \subset \omega$ be as in Theorem 3.1. Since $\iota_E(\omega') > 0$, Theorem 13.7 from BLPS (1998) constructs a percolation $\omega'' \subset \omega'$ such that a.s. ω'' has all components with infinitely many ends and the distribution of (ω'', ω) is Γ-invariant. By Lemma 3.8, there is a forest $\mathfrak{F}' \subset \omega''$ such that some tree in $\mathfrak{F}'$ has infinitely many ends a.s. and the distribution of $(\mathfrak{F}', \omega)$ is Γ-invariant. Let $\mathfrak{F}''$ be the union of the infinite components of $\mathfrak{F}'$. By Lemma 3.7, $\mathbf{E}[\deg_{\mathfrak{F}''} o \mid o \in \mathfrak{F}''] > 2$. The proof is completed as for Theorem 3.6. □

Applying Theorem 3.10 to the case where $\omega = G$ a.s., we obtain an invariant random forest in G with $\iota_E(\mathfrak{F}) > 0$. This is related to the result of Benjamini and Schramm (1997) which says that every bounded-degree graph with $\iota_E > 0$ contains a tree T with $\iota_E(T) > 0$. In fact, the latter result can be used to extend our theory to the non-transitive setting as follows.

Corollary 3.11. (The Non-Transitive Case) *Let G be a graph of bounded degree with $\iota_E(G) > 0$. Then there is some $p_0 < 1$ such that p-Bernoulli bond percolation on G has a subgraph with $\iota_E > 0$ a.s. whenever $p > p_0$.*

Proof. By the result of Benjamini and Schramm (1997) mentioned above, there is a tree $T \subset G$ with $\iota_E(T) > 0$. Let T_3 be the 3-regular tree. There is a map ϕ that takes $V(T_3)$ into $V(T)$, takes every edge $e = [v, u] \in E(T_3)$ to a path of bounded length $\phi(e)$ in T joining $\phi(v)$ to $\phi(u)$, and when $e, e' \in E(T_3)$ are distinct, the corresponding paths $\phi(e), \phi(e')$ are edge-disjoint. Consequently, p-Bernoulli percolation on T can be pulled back via ϕ to a bond percolation ω on T_3 in which the events $\{e \in \omega\}$ $(e \in E(T_3))$

are mutually independent. Moreover, $\mathbf{P}[e \in \omega] \geqslant 1-k(1-p)$, where k is the maximum length of a path $\phi(e')$, $e' \in E(T_3)$. Consequently, ω dominates $(1-k(1-p))$-Bernoulli bond percolation ω' on T_3. By Theorem 3.5, when $3(1-k(1-p)) > 2$, there is with positive probability, and therefore a.s., a subgraph $\omega'' \subset \omega'$ with $\iota_E(\omega'') > 0$. Now $\phi(\omega'')$ is the required subgraph of ω. □

4. Speed and Transience.

In this section, we prove that in many cases, simple random walk on the infinite components of invariant percolation on a nonamenable transitive graph G has positive speed.

Let ω be a percolation on G. It will be useful to consider **delayed simple random walk** $Z = Z^\omega$ on ω, defined as follows. Set $Z(0) := o$, where $o \in V(G)$ is some fixed base point. If $n \geqslant 0$, conditioned on $\langle Z(0), \ldots, Z(n)\rangle$ and ω, let $Z'(n+1)$ be chosen from $Z(n)$ and its neighbors in $E(G)$ with equal probability. Set $Z(n+1) := Z'(n+1)$ if the edge $[Z(n), Z'(n+1)]$ belongs to ω; otherwise, let $Z(n+1) := Z(n)$.

For any sequence $\langle z(0), z(1), \ldots\rangle$, let $\mathcal{S}z$ be the shifted sequence $\langle \mathcal{S}z(0), \mathcal{S}z(1), \ldots\rangle$ defined by $\mathcal{S}z(n) := z(n+1)$ and let $\mathcal{S}(z,\omega) := (\mathcal{S}z, \omega)$. For $\gamma \in \Gamma$, we set

$$\gamma(w,\omega) := (\gamma w, \gamma\omega),$$

where $(\gamma w)(n) := \gamma(w(n))$.

The following lemma is from Lyons and Schramm (1998); it generalizes similar lemmas in Lyons and Peres (1998) and in Häggström (1997).

Lemma 4.1. **(Stationarity of Random Walk)** *Let G be a graph with a transitive unimodular closed automorphism group $\Gamma \subset \mathrm{Aut}(G)$. Let $o \in V(G)$ be some base point. Let $(\mu, 2^{E(G)})$ be a Γ-invariant percolation process on G; let ω denote the percolation configuration. Let $\widehat{\mu}_o$ be the joint law of ω and delayed simple random walk on ω, as defined above. Then $\widehat{\mu}_o[\mathcal{A}] = \widehat{\mu}_o[\mathcal{S}\mathcal{A}]$ for every Γ-invariant $\mathcal{A} \subset V^{\mathbb{Z}} \times \Omega$. In other words, the restriction of $\widehat{\mu}_o$ to the Γ-invariant σ-field is $\mathcal{S}$-stationary.*

Let μ' be the measure on subgraphs $\omega \subset G$ whose Radon-Nikodym derivative with respect to μ is $\deg_\omega o/\mathbf{E}_\mu[\deg_\omega o]$. Let $\widehat{\mu}'_o$ be the joint law of ω and simple random walk (non-delayed) on ω starting at o. Then the restriction of $\widehat{\mu}'_o$ to the Γ-invariant σ-field is $\mathcal{S}$-stationary.

Lemma 4.2. **(Speed Exists and is Not Random)** *Let G be a graph with a transitive unimodular closed automorphism group $\Gamma \subset \mathrm{Aut}(G)$. Let ω be a Γ-invariant percolation on G. Then the speed Λ of delayed simple random walk on ω exists and is an ω-measurable random variable (possibly zero).*

If ω has indistinguishable components and is ergodic, then, conditioned on $|K(o)| = \infty$, Λ is equal a.s. *to a constant.*

The same statements hold for simple random walk in place of delayed simple random walk.

Proof. Let $f_n(\hat{Z}) := \mathrm{dist}_G(o, Z(n))$. Then

$$f_{n+m}(\hat{Z}) \leqslant f_n(\hat{Z}) + f_m(\mathcal{S}^n \hat{Z})$$

by the triangle inequality. Consequently, the Subadditive Ergodic Theorem shows that the speed

$$\Lambda = \Lambda(\hat{Z}) = \lim_{n\to\infty} f_n(\hat{Z})/n = \lim_{n\to\infty} \mathrm{dist}_G\big(o, Z(n)\big)/n$$

exists a.s.

To show that the speed $\Lambda(\hat{Z})$ depends only on ω and not on the path of the random walk a.s., define $F(\hat{Z})$ to be the variance, conditioned on ω, of the speed of an independent random walk starting from $Z(0)$. By Lévy's 0-1 Law, $F(\mathcal{S}^n \hat{Z})$ converges to zero a.s. But by stationarity, the distribution of $F(\mathcal{S}^n \hat{Z})$ is the same for all n. Hence, it is 0.

The statement concerning indistinguishable components is a consequence of the definition.

The same proof applies to simple random walk since the measures μ and μ' of Lemma 4.1 are mutually absolutely continuous. □

Our main tool to convert the geometric information of Section 3 to probabilistic information is the following:

Theorem 4.3. (Speed When There is a Subgraph with $\iota_E > 0$) *Let G be a graph with a transitive unimodular closed automorphism group $\Gamma \subset \mathrm{Aut}(G)$. Let $\omega' \subset \omega$ be percolations on G such that the distribution of the pair (ω', ω) is Γ-invariant. Suppose that $\omega' \neq \emptyset$ and $\iota_E(\omega') > 0$* a.s. *Then simple random walk on ω has positive speed* a.s. *on the event that $K(o) \cap \omega' \neq \emptyset$.*

Proof. Let V^* be the vertices of ω' that are in the ω-component of o, and let Z be delayed simple random walk on ω starting at o. Note that given ω, Z is reversible with uniform stationary distribution. Given ω and ω' with $o \in \omega'$, there is an induced walk Z^* on V^* defined as follows. Set $t_0 := 0$. Since the transformed $\hat{Z}$ is stationary (Lemma 4.1), the Poincaré recurrence theorem (see, e.g., Petersen (1983), p. 34) shows that conditioned on $o \in \omega'$, there is a.s. some first time $t_1 > 0$ such that $Z(t_1) \in \omega'$. (Strictly speaking, Lemma 4.1 does not apply when there is an extra "scenery" ω', but the lemma extends easily to this situation; see Lyons and Schramm (1998).) Inductively, for $k > 0$, let t_k be the first time $t > t_{k-1}$ such that $Z(t) \in \omega'$.

Define $Z^*(k) := Z(t_k)$. Then given ω and ω' with $o \in \omega'$, Z^* is just the Markov chain Z induced on the states V^*. In particular, it is reversible with the same stationary distribution on V^*, i.e., uniform.

We claim that, given ω and ω' with $o \in \omega'$, the spectral radius $\rho(Z^*)$ is less than 1 a.s. Given two vertices $u^*, v^* \in V^*$, let $p^*(u^*, v^*)$ denote the transition probability of the Markov chain Z^*. Let G^* be the graph whose vertices are V^* and whose edges $[u^*, v^*]$ are those pairs with $p^*(u^*, v^*) > 0$. Note that there is some positive lower bound $c > 0$ for $p^*(u^*, v^*)$ whenever $[u^*, v^*] \in \omega'$. Consequently,

$$\inf \left\{ \frac{1}{|K^*|} \sum_{e^* \in \partial_E K^*} p^*(e^*) : K^* \subset V(G^*) \text{ is finite} \right\} \geqslant c\iota_E(\omega') > 0 \,.$$

Since the stationary distribution is uniform, this implies that $\rho(Z^*) < 1$ a.s. (see, e.g., Kaimanovich (1992)), as claimed.

Fix any $\rho_0 < 1$ such that $\rho(Z^*) < \rho_0$ with positive probability, and let $\mathcal{A}$ be the event that $o \in \omega'$ and $\rho(Z^*) < \rho_0$. Then, for some $\zeta < 1$ and all $v \in V(G)$,

$$\mathbf{P}\big[Z(t_k) = v \mid \omega, \omega'\big] \leqslant \zeta^k \quad \text{on } \mathcal{A} \,,$$

which gives

$$\mathbf{P}\big[Z(t_k) = v \mid \mathcal{A}\big] \leqslant \zeta^k \,. \tag{4.1}$$

Since the number of vertices $v \in V(G)$ with $\mathrm{dist}_G(o, v) < r$ is bounded by $(\deg_G o)^{r+1}$, by summing (4.1) over all such vertices, we get

$$\mathbf{P}\Big[\mathrm{dist}_G\big(o, Z(t_k)\big) < r \mid \mathcal{A}\Big] \leqslant \zeta^k (\deg_G o)^{r+1}.$$

Let β be such that $(\deg_G o)^\beta = 1/\zeta$, and choose $r := (\beta/2)k - 1$. Then

$$\mathbf{P}\Big[\mathrm{dist}_G\big(o, Z(t_k)\big) < \beta k/2 - 1 \mid \mathcal{A}\Big] \leqslant \zeta^{k/2}. \tag{4.2}$$

By the Borel-Cantelli lemma, it follows that

$$\liminf_{k \to \infty} \mathrm{dist}_G\big(o, Z(t_k)\big)/k \geqslant \beta/2$$

a.s. on $\mathcal{A}$. Also, the ergodic theorem ensures that $\lim t_k/k < \infty$ a.s., whence

$$\liminf_{k \to \infty} \mathrm{dist}_G\big(o, Z(t_k)\big)/t_k > 0$$

a.s. on $\mathcal{A}$ and hence a.s. when $K(o) \cap \omega' \neq \emptyset$. This shows that the speed of Z is positive a.s. when $K(o) \cap \omega' \neq \emptyset$ by Lemma 4.2. By the obvious coupling of delayed random walk and simple random walk, it follows that also the speed of simple random walk is positive a.s. when $K(o) \cap \omega' \neq \emptyset$.

□

Theorem 4.4. (Speed) *Let G be a graph with a transitive unimodular closed automorphism group $\Gamma \subset \mathrm{Aut}(G)$, and suppose that $\iota_E(G) > 0$. Let ω be a Γ-invariant percolation on G. Then simple random walk on some infinite cluster of ω has positive speed with positive probability in each of the following cases:*

(i) *ω is Bernoulli percolation that has infinite components a.s.;*
(ii) *ω has a unique infinite cluster a.s.;*
(iii) *ω has a cluster with at least three ends with positive probability;*
(iv) $\mathbf{E}[\deg_\omega o \mid o \in \omega] > \alpha(G)$.

Proof. This follows from Theorems 3.9 and 4.3. □

In case G is a tree, (iii) and (iv) of this theorem were established by Häggström (1997).

In case the percolation is ergodic and has indistinguishable components, like Bernoulli percolation, we have the stronger conclusion that simple random walk has positive speed on every infinite component a.s.

Remark 4.5. It does not suffice in Theorem 4.4 to drop in (i) the assumption that ω is Bernoulli. For example, if ω is the wired uniform spanning forest (see Section 5), then every component is a tree with one end (BLPS (1998)), whence is recurrent.

In order to derive additional consequences of Theorem 4.4, we now extend Theorem 1.6:

Lemma 4.6. *Let G be a graph with a transitive unimodular closed automorphism group $\Gamma \subset \mathrm{Aut}(G)$. Let ω be a Γ invariant bond percolation on G. The following are equivalent:*

(i) *the speed of simple random walk $X(t)$ on $K(o)$ is zero a.s. in the G-metric:*
$$\lim_{t\to\infty} \mathrm{dist}_G\big(X(t)\big)/t = 0\,;$$
(ii) *the speed of simple random walk $X(t)$ on $K(o)$ is zero a.s. in the ω-metric:*
$$\lim_{t\to\infty} \mathrm{dist}_\omega\big(X(t)\big)/t = 0\,;$$
(iii) *the asymptotic entropy of simple random walk on $K(o)$ is zero a.s.;*
(iv) *there are no nonconstant bounded harmonic functions on $K(o)$ a.s.*

Proof. Because of Lemma 4.1, the equivalence of (iii) and (iv) follows from Kaimanovich and Woess (1998). Clearly, (ii) implies (i). We show that (i) implies (iii) implies (ii).

Assume (i). Fix ω such that the speed on $K(o)$ is zero. Let μ_t^ω denote the law of $X(t)$ on $K(o)$. Let B_r denote the ball of radius $\lfloor r \rfloor$ in G centered at o. Given $\epsilon > 0$, choose t_0 large enough that for all $t \geqslant t_0$, we have

$\mu_t^\omega(B_{t\epsilon}) \geqslant 1-\epsilon$. Let $D := \deg_G o$. Then for $t \geqslant t_0$, concavity of log gives the inequality

$$\sum_{x \in B_{t\epsilon}} -\mu_t^\omega(x) \log \mu_t^\omega(x) \leqslant \mu_t^\omega(B_{t\epsilon}) \log\big(|B_{t\epsilon}|/\mu_t^\omega(B_{t\epsilon})\big) \leqslant \log\big(D^{t\epsilon}/(1-\epsilon)\big)\,.$$

Similarly,

$$\sum_{x \notin B_{t\epsilon}} -\mu_t^\omega(x) \log \mu_t^\omega(x) = \sum_{x \in B_t - B_{t\epsilon}} -\mu_t^\omega(x) \log \mu_t^\omega(x) \leqslant \epsilon \log(D^t/\epsilon)\,.$$

Since this holds for all $t \geqslant t_0$ and ϵ was arbitrary, (iii) follows.

Now assume that (ii) does not hold. Let $\mathcal{A}_\ell$ be the event that the speed is at least ℓ, and note that $\mathcal{A}_\ell$ is ω measurable, by Lemma 4.2. Then by the famous bound of Varopoulos (1985) and Carne (1985), we have

$$\lim_{t\to\infty} -\frac{1}{t} \log \mu_t^\omega\big(X(t)\big) \geqslant \ell^2/2$$

on $\mathcal{A}_\ell$. In other words, (iii) does not hold. □

Corollary 4.7. *Let G be a graph with a transitive unimodular closed automorphism group $\Gamma \subset \mathrm{Aut}(G)$. Suppose that G is nonamenable. Let ω be any Γ-invariant, ergodic, insertion tolerant percolation that has more than one infinite component a.s. Then every infinite component of ω admits nonconstant bounded harmonic Dirichlet functions.*

Proof. By Theorem 2.2 and ergodicity, it suffices to establish the existence of nonconstant bounded harmonic Dirichlet functions on $K(o)$ with positive probability.

We know from insertion tolerance and ergodicity that there are infinitely many infinite components a.s. By insertion tolerance again, we also have that some, hence all, infinite components have at least three ends. By Theorem 4.4, all infinite components are transient. By insertion tolerance, it follows that with positive probability, $K(o)$ has a finite subset K whose removal breaks $K(o)$ into at least two transient components. In such a case, $K(o)$ has nonconstant bounded harmonic Dirichlet functions (e.g., the probability that a simple random walk starting at v eventually stays in a fixed transient component of $K(o) - K$ is such, as a function of v). See Soardi (1994), Theorems 4.20 and 3.73. This establishes our goal. □

In order to prove transience in certain amenable cases, we shall use:

Lemma 4.8. (Transience of Big Trees) *If T is any locally finite tree with $p_c(T) < 1$, then simple random walk is transient on T.*

Proof. By Lyons (1990), the branching number of T is $1/p_c(T)$ and this is the critical value for transience of biased random walk on T. Since this is larger than 1, it follows that, in particular, simple random walk is transient. □

For a Cayley graph G, let ζ_n be the number of elements of G at distance n from o. It is evident that $\langle \zeta_n \rangle$ is submultiplicative, whence the **growth rate** $\mathrm{gr}(G) := \lim \zeta_n^{1/n} = \inf \zeta_n^{1/n}$ exists.

Theorem 4.9. (Transience Above the Reciprocal Growth Rate) *Let G be a Cayley graph with $\mathrm{gr}(G) > 1$, and let $p \in (1/\mathrm{gr}(G), 1)$. Then simple random walk is transient on every infinite cluster of p-Bernoulli percolation a.s.*

Proof. Let ω be p-Bernoulli percolation. By ergodicity and indistinguishability of components, it suffices to prove transience of $K(o)$ with positive probability. As shown in Lyons (1995), there is a tree $T \subset G$ with $p_c(T) = 1/\mathrm{gr}(G)$. This means that the component ω' of o in $\omega \cap T$ has $p_c(\omega') < 1$ with positive probability, whence by Lemma 4.8, ω is transient with positive probability. By Rayleigh monotonicity, the same is true of the component of o in G. □

The following conjecture would imply Conjecture 1.5 (by taking $\omega := G$ and using Lemma 4.6):

Conjecture 4.10. (Monotonicity of Entropy) *Let G be a graph with a transitive unimodular closed automorphism group $\Gamma \subset \mathrm{Aut}(G)$. Let ω and ω' be two Γ-invariant percolations on G such that $\omega' \subseteq \omega$. Then the asymptotic entropy of delayed simple random walk on ω' is at most the asymptotic entropy of delayed simple random walk on ω.*

The following conjecture for finite graphs can be shown to imply Conjecture 4.10.

Conjecture 4.11. *Let G be a finite graph and $C : E(G) \to \mathbb{R}^+$. Consider the continuous-time (reversible) Markov chain $\langle X(t) \rangle$ on $V(G)$ whose transition rate from u to v is $C(u,v)$. Let $h_t(v, C)$ be the entropy of $X(t)$ when $X(0) = v$ and $h_t(C) := \sum_{v \in V(G)} h_t(v, C)$. Then for all t, given two functions C' and C with $C'(e) \leqslant C(e)$ for all $e \in E(G)$, we have $h_t(C') \leqslant h_t(C)$.*

Here is an equivalent formulation of this conjecture. Given a matrix B, let $H(B)$ be the sum of $-b_{i,j} \log b_{i,j}$ over all entries $b_{i,j}$ of the matrix. Let $\mathcal{A}_n$ be the space of $n \times n$ real symmetric matrices with non-negative off-diagonal terms and with each row summing to zero. Then a reformulation of Conjecture 4.11 is that $H(\exp A)$ is (weakly) monotone increasing in the off-diagonal entries of A, where A ranges in $\mathcal{A}_n$.

5. Harmonic Dirichlet Functions.

In this section, we study the existence of nonconstant harmonic Dirichlet functions on percolation components.

We first describe the spanning forest measures we use. A **spanning tree** of a finite graph is a subgraph without cycles that is connected and includes every vertex of the graph. Motivated by some questions of R. Lyons, Pemantle (1991) showed that if an infinite graph G is exhausted by finite subgraphs G_n, then the uniform distributions on the spanning trees of G_n converge weakly to a measure supported on spanning forests* of G. We call this the **free uniform spanning forest** (FSF), since there is another natural construction where the exterior of G_n is identified to a single vertex ("wired") before passing to the limit. This second construction, which we call the **wired uniform spanning forest** (WSF), was implicit in Pemantle's paper and was made explicit by Häggström (1995). Both measures are concentrated on the set of forests, all of whose trees are infinite. See BLPS (1998) or Lyons (1998) for an exposition and more details. For convenience, we will use the symbols FSF and WSF also for the uniform measure on spanning trees of a finite graph. For the proof of Theorem 5.7, we shall have need of one more measure on infinite transient graphs, the **oriented wired spanning forest**, denoted OWSF. We refer to BLPS (1998) for its definition. For our purposes, it is enough to know that it is the same as WSF, except that each edge in the forest is oriented in such a way that there is exactly one outgoing edge from each vertex. All these measures, FSF, WSF, and OWSF, are invariant under Aut(G). We typically denote the random spanning forest by $\mathfrak{F}$.

The following three lemmas are taken from BLPS (1998):

Lemma 5.1. **($\mathcal{O}_{\mathbf{HD}}$ Criterion)** *For any (connected) graph G, we have* FSF $=$ WSF *iff $G \in \mathcal{O}_{\mathbf{HD}}$.*

Write $\mathbf{E}^G_{\mathsf{WSF}}$, $\mathbf{E}^G_{\mathsf{FSF}}$ for expectation with respect to the random spanning forests on G.

Lemma 5.2. **(Domination)** *For any graph G, we have $\mathbf{E}^G_{\mathsf{FSF}}[\deg_{\mathfrak{F}} v] \geqslant \mathbf{E}^G_{\mathsf{WSF}}[\deg_{\mathfrak{F}} v]$ for every $v \in V$, with equality for every v iff* FSF $=$ WSF.

Lemma 5.3. (WSF**-Expected Degree)** *In any infinite transitive graph G, the* WSF*-expected degree of every vertex is 2.*

The following lemma is from BLPS (1999):

---

* In graph theory, "spanning forest" usually means a maximal subgraph without cycles, i.e., a spanning tree in each connected component. We mean, instead, a subgraph without cycles that contains every vertex.

Lemma 5.4. (Small Trees and Expected Degree) *Let Γ be a closed unimodular subgroup of $\mathrm{Aut}(G)$ that acts transitively on G and let ω be the configuration of a Γ-invariant percolation on G. Fix a vertex o. Let F_o be the event that $K(o)$ is an infinite tree with finitely many ends, and let F'_o be the event that $K(o)$ is a finite tree.*

(i) *If $P[F_o] > 0$, then* $\mathbf{E}\big[D(o) \bigm| F_o\big] = 2$.

(ii) *If $P[F'_o] > 0$, then* $\mathbf{E}\big[D(o) \bigm| F'_o\big] < 2$.

Fix any base point $o \in V(G)$ and let $\mathcal{A}_o$ be the event that $K(o)$ is infinite. Let $\mathbf{E}$ refer to the probability measure of the percolation. Extending the notation above, we write $\mathbf{E}^{\omega}_{\mathsf{FSF}}$ and $\mathbf{E}^{\omega}_{\mathsf{WSF}}$ for expectation with respect to the free and wired spanning forest measures on ω (given ω).

Theorem 5.5. ($\mathcal{O}_{\mathbf{HD}}$ Stability when Amenable) *Let G be an amenable graph with a transitive automorphism group $\Gamma \subset \mathrm{Aut}(G)$ and ω a Γ-invariant percolation. Then a.s. every component of ω is in $\mathcal{O}_{\mathbf{HD}}$.*

Proof. We show that the measures FSF and WSF coincide on $K(o)$ a.s. given $\mathcal{A}_o$. By the argument in Burton and Keane (1989), a.s. no component in any invariant percolation on G can have more than 2 ends. Applying this to the percolations given by taking the FSF or the WSF of each component (independently for each component) of ω in conjunction with Lemma 5.4, we obtain that

$$\mathbf{E}\big[\mathbf{E}^{\omega}_{\mathsf{WSF}}[\deg_{\mathfrak{F}} o] \bigm| \mathcal{A}_o\big] = 2 = \mathbf{E}\big[\mathbf{E}^{\omega}_{\mathsf{FSF}}[\deg_{\mathfrak{F}} o] \bigm| \mathcal{A}_o\big]\,.$$

Hence, the result follows from Lemma 5.2. □

Lemma 5.6. (Expected Degree for Recurrent Trees) *Let G be a graph with a transitive unimodular closed automorphism group $\Gamma \subset \mathrm{Aut}(G)$. Let ω be a Γ-invariant random forest in G. Suppose that a.s. all components of ω are recurrent. Then* $\mathbf{E}[\deg_{\omega} o] \leqslant 2$.

Proof. This follows from Lemmas 3.7, 5.4, and 4.8. □

Theorem 5.7. ($\neg\mathcal{O}_{\mathbf{HD}}$ Stability when High Marginals) *Let G be a graph with a transitive unimodular closed automorphism group $\Gamma \subset \mathrm{Aut}(G)$. If $G \notin \mathcal{O}_{\mathbf{HD}}$, then there is some $p_0 < 1$ such that for every Γ-invariant bond percolation ω with $\inf_{e\in E} \mathbf{P}[e \in \omega] > p_0$, some component of ω is not in $\mathcal{O}_{\mathbf{HD}}$ with positive probability. If ω is ergodic and has indistinguishable components, then the same hypotheses imply that a.s., no infinite component of ω is in $\mathcal{O}_{\mathbf{HD}}$.*

Proof. We show that $p_0 := 2/\mathbf{E}^{G}_{\mathsf{FSF}}[\deg_{\mathfrak{F}} o]$ works.

First, $p_0 < 1$ by Lemmas 5.1, 5.2, and 5.3. Let $\mathcal{T}_o$ be the event that $K(o)$, the component of o in ω, is transient. We claim that

$$\mathbf{P}[\mathcal{T}_o] > 0 \tag{5.1}$$

and

$$\mathbf{E}\big[\mathbf{E}^{\omega}_{\mathsf{WSF}}[\deg_{\mathfrak{F}} o] \mid \mathcal{T}_o\big] = 2 < \mathbf{E}\big[\mathbf{E}^{\omega}_{\mathsf{FSF}}[\deg_{\mathfrak{F}} o] \mid \mathcal{T}_o\big]\,. \tag{5.2}$$

This suffices for the first statement by Lemma 5.2. The second statement then follows by ergodicity.

Let $\mathfrak{F}'$ be the union of components of $\mathfrak{F}$ that are contained in recurrent components of ω. Lemma 5.6 implies that if $\mathbf{P}[\mathcal{T}_o] < 1$, then $\mathbf{E}^{\omega}_{\mathsf{FSF}}[\deg_{\mathfrak{F}} o \mid o \in \mathfrak{F}'] \leqslant 2$, which means $\mathbf{E}^{\omega}_{\mathsf{FSF}}[\deg_{\mathfrak{F}} o \mid \neg\mathcal{T}_o] \leqslant 2$. Consequently, (5.1) and the inequality in (5.2) will be established once we prove

$$\mathbf{E}\big[\mathbf{E}^{\omega}_{\mathsf{FSF}}[\deg_{\mathfrak{F}} o]\big] > p_0 \mathbf{E}^{G}_{\mathsf{FSF}}[\deg_{\mathfrak{F}} o] = 2\,.$$

Now

$$\mathbf{E}^{\omega}_{\mathsf{FSF}}[\deg_{\mathfrak{F}} o] = \sum_{x \sim o} \mathbf{P}^{\omega}_{\mathsf{FSF}}\big[[o,x] \in \mathfrak{F}\big] = \sum_{\substack{x \sim o \\ [o,x] \in \omega}} \mathbf{P}^{\omega}_{\mathsf{FSF}}\big[[o,x] \in \mathfrak{F}\big]\,. \tag{5.3}$$

Let B_n be the ball of radius n centered at o in G. By Kirchhoff's theorem and Rayleigh's monotonicity principle (see Lyons and Peres (1998) or BLPS (1998)), for each n and each $e \in \omega$,

$$\mathbf{P}^{\omega \cap B_n}_{\mathsf{FSF}}[e \in \mathfrak{F}] \geqslant \mathbf{P}^{B_n}_{\mathsf{FSF}}[e \in \mathfrak{F}]\,.$$

Taking a limit as $n \to \infty$, we obtain

$$\mathbf{P}^{\omega}_{\mathsf{FSF}}[e \in \mathfrak{F}] \geqslant \mathbf{P}_{\mathsf{FSF}}[e \in \mathfrak{F}]$$

by the definition of the FSF, whence (5.3) gives

$$\mathbf{E}^{\omega}_{\mathsf{FSF}}[\deg_{\mathfrak{F}} o] \geqslant \sum_{\substack{x \sim o \\ [o,x] \in \omega}} \mathbf{P}^{G}_{\mathsf{FSF}}[e \in \mathfrak{F}]\,.$$

Taking expectation, we obtain

$$\begin{aligned}
\mathbf{E}\big[\mathbf{E}^{\omega}_{\mathsf{FSF}}[\deg_{\mathfrak{F}} o]\big] &\geqslant \mathbf{E}\Big[\sum_{x \sim o} \mathbf{1}_{[o,x] \in \omega} \mathbf{P}^{G}_{\mathsf{FSF}}\big[[o,x] \in \mathfrak{F}\big]\Big] \\
&= \sum_{x \sim o} \mathbf{P}\big[[o,x] \in \omega\big] \mathbf{P}^{G}_{\mathsf{FSF}}\big[[o,x] \in \mathfrak{F}\big] \\
&> \sum_{x \sim o} p_0 \mathbf{P}^{G}_{\mathsf{FSF}}\big[[o,x] \in \mathfrak{F}\big] \\
&= p_0 \mathbf{E}^{G}_{\mathsf{FSF}}[\deg_{\mathfrak{F}} o]\,,
\end{aligned}$$

as desired.

For the equality in (5.2), we use the oriented wired spanning forest, OWSF, on each transient component of ω, chosen independently on each component. Let $\varphi(x, y)$ be the probability that ($K(x)$ is transient and that) $[x, y]$ belongs to the oriented wired spanning forest of ω. Since OWSF is Γ-invariant, φ is invariant under the diagonal action of Γ, whence the Mass-Transport Principle says that

$$\sum_x \varphi(o, x) = \sum_x \varphi(x, o) .$$

The left-hand side is the expected outdegree of o, which is $\mathbf{P}[\mathcal{T}_o]$. Hence, the right-hand side, the expected in-degree of o, is also $\mathbf{P}[\mathcal{T}_o]$. This shows that $\mathbf{E}\big[\mathbf{E}^{\omega}_{\mathsf{WSF}}[\deg_{\mathfrak{F}} o] \mid \mathcal{T}_o\big] = 2$. □

Example 5.8. It does not suffice in Theorem 5.7 to assume merely that the components of ω are infinite. For example, if ω is given by the WSF, then every component is a tree with one end (BLPS (1998)), whence is recurrent and in $\mathcal{O}_{\mathbf{HD}}$. However, as stated in Conjecture 1.11, we believe that this *is* sufficient for Bernoulli percolation.

Example 5.9. The hypothesis that Γ be unimodular cannot be omitted in Theorem 5.7. For example, let G be a regular tree of degree 3 and ξ be an end of G. Let Γ be the group of automorphisms of G that fix ξ. Let H_n, $n \in \mathbb{Z}$, be the horocycles with respect to ξ. (More precisely, fix a base point $o \in V$, let $\langle v_m \rangle$ be a sequence converging to ξ. Then a vertex v is in H_n iff $\operatorname{dist}(v_m, v) - \operatorname{dist}(v_m, o) = n$ for all but finitely many m.) To define ω, we first define a percolation η. Given any $p_0 < 1$, for each n independently, let all the edges joining H_n to H_{n+1} be in η with probability p_0. Each component of η is a finite tree a.s. For each component K of η, let $n(K)$ be the largest n such that $K \cap H_n \neq \emptyset$. Choose an edge joining $K \cap H_{n(K)}$ to $H_{n(K)+1}$ at random uniformly among all such edges and independently for each K; let η' be the set of the chosen edges (over all K). Now let $\omega := \eta \cup \eta'$. Each component of ω is a tree with exactly one end, so is recurrent and in $\mathcal{O}_{\mathbf{HD}}$. Yet $G \notin \mathcal{O}_{\mathbf{HD}}$.

Question 5.10. Does Theorem 5.7 hold for Bernoulli percolation when the unimodularity assumption is omitted?

6. Anchored Expansion and Stability.

Cheeger's inequality relates the isoperimetric constant, which is geometric, to the spectral radius, which governs the exponential decay of return probabilities of simple random walk or Brownian motion. We discuss a geometric constant that we hope can replace the isoperimetric constant in graphs and manifolds that are not uniformly expanding.

Consider, for example, the hyperbolic space $\mathbb{H}^n$ and perturb the metric on an extremely sparse sequence of balls with radii growing very slowly to infinity; for instance, pick the center of the n-th ball at distance e^n from a fixed origin, and let $\log n$ be its radius. If we modify the metric inside these balls so that it is flat on sub-balls of half the radius, we get a manifold with zero isoperimetric constant. Many properties of M (such as the existence of nonconstant bounded harmonic functions or the speed of Brownian motion), are, however, unchanged from $\mathbb{H}^n$.

Definition 6.1. Fix some base point $o \in G$. Call

$$\iota_E^*(G) := \lim_{n\to\infty} \inf \left\{ \frac{|\partial_E S|}{|S|} : o \in S \subset V(G),\ S \text{ is connected},\ n \leqslant |S| < \infty \right\}$$

the **anchored expansion constant** of G. Note that $\iota_E^*(G)$ is independent of the choice of the base point o and that $\iota_E^*(G) \geqslant \iota_E(G)$.

By attaching a sequence of paths of length $1, 2, \ldots$ at a very sparse sequence of vertices of a binary tree, we get an example of a graph G for which $\iota_E^*(G) > \iota_E(G) = 0$.

Isoperimetric quantities such as this, where the set S is required to contain a fixed vertex o, were introduced in greater generality by Thomassen (1992). He used the word "rooted" instead of "anchored".

Benjamini and Schramm (1996) show that $\iota_E(G)$ is related to $p_c(G)$; in fact, their proof of Thm. 2 shows that $p_c(G) \leqslant (1 + \iota_E^*(G))^{-1}$.

When the isoperimetric constant $\iota_E(G)$ of a bounded degree (not necessarily transitive) graph is positive, Dodziuk's (1984) discrete version of Cheeger's inequality gives an upper bound $\overline{\rho} < 1$ for the spectral radius $\rho(G)$, where $\overline{\rho}$ depends on $\iota_E(G)$ and the maximum degree in G. In turn, this implies that lim inf speed Λ^- of simple random walk X starting at a base point $o \in V(G)$ is positive almost surely.

By analogy, we make the following

Conjecture 6.2. *Let G be a bounded degree graph with $\iota_E^*(G) > 0$. Then $\Lambda^- > 0$ with positive probability.*

It might even be the case that $\iota_E^*(G) > 0$ implies $\Lambda^- > 0$ a.s. In fact, after seeing an earlier draft of this paper, Virag (1998) proved that the lim sup speed is positive when G is a tree of bounded degree with $\iota_E^*(G) > 0$.

He also has shown that whenever $\iota_E^*(G) > 0$, there is a subgraph G' of G with $\iota_E(G') > 0$.

Thomassen (1992) has shown that if a graph satisfies a weaker type of anchored isoperimetric inequality, then it is transient; Conjecture 6.2 has a stronger hypothesis and a stronger conclusion.

The motivation for looking at $\iota_E^*(G)$ is that ι_E^* is more stable than ι_E under random perturbations of G. For example, let G be an infinite graph of bounded degree and pick a probability distribution $\mathbf{P}$ on the strictly positive integers. Replace each edge $e \in G$ by a path of length L_e, where L_e is distributed according to $\mathbf{P}$, and all L_e $(e \in E(G))$ are independent. Let $G^{\mathbf{P}}$ denote the random graph obtained in this way. If $\mathbf{P}$ has a bounded support, then $\iota_E(G) > 0$ implies $\iota_E\left(G^{\mathbf{P}}\right) > 0$, while if $\mathbf{P}$ has unbounded support then, almost surely, $\iota_E\left(G^{\mathbf{P}}\right) = 0$.

Question 6.3. Does $\iota_E(G) > 0$ imply that $\iota_E^*\left(G^{\mathbf{P}}\right) > 0$ a.s. when $\mathbf{P}$ is the geometric distribution on the positive integers? What about other distributions $\mathbf{P}$ with finite mean? Can this be settled in the case where G is a regular tree?

It is not hard to construct $\mathbf{P}$ and G such that $\iota_E(G) > 0$ while $\iota_E^*\left(G^{\mathbf{P}}\right) = 0$ a.s. (Take G to be a binary tree and $\mathbf{P}$ to have a fat tail.)

Lyons, Pemantle, and Peres (1995) proved that simple random walk on a random perturbation of any regular tree, with $\mathbf{P}$ the geometric distribution, has positive speed almost surely. More generally, simple random walk has positive speed on every supercritical Galton-Watson tree a.s. given nonextinction.

Question 6.4. Is $\iota_E^*(T) > 0$ a.s. for supercritical Galton-Watson trees given nonextinction?

Question 6.5. If $\iota_E(G) > 0$ and ω is Bernoulli percolation on G, must every infinite component K of ω have $\iota_E^*(K) > 0$ a.s.?

After seeing an earlier draft of this paper, answers to the last three questions have largely been provided by Chen and Peres (1998). That is, Questions 6.3 and 6.4 were answered completely and Question 6.5 was answered affirmatively for p-Bernoulli percolation when p is sufficiently close to 1.

References.

Babson, E. and Benjamini, I. (1999). Cut sets and normed cohomology with applications to percolation, *Proc. Amer. Math. Soc.*, 589–597.

Benjamini, I., Lyons, R., Peres, Y., and Schramm, O. (1999). Group-invariant percolation on graphs, *Geom. Funct. Anal.*, to appear.

Benjamini, I., Lyons, R., Peres, Y., and Schramm, O. (1998). Uniform spanning forests, *preprint.*

Benjamini, I., Pemantle, R. and Peres, Y. (1998). Unpredictable paths and percolation, *Ann. Probab.* **26**, 1198–1211.

Benjamini, I. and Schramm, O. (1996a). Harmonic functions on planar and almost planar graphs and manifolds, via circle packings, *Invent. Math.* **126**, 565–587.

Benjamini, I. and Schramm, O. (1996b). Percolation beyond $\mathbb{Z}^d$, many questions and a few answers, *Electronic Commun. Probab.* **1**, 71–82.

Benjamini, I. and Schramm, O. (1997). Every graph with a positive Cheeger constant contains a tree with a positive Cheeger constant, *Geom. Funct. Anal.* **7**, 403–419.

Burton, R. M. and Keane, M. (1989). Density and uniqueness in percolation, *Commun. Math. Phys.* **121**, 501–505.

Carne, T. K. (1985). A transmutation formula for Markov chains, *Bull. Sci. Math.* (2) **109**, 399–405.

Chen, D. and Peres, Y. (1998). Anchored expansion, percolation, and speed, *preprint.*

De Masi, A., Ferrari, P. A., Goldstein, S., and Wick, W. D. (1989). An invariance principle for reversible Markov processes. Applications to random motions in random environments, *J. Statist. Phys.* **55**, 787–855.

Derriennic, Y. (1980). Quelques applications du théorème ergodique sous-additif, *Astérisque* **74**, 183–201.

Dodziuk, J. (1984). Difference equations, isoperimetric inequality, and transience of certain random walks, *Trans. Amer. Math. Soc.* **284**, 787–794.

Grimmett, G. R. (1989). *Percolation.* Springer, New York.

Grimmett, G. R., Kesten, H. and Zhang, Y. (1993). Random walk on the infinite cluster of the percolation model, *Probab. Theory Relat. Fields* **96**, 33–44.

Grimmett, G. R. and Newman, C. M. (1990). Percolation in ∞+1 dimensions, in *Disorder in Physical Systems*, G. R. Grimmett and D. J. A. Welsh (editors), pp. 219–240. Clarendon Press, Oxford.

Häggström, O. (1995). Random-cluster measures and uniform spanning trees, *Stoch. Proc. Appl.* **59**, 267–275.

Häggström, O. (1997). Infinite clusters in dependent automorphism invariant percolation on trees, *Ann. Probab.* **25**, 1423–1436.

Häggström, O. and Peres, Y. (1998). Monotonicity of uniqueness for percolation on Cayley graphs: all infinite clusters are born simultaneously, *Probab. Th. Rel. Fields*, to appear.

Holopainen, I. and Soardi, P. M. (1997). p-harmonic functions on graphs and manifolds, *Manuscripta Math.* **94**, 95–110.

Kaimanovich, V. A. (1990). Boundary and entropy of random walks in random environment, in *Probability theory and mathematical statistics*, Vol. I (Vilnius, 1989), 573–579, "Mokslas", Vilnius, 1990.

Kaimanovich, V. A. (1992). Dirichlet norms, capacities and generalized isoperimetric inequalities for Markov operators, *Potential Anal.* **1**, 61–82.

Kaimanovich, V. A. and Vershik, A. M. (1983). Random walks on discrete groups: boundary and entropy, *Ann. Probab.* **11**, 457–490.

Kaimanovich, V. A. and Woess, W. (1998). The Poisson boundary of quasi-transitive graphs, *in preparation.*

Kanai, M. (1986). Rough isometries and the parabolicity of riemannian manifolds, *J. Math. Soc. Japan* **38**, 227–238.

Kesten, H. (1959). Symmetric random walks on groups, *Trans. Amer. Math. Soc.* **92**, 336–354.

Lyons, R. (1990). Random walks and percolation on trees, *Ann. Probab.* **18**, 931–958.

Lyons, R. (1995). Random walks and the growth of groups, *C. R. Acad. Sci. Paris* **320**, 1361–1366.

Lyons, R. (1998). A bird's-eye view of uniform spanning trees and forests, in *Microsurveys in Discrete Probability*, D. Aldous and J. Propp (editors). AMS, Providence, pp. 135–162.

Lyons, R., Pemantle, R. and Peres, Y. (1995). Ergodic theory on Galton-Watson trees: speed of random walk and dimension of harmonic measure, *Ergodic Theory Dynamical Systems* **15**, 593–619.

Lyons, R. and Peres, Y. (1998). *Probability on Trees and Networks.* Cambridge University Press, in preparation. Current version available at http: //php.indiana.edu/~rdlyons.

Lyons, R. and Schramm, O. (1998). Indistinguishability of percolation clusters, *Ann. Probab.*, to appear.

Lyons, T. (1987). Instability of the Liouville property for quasi-isometric Riemannian manifolds and reversible Markov chains, *J. Diff. Geom.* **26**, 33–66.

Medolla, G. and Soardi, P. M. (1995). Extension of Foster's averaging formula to infinite networks with moderate growth, *Math. Z.* **219**, 171–185.

Pemantle, R. (1991). Choosing a spanning tree for the integer lattice uniformly, *Ann. Probab.* **19**, 1559–1574.

Petersen, K. (1983). *Ergodic Theory.* Cambridge University Press, Cambridge.

Saloff-Coste, L. (1992). A note on Poincare, Sobolev, and Harnack inequalities, *Internat. Math. Res. Notices*, no. 2, 27–38.

Soardi, P. M. (1993). Rough isometries and Dirichlet finite harmonic functions on graphs, *Proc. Amer. Math. Soc.* **119**, 1239–1248.

Soardi, P. M. (1994). *Potential Theory on Infinite Networks.* Springer, Berlin.

Soardi, P. M. and Woess, W. (1990). Amenability, unimodularity, and the spectral radius of random walks on infinite graphs, *Math. Z.* **205**, 471–486.

Thomassen, C. (1992). Isoperimetric inequalities and transient random walks on graphs, *Ann. Probab.* **20**, 1592–1600.

Varopoulos, N. Th. (1985). Long range estimates for Markov chains, *Bull. Sci. Math.* (2) **109**, 225–252.

Virag, B. (1998). Personal communication.

Woess, W. (1994). Random walks on infinite graphs and groups — a survey on selected topics, *Bull. London Math. Soc.* **26**, 1–60.

Itai Benjamini
Mathematics Department,
The Weizmann Institute of Science,
Rehovot 76100, Israel
itai@wisdom.weizmann.ac.il
http://www.wisdom.weizmann.ac.il/~itai/

Russell Lyons
Department of Mathematics,
Indiana University,
Bloomington, IN 47405-5701, USA
rdlyons@indiana.edu
http://php.indiana.edu/~rdlyons/

Oded Schramm
Mathematics Department,
The Weizmann Institute of Science,
Rehovot 76100, Israel
schramm@wisdom.weizmann.ac.il
http://www.wisdom.weizmann.ac.il/~schramm/

Twist surfaces

Robert Brooks

In this paper, we consider the problem: what kinds of Riemann surfaces S will have first eigenvalue $\lambda_1(S)$ large? By large, we mean that there is some positive lower bound for S independent of the genus.

There are a number of constructions of surfaces with λ_1 large, the main example being the quotients of the hyperbolic plane by the congruence subgroups of $PSL(2,\mathbb{Z})$. Most of the known examples rely on fairly heavy number-theoretic or representation-theoretic machinery. The construction of [SX] replaces much of the representation theory with geometric arguments, but still only applies to surfaces with some special types of symmetries. On the other hand, it is our belief that "large first eigenvalue" should be a much more common phenomenon than this, although it seems to be difficult to pinpoint good geometric criteria which will allow for this.

The technique we will employ involves thinking of a Riemann surface in terms of a graph. Then a surface will have large first eigenvalue if and only if the corresponding graph has large first eigenvalue.

The novelty of our approach is that we will associate a graph to a surface in such a way that many surfaces will share the same graph. The surfaces which share the same graph, which we will call *twist surfaces*, will in general have very little in common with one another, except for the fact that they will have a common lower bound for the first eigenvalue. They will not even have the same genus! In this way, we may build from arithmetic examples large collections of very non-arithmetic examples which have their first eigenvalue bounded from below.

Our thinking on this question was inspired by the theorem of Joel Friedman [F], who showed that, for k even, a "typical" k-regular finite graph is close to being Ramanujan. One would hope, arguing by analogy, that a "typical" Riemann surface has first eigenvalue close to $1/4$. While the present results are far from that goal, we hope to emphasize the role that finite graph theory can play in constructing fairly exotic surfaces whose spectral theory is nonetheless under good control.

An important part of our argument revolves around the use of the Ahlfors-Schwarz Lemma. We believe that this could be a powerful tool in the relationship between graph theory and geometry, and hope the reader finds this discussion of independent interest.

The results of this paper will be more qualitative than quantitative. Our main intent here is to show that a graph-theoretic approach to these questions leads in a natural way to phenomena which would be hard to imagine any other way. We intend to give a more detailed account of our results elsewhere.

Acknowledgements: It is a pleasure to thank the organizers of the Conference on Discrete Potential Theory and Random Walks for their invitation to participate in this conference and to write this article.

We would also like to thank Gil Kalai and Eran Makover for helpful discussions.

1 The Marriage Lemma.

We begin this section with the following problem: given a compact manifold M_0, describe the behavior of the first eigenvalue of the Laplacian $\lambda_1(M_i)$ as M_i ranges over finite coverings of M_0.

The result of [SGTC] in some sense gives a complete answer to this problem. To state it, let us recall that a finite covering M_i of M_0 is described by a finite index subgroup $\pi_i = \pi_1(M_i)$ of the fundamental group $\pi = \pi_1(M_0)$ of M_0.

Picking a symmetric set of generators of π, we may associate to the pair $<\pi, \pi_i>$ a graph $\Gamma_i = \Gamma(\pi, \pi_i)$ called the Schreier graph of π/π_i, as follows:

(i) the vertices of Γ_i are the cosets of π/π_i.

(ii) two vertices v_1 and v_2 are joined by one edge for each j such that $g_j v_1 = v_2$.

Clearly, Γ_i is a k-regular graph.

We may then associate to Γ_i the first eigenvalue $\lambda_1(\Gamma_i)$ of the graph Γ_i. We then have:

Theorem 1.1 ([SGTC]) $\lambda_1(M_i) \to 0$ *if and only if* $\lambda_1(\Gamma_i) \to 0$.

The proof of [SGTC] passes through the Cheeger constants $h(M_i)$ and $h(\Gamma_i)$, where $h(M)$ is defined by

$$h(M) = \inf_H \frac{\text{area}(H)}{\min(\text{vol}(A), \text{vol}(B))},$$

where H runs over hypersurfaces dividing M into two pieces A and B, and where "area" denotes $(n-1)$-dimensional volume, while "vol" denotes n-dimensional volume. $h(\Gamma)$ for Γ a graph is defined analogously.

One shows that

$$c_1 h(\Gamma_i) \leq h(M_i) \leq c_2 h(\Gamma_i),$$

where c_1 and c_2 are geometric constants depending on a choice for fundamental domain of M_0. To pass to bounds for λ_1, we make use of the bounds of Cheeger [Ch] and Buser [Bu], which say that

$$(1/4)h^2 \leq \lambda_1 \leq c_3 h + c_4 h^2,$$

for some constants c_3 and c_4 which depend only on a lower bound for the Ricci curvature of M_0.

A proof of Theorem 1.1 which does not pass through the Cheeger constant may be found in [Bur].

Theorem 1.1 leaves open the problem of deciding when $\lambda_1(\Gamma_i)$ is bounded away from 0. The following give some methods for determining when this is the case:

(i) Let π be a group with Kazhdan's Property T. Then there is a lower bound c for $\lambda_1(\Gamma_i)$ for all finite-index subgroups π_i of π.

(ii) Let π be the group $PSL(2, \mathbb{Z})$, and let π_p be the p-th congruence subgroup

$$\pi_p = \left\{ \begin{pmatrix} a & b \\ c & d \end{pmatrix} \equiv \pm \begin{pmatrix} 1 & 0 \\ 0 & 1 \end{pmatrix} (\mod p) \right\}.$$

Then $\lambda_1(\Gamma_p)$ is bounded uniformly away from 0.

This is a reformulation of Selberg's 3/16 Theorem [Sel] (without the constant 3/16).

From this description, it would appear that a lower bound for $\lambda_1(M_i)$ would be heavily dependent on number-theoretic or representation-theoretic properties of groups. On the other hand, it is our contention here that this is not really the case. To that end, let us ask the question: how does one construct coverings of M_0 with λ_1 large which do not have such a description?

We first observe that in passing from the pair (π, π_i) to the graph Γ_i, one loses a lot of information. Let us ask the question: given the graph Γ_i, in how many ways can we describe it as the Schreier graph corresponding to (π, π_j) for some possibly very different group π_j? Since (π, π_i) and (π, π_j) share the same Schreier graph, it follows from Theorem 1.1 that the correponding manifolds share the same lower bound for λ_1, even though they may be quite different manifolds.

The extra information which is lost in passing from the pair (π, π_i) to the graph Γ_i may be described as follows:

Definition 1.1 *Let Γ be a k-regular graph. A* coloring of type (r,s) *of* Γ, *with $2r+s=k$, is an assignment to each edge of Γ either*

(i) an integer in the set $\{1,\dots,r\}$ and a direction,

or

(ii) an integer in the set $\{r+1,\dots,r+s\}$,

such that each vertex has precisely one outgoing edge and one incoming edge of each color $\{1,\dots,r\}$ and one edge of each color in $\{r+1,\dots,r+s\}$.

It is easily seen that writing $\Gamma_i = \Gamma(\pi,\pi_i)$ describes a coloring of Γ_i. One takes the generating set and writes it in asymmetrical form as r generators $g_1,\dots,g_r$ such that $g_i \neq g_i^{-1}$, and s generators $g_{r+1},\dots,g_{r+s}$ with $g_{r+j} = g_{r+j}^{-1}$. Then an edge given by left-multiplication by g_i is given the color i.

It is not very difficult to see that the converse to this holds: given a coloring of the graph Γ_i, let g_i be the map of the vertices of Γ described as follows: if $i \leqslant r$, then g_i is the map which takes each vertex x of Γ_i to the vertex of Γ_i joining x along an outgoing edge of color i. If $j > r$, then g_j sends each vertex to the vertex adjoining it along an edge of color j.

If we denote by π^* the group generated by the g_j's, and π_0^* the stabilizer of a vertex, then clearly

$$\Gamma_i = \Gamma(\pi^*,\pi_0^*).$$

In general, we have to impose an extra condition to obtain from this a covering M_j of our given manifold M_0. The extra condition is that the map $g_i \to \pi^*$ extend to a homomorphism $\pi_1(M_0) \to \pi^*$. This will be the case automatically if$\pi_1(M)$ is a free group, but if it is not, it may well be difficult to establish whether this holds for a given coloring.

Geometrically, one can visualize the new covering of M_0 in the following way: the graph Γ_i describes how to build a covering of M_0 by instructing us how to glue copies of a fundamental domain together. When we change the coloring of the graph, we retain the information of which copies of the fundamental domain are glued together, but we change the faces along which they are glued together.

The point now is that it is a fairly simple matter to obtain from a given coloring many new colorings. In fact, we will see that this is essentially a combinatorial problem that can be solved with relative ease and flexibility. The main tool is the classical Marriage Lemma:

Lemma 1.1 (Marriage Lemma) *Let A and B be two sets of the same size, and for each element x of A, there is a subset B_x of B of "suitable matches"* *for x. Then there is a 1-to1 function $f : A \to B$ which assigns to each x one of its suitable matches, if and only if, for each subset C of Y, the set D of elements of A who have a suitable match lying in C has cardinality at least as large as C.*

The proof is inductive: one makes arbitrary choices, at each stage eliminating a given choice as a suitable match for other elements of A, until one hits the case where for some C, the set D has the same cardinality as C. One then assigns a matching of C by the inductive hypothesis.

It is easily seen that the Marriage Lemma gives us many ways of finding a coloring of type $(0, k)$ for any k-regular bipartite graph, by letting A and B be the two sets of vertices, and the "suitable matches" be the pairs of vertices joined by an edge. The condition in the Marriage Lemma is easily seen to be satisfied: if C is any subset of B, then there are $k\#(C)$ edges leading from it, which must link up with at least $(1/k)k\#(C)$ vertices in A. After one matching of A with B is made, we assign all the edges involved a color, remove them to obtain a $(k-1)$-regular graph, and proceed inductively.

With a little more work, we may show:

Theorem 1.2 *Let Γ be any k-regular graph, with k even and > 2. Then there are a large number of colorings of type $(k/2, 0)$ of Γ.*

Of course, any connected 2-regular graph has precisely two colorings.

The proof begins by decomposing Γ into a disjoint union of closed paths. This is a simple classical construction, with a large number of possible solutions. By choosing arbitrarily a direction for each closed path, we may put a direction on each edge, so that each vertex has $k/2$ incoming and $k/2$ outgoing edges.

Now let Γ' be the bipartite graph whose vertices consist of two sets A and B, each equal to the vertex set of Γ. A vertex x of A is joined to a vertex y in B if and only if there is an edge in Γ which is outgoing at x and incoming at y. It is clear that a coloring of Γ' of type $(0, k/2)$ gives a coloring of Γ of type $(k/2, 0)$.

We remark that a coloring of type $(k/2, 0)$ is the same thing as presenting Γ as a covering of the k-regular graph with one vertex. In this way, we can give a quick proof of Leighton's Theorem [Le], that the obvious necessary condition for two graphs to have a common finite cover, in the special case when the two graphs are k-regular, with k even. This was first proved, in a somewhat different way which is also valid for k odd, in [AG]. Their proof also makes use of the Marriage Lemma.

One may establish the k-regular case, k odd, of the theorem of [AG] by first passing to the bipartite double coverings of the graphs involved. Note that the "obvious necessary condition" mentioned above is always satisfied when both graphs are k-regular.

For a geometrically-oriented discussion of these and related questions, we refer the reader to the paper [BK].

2 Riemann Surfaces and Graphs.

In this section, we discuss some relationships between Riemann surfaces and graphs. For a more complete discussion of this topic, see [RSG].

Let Γ be a graph which is a triangulation of an oriented surface. We may describe the triangulation and orientation conditions purely combinatorially, see [RSG] and [PS]. Furthermore, we may read off the genus of the surface from the formula

$$g = E/6 - V/2 + 1,$$

where E is the number of edges and V the number of vertices.

We now want to build Riemann surfaces out of Γ. To do this, we consider the marked ideal triangle T, shown below. We have placed "tick marks" at the points i, $i+1$, and $\frac{i+1}{2}$.

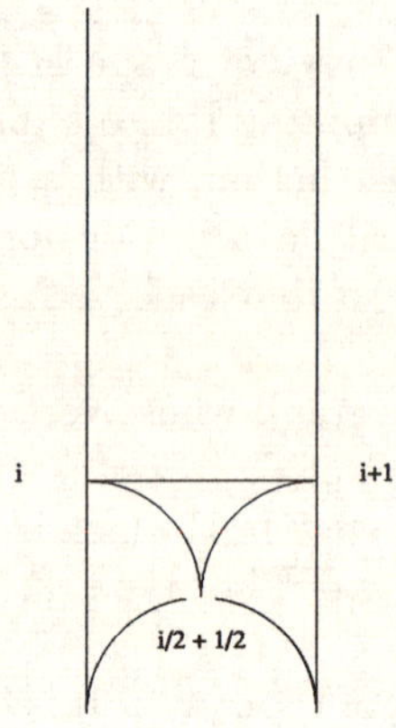

Figure 1: The marked ideal triangle

The idea now is to paste in one copy of T for each face of the triangulation. To do this, we have to describe how to glue together two faces which share an edge. To that end, let f be some function on the edges. Then if T_1 and T_2 are two copies of T which share the edge e, we will glue T_1 to T_2 along e by placing the tick marks along the sides at distance $f(e)$. We can place a sign on the distance by choosing a direction for the edge, and declaring T_1 and T_2

to be positively ordered if, when passing from T_1 to T_2, the direction of the edge points to the left. Then the distance between the tick marks is positive if the tick mark for T_2 is to the left of the tick mark for T_1.

We will call the resulting surface $S_f^O(\Gamma)$.

In general, the surface $S_f^O(\Gamma)$ will not be complete, but there is a simple criterion for completeness:

Lemma 2.1 *The surface S_f^O is complete if and only if, for each vertex v of Γ, the sum over incoming edges is zero:*

$$\sum_{e:v\varepsilon e} f(e) = 0.$$

To see this, we observe that if $\sum_{e:v\varepsilon e} f(e) \neq 0$, then one may travel along horocycles around the cusp. After each cycle, one is some constant closer to the cusp, but the length of the horocycle decreases geometrically. Thus, one can get to infinity in finite time.

Let us consider this construction for the simplest case, where we glue two triangles together to get a punctured torus:

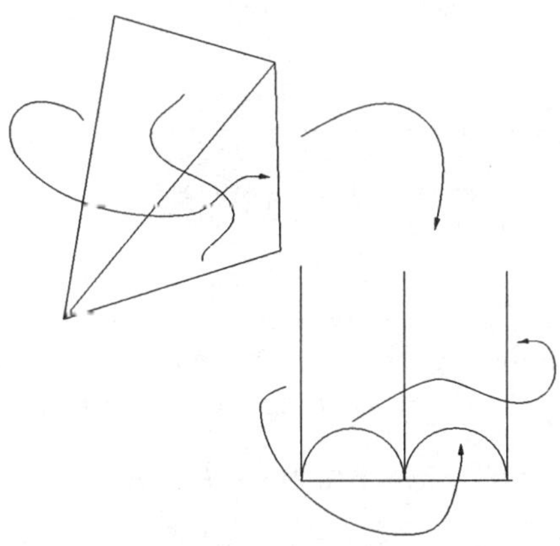

Figure 2: Building a torus

Here, Γ has one vertex, three edges, and two faces. The space of f's which give a complete metric is two-dimensional— one dimension for each eadge, and one relation for the completeness condition at the vertex. Since each edge contributes twice to the vertex condition, one needs to check that the two contributions enter with the same sign.

Recalling that the Teichmüller space of the once-punctured α torus has two real dimensions, one may check easily that the map

$$f \to S_f(\Gamma)$$

is an isomorphism from such f to the Teichmüller space.

One may realize the two real parameters in the following way: in Figure 3 below, we have glued two ideal triangles together along the three possible sides. The distance between the tick marks is reflected in the cross-ratio of the four points at infinity. Taking two of these cross-ratios as parameters, and observing that the completeness condition prescribes the third cross-ratio once the first two are known, gives us the desired two real parameters.

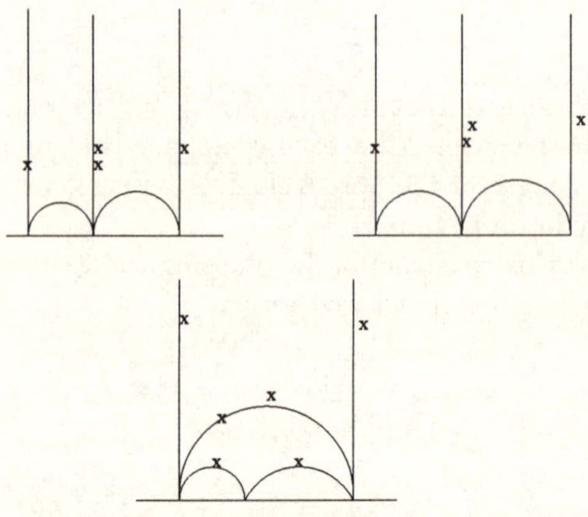

Figure 3: The three cross-ratios

The fact that we get an isomorphism here is not an accident. For a general graph Γ, the space of f's which give rise to a complete metric has dimension $E - V$. The Teichmüller space of a surface of genus g with V punctures is isomorphic to $\mathbb{R}^{6g-6+2V}$. From our formula for g, we see that

$$6g - 6 + 2V = E - V.$$

One may then show:

Theorem 2.1 *For any graph Γ which is the triangulation of a surface of genus g, the map $f \to S_f^O(\Gamma)$ gives an isomorphism from $\mathbb{R}^{E-V}$ to the Teichmüller space of surfaces of genus g with V punctures.*

Question: We observe that the Teichmüller space has a natural complex structure. Is it possible to see the complex structure on $\mathbb{R}^{E-V}$ in these coordinates?

Now let S^O be a Riemann surface with punctures. We may associate to S^O a compact Riemann surface, denoted S^C, in a canonical way, which is easily described conformally. The neighborhood of each puncture is conformally equivalent to a punctured disk. By replacing each punctured disk with an unpunctured disk, we obtain the desired compactification.

We now apply this construction to the surfaces $S_0^O(\Gamma)$, where the subscript "0" signifies that we choose the function f to be the zero function. We then have:

Theorem 2.2 *As Γ runs over all graphs which are triangulations of a surface of genus g, the surfaces $S_0^C(\Gamma)$ run over a dense subset of the moduli space of surfaces of genus g.*

The proof of this theorem is essentially due to Belyi [Bel], see also [PS]. The surfaces $S_0^C(\Gamma)$ are like a dense set of "rational points" (being countable in number) in the moduli space.

This theorem should be compared the the theorem of [CPCE], which gives an alternative method to assign to the set of triangulations of a surface a dense set of Riemann surfaces in the moduli space, based on circle packings. The surfaces $S^O(\Gamma)$ and $S^C(\Gamma)$ which arise in this way have a number of alternative characterizations, which shed some light on this comparison:

Theorem 2.3 *The surfaces $S^O(\Gamma)$ are characterized by each of the following properties:*

(1a) They are finite coverings of the modular surface $\mathbb{H}^2/PSL(2,\mathbb{Z})$.

(1b) They carry a "horocyclic circle packing," that is, for each cusp there is a horocycle, and the region interior to all the horocycles consists of a finite union of horocyclic triangles.

The surfaces $S^C(\Gamma)$ are characterized by each of the following properties:

(2a) They are branched coverings of S^2, branched at three values.

(2b) They are defined by algebraic equations whose coefficients lie in a number field.

The equivalence of (2a) and (2b) in the theorem is precisely Belyi's Theorem [Bel].

Question: If one replaced "horocyclic circle packing" with "circle packing" in (1b), one would obtain the dense set of surfaces of [CPCE]. Are the two mappings of the set of triangulations of surfaces to compact surfaces coarse quasi-isometrically equivalent? Is there a proof of Theorem 2.2 which shows directly how to approximate a given surface by surfaces $S^C(\Gamma)$, similar to the continued fraction argument of [CPCE]?

From part 1(a) above, we see that each graph Γ which is a triangulation of a surface gives rise to a surface $S^O(\Gamma)$ which is a covering of $\mathbb{H}^2/PSL(2,\mathbb{Z})$. This covering is in turn described by a graph, which we will denote by Γ^*, and call the *dual graph* of Γ.

The graph Γ^* has the following explicit description in terms of Γ:

(i) The vertices of Γ^* are the triangular faces of Γ

(ii) Two vertices of Γ^* are joined by an edge if and only if the corresponding faces share an edge.

To visualize the graph Γ^*, we choose the fundamental domain for the group $PSL(2,\mathbb{Z})$ shown in Figure 4 below. It is not the usual fundamental domain for $PSL(2,\mathbb{Z})$. Instead, we have taken the usual fundamental domain for $PSL(2,\mathbb{Z})$ and cut it in half, moving the left half distance 1 to the right, and then gluing it to the right half.

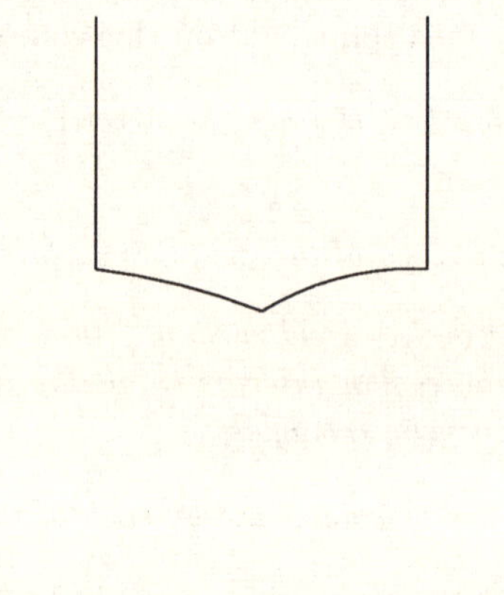

Figure 4: The fundamental domain

With this choice of fundamental domain, we observe that three copies of the fundamental domain fit together to make one ideal triangle. The graph Γ^* will then be the trivalent graph which is given in each triangle by the three glue lines coming out of the center of the triangle.

We illustrate this process in Figure 5 below, where we see the cube realized as the dual graph to the octahedron.

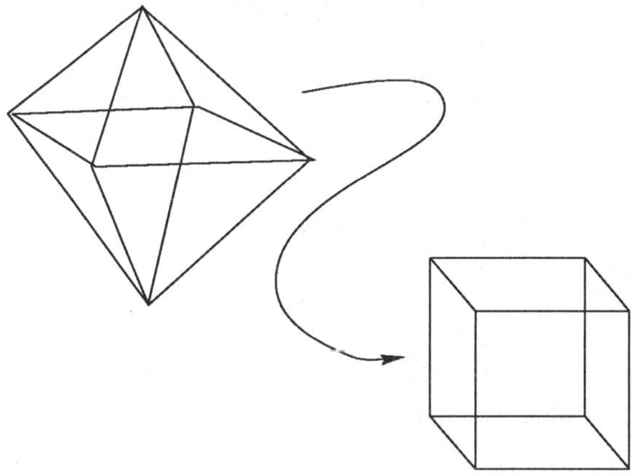

Figure 5: Constructing the dual graph Γ^*

3 The Ahlfors-Schwarz Lemma.

We begin this section with the following question: Given a punctured surface S^O, with a complete, finite-area hyperbolic metric, we have seen how to associate to it a compact surface S^C. We may place on S^C its constant curvature metric. What is the relationship between the geometry of S^O and S^C?

It is not difficult to convince oneself that in general there need not be any relationship whatsoever. For instance, if S^O is a sphere with n punctures, $n \geq 3$, then S^C will not carry any metric of curvature -1.

However, when Γ is a k-regular graph, $k \geq 7$, then it is fairly easy to describe the geometry of $S^C(\Gamma)$ in terms of the geometry of $S^O(\Gamma)$. One simply replaces the ideal hyperbolic triangles with equilateral hyperbolic triangles with angles equal to $2\pi/k$. It is easily checked that the tick-mark condition is automatically fulfilled in this case.

When Γ is not necessarily regular, then the process of building $S^C(\Gamma)$ from triangles would seem to be complicated. One has to choose the angles at the vertices in such a way that all the vertex angles add up to 2π, and so that the corresponding sides have equal length — so far this is a fairly routine exercise in hyperbolic trigonometry. But when one imposes the tick-mark condition, the problem would appear to become fairly transcendental, and one loses control.

We will see below that the Ahlfors-Schwarz Lemma gives us a powerful technique to establish that two conformally equivalent metrics are actually close. Here is a version of the Ahlfors-Schwarz Lemma, taken from [GA], which we will find useful:

Lemma 3.1 *Let ds_1 and ds_2 be two complete, conformally equivalent metrics on a surface S, with curvature κ_1 and κ_2 respectively.*

Suppose that

$$\sup_S \kappa_1 \leqslant \inf_S \kappa_2 < 0.$$

Then

$$ds_1 \leqslant ds_2$$

everywhere on S.

We then have as a corollary:

Corollary 3.1 *Suppose that*

$$C \sup_S \kappa_2 \leqslant \kappa_1 \leqslant \frac{1}{C} \inf_S \kappa_2 < 0.$$

Then

$$\frac{1}{C} ds_2 \leqslant ds_1 \leqslant C ds_2.$$

In other words, for conformally equivalent metrics of negative curvature on surfaces, curvature closness implies metric closeness.

We will apply this to radially symmetric metrics conformally equivalent to the standard metric on the hyperbolic disk – that is, metrics of the form

$$ds^2 = f^2(r)[dr^2 + \sinh^2(r) d\theta^2].$$

The curvature of this metric is given by the formula

$$\kappa(r) = \frac{-[(\frac{f'}{f})' + 1 + \frac{f'}{f} \coth(r)]}{f^2}.$$

When we set the curvature of this metric equal to -1, we get two solutions: the solution

$$f_1(r) = 1$$

and the solution

$$f_2(r) = \frac{1}{\sinh(r) \log(\coth(r/2))},$$

corresponding to the hyperbolic metrics on the disk and punctured disk respectively.

The following lemma is a fairly simple exercise in cutting and pasting solutions of differential equations, see [PS] for details:

Lemma 3.2 *For all ε, there exists an R and a function f_R such that $f_R(r) = f_2(r)$ for $r > R$, $f_R = f_1$ for r small, and the curvature κ_R lies between $-(1+\varepsilon)$ and $-(1-\varepsilon)$.*

We now proceed as follows: if p is a cusp of a Riemann surface S^O, we will say that the cusp has length l if there is a cusp neighborhood of p with a horocycle of length l. We observe that if $S^O = S^O(\Gamma)$, then the length of the cusp p is at least the number of edges which meet the vertex corresponding to p.

By the lemma, for each ε there is an l such that, if the length of the cusp is at least l, then we may replace the metric in the cusp neighborhood with a conformally equivalent metric which has curvature everywhere between $-(1+\varepsilon)$ and $-(1-\varepsilon)$, and which furthermore extends over the cusp. If each cusp of S^0 is sufficiently large, then we may do this on all the cusps to obtain a metric on S^C which first of all has curvature everywhere close to -1, and secondly agrees with the hyperbolic metric on S^0. Applying the Ahlfors-Schwarz Lemma, we see that this metric must be close to the hyperbolic metric on S^C.

Playing with this idea, we can show (see [PS]):

Theorem 3.1 *For all l sufficiently large, there is a constant $C(l)$ such that the Cheeger constants $h(S^O)$ and $h(S^C)$ satisfy*

$$\frac{1}{C(l)}h(S^C) \leqslant h(S^O) \leqslant C(l)h(S^C)$$

whenever all the cusp neighborhoods have length $\geq l$.

Furthermore, $C(l) \to 1$ as $l \to \infty$.

As a consequence, we have that $\lambda_1(S^O)$ and $\lambda_1(S^C)$ are bounded in terms of one another.

Based on suggestions made to us by Laurent Saloff-Coste and Yves Colin de Verdiere, we believe that we can show that

$$\frac{1}{C(l)}\lambda_1(S^C) \leqslant \lambda_1(S^O) \leqslant C(l)\lambda_1(S^C)$$

for some constant $C(l) \to 1$ as $l \to \infty$, but at this writing our proof is not complete.

4 Twisting.

As a consequence of the last section, we have:

Theorem 4.1 *There are constants $C(d)$ for d sufficiently large, with the following property: Suppose that Γ is a graph which is the triangulation of a surface, with Γ^* the dual graph. Suppose that each vertex of Γ has at least d edges. Then*

$$\lambda_1(S^C(\Gamma)) \geqslant C(d)[h(\Gamma^*)]^2.$$

Proof: In effect, Theorem 3.1 tells us that $h(S^C)$ is close to $h(S^0)$. But Theorem 1.1 tells us that $h(S^O)$ is controlled by $h(\Gamma^*)$.

We would like to employ Theorem 4.1 in order to construct Riemann surfaces with large λ_1. According to the theorem, we must proceed in two steps: first of all, we must find trivalent graphs Γ^* with large λ_1. Secondly, we must find graphs Γ whose duals are Γ^*, and which have large valence.

The main point here is that many graphs Γ may correspond to the same Γ^*. So even if the construction of Γ^* is somewhat standard, the resulting Γ's have the potential of being quite exotic. As long as the valences of these graphs are large, they will all share a common lower bound for λ_1.

We will call a graph Γ_2 which shares the same Γ^* with Γ_1 a *twist* of Γ_1. We will call the corresponding surface $S^C(\Gamma_2)$ a twist surface of Γ_1.

There are a number of known constructions of trivalent graphs with large λ_1, all of which have a fairly heavy number-theoretic or representation-theoretic component. Here is my favorite, taken from [PS]:

For each prime number p, let π_p = the p-th Platonic graph be the graph whose vertices are

$$\left\{ \begin{pmatrix} a \\ b \end{pmatrix} \varepsilon\ \mathbb{Z}/p \times \mathbb{Z}/p - (0,0) \right\},$$

with

$$\begin{pmatrix} a \\ b \end{pmatrix} \equiv \begin{pmatrix} -a \\ -b \end{pmatrix}.$$

The edges of π_p are given by

$$\begin{pmatrix} a \\ b \end{pmatrix} \sim \begin{pmatrix} c \\ d \end{pmatrix}$$

if and only if

$$\det \begin{pmatrix} a & c \\ b & d \end{pmatrix} \equiv \pm 1 \pmod p.$$

The dual graph π_p' has vertices which are described by triples

$$\left\{ \begin{pmatrix} a \\ b \end{pmatrix}, \begin{pmatrix} a' \\ b' \end{pmatrix}, \begin{pmatrix} a'' \\ b'' \end{pmatrix} \right\}$$

with

$$\pm\det\begin{pmatrix} a & a' \\ b & b' \end{pmatrix} \equiv \pm\det\begin{pmatrix} a & a'' \\ b & b'' \end{pmatrix} \equiv \pm\det\begin{pmatrix} a' & a'' \\ b' & b'' \end{pmatrix} \equiv \pm 1 \pmod{p}.$$

Two vertices are joined by an edge if and only if the two sets have two members in common.

When p is not a prime, there is an analogous construction, see [PS] or [BFK] for details, and also for a discussion of the geometry of these graphs.

It is easily seen that $S^O(\pi_p)$ is the quotient of $\mathbb{H}^2$ by the p-th congruence subgroup. The fact that $\lambda_1(\pi_p')$ is bounded from below independent of p is more-or-less equivalent to the Selberg 3/16 Theorem [Se].

Now, we ask the question: what are the various twists of the graphs π_p and of the surfaces $S^C(\pi_p)$? More generally, given a graph Γ which is the triangulation of a surface, how may we describe the twists of Γ and of $S^C(\Gamma)$?

As one would expect from the discussion of §1, the problem should be somewhat akin to finding all the different colorings of a graph. Here is a precise statement:

Suppose that Γ is a graph which is the triangulation of an oriented surface. Then the graph Γ^* inherits from Γ the additional structure of a cyclic ordering of the vertices at each point. To see this, we note that each vertex corresponds to an ideal triangle, and the edges correspond to the sides of the triangle. The cyclic ordering then means that we traverse the edges of the triangle in the clockwise direction.

Now suppose we reverse the cyclic ordering at a certain number of vertices of Γ^*. This will correspond the gluing in the ideal triangle with orientation reversed— in other words, we rotate the edge of the second triangle about the tick mark.

We will now try to construct a graph Γ' which is again the triangulation of an oriented surface, with $(\Gamma')^* = \Gamma^*$, but with the induced cyclic ordering on the vertices of Γ^* agreeing with the new cyclic ordering on the vertices of Γ^*.

The idea to construct Γ' is as follows: each vertex of Γ corresponds to a path in Γ^* which consists of always making a left-hand turn whenever one comes to a vertex. Conversely, a path in Γ^* which consists of left-hand turns corresponds to traveling around a vertex in Γ.

Changing the orientation at a vertex corresponds to changing left-hand turns into right-hand turns, and *vice versa*.

Γ' may then be described as the graph whose vertices consist of all paths making only left-hand turns in the new orientation on Γ^*. Two vertices are joined by an edge if and only if the corresponding paths share an edge.

In order to build the surface $S^O(\Gamma')$, we glue together, for each left-hand turn path of length n, n copies of the fundamental domain for $PSL(2, \mathbb{Z})$

of Figure 4. We then glue together two edges of copies of the fundamental domain whenever the edge appears in two different left-hand-turn paths. In particular, there is no obstruction to building the covering $S^O(\Gamma')$ for any choice of change of orientation, in contrast to the situation described in §1.

We illustrate this process with the graph π_4, which is the octahedron. The dual graph in this case is then π_4', which is the cube.

If we reverse the orientations at the four vertices circled in Figure 6 below, then there are precisely four paths consisting of all left-hand turns. The resulting surface $S^C(\Gamma')$ will then have eight faces, corresponding to the eight vertices of the cube, twelve edges corresponding to the twelve edges of the cube, and four vertices corresponding to the four all-left-hand turns. It will thus have genus 1.

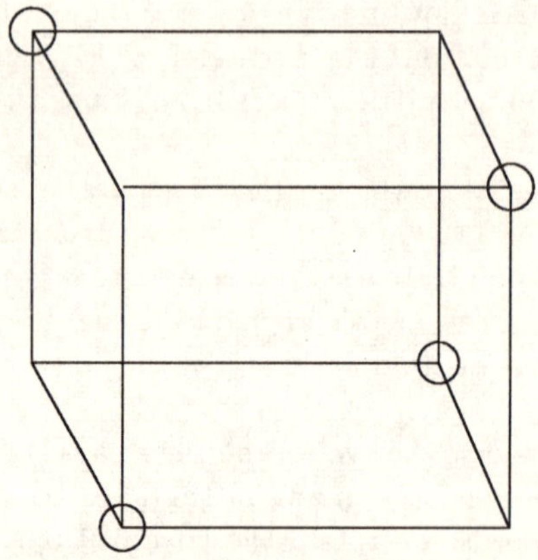

Figure 6: Reversing orientations on the cube

It is a pleasant exercise to check that conformally $S^C(\Gamma')$ is the equilateral torus.

In Figure 7, we illustrate a number of new orientations on the cube which give rise to new surfaces. The reader is invited to try to calculate their conformal structures. As a hint, we remark that the first two are tori, while the third is a surface of genus 2.

In Figure 8, we also give a new orientation to the dodecahedron π_5, which thus describes some new surface.

Playing this game for π_p, p large, we construct large numbers of twists (roughly speaking, exponentially many in p^3) of the surfaces $S^C(\pi_p)$. If we can show that all of these surfaces have large cusps (for sufficiently large p), then we will have constructed very large families of surfaces with large λ_1, since a lower bound for λ_1 is provided by a common lower bound for $\lambda_1(\pi_p')$,

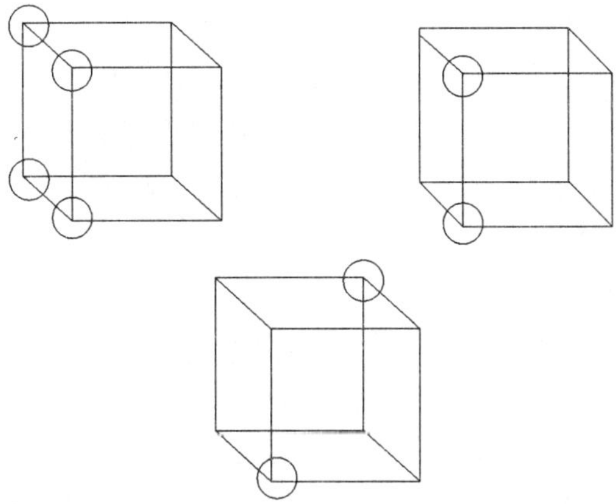

Figure 7: Three twist surfaces

which in turn is provided by the Selberg Theorem.

But it is easy to show that all the cusps are large. Indeed, the length of a cusp is precisely the length of the corresponding closed path in Γ^*. But it is not difficult to see that the length of a closed path in π'_p must be at least (const) $\log(p)$ for some (const) independent of p. Thus the lengths of the cusps grow at least logarithmically in p, and the construction is complete.

We observe that the various twists of the surfaces $S^C(\pi_p)$ appear to have very little in common, outside of their common dual graphs π'_p. Indeed, as we have seen by example, the genus of a twist may well be different from the genus of the original surface.

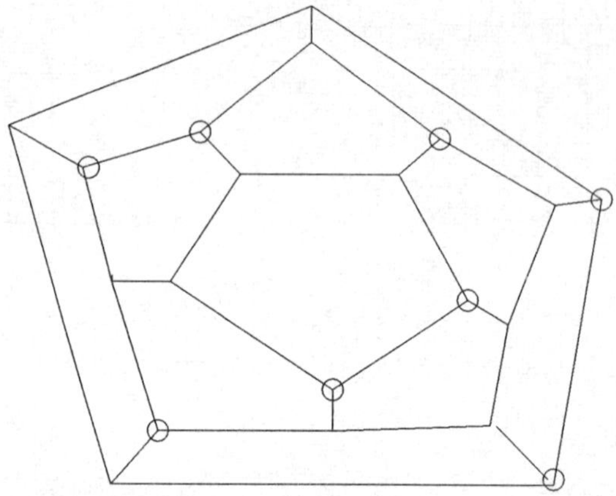

Figure 8: Reversing orientations on the dodecahedron

References

[Ah] L. Ahlfors, *An Extension of Schwarz's Lemma,* Trans. AMS 43 (1938), pp. 359-364.

[AG] D. Angluin and A Gardiner, *Finite Common Coverings of Pairs of Graphs,* J. Comb. Theory B 30 (1981), pp. 184-187.

[BK] H. Bass and R. Kulkarni, *Uniform Tree Lattices,* JAMS 3 (1990), pp. 843-902.

[Be] A. Belyi, *On Galois Extensions of a Maximal Cyclotomic Polynomial,* Izvestia 14 (1980), pp. 247-256.

[CPCE] R. Brooks, *Circle Packings and Co-compact Extensions of Kleinian Groups,* Inv. Math. 86 (1986), pp. 461-469.

[GA] R. Brooks, *Some Geometric Aspects of the Work of Lars Ahlfors,* to appear in Proc. Ahlfors Mem. Lect.

[PS] R. Brooks, *Platonic Surfaces,* to appear in Comm. Math. Helv.

[RSG] R. Brooks, Some Relations between Graph Theory and Riemann Surfaces, Isr. Math. Conf. Proc. 11 (1997), pp. 61-73.

[SGTC] R. Brooks, *The Spectral Geometry of a Tower of Coverings,* J. Diff. Geom. 23 (1985), pp. 97-107.

[BFK] R. Brooks, H. Farkas, and I. Kra, *Number Theory, Theta Identities, and Modular Curves,* Contemp. Math. 201 (1997), pp. 125-154.

[Bur] M. Burger, *Spectre du Laplacien, Graphes, et Topologie de Fell,* Comm. Math. Helv. 63 (1988), pp. 226-252.

[Bu] P. Buser, *A Note on the Isoperimetric Constant,* Ann. Sci. Ec. Norm. Sup. 15 (1982), pp. 213-230.

[Ch] J. Cheeger, *A Lower Bound for the Smallest Eigenvalue of the Laplacian,* in Gunning (ed), Problems in Analysis, Princeton University Press (1970), pp. 195-199.

[F] J. Friedman, *On the Second Eigenvalue and Random Walk in Random d-regular Graphs,* Combinatorica 11 (1991), pp. 331-362.

[Le] F. Leighton, *Finite Common Coverings of Graphs,* J. Comb. Th. B 33 (1982), pp. 231-238.

[SX] P. Sarnak and X. Xue, *Bounds for Multiplicities of Automorphic Representations,* Duke Math. J. 64 (1991), pp. 207-227.

[Se] A. Selberg, *On the Estimation of Fourier Coefficients of Modular Forms,* Proc. Symp. Pure Math. VIII (1965), pp. 1-15.

Robert Brooks
Department of Mathematics
The Technion, Haifa 32000, Israel
email: `rbrooks@tx.technion.ac.il`
http:// www.math.technion.ac.il/~rbrooks

Harmonic functions on buildings of type $\tilde{A}_n$

Donald I. Cartwright

1. Introduction.

For each function f on the vertex set of a graph, and for each vertex x, let $(Af)(x)$ be the average value of $f(y)$ as y runs through the set of neighbours of x. If $\lambda \in \mathbb{C}$, a function h is called *λ-harmonic* if $Ah = \lambda h$, and simply *harmonic* if this holds for $\lambda = 1$.

In this paper, I would like to discuss λ-harmonic functions on the vertex set $\mathfrak{X}^0$ of an affine building $\mathfrak{X}$. To keep the discussion as concrete as possible, I shall restrict my attention to the simplest of these buildings, those of type $\tilde{A}_n$. In such a building, each vertex x has a *type* $\tau(x) \in \{0, 1, \ldots, n\}$ and it is natural to define not just one, but n averaging operators A_r, $r = 1, \ldots, n$. Here $(A_r f)(x)$ is the average value of $f(y)$ as y runs through the set of neighbours of x of type $\tau(x) + r \bmod n+1$. Now if $\lambda = (\lambda_1, \ldots, \lambda_n)$ is an n-tuple of complex numbers, we call a function h *λ-harmonic* if $A_r h = \lambda_r h$ for $r = 1, \ldots, n$, and simply *harmonic* if this holds for $\lambda_1 = \cdots = \lambda_n = 1$.

We shall mostly have in mind the case $n \geq 2$ below. If $n = 1$, an $\tilde{A}_n$ building is a tree in which each vertex has at least three neighbours. If we make the assumption that this tree is homogeneous, with each vertex having exactly $q + 1$ neighbours, then what we say below is true also for $n = 1$, though a separate discussion for that case would be much easier.

The operators A_r commute, and generate an algebra $\mathcal{A}$ which is spanned by more general averaging operators A_k, $k \in \mathbb{N}^n$ (Theorem 3.1). This is basically a reformulation of a special case of a well known fact for buildings arising from a p-adic reductive group G. In that situation there is a compact open subgroup K of G, and (see [5, Proposition 2.4] for the case $n = 2$) $\mathcal{A}$ is isomorphic to the convolution algebra of compactly supported bi-K-invariant functions, known to be commutative (see [13]).

To our building $\mathfrak{X}$ there is associated a "boundary" Ω, which is a totally disconnected compact Hausdorff space. This is the "spherical building at infinity" of $\mathfrak{X}$, and consists of equivalence classes of sectors (see §4). For each $\omega \in \Omega$ and each vertex x, there is a unique sector $S^x(\omega)$ in the class ω having "base vertex" x. Given $x, y \in \mathfrak{X}^0$ and $\omega \in \Omega$, one can define n integers $m_i(x, y; \omega)$, $i = 1, \ldots, n$, which give the coordinates of the vertices of $S^y(\omega)$ in terms of those of $S^x(\omega)$. Given $n+1$ complex numbers $s_1, \ldots, s_{n+1}$ whose product is 1, and given a base vertex o, we define the "Poisson kernel" by

$$h_{o,x}(\omega) = \frac{s_1^{m_1+\cdots+m_n} \cdots s_{n-1}^{m_{n-1}+m_n} s_n^{m_n}}{q^{\rho(m)}}, \tag{1.1}$$

where $m_i = m_i(o, x; \omega)$, $i = 1, \ldots, n$, and where

$$\rho(k) = \frac{1}{2} \sum_{i=1}^{n} i(n+1-i)k_i \tag{1.2}$$

for n-tuples $k = (k_1, \ldots, k_n)$ of integers. This definition is motivated by the "unramified principal series" representations of $GL(n+1, F)$, where F is a local field (see §5). It turns out (Proposition 4.1) that for any fixed ω, $h(x) = h_{o,x}(\omega)$ defines a λ-harmonic function for $\lambda = (\lambda_1, \ldots, \lambda_n)$, where

$$\lambda_r = \frac{q^{r(n+1-r)/2}}{M_{n+1,r}} \sigma_r(s_1^{-1}, \ldots, s_{n+1}^{-1}), \tag{1.3}$$

where σ_r denotes the rth elementary symmetric function, and where $M_{n+1,r}$ is as in Lemma 2.5 below. In the special case

$$s_1 = \frac{1}{q^{\frac{n}{2}}}, s_2 = \frac{q}{q^{\frac{n}{2}}}, \cdots, s_{n+1} = \frac{q^n}{q^{\frac{n}{2}}}, \tag{1.4}$$

one shows that $\lambda_1 = \cdots = \lambda_n = 1$, and that

$$h_{o,x}(\omega) = q^{-\sum_{i=1}^{n} i(n+1-i)m_i} = q^{-2\rho(m)}. \tag{1.5}$$

In this case, the Poisson kernel has an added significance. There are certain natural regular Borel probability measures ν_x defined on Ω (see §4), and

$$h_{o,x}(\omega) = \frac{d\nu_x}{d\nu_o}(\omega), \tag{1.6}$$

the Radon-Nikodym derivative of ν_x with respect to ν_o.

For general $s_1, \ldots, s_{n+1}$, the Poisson kernel belongs to the space $C^\infty(\Omega)$ of locally constant functions on Ω. Given any finitely additive measure μ on the algebra of open and closed subsets of Ω, the integral

$$h(x) = \int_\Omega h_{o,x}(\omega)\, d\mu(\omega) \tag{1.7}$$

makes sense, and defines a λ-harmonic function h for λ as in (1.3).

The following result is a reformulation of a result of S. Kato [11, Theorem 3.2, 12, Theorem 3.7] (which is stated without any reference to buildings). Kato's result applies to any building arising from a p-adic reductive group, and so is basically much more general. But it is not applicable to all $\tilde{A}_n$ buildings when $n = 1$ (because q need not be a power of a prime), and when $n = 2$ (when "non-classical" $\tilde{A}_n$ buildings exist for more complicated reasons). When $n \geq 3$ it is applicable by a result of Tits [18, p. 137]. The following result is due to Mantero and Zappa when $n = 1$ and $n = 2$ (see [9] for the case $n = 1$ and [16] for the case $n = 2$; see also [15] for the announcement, and [17] for the proof of the other cases of affine buildings not covered by Kato's hypotheses).

Theorem 1.1. *Suppose that $\lambda = (\lambda_1, \ldots, \lambda_n)$ is any n-tuple of complex numbers. Choose $n+1$ complex numbers $s_1, \ldots, s_{n+1}$ whose product is 1 and such that (1.3) holds for each r. Assume that $qs_j \neq s_i$ for all $1 \leq i < j \leq n+1$. Then each λ-harmonic function h on $\mathfrak{X}^0$ can be expressed as an integral (1.7) for a unique finitely additive measure μ. In particular, taking the s_j's as in (1.4), each harmonic function on $\mathfrak{X}^0$ has an integral representation (1.7) for a unique μ.*

Note that for any λ, numbers $s_1, \ldots, s_{n+1}$ with product 1 can be found such that (1.3) holds. Any permutation of these numbers will of course also work, and so if they are ordered so that $|s_1| \leqslant \cdots \leqslant |s_{n+1}|$, then $qs_j \neq s_i$ for all $1 \leq i < j \leqslant n+1$.

One of my aims in this paper was to prove Theorem 1.1 for all n using a method I found in a paper of Gerardin [10] (in the case $n = 1$ and $\lambda = 1$). Starting from a λ-harmonic function h on $\mathfrak{X}^0$, the idea is to define for each vertex o and each $k \in \mathbb{N}^n$ a functional J_k^o on $C^\infty(\Omega)$ by

$$J_k^o(\varphi) = \sum_{v \in S_k(o)} h(v) \int_{\Omega_o(v)} \varphi \, d\nu_o, \tag{1.8}$$

where $S_k(o)$ is the set of vertices which have "coordinates" k with respect to o (see §2), and where $\Omega_o(v)$ is the set of $\omega \in \Omega$ such that v is a vertex of $S^o(\omega)$. One then seeks a family $\{c_\ell : \ell \in F\}$ of constants, where $F \subset \mathbb{Z}^n$ is finite, such that

$$T_k^o(\varphi) = C q^{\rho(k)} s_1^{k_1 + \cdots + k_n} \cdots s_n^{k_n} \left(\sum_{\ell \in F} c_\ell q^{\rho(\ell)} J_{k+\ell}^o(\varphi) \right) \tag{1.9}$$

is independent of $k = (k_1, \ldots, k_n)$ once all the k_i are large enough (how large depending on φ). Write $T^o(\varphi)$ for the common value of $T_k^o(\varphi)$, k large. Then

$$T^x(\varphi) = T^o(h_{o,x}\varphi) \tag{1.10}$$

for all vertices o and x and all $\varphi \in \mathcal{C}^\infty(\Omega)$ (Lemma 6.2). The constant C is a normalization factor, chosen so that

$$T^o(\mathbf{1}) = h(o), \tag{1.11}$$

where $\mathbf{1}$ is the function taking the constant value 1 on Ω. Thus by (1.10),

$$h(x) = T^x(\mathbf{1}) = T^o(h_{o,x}\mathbf{1}) = T^o(h_{o,x}).$$

The linear functionals $T : \mathcal{C}^\infty(\Omega) \to \mathbb{C}$ are obviously in one to one correspondence with the finitely additive measures μ on the algebra of open and closed subsets of Ω. So if μ corresponds to T^o, then (1.7) holds. The uniqueness part of Theorem 1.1 is clearly equivalent to the assertion that

$$\mathcal{C}^\infty(\Omega) = \text{ the linear span of the functions } \omega \mapsto h_{o,x}(\omega), x \in \mathfrak{X}^0. \tag{1.12}$$

Unfortunately I have only succeeded in carrying out this program for $n = 1, 2$, (see §7), thus re-proving results of Mantero and Zappa. For general n, I give constants (found with some help from Tim Steger) which I conjecture to "work", and give a very concrete conjecture about a system of recurrence relations which would at least give the existence part of Theorem 1.1 (see §6). In terms of unramified principal series representations π_s of $GL(n+1, F)$, F a local field, (1.12) is equivalent to the assertion that $\mathbf{1}$ is a cyclic vector for π_s. Note that π_s is known to be irreducible when the s_i's are distinct and $qs_j \neq s_i$ for all $1 \leq i, j \leq n+1$ [8, Proposition 3.5(b)]. In the case of harmonic functions at least, it would be desirable to prove that if h is positive, then so is the measure μ in (1.7). The above method of finding μ does not seem very appropriate when trying to prove this. See §7 for the case $n = 1$ however.

2. $\tilde{A}_n$ buildings.

Buildings are simplicial complexes which are made by "gluing together" Coxeter complexes, which are formed from Coxeter groups. Some Coxeter groups arise as groups generated by reflections in hyperplanes. See [1] or [2, Chapters I and VI] for a general discussion. We consider here only the case of the "$\tilde{A}_n$" Coxeter group and complex.

In $\mathbb{R}^{n+1}$ consider the family of hyperplanes $H_{i,j;k} = \{(x_1, \ldots, x_{n+1}) : x_i - x_j = k\}$, where $1 \leq i \neq j \leq n+1$ and $k \in \mathbb{Z}$. The reflection in the hyperplane $H_{i,j;k}$ is the affine map $x \mapsto r_{i,j}x + k(e_i - e_j)$, where $e_1, \ldots, e_{n+1}$ is the usual basis of $\mathbb{R}^{n+1}$ and where $r_{i,j}x$ is x with its ith and jth coordinates interchanged. These reflections generate a group W of affine transformations, which is the semi-direct product of the group S_{n+1} of permutations of $\{1, \ldots, n+1\}$, and the group of translations $x \mapsto x + c$, where $c = (c_1, \ldots, c_{n+1}) \in \mathbb{Z}^{n+1}$ and $\sum_{i=1}^{n+1} c_i = 0$. Here S_{n+1} acts on $\mathbb{R}^{n+1}$ in the natural way. One of the connected components of the complement in $\mathbb{R}^{n+1}$ of the union of all the hyperplanes is

$$C_0^\circ = \{(x_1, \ldots, x_{n+1}) : x_1 < x_2 < \cdots < x_{n+1} < x_1 + 1\},$$

and the other connected components are in one to one correspondence with the elements of W ($w(C_0^\circ) \leftrightarrow w$). The group W is generated by the reflections $s_1, \ldots, s_{n+1}$ in the "walls" $H_{1,2;0}, \ldots, H_{n,n+1;0}, H_{n+1,1;1}$ of C_0°, and $s_i \circ s_j$ has order 3 if $j = i \pm 1 \bmod n+1$, and order 2 otherwise.

Any translation $x \mapsto x + c$, where $c \in \mathbb{Z}^{n+1}$, even if it is not in W, permutes the hyperplanes $H_{i,j;k}$ and hence the connected components of the complement of their union.

The hyperplanes $H_{i,j;k}$ do not satisfy one of the conditions needed for the general theory of [1], because the intersection of the hyperplanes $H_{i,j;0}$ passing through the origin is not $\{0\}$, but rather $\mathbb{R}\mathbf{1}$, where $\mathbf{1} = (1, 1, \ldots, 1)$. We get around this by forming $E = \mathbb{R}^{n+1}/\mathbb{R}\mathbf{1}$ (alternatively, we could have worked in $(\mathbb{R}\mathbf{1})^\perp = \{(x_i) \in \mathbb{R}^{n+1} : \sum_{i=1}^{n+1} x_i = 0\}$). We can write $(x_1, \ldots, x_{n+1}) = x_1\mathbf{1} + \sum_{i=1}^{n}(x_{i+1} - x_i)\mathbf{b}_i$, where $\mathbf{b}_i = (0, \ldots, 0, 1, \ldots, 1)$, in which there are i 0's. Let $\varphi : \mathbb{R}^{n+1} \to E$ be the natural map. Then $\varphi(C_0^\circ)$ is a simplex C_0, the interior of the convex hull of the origin and the images $b_1, \ldots, b_n$ of $\mathbf{b}_1, \ldots, \mathbf{b}_n$ in E. The group W acts as a group of affine transformations of E. The other connected components in $\mathbb{R}^{n+1}$ project onto other simplices, which are in one to one correspondence with the elements of W. We obtain a simplicial complex Σ, whose vertices are the images in E of the elements of $\mathbb{Z}^{n+1}$, and whose maximal simplices, called *chambers*, are the $\{w(0), w(b_1), \ldots, w(b_n)\}$, $w \in W$. The maximal simplices containing the origin are the images of C_0 under the elements of S_{n+1}. Another way of thinking of Σ is as follows: Form $\mathbb{Z}^{n+1}/\mathbb{Z}\mathbf{1}$, and call "vertices" $\mathbf{a} + \mathbb{Z}\mathbf{1}$ and $\mathbf{b} + \mathbb{Z}\mathbf{1}$ *adjacent* if, for some representatives $\mathbf{a} = (a_j)$ and $\mathbf{b} = (b_j)$ of these vertices, $a_j \le b_j \le a_j + 1$ for all j. A simplex is then a set of vertices, any two of which are adjacent.

We define the *type* $\tau(x) \in \{0, 1, \ldots, n\}$ of each vertex x of Σ by writing $\tau(x) = i$ if x is the image of an $n+1$-tuple $(j_1, \ldots, j_{n+1})$ of integers such that $\sum_{\nu=1}^{n+1} j_\nu = n+1-i \bmod n+1$. We have used this definition of $\tau(x) = i$ rather than the perhaps more natural requirement that $\sum_{\nu=1}^{n+1} j_\nu = i \bmod n+1$, because it makes some formulas a little neater later. For example, the vertex b_i defined above is of type i. Two chambers are said to be *i-adjacent* if they differ only in their vertex of type i. A *gallery* is a finite sequence $D_0, \ldots, D_N$ of chambers, such that each D_k is i_k-adjacent to D_{k-1} for $k = 1, \ldots, N$, for some $i_1, \ldots, i_N$. The word $i_1 \cdots i_N$ is called the *type* of the gallery.

Any translation $x \mapsto x + (k_1 b_1 + \cdots + k_n b_n)$, where $k_1, \ldots, k_n \in \mathbb{Z}$, maps vertices of Σ to vertices of Σ, and maps simplices to simplices, though it may not preserve types.

If we consider only the hyperplanes $H_{i,j;k}$ which pass through the origin, i.e., for which $k = 0$, then the corresponding reflections generate S_{n+1}. One of the connected components of the complement in $\mathbb{R}^{n+1}$ of these hyperplanes is

$$S_0^\circ = \{(x_1, \ldots, x_{n+1}) : x_1 < x_2 < \cdots < x_{n+1}\},$$

and the other connected components are in one to one correspondence with the elements of S_{n+1}. The image in E of S_0° is the simplicial cone

$$S_0 = \{t_1 b_1 + \cdots + t_n b_n : t_1, \ldots, t_n > 0\}.$$

Let $\mathcal{S}_0$ denote the subcomplex of Σ whose vertices are the vertices of Σ in the closure $\overline{S}_0$ of S_0. That is, $\mathcal{S}_0$ consists of the points $k_1 b_1 + \cdots + k_n b_n$, where $k_1, \ldots, k_n \in \mathbb{N}$. Any vertex x of Σ, being the image of some $n+1$-tuple of integers, can be moved into $\mathcal{S}_0$ by application of some $w \in S_{n+1}$. Then $x = k_1 b_1 + \cdots + k_n b_n$, where $k_1, \ldots, k_n \in \mathbb{N}$, and we refer to $k_1, \ldots, k_n$ or $k = (k_1, \ldots, k_n)$ as the *coordinates* of x with respect to 0.

Let $\mathcal{E}_r$ (respectively, $\mathcal{E}'_r$) denote the set of $n+1$-tuples $\epsilon = (\epsilon_1, \ldots, \epsilon_{n+1})$ of 0's and 1's, in which $\epsilon_i = 0$ (respectively, $\epsilon_i = 1$) for exactly r i's (where $1 \leq r \leq n$). Of course $\mathcal{E}'_r = \mathcal{E}_{n+1-r}$. Let C_0^* be the chamber in Σ consisting of the images in E of $0 = (0, \ldots, 0), (1, 0, \ldots, 0), \ldots, (1, \ldots, 1, 0)$.

Lemma 2.1. *Let $\delta \in \mathcal{E}'_r$. Amongst the galleries in Σ from C_0^* to a chamber C containing δ, the shortest ones have length*

$$N_r^*(\delta) = -\frac{r(r+1)}{2} + \sum_{i=1}^{n+1} i\delta_i\,. \tag{2.1}$$

Proof. Let $\delta_{\ell_1} = \cdots = \delta_{\ell_r} = 1$, where $1 \leqslant \ell_1 < \cdots < \ell_r \leq n+1$. The vertices $x \neq 0$ of Σ lying in a chamber with 0 are the images in E of the non-constant $n+1$-tuples of 0's and 1's. These correspond in an obvious way with proper non-empty subsets of $\{1, \ldots, n+1\}$. The chambers C in Σ containing 0 are in 1–1 correspondence with the permutations w of $\{1, \ldots, n+1\}$, with $w \in S_{n+1}$ corresponding to

$$C = \{\{w(1)\}, \{w(1), w(2)\}, \ldots, \{w(1), \ldots, w(n)\}\} = wC_0^*.$$

Thus the chambers containing 0 and δ correspond to the permutations $w \in S_{n+1}$ such that $w(\{1, \ldots, r\}) = \{\ell_1, \ldots, \ell_r\}$. Two chambers are then $n+$

$1-i$-adjacent if they differ only in their component subsets of size i, and so are of the form $wC_0^*, ws_iC_0^*$, where s_i is the transposition $(i, i+1)$. So the length of a minimal gallery from C_0^* to wC_0^* is the length $\ell(w)$ of w with respect to the generators $s_1, \ldots, s_n$ of S_{n+1}. This equals the number of pairs (i, j) such that $1 \leq i < j \leq n+1$ and $w(i) > w(j)$ [21, Cor 9.26, Ex. 9.9]. Amongst the $w \in S_{n+1}$ such that $w(\{1, \ldots, r\}) = \{\ell_1, \ldots, \ell_r\}$, it is easy to see that $\ell(w)$ is minimized by taking $w(i) = \ell_i$ for $i = 1, \ldots, r$, and then $(w(r+1), \ldots, w(n+1))$ to be the numbers of $\{1, \ldots, n+1\} \setminus \{\ell_1, \ldots, \ell_r\}$ in increasing order, and that $\ell(w)$ equals $N_r^*(\delta)$ for this w. □

If $k = (k_1, \ldots, k_n) \in \mathbb{N}^n$, form the $n+1$-tuple $(j_1, \ldots, j_{n+1})$, where $j_1 = 0$ and $j_r = k_1 + \cdots + k_{r-1}$ for $r = 2, \ldots, n+1$. Then $j_1 \leqslant \cdots \leq j_{n+1}$. If $\epsilon \in \mathcal{E}_r$, form the $n+1$-tuple $j + \epsilon = (j_1 + \epsilon_1, \ldots, j_{n+1} + \epsilon_{n+1})$. We want to calculate the coordinates of the image of this point in E. Let $\Delta\epsilon$ be the n-tuple of integers defined by $(\Delta\epsilon)_i = \epsilon_{i+1} - \epsilon_i$ for $1 \leq i \leq n$. If the $n+1$-tuple $j+\epsilon$ forms an increasing sequence, or equivalently, if $k_i + (\Delta\epsilon)_i \geq 0$ for all i (written $k + \Delta\epsilon \geq 0$), then the coordinates are the numbers $k_i + (\Delta\epsilon)_i$, $i = 1, \ldots, n$. This is automatic when $k_i > 0$ for all i, but we need to consider arbitrary $k \in \mathbb{N}^n$, and so we need the following lemma, the obvious proof of which is omitted.

Lemma 2.2. *Let $k = (k_1, \ldots, k_n) \in \mathbb{N}^n$, and let $\{i : k_i \geq 1\} = \{i_1, \ldots, i_t\}$, where $1 \leq i_1 < \cdots < i_t \leq n$. Also, write $i_0 = 0$ and $i_{t+1} = n+1$. Let $\epsilon \in \mathcal{E}_r$ and suppose that for $\nu = 1, \ldots, t+1$, $\epsilon_i = 0$ for exactly r_ν of the $i \in \{i_{\nu-1}+1, \ldots, i_\nu\}$ (thus $0 \leq r_\nu \leqslant i_\nu - i_{\nu-1}$ for each ν, and $r_1 + \cdots + r_{t+1} = r$). Then $k + \Delta\epsilon \geq 0$ if and only if for each ν, the sequence $(\epsilon_{i_{\nu-1}+1}, \ldots, \epsilon_{i_\nu})$ is made up of r_ν 0's followed by $i_\nu - i_{\nu-1} - r_\nu$ 1's.*

So if $\delta \in \mathcal{E}_r$ and if for $\nu = 1, \ldots, t+1$, $\delta_i = 0$ for exactly r_ν of the $i \in \{i_{\nu-1}+1, \ldots, i_\nu\}$, then there is a permutation $w \in S_{n+1}$ fixing $(j_1, \ldots, j_{n+1})$ and moving δ to the ϵ of Lemma 2.2.

Now let $\mathfrak{X}$ be a thick locally finite building of type $\tilde{A}_n$. In particular, $\mathfrak{X}$ is a simplicial complex in which the maximal simplexes (chambers) have $n+1$ vertices. Each vertex x of $\mathfrak{X}$ has a type $\tau(x) \in \{0, 1, \ldots, n\}$, so that each chamber has one vertex of each type. We define i-adjacency as for the Coxeter complex. Assuming that $n \geq 2$, there is an integer $q \geq 2$ such that for each chamber C and each $i \in \{0, 1, \ldots, n\}$, there are exactly q chambers $C' \neq C$ which are i-adjacent to C. If $n = 1$, then we shall assume that the tree $\mathfrak{X}$ is homogeneous, with each vertex having exactly $q+1$ neighbours. There is a family of subcomplexes called *apartments*, each of which is isomorphic with the $\tilde{A}_n$ Coxeter complex described above, and the isomorphism can be chosen to be type-preserving. Given any two simplexes of $\mathfrak{X}$, there is an apartment containing them both. The next result is well-known.

Lemma 2.3. *Let a, b be vertices of $\mathfrak{X}$, and let A be an apartment containing a and b. Let $\varphi : A \to \Sigma$ be a type-preserving isomorphism such that $\varphi(a) = 0$ and $\varphi(b) = k_1 b_1 + \cdots + k_n b_n$, where $k_1, \ldots, k_n \in \mathbb{N}$. Then $k_1, \ldots, k_n$ do not depend on the particular apartment A containing a and b.*

Proof. Suppose that a, b also belong to some other apartment A', and that $\varphi' : A' \to \Sigma$ is a type-preserving isomorphism mapping a to 0 and b to $k_1' b_n + \cdots + k_n' b_n$, where $k_1', \ldots, k_n' \in \mathbb{N}$. There is a type-preserving isomorphism $\psi : A \to A'$ which fixes a and b [2, pp. 76, 78]. Thus $\alpha = (\varphi')^{-1} \circ \psi \circ \varphi : \Sigma \to \Sigma$ is a type-preserving isomorphism which fixes 0 and maps $k_1 b_n + \cdots + k_n b_n$ to $k_1' b_n + \cdots + k_n' b_n$. Hence α is induced from an element of the Weyl group W defined at the beginning of this section. Thus α must correspond to a translation $x \mapsto x + c$ in $\mathbb{R}^{n+1}$, where $\sum_i c_i = 0$, followed by a permutation w of $\{1, \ldots, n+1\}$, acting on $\mathbb{R}^{n+1}$. Since α fixes 0, it corresponds simply to a permutation w. Since α maps $k_1 b_n + \cdots + k_n b_n$ to $k_1' b_n + \cdots + k_n' b_n$, w must map the $n+1$-tuple j defined before Lemma 2.2 using the k_i's, to the $n+1$-tuple j' defined using the k_i''s. But both j and j' are increasing sequences, and so $wj = j'$ implies that $j' = j$ (and that w permutes each set $\{i_{\nu-1} + 1, \ldots, i_\nu\}$, in the notation of Lemma 2.2). Hence $k_i = k_i'$ for each i. □

Let a, b and $k = (k_1, \ldots, k_n)$ be as in Lemma 2.3. Then we write $b \in S_k(a)$. For example, $S_{e_r}(a)$ is the set of vertices b of $\mathfrak{X}$ such that $d(a, b) = 1$ and $\tau(b) = \tau(a) + r \bmod n + 1$.

Later we shall need the idea of the *convex hull* of a set X of vertices contained in some apartment A. A half-apartment or root in Σ is defined to be the image of a half-space $\{x \in \mathbb{R}^{n+1} : x_i - x_j \leq k\}$ in $\mathbb{R}^{n+1}$. If $\varphi : A \to \Sigma$ is an isomorphism, then a root in A is the preimage of a root in Σ. The convex hull of $X \subset A$ is defined to be the intersection of the roots of A containing X. This depends only on X and not on the particular apartment A containing X. In particular, if a, b and φ are as in Lemma 2.3, then the convex hull $[a, b]$ of $\{a, b\}$ has as its set of vertices the points $\varphi^{-1}(k_1' b_1 + \cdots + k_n' b_n)$, where $0 \leq k_i' \leq k_i$ for all i.

Below, if $\delta \in \mathcal{E}_r = \mathcal{E}'_{n+1-r}$, we write $N(\delta)$ for $N^*_{n+1-r}(\delta)$.

Lemma 2.4. *Let a, b be vertices of $\mathfrak{X}$, let $b \in S_k(a)$, where $k \in \mathbb{N}^n$, and let $1 \leq r \leq n$. Then the n-tuples $\ell \in \mathbb{N}^n$ for which $S_{e_r}(b) \cap S_\ell(a) \neq \emptyset$ are those of the form $\ell = k + \Delta\epsilon$, where $\epsilon \in \mathcal{E}_r$ and $k + \Delta\epsilon \geq 0$. In this case,*

$$|S_{e_r}(b) \cap S_\ell(a)| = \sum_{\substack{\delta \in \mathcal{E}_r : \\ \delta^* = \epsilon}} q^{N(\delta)}, \tag{2.2}$$

where $\delta^* = \epsilon$ means that $w\delta = \epsilon$ for some $w \in S_{n+1}$ which fixes $(j_1, \ldots, j_{n+1})$ (so, in the notation of Lemma 2.2, for $\nu = 1, \ldots, t+1$, $\delta_i = 0$ for exactly r_ν of the $i \in \{i_{\nu-1} + 1, \ldots, i_\nu\}$).

Proof. Let A be an apartment containing a and b, let $\varphi : A \to \Sigma$ be a type-preserving isomorphism such that $\varphi(a) = 0$ and $\varphi(b) = k_1 b_1 + \cdots + k_n b_n$. We shall identify A and Σ, and suppress any further mention of φ. Let $\mathcal{S}_0$ denote the sector in Σ defined before Lemma 2.1. The "opposite sector" is obtained by applying the permutation $w_0 : (x_1, \dots, x_{n+1}) \mapsto (x_{n+1}, \dots, x_1)$ to $\mathcal{S}_0$, and consists of the vertices $-(k'_1 b_1 + \cdots + k'_n b_n)$, where $k'_1, \dots, k'_n \in \mathbb{N}$, because $w_0(\mathbf{b}_i) = \mathbf{1} - \mathbf{b}_{n+1-i}$, so that $w_0(b_i) = -b_{n+1-i}$. Notice that $w_0(C_0) = C_0^*$, defined before Lemma 2.1.

Let $\mathcal{S}$ denote the sector $b - \mathcal{S}_0$ obtained by translating by b in A the sector opposite $\mathcal{S}_0$. Let ρ be the retraction $\rho_{\mathcal{S},A}$ of $\mathfrak{X}$ onto A [18, p. 121]. If $u \in S_{e_r}(b)$, then clearly $\rho(u)$ must be the image u_δ in $E = \mathbb{R}^{n+1}/\mathbb{R}\mathbf{1}$ of $j + \delta$ for some $\delta \in \mathcal{E}_r$, where $j \in \mathbb{N}^{n+1}$ was defined before Lemma 2.2.

Now fix $\delta \in \mathcal{E}_r$. We next show that $\rho(u) = u_\delta$ for exactly $q^{N(\delta)}$ u's in $S_{e_r}(b)$. Let $u \in S_{e_r}(b)$, and suppose that $\rho(u) = u_\delta$. Then u and b lie in a common chamber C. Note that there is an apartment A' containing C and $\mathcal{S}$. For by Lemma 9.4 of [18], there is a subsector $c - \mathcal{S}_0$ of $\mathcal{S}$ lying in an apartment A' with C. But $b \in C \subset A'$ and so we can translate the sector $c - \mathcal{S}_0$ by $b - c$ in A' obtaining a sector $\mathcal{S}' \subset A'$ parallel to $\mathcal{S}$ and with basepoint b. Hence $\mathcal{S}' = \mathcal{S}$ by Lemma 9.7 of [18]. Hence $C, \mathcal{S} \subset A'$. Let $C_0^b = b + C_0^*$ be the base chamber of $\mathcal{S}$. Then $\rho(u) = \rho_{C_0^b,A}(u)$ [18, p. 121]. Also, there is an isomorphism $\psi : A' \to A$ fixing each vertex of $\mathcal{S}$, and $\rho_{C_0^b,A}$ agrees with ψ on A'. So $\psi(u) = u_\delta$. By Lemma 2.1, there is a gallery in Σ of length $N(\delta)$ from C_0^* to a chamber containing δ. Translating by b, we get a gallery γ_δ of length $N(\delta)$ from C_0^b to a chamber C^δ containing u_δ. By Lemma 2.1, C^δ is the unique chamber of $\mathfrak{X}$ containing u_δ which is nearest C_0^b (see [18, Cor. 3.9]). Let f_δ denote the type of γ_δ. Applying ψ^{-1} to γ_δ, we get a gallery of type f_δ from C_0^b to a chamber containing u.

In $\mathfrak{X}$ there are $q^{N(\delta)}$ galleries of type f_δ starting from C_0^b, because for each chamber C and each index i, there are exactly q chambers $C' \neq C$ such that C and C' are i-adjacent. We claim that there is a one to one correspondence between these galleries and the vertices $u \in S_{e_r}(b)$ for which $\rho(u) = u_\delta$.

We have seen above that if $u \in S_{e_r}(b)$ and $\rho(u) = u_\delta$, then there is gallery of type f_δ starting from C_0^b and ending in a chamber containing u. Let $\gamma' : C'_0 = C_0^b, C'_1, \dots, C'_N$ be any gallery of type f_δ starting from C_0^b. There is an apartment A' containing C_0^b and C'_N [2, p. 76]. Now γ' is minimal [18, 3.1 (iv)], and so A' contains all the C'_j's [18, Theorem 3.8]. By [2, p. 76], there is an isomorphism $\psi' : A' \to A$ fixing each vertex of C_0^b. Thus the $\psi'(C'_j)$'s form a gallery in A starting from C_0^b and having type f_δ. This gallery is therefore identical with γ_δ [4, Lemma 2.9(a)]. Let $u' = {\psi'}^{-1}(u_\delta)$. Then $u_\delta \in C^\delta = \psi'(C'_N)$, and so $u' \in C'_N$. Also, $\rho(u') = \rho_{C_0^b,A}(u') =$

$\psi'(u') = u_\delta$. Note that C'_N is the unique chamber in $\mathfrak{X}$ containing u' which is nearest C_0^b. For if C^* is the chamber containing u' such $d(C^*, C_0^b)$ is minimal, then C^* is in A' [18, Theorem 3.8], and so $d(C^*, C_0^b) = d(\psi'(C^*), C_0^b) \geq d(C^\delta, C_0^b) = N = d(C'_N, C_0^b)$. Now let $\gamma' : C'_0 = C_0^b, C'_1, \ldots, C'_N$ and $\gamma'' : C''_0 = C_0^b, C''_1, \ldots, C''_N$ be two galleries of type f_δ starting from C_0^b. Let u' and u'' be the vertices of C'_N and C''_N, respectively, which are mapped to u_δ by ρ. We claim that if $u' = u''$, then $\gamma' = \gamma''$, i.e., $C'_j = C''_j$ for each j. But we saw above that C'_N, respectively C''_N, is the unique chamber containing u', respectively u'', which is nearest C_0^b. Thus $u' = u''$ implies that $C'_N = C''_N$. Now let A'' be an apartment containing C_0^b and $C'_N = C''_N$. Then γ' and γ'' are reduced galleries from C_0^b to C'_N, and so lie in A'' [18, Theorem 3.8]. But in A'' there is only one gallery of type f_δ starting from C_0^b [4, Lemma 2.9(a)], and so $\gamma' = \gamma''$.

Now if $u \in S_{e_r}(b)$ and $\rho(u) = u_\delta$, and if $\delta^* = \epsilon$, then $u \in S_{k+\Delta\epsilon}(a)$. For there is a permutation $w \in S_{n+1}$ such that $w(u_\delta) = u_\epsilon \in \mathcal{S}_0$, which has coordinates $k_i + (\Delta\epsilon)_i$, $i = 1, \ldots, n$, with respect to a. As we saw above, there is an apartment A' containing u and $\mathcal{S}_0$ and there is an isomorphism $\psi : A' \to A$ fixing each vertex of $\mathcal{S}_0$. The isomorphism $\psi^{-1} w \psi$ of A' fixes a and maps u to $\psi^{-1}(u_\epsilon) \in \psi^{-1}(\mathcal{S}_0)$. So the coordinates of u with respect to a are the same as those of u_ϵ. Formula (2.2) is now clear. □

Lemma 2.5. *For $r = 0, 1, \ldots, n+1$, let*

$$M_{n+1,r} = \frac{(q^{n+1}-1)\cdots(q-1)}{(q^r-1)\cdots(q-1)(q^{n+1-r}-1)\cdots(q-1)} = M_{n+1,n+1-r}. \tag{2.3}$$

Then

$$M_{n+1,r} = \sum_{\epsilon \in \mathcal{E}_r} q^{N(\epsilon)}. \tag{2.4}$$

Proof. It is well known (see, for example, [19]) that $M_{n+1,r}$ is the number $|\Pi_r(\mathbf{V})|$ of r-dimensional subspaces in an $n+1$ dimensional vector space $\mathbf{V}$ over the field $\mathbb{F}_q$ of order q. It is also well-known (see [18, Theorem 3.5 and p. 95]) that for any vertex b of $\mathfrak{X}$, the set of vertices at distance 1 from b is isomorphic to a projective space, and hence is isomorphic to the space $|\Pi(\mathbf{V})|$ of nonzero proper subspaces of V (except when the projective space is nondesarguesian, which can only happen if $n = 2$). Under this isomorphism, neighbours of b of type $\tau(b) + r \bmod n+1$ correspond to r-dimensional subspaces of $\mathbf{V}$.

Take any $b \in S_{k_1,\ldots,k_n}(a)$, where $k_1, \ldots, k_n > 0$. Then $k + \Delta\epsilon \geq 0$ for all $\epsilon \in \mathcal{E}_r$, and there is only one term in the sum (2.2). Hence

$$M_{n+1,r} = |S_{e_r}(b)| = \sum_{\epsilon \in \mathcal{E}_r} |S_{e_r}(b) \cap S_{k+\Delta\epsilon}(a)| = \sum_{\epsilon \in \mathcal{E}_r} q^{N(\epsilon)}. \qquad \square$$

Formula (2.4) can be checked directly if $n = 2$. □

Remark. We can also prove Lemma 2.5 directly, as follows (see [14, p. 26]). If we expand the product

$$(1+qx)(1+q^2x)\cdots(1+q^{n+1}x),$$

then for $r = 0, \ldots, n+1$, the coefficient of x^r equals

$$\sum_{\epsilon\in\mathcal{E}'_r} q^{\sum_{i=1}^{n+1} i\epsilon_i} = \sum_{\epsilon\in\mathcal{E}'_r} q^{N_r^*(\epsilon)+r(r+1)/2}$$

(in the notation of Lemma 2.1). On the other hand, writing

$$(1+qx)(1+q^2x)\cdots(1+q^{n+1}x) = \sum_{r=0}^{n+1} C_{n+1,r}x^r,$$

it is obvious that the coefficients satisfy the recurrence

$$C_{n+1,k} = C_{n,k} + q^{n+1}C_{n,k-1} \quad \text{for } k = 0, \ldots, n+1,$$

where we set $C_{n,n+1} = C_{n,-1} = 0$. The same recurrence is satisfied by the numbers $M_{n+1,k}q^{k(k+1)/2}$, and as

$$(1+qx)(1+q^2x)\cdots(1+q^{n+1}x) = \sum_{k=0}^{n+1} M_{n+1,k}q^{k(k+1)/2}x^k \tag{2.5}$$

holds for $n = 2$, it holds for all $n \geq 2$. Hence (2.4) holds. □

Corollary 2.6. *With notation as in Lemma 2.2, the sum (2.2) equals*

$$q^{r(n+1)-r^2/2}\prod_{\nu=1}^{t+1} q^{r_\nu(r_\nu-2i_\nu)/2}M_{i_\nu-i_{\nu-1},r_\nu}\,. \tag{2.6}$$

Proof. If $\delta^* = \epsilon$, then $\delta_i = 0$ for exactly r_ν i's in $\{i_{\nu-1}+1, \ldots, i_\nu\}$ for $\nu = 1, \ldots, t+1$. Let $\mathcal{E}_{k,r}$ be defined as was $\mathcal{E}_r$, except that it is a set of k-tuples, not $n+1$-tuples. Think of δ as the concatenation of $\delta^1, \ldots, \delta^{t+1}$, where $\delta^\nu \in \mathcal{E}_{i_\nu - i_{\nu-1}, r_\nu}$ for each ν. Then

$$\begin{aligned}
N(\delta) &+ \frac{(n+1-r)(n+2-r)}{2} \\
&= \sum_{i=1}^{n+1} i\delta_i \\
&= \sum_{\nu=1}^{t+1} \sum_{i=i_{\nu-1}+1}^{i_\nu} i\delta_i \\
&= \sum_{\nu=1}^{t+1} \Big(\sum_{i=i_{\nu-1}+1}^{i_\nu} (i - i_{\nu-1})\delta_i \Big) + \sum_{\nu=1}^{t+1} i_{\nu-1} \Big(\sum_{i=i_{\nu-1}+1}^{i_\nu} \delta_i \Big) \\
&= \sum_{\nu=1}^{t+1} \Big(N(\delta^\nu) + \frac{(i_\nu - i_{\nu-1} - r_\nu)(i_\nu - i_{\nu-1} - r_\nu + 1)}{2} \Big) \\
&\qquad + \sum_{\nu=1}^{t+1} i_{\nu-1}(i_\nu - i_{\nu-1} - r_\nu) \\
&= \sum_{\nu=1}^{t+1} \Big(N(\delta^\nu) + \frac{1}{2}(i_\nu^2 - i_{\nu-1}^2 + i_\nu - i_{\nu-1} + r_\nu^2 - r_\nu - 2r_\nu i_\nu) \Big) .
\end{aligned}$$

Hence

$$N(\delta) = (n+1)r - \frac{r^2}{2} + \sum_{\nu=1}^{t+1} \Big(N(\delta^\nu) + \frac{r_\nu(r_\nu - 2i_\nu)}{2} \Big) .$$

Summing over the $\delta \in \mathcal{E}_r$ such that $\delta^* = \epsilon$ amounts to summing over all $\delta^\nu \in \mathcal{E}_{i_\nu - i_{\nu-1}, r_\nu}$ for $\nu = 1, \ldots, t+1$. Hence the result follows from Lemma 2.5. □

Proposition 2.7. *Let x be any vertex of Δ, let $k = (k_1, \ldots, k_n) \in \mathbb{N}^n$, and let $\{i : k_i \geq 1\} = \{i_1, \ldots, i_t\}$. Then*

$$|S_{k_1,\ldots,k_n}(x)| = q^{-\sum_{\nu=1}^{t} i_\nu(i_{\nu+1} - i_\nu)} \begin{bmatrix} n+1 \\ i_1 - i_0, \ldots, i_{t+1} - i_t \end{bmatrix}_q \cdot q^{\sum_{i=1}^{n} i(n+1-i)k_i} ,$$

where $\Big[\cdots\Big]_q = [n+1]_q / ([i_1 - i_0]_q \cdots [i_{t+1} - i_t]_q)$, and $[k]_q = (q^k - 1) \cdots (q-1)$.

Proof. The proof is by induction on $k_1 + \cdots + k_n$, the induction step being proved firstly using the fact that if $b \in S_k(a)$, where $k = (k_1, \ldots, k_n)$, and if $i_{\lambda-1} \leq r \leq i_\lambda$, then

$$|S_{e_r}(b) \cap S_{k+e_r}(a)| = q^{-(r-i_{\lambda-1})(i_\lambda - r)} M_{i_\lambda - i_{\lambda-1}, r - i_{\lambda-1}} q^{r(n+1-r)} . \quad (2.7)$$

This is a special case of Lemma 2.4 and Corollary 2.6 (taking $\epsilon = (0, \dots, 0, 1, \dots, 1)$). Secondly, if $u \in S_{k+e_r}(a)$, then there is a unique $b = u^- \in S_k(a)$ such that $u \in S_{e_r}(b)$. This follows from Lemma 2.4 (with r replaced by $n+1-r$, k by $k+e_r$, and with $\epsilon = (1, \dots, 1, 0, \dots, 0)$ (r 1's)), which shows that $|S_{e_{n+1-r}}(u) \cap S_k(a)| = 1$. Thus

$$\begin{aligned} |S_{k+e_r}(a)| &= \sum_{b \in S_k(a)} |\{u \in S_{k+e_r}(a) : u^- = b\}| \\ &= \sum_{b \in S_k(a)} |S_{e_r}(b) \cap S_{k+e_r}(a)|, \end{aligned}$$

which is $|S_k(a)|$ times the number on the right in (2.7). □

Corollary 2.8. *If $k = (k_1, \dots, k_n) \in \mathbb{N}^n$, then*

$$N_k = N_{k_1, \dots, k_n} = |S_{k_1, \dots, k_n}(x)|$$

is independent of the vertex x of Δ. Moreover,

$$N_{k_1, \dots, k_n} = N_{k_n, \dots, k_1}. \tag{2.8}$$

Of course, things are simpler when $k_i > 0$ for all i. For then $t = n$ and $i_\nu = \nu$ for all ν:

Corollary 2.9. *If $k = (k_1, \dots, k_n) \in \mathbb{N}^n$ and $k_i \geq 1$ for all i, then*

$$N_k = q^{-n(n+1)/2} \prod_{i=1}^{n+1} \left(\frac{q^i - 1}{q - 1} \right) \cdot q^{2\rho(k)}. \tag{2.9}$$

3. The averaging operators A_k.

Let $\mathfrak{X}^0$ denote the vertex set of our $\tilde{A}_n$ building $\mathfrak{X}$. Let $\mathcal{C}(\mathfrak{X})$ denote the space of all functions $f : \mathfrak{X}^0 \to \mathbb{C}$, and let $\mathcal{C}_c(\mathfrak{X})$ denote the space of finitely supported functions $f : \mathfrak{X}^0 \to \mathbb{C}$. If $f \in \mathcal{C}_c(\mathfrak{X})$ and $F \in \mathcal{C}(\mathfrak{X})$, we define

$$\langle F, f \rangle = \langle f, F \rangle = \sum_{x \in \mathfrak{X}^0} f(x) F(x).$$

If $k = (k_1, \dots, k_n) \in \mathbb{N}^n$, define $A_k : \mathcal{C}(\mathfrak{X}) \to \mathcal{C}(\mathfrak{X})$ by

$$(A_k F)(x) = \frac{1}{N_k} \sum_{y \in S_k(x)} F(y). \tag{3.1}$$

Notice that A_k maps $\mathcal{C}_c(\mathfrak{X})$ into itself. Also, if $f \in \mathcal{C}_c(\mathfrak{X})$ and $F \in \mathcal{C}(\mathfrak{X})$ or vice versa, then by (2.8) and because $b \in S_k(a)$ if and only if $a \in S_{k'}(b)$,

$$\langle f, A_k F\rangle = \langle A_{k'} f, F\rangle\,, \tag{3.2}$$

where $k' = (k_n, \dots, k_1)$. If $\delta_x(y) = 1$ when $y = x$ and 0 otherwise, then

$$A_k \delta_x = N_k^{-1} \mathbf{1}_{S_{k'}(x)}\,. \tag{3.3}$$

Theorem 3.1. *The linear span of the operators A_k, $k \in \mathbb{N}^n$, is a commutative algebra $\mathcal{A}$ generated by the operators $A_r = A_{e_r}$, $r = 1, \dots, n$, and the identity operator A_0. The operators A_k satisfy the relations*

$$A_k A_r = \frac{1}{N_{e_r}} \sum_{\substack{\epsilon \in \mathcal{E}_r:\\ k+\Delta\epsilon \geq 0}} |S_{e_r}(b) \cap S_{k+\Delta\epsilon}(a)| A_{k+\Delta\epsilon}\,, \tag{3.4}$$

where a, b are any vertices such that $b \in S_k(a)$.

Proof. We saw in Lemma 2.4 that for any $k, m \in \mathbb{N}^n$ and any $r \in \{1, \dots, n\}$, if $b \in S_k(a)$, the number $\alpha_{m,r;k} = |S_{e_r}(b) \cap S_m(a)|$ does not depend on the particular vertices a and b, and is non-zero only if $m = k + \Delta\epsilon$ for some $\epsilon \in \mathcal{E}_r$ (such that $k + \Delta\epsilon \geq 0$). For any vertices t, y,

$$\begin{aligned}
N_k(A_k A_{e_r} \delta_t)(y) &= \sum_{x \in S_k(y)} (A_{e_r}\delta_t)(x) \\
&= N_{e_r}^{-1} \sum_{x \in S_k(y)} \mathbf{1}_{S_{e_{n+1-r}}(t)}(x) \quad \text{by (3.3)} \\
&= N_{e_r}^{-1} |S_{e_{n+1-r}}(t) \cap S_k(y)| \\
&= N_{e_r}^{-1} \sum_{m \in \mathbb{N}^n} \alpha_{k,n+1-r;m} \mathbf{1}_{S_m(y)}(t) \\
&= N_{e_r}^{-1} \sum_{m \in \mathbb{N}^n} \alpha_{k,n+1-r;m} \mathbf{1}_{S_{m'}(t)}(y) \\
&= N_{e_r}^{-1} \sum_{m \in \mathbb{N}^n} \alpha_{k,n+1-r;m} N_m(A_m\delta_t)(y) \quad \text{by (3.3)}
\end{aligned} \tag{3.5}$$

Hence

$$A_k A_{e_r} \delta_t = \sum_{m \in \mathbb{N}^n} c_{k,e_r;m} A_m \delta_t \tag{3.6}$$

for

$$c_{k,e_r;m} = N_m \alpha_{k,n+1-r;m} / (N_{e_r} N_k)\,. \tag{3.7}$$

It follows easily from (3.2) that in (3.6), δ_t can be replaced by any $F \in \mathcal{C}(\mathfrak{X})$, and so $A_k A_{e_r} = \sum_{m\in\mathbb{N}^n} c_{k,e_r;m} A_m$. Also,

$$\begin{aligned} N_k(A_k A_{e_r}\delta_t)(y) &= \sum_{x\in S_k(y)} (A_{e_r}\delta_t)(x) \\ &= \sum_{x\in S_k(y)} \langle A_{e_r}\delta_t, \delta_x\rangle \\ &= \sum_{x\in S_k(y)} \langle \delta_t, A_{e_{n+1-r}}\delta_x\rangle \\ &= N_{e_r}^{-1} \sum_{x\in S_k(y)} \langle \delta_t, \mathbf{1}_{S_{e_r}(x)}\rangle, \end{aligned}$$

and so if we sum this over $t \in S_m(y)$, we obtain

$$\begin{aligned} N_k \sum_{t\in S_m(y)} (A_k A_{e_r}\delta_t)(y) &= N_{e_r}^{-1} \sum_{x\in S_k(y)} |S_{e_r}(x) \cap S_m(y)| \\ &= N_k \alpha_{m,r;k}/N_{e_r} \end{aligned} \tag{3.8}$$

But from (3.5) and (3.7) we know that if $t \in S_m(y)$, then

$$N_k(A_k A_{e_r}\delta_t)(y) = \frac{\alpha_{k,n+1-r;m}}{N_{e_r}} = \frac{N_k c_{k,e_r;m}}{N_m}.$$

So the sum over $t \in S_m(y)$ of $N_k(A_k A_{e_r}\delta_t)(y)$ is also equal to $N_k c_{k,e_r;m}$. Comparing this with (3.8), we see that $c_{k,e_r;m} = \alpha_{m,r;k}/N_{e_r}$. This means that (3.4) holds.

We now show that each A_k, $k \in \mathbb{N}^n$, is a polynomial in $A_1, \ldots, A_n$. Write $|k| = k_1 + \cdots + k_n$ if $k = (k_1, \ldots, k_n)$. Notice that if $\epsilon \in \mathcal{E}_r$, then $|k + \Delta\epsilon| = |k| + \epsilon_{n+1} - \epsilon_1 \leqslant |k| + 1$, with equality if and only if $\epsilon_{n+1} = 1$ and $\epsilon_1 = 0$. Let $\mathcal{B}$ denote the algebra of operators generated by $A_0, A_1, \ldots, A_n$. We show that $A_k \in \mathcal{B}$ by induction on $|k|$. Clearly $A_k \in \mathcal{B}$ if $|k| \leq 1$. Suppose that $k = (k_1, \ldots, k_n) \in \mathbb{N}^n$ and that we have shown that $A_{\tilde{k}} \in \mathcal{B}$ for all $\tilde{k} \in \mathbb{N}^n$ such that $|\tilde{k}| < |k|$. Let r be the first index i such that $k_i > 0$, and let $\tilde{k} = (0, \ldots, 0, k_r - 1, k_{r+1}, \ldots, k_n)$. Then $|\tilde{k}| < |k|$, and so $A_{\tilde{k}} \in \mathcal{B}$. Therefore $A_{\tilde{k}} A_r \in \mathcal{B}$. But (3.4) tells us that $A_{\tilde{k}} A_r$ is a linear combination of operators $A_{\tilde{k}+\Delta\epsilon}$, where $\epsilon \in \mathcal{E}_r$ and $\tilde{k} + \Delta\epsilon \geq 0$. The condition $\tilde{k} + \Delta\epsilon \geq 0$ implies that $\epsilon_1 = \cdots = \epsilon_r$. If $\epsilon_1 = \cdots = \epsilon_r = 1$, then $|\tilde{k} + \Delta\epsilon| = |\tilde{k}| + \epsilon_{n+1} - \epsilon_1 \leqslant |\tilde{k}| < |k|$, and so $A_{\tilde{k}+\Delta\epsilon} \in \mathcal{B}$ by the induction hypothesis. If $\epsilon_1 = \cdots = \epsilon_r = 0$, then ϵ must be $(0, \ldots, 0, 1, \ldots, 1)$, and so $\tilde{k} + \Delta\epsilon = k$. Hence (3.4) (with $\tilde{k}$ in place of k), has the form $A_{\tilde{k}} A_r = cA_k + \sum_{\ell\in\mathbb{N}^n:|\ell|<|k|} c_\ell A_\ell$, where $c > 0$. It follows that $A_k \in \mathcal{B}$.

Conversely, each polynomial in $A_1, \ldots, A_n$ is a linear combination of operators A_k. For by an obvious induction on d (using (3.4)), any monomial $A_{i_1} \cdots A_{i_d}$, where $i_1, \ldots, i_d \in \{1, \ldots, n\}$, is such a linear combination. Thus $\mathcal{A} = \mathcal{B}$.

It remains to show that $\mathcal{A}$ is commutative. To do this, we verify that $A_s A_r = A_r A_s$ for $1 \le r, s \le n$. Take $k = e_s$ in Lemma 2.4. Then in the notation of Lemma 2.2 we have $t = 1$, $i_0 = 0$, $i_1 = s$, and $i_2 = n+1$. The only $\epsilon \in \mathcal{E}_r$ for which $k + \Delta\epsilon \ge 0$ are the ones made up of r_1 0's, then $s - r_1$ 1's, then r_2 0's and finally $n+1-s-r_2$ 1's, where $0 \le r_1 \le s$, $0 \le r_2 \le n+1-s$ and $r_1 + r_2 = r$. For such an ϵ, we have $k + \Delta\epsilon = e_{r_1} + e_{s+r_2}$. This is true even if $r_1 = 0$ or $r_2 = n+1-s$, provided that we agree that $e_0 = e_{n+1} = 0$. To express this more symmetrically, we set $\alpha = r_1$ and $\beta = s + r_2$. Then $0 \leqslant \alpha \leqslant \min\{r, s\}$, $\max\{r, s\} \leqslant \beta \le n+1$ and $\alpha + \beta = r + s$. Conversely, any pair (α, β) satisfying these conditions comes from a pair of numbers r_1, r_2 satisfying the above conditions. Hence

$$A_r A_s = \sum_{\substack{(\alpha,\beta):\alpha+\beta=r+s \text{ and} \\ 0 \leqslant \alpha \le \min\{r,s\}, \max\{r,s\} \le \beta \le n+1}} C_{r,s;\alpha,\beta} A_{e_\alpha + e_\beta},$$

where we read off the coefficients from (2.6). After a little algebra, we find that

$$C_{r,s;\alpha,\beta} = q^{\alpha(n+1-\beta)} \begin{bmatrix} r, s, n+1-r, n+1-s \\ \alpha, n+1-\beta, r-\alpha, s-\alpha, n+1 \end{bmatrix}_q,$$

where $[\cdots]_q = [r]_q[s]_q[n+1-r]_q[n+1-s]_q/([\alpha]_q[n+1-\beta]_q[r-\alpha]_q[s-\alpha]_q[n+1]_q)$ with notation as in Proposition 2.7. Since $C_{r,s;\alpha,\beta}$ is thus symmetric in r and s, we see that $A_r A_s = A_s A_r$. □

4 The boundary Ω of $\mathfrak{X}$.

A subcomplex S of $\mathfrak{X}$ is called a *sector* if there is an apartment A containing S and an isomorphism $\varphi : A \to \Sigma$ such that $\varphi(S)$ is the subcomplex $\mathcal{S}_0$ defined before Lemma 2.1. The vertex of S mapped to 0 by φ is called the *base vertex* of S. The chamber mapped to C_0 is called the *base chamber* of S. If S and S' are sectors in $\mathfrak{X}$ and if $S' \subset S$, then we call S' a *subsector* of S. In this case, if A is an apartment containing S, and $\varphi : A \to \Sigma$ an isomorphism such that $\varphi(S) = \mathcal{S}_0$, then $\varphi(S') = v + \mathcal{S}_0 = \{k_1 b_1 + \cdots + k_n b_n : k_i \in \mathbb{N} \text{ and } k_i \ge v_i \text{ for all } i\}$, for some $v = v_1 b_1 + \cdots + v_n b_n \in \mathcal{S}_0$. This is because the only automorphisms f of Σ such that $f(\mathcal{S}_0) \subset \mathcal{S}_0$ are the translations $x \to v + x$, where $v \in \mathcal{S}_0$ (see [18, Lemma 9.2]). We call two sectors in $\mathfrak{X}$ *equivalent* if they contain a common subsector. The *boundary* Ω of $\mathfrak{X}$ is the set of equivalence classes of sectors. We also call Ω the "spherical building at infinity" associated with $\mathfrak{X}$ [18,

§9.3]. Given any vertex x of $\mathfrak{X}$ and any $\omega \in \Omega$, there is a unique sector $S^x(\omega)$ in $\mathfrak{X}$ in the equivalence class ω and having base vertex x [18, Lemma 9.7]. If A is any apartment containing $S^x(\omega)$, there is an isomorphism $\varphi : A \to \Sigma$ mapping x to 0 and $S^x(\omega)$ to $\mathcal{S}_0$. If $k = (k_1, \ldots, k_n) \in \mathbb{N}^n$, we define $S^x_{k_1,\ldots,k_n}(\omega)$ to be the vertex of $S^x(\omega)$ mapped by φ to $k_1 b_1 + \cdots + k_n b_n$. This does not depend on the particular apartment A containing $S^x(\omega)$, nor on the particular isomorphism $\varphi : A \to \Sigma$ mapping x to 0 and $S^x(\omega)$ to $\mathcal{S}_0$. For suppose that A' is another apartment containing $S^x(\omega)$, and that $\varphi' : A' \to \Sigma$ is an isomorphism mapping x to 0 and $S^x(\omega)$ to $\mathcal{S}_0$. There is an isomorphism $\psi : A' \to A$ which fixes each vertex of $S^x(\omega)$. Then $\varphi \circ \psi$ and φ' are two isomorphisms $A' \to \Sigma$ mapping $S^x(\omega)$ to $\mathcal{S}_0$. They must coincide, because the identity is the only isomorphism of Σ which maps $\mathcal{S}_0$ onto $\mathcal{S}_0$. So if $\varphi(y) = k_1 b_1 + \cdots + k_n b_n$, then $\varphi'(y) = \varphi(\psi(y)) = \varphi(y)$. So $S^x_{k_1,\ldots,k_n}(\omega)$, defined using A and φ, is the same as $S^x_{k_1,\ldots,k_n}(\omega)$, defined using A' and φ'.

If x and y are vertices of $\mathfrak{X}$, and if $\omega \in \Omega$, then $S^x(\omega)$ and $S^y(\omega)$ contain a common subsector. If z is the base vertex of this sector, then $z = S^x_{k_1,\ldots,k_n}(\omega)$ and $z = S^y_{k'_1,\ldots,k'_n}(\omega)$ for some $k = (k_1, \ldots, k_n)$ and $k' = (k'_1, \ldots, k'_n) \in \mathbb{N}^n$. It follows (see [6, Lemma 2.1]) that setting $m_i = m_i(x, y; \omega) = k'_i - k_i$ for $i = 1, \ldots, n$, then

$$S^x_{j_1,\ldots,j_n}(\omega) = S^y_{j_1+m_1,\ldots,j_n+m_n}(\omega)$$

for all sufficiently large $j_1, \ldots, j_n \in \mathbb{N}$ (in fact, such that $j_i \geq k_i$ for each i).

There is a natural topology on Ω, which we now describe. Let $o, x \in \mathfrak{X}^0$ with $x \in S_k(o)$, where $k \in \mathbb{N}^n$. Let $\Omega_o(x)$ denote the set of $\omega \in \Omega$ such that $S^o(\omega)$ contains x (equivalently, such that $x = S^o_k(\omega)$). If we fix $k \in \mathbb{N}^n$, then

$$\Omega = \bigcup_{x \in S_k(o)} \Omega_o(x),$$

a partition of Ω into N_k sets. The topology on Ω is the one for which the sets $\Omega_o(x)$, o fixed x varying, form a base of open sets. Arguing as in [6, Lemma 2.5], we see that the topology does not depend on o.

Now let $x \in S_{e_r}(o)$. Then there is a sequence $\epsilon = (\epsilon_1, \ldots, \epsilon_{n+1})$ in $\mathcal{E}_r$ such that $m_i(o, x; \omega) = \epsilon_i - \epsilon_{i+1}$ for all i. For there is an apartment A containing $S = S^o(\omega)$ and x (see the proof of Lemma 2.4). Let $\varphi : A \to \Sigma$ be an isomorphism mapping o to 0 and S to $\mathcal{S}_0$. Then $\varphi(x) = (\epsilon_2 - \epsilon_1)b_1 + \cdots + (\epsilon_{n+1} - \epsilon_n)b_n$ for some $\epsilon \in \mathcal{E}_r$. The vertices $(\epsilon_2 - \epsilon_1 + k_1)b_1 + \cdots + (\epsilon_{n+1} - \epsilon_n + k_n)b_n$, where $k = (k_1, \ldots, k_n) \in \mathbb{N}^n$, form a sector in Σ with base vertex $\varphi(x)$ which is equivalent to $\mathcal{S}_0$, and so their preimages under φ form a sector in A with base point x which is equivalent to S. Hence this sector is $S^x(\omega)$. Moreover, provided that $k_i, k_i + \epsilon_i - \epsilon_{i+1} \geq 0$ for all i,

$$S^x_{k_1,\ldots,k_n}(\omega) = S^o_{k_1+\epsilon_2-\epsilon_1,\ldots,k_n+\epsilon_{n+1}-\epsilon_n}(\omega).$$

Thus $m_i(o,x;\omega) = \epsilon_i - \epsilon_{i+1}$ for all i.

We write $\omega \in \Omega_o(\epsilon, x)$ if $m_i(o,x;\omega) = \epsilon_i - \epsilon_{i+1}$ for all i.

Given $\epsilon \in \mathcal{E}_r$, the length $N'(\epsilon)$ of a shortest gallery from C_0 to a gallery containing ϵ is the same as the length of a shortest gallery from C_0^* to a gallery containing $\mathbf{1} - \epsilon \in \mathcal{E}'_r$, and so by Lemma 2.1,

$$N'(\epsilon) = -\frac{r(r+1)}{2} + \sum_{i=1}^{n+1} i(1-\epsilon_i) = \frac{(n+1)(n+2)}{2} - \frac{r(r+1)}{2} - \sum_{i=1}^{n+1} i\epsilon_i\,.$$

The proof of Lemma 2.4 shows that, given $\omega \in \Omega$, there are $q^{N'(\epsilon)}$ neighbours y of o such that $\omega \in \Omega_o(\epsilon, y)$.

Proposition 4.1. *Fix a vertex o and a boundary point $\omega \in \Omega$, and define $f : \mathfrak{X}^0 \to \mathbb{C}$ by $f(x) = h_{o,x}(\omega)$, as defined in (1.1). Then*

$$A_r f = \lambda_r f \quad \text{for } r = 1, \dots, n,$$

where the λ_r are as in (1.3).

Proof. It is easy to see that for all vertices x, y, z of $\mathfrak{X}$, $m_i(x,y;\omega) = m_i(x,z;\omega) + m_i(z,y;\omega)$, and hence

$$h_{x,y}(\omega) = h_{x,z}(\omega)h_{z,y}(\omega). \tag{4.2}$$

Thus

$$N_{e_r}(A_r f)(x) = \sum_{y\in S_{e_r}(x)} h_{o,y}(\omega) = f(x) \sum_{y\in S_{e_r}(x)} h_{x,y}(\omega).$$

This last sum is

$$\sum_{\epsilon\in\mathcal{E}_r} \left(\sum_{\substack{y:d(x,y)=1 \text{ and} \\ \omega\in\Omega_x(\epsilon,y)}} h_{x,y}(\omega) \right).$$

For each y such that $\omega \in \Omega_x(\epsilon, y)$,

$$\begin{aligned} h_{x,y}(\omega) &= \frac{s_1^{\epsilon_1-\epsilon_2} \cdots s_n^{\epsilon_n-\epsilon_{n+1}}}{q^{\frac{1}{2}\sum_{i=1}^n i(n+1-i)(\epsilon_i-\epsilon_{i+1})}} \\ &= \frac{s_1^{\epsilon_1} \cdots s_n^{\epsilon_n} s_{n+1}^{\epsilon_{n+1}}}{q^{(n+2)(n+1-r)/2-\sum_{i=1}^{n+1} i\epsilon_i}} \end{aligned} \tag{4.3}$$

Hence

$$\begin{aligned} N_{e_r}(A_r f)(x) &= \sum_{\epsilon\in\mathcal{E}_r} q^{N'(\epsilon)} \frac{s_1^{\epsilon_1} \cdots s_n^{\epsilon_n} s_{n+1}^{\epsilon_{n+1}}}{q^{(n+2)(n+1-r)/2-\sum_{i=1}^{n+1} i\epsilon_i}} \\ &= q^{r(n+1-r)/2} \sum_{\epsilon\in\mathcal{E}_r} s_1^{\epsilon_1} \cdots s_n^{\epsilon_n} s_{n+1}^{\epsilon_{n+1}} \\ &= q^{r(n+1-r)/2} \sigma_{n+1-r}(s_1, \dots, s_{n+1}) \\ &= q^{r(n+1-r)/2} \sigma_r(s_1^{-1}, \dots, s_{n+1}^{-1}). \quad \square \end{aligned}$$

Given $\lambda_1, \ldots, \lambda_n \in \mathbb{C}$, we can find numbers $s_1, \ldots, s_{n+1}$ such that $s_1 \cdots s_{n+1} = 1$ and so that (1.3) holds, namely the roots of the polynomial $1 + \lambda'_1 X + \cdots + \lambda'_n X^n + X^{n+1}$, where $\lambda'_r = (-1)^{n+1-r} M_{n+1,r} q^{-r(n+1-r)/2} \lambda_r$. In particular, using (2.5), it is easy to see that the choice (1.4) of the s_i's gives $\lambda_1 = \cdots = \lambda_n = 1$, and that (1.5) holds.

For each vertex o, there is a regular Borel probability measure ν_o on Ω such that each of the sets $\Omega_o(v)$, $v \in S_k(o)$, have measure $1/N_k$. Using (2.9) and arguing as in [6, Lemma 2.5], we see that (1.6) holds.

5. Motivation from representation theory.

For each integer $n \geq 1$ and each non-archimedean local field F, one can construct a building $\mathfrak{X}_F$ of type $\tilde{A}_n$. For details, see [20]. Let $\mathcal{O}$ be the valuation ring of F. Let $\mathrm{ord} : F^\times \to \mathbb{Z}$ be the valuation on F, which we assume surjective, and we let ϖ be an element of F such that $\mathrm{ord}(\varpi) = 1$. As usual, we set $\mathrm{ord}(0) = \infty$. Let q denote the order of the residual field $\mathcal{O}/\varpi\mathcal{O}$ of F. We let $|x| = q^{-\mathrm{ord}(x)}$ denote the normalized absolute value on F corresponding to ord.

Let V denote the space F^{n+1} of column vectors of length $n+1$ with entries in F. A *lattice* in V is a subset of V of the form $\{a_1 v_1 + \cdots + a_{n+1} v_{n+1} : a_1, \ldots, a_{n+1} \in \mathcal{O}\}$, where $v_1, \ldots, v_{n+1}$ is a basis of V. Call two lattices L and L' *equivalent* if $L' = cL$ for some $c \in F^\times$. Let $[L]$ denote the equivalence class of L. We call two lattice classes $[L]$ and $[L']$ *adjacent* if representatives L and L' can be found such that $\varpi L \subsetneqq L' \subsetneqq L$.

The building $\mathfrak{X}_F$ is a simplicial complex which has the set of equivalence classes of lattices as its vertex set. A simplex of $\mathfrak{X}_F$ is a set $\{[L_1], \ldots, [L_r]\}$ of vertices, any two of which are adjacent.

The apartments in $\mathfrak{X}_F$ are the subcomplexes A of $\mathfrak{X}_F$ constructed as follows. Let $v_1, \ldots, v_{n+1}$ be a fixed basis of V. If $r_1, \ldots, r_{n+1}$ are integers, let $L_{r_1,\ldots,r_{n+1}}$ be the lattice $\varpi^{r_1}\mathcal{O}v_1 + \cdots + \varpi^{r_{n+1}}\mathcal{O}v_{n+1}$. We let A consist of all the lattice classes $[L_{r_1,\ldots,r_{n+1}}]$, where $r_1, \ldots, r_{n+1} \in \mathbb{Z}$.

Consider the natural action of $G = GL(n+1, F)$ on V. Clearly, G acts on the set of lattices and on the set of lattice classes. This action is transitive. Similarly, G acts transitively on the set of apartments and on the set of equivalence classes of sectors (if $\omega \in \Omega$ is the equivalence class of a sector S, then we define $g\omega$ to be the class of gS). It is clear from the definitions that the numbers $m_\nu(x, y; \omega)$ satisfy $m_\nu(gx, gy; g\omega) = m_\nu(x, y; \omega)$ for all vertices x, y and for all $\omega \in \Omega$.

The stabilizers of a given vertex, chamber or sector class are important subgroups of G. Let $e_1, \ldots, e_{n+1}$ be the usual basis of V. Then $L_{r_1,\ldots,r_{n+1}}$, as defined above, is here the set of vectors $x = (x_i) \in V$ such that $\mathrm{ord}(x_i) \geq r_i$ for each i. Write L_0 in place of $L_{0,\ldots,0}$. Let A_0 denote the apartment defined using the basis $e_1, \ldots, e_{n+1}$. Let S_0 denote the sector of A_0 consisting of the lattice classes $[L_{i_1,\ldots,i_{n+1}}]$, where $i_1 \leqslant \ldots \leq i_{n+1}$. The base vertex of S_0 is $o = [L_0]$.

The stabilizer in G of the sector class ω_0 of S_0 is the subgroup P consisting of upper-triangular matrices g in G [20, (8.1)]. In fact, if $g = (g_{i,j}) \in G$ is upper-triangular, and if $\ell_i = \mathrm{ord}(g_{i,i})$ for each i, then $g(L_{r_1,\ldots,r_{n+1}}) = L_{r_1+\ell_1,\ldots,r_{n+1}+\ell_{n+1}}$ holds provided that the numbers $r_{i+1} - r_i$ are sufficiently large, for each i. It follows that $m_i(o, go; \omega_0) = \ell_i - \ell_{i+1}$ for $i = 1, \ldots, n$.

It is routine to verify that

$$\begin{aligned} J(f) &= \int \cdots \int \frac{f((g_{i,j}))}{|g_{1,1}|^{n+1} \cdots |g_{n+1,n+1}|} \, dg_{1,1} dg_{1,2} \cdots dg_{n+1,n+1} \\ &= \int \cdots \int \frac{f((g_{i,j}))}{\prod_{i=1}^{n+1} |g_{i,i}|^{n+2-i}} \prod_{\substack{i,j: \\ 1 \leq i < j \leq n+1}} dg_{i,j} \end{aligned}$$

(for compactly supported continuous functions on P) defines a left Haar integral for P. Here the integral is an iterated integral, with $dg_{i,j}$ referring to a Haar measure μ on the additive group of F, the (i,j)-th integral being over F if $i < j$ and over $F \setminus \{0\}$ if $i = j$. The key point is that $\int_F f(ax)\, d\mu(x) = |a|^{-1} \int_F f(x)\, d\mu(x)$ for compactly supported continuous functions f on F and non-zero $a \in F$.

As usual, we write $\int_P f(g)\, dg$ in place of $J(f)$. It is also routine to verify that the modular function $\delta = \delta_P$ of P (that is, the function δ such that $\int_P f(gp)\, dg = \delta(p)^{-1} \int_P f(g)\, dg$ for $p \in P$) is given by

$$\delta((p_{i,j})) = |p_{1,1}|^{-n} |p_{2,2}|^{-n+2} \cdots |p_{n+1,n+1}|^n = \prod_{i=1}^{n+1} |p_{i,i}|^{2i-n-2}.$$

Suppose that $\mathrm{ord}(p_{i,i}) = \ell_i$ for each i. Then $|p_{i,i}| = q^{-\ell_i}$, and so $\delta(p) = q^N$ for

$$N = -\sum_{i=1}^{n+1} \ell_i(2i - n - 2) = \sum_{i=1}^{n+1} (\ell_1 - \ell_i)(2i - n - 2).$$

Writing $\ell_1 - \ell_{j+1} = m_1 + \cdots + m_j$, we can express N in terms of the m_j's, and we obtain the formula

$$\delta(p) = q^{\sum_{j=1}^{n} j(n+1-j)m_j} = q^{2\rho(m)}, \tag{5.1}$$

where $m_j = m_j(o, po; \omega_0)$ for each j.

The representations π_s of G.

Suppose that $s = (s_1, \ldots, s_{n+1})$ is an $n+1$-tuple of non-zero complex numbers satisfying $s_1 s_2 \cdots s_{n+1} = 1$. The map

$$\chi_s : p = (p_{i,j}) \mapsto s_1^{\operatorname{ord} p_{1,1}} s_2^{\operatorname{ord} p_{2,2}} \cdots s_{n+1}^{\operatorname{ord} p_{n+1,n+1}}$$

is a continuous homomorphism of P into the multiplicative group $\mathbb{C}^\times$ of non-zero complex numbers which is trivial on the center $Z = \{\lambda I : \lambda \in F^\times\}$ of P. Moreover, χ_s is trivial on the subgroup M° of P consisting of matrices $p = (p_{i,j}) \in P$ for which $p_{i,i} \in \mathcal{O}^\times$ for all i.

We can form the space $\mathcal{I}_s$ of locally constant functions $f : G \to \mathbb{C}$ such that

$$f(pg) = \chi_s(p)\sqrt{\delta(p^{-1})}\, f(g)$$

for all $g \in G$ and $p \in P$. This space is invariant under right translations, and so we get a representation ρ_s of G on $\mathcal{I}_s$ by setting $(\rho_s(g)f)(g') = f(g'g)$. These representations, for varying $s = (s_1, \ldots, s_{n+1})$, make up what is called the *unramified principal series* of representations of G (see [3]).

There is another way of looking at these representations which is more appropriate to our present purposes. Let $\mathcal{C}^\infty(\Omega)$ denote the space of locally constant functions on Ω. For $s = (s_1, \ldots, s_{n+1})$, satisfying $s_1 s_2 \cdots s_{n+1} = 1$ as above, for $g \in G$ and for $F \in \mathcal{C}^\infty(\Omega)$, define $\pi_s(g)F : \Omega \to \mathbb{C}$ by

$$(\pi_s(g)F)(\omega) = \frac{s_1^{m_1+\cdots+m_n} \cdots s_{n-1}^{m_{n-1}+m_n} s_n^{m_n}}{q^{\rho(m)}} F(g^{-1}\omega),$$

where $m_i = m_i(o, go; \omega)$ for each i. Thus $h_{o,go} = \pi_s(g)\mathbf{1}$.

Proposition 5.1. *Define $T : \mathcal{C}^\infty(\Omega) \to \mathcal{I}_s$ by $(TF)(g) = (\pi_s(g)F)(\omega_0)$. Then T is a linear bijection which intertwines π_s and ρ_s.*

Proof. Using $m_i(o, po; \omega_0) = \ell_i - \ell_{i+1}$ for $p = (p_{i,j}) \in P$, where $\ell_i = \operatorname{ord}(p_{i,i})$, it is easy to see that $(\pi_s(p)F)(\omega_0) = \chi_s(p)\sqrt{\delta(p^{-1})}\, F(\omega_0)$ for $F \in \mathcal{C}^\infty(\Omega)$, and then that $TF \in \mathcal{I}_s$. Using the fact that G acts transitively on Ω, with P being the stabilizer of ω_0, the rest of the proof is routine. □

6. The functionals J_k.

Let $\lambda = (\lambda_1, \ldots, \lambda_n) \in \mathbb{C}^n$, and let h be a λ-harmonic function on the vertex set of $\mathfrak{X}$. For any vertex o of $\mathfrak{X}$ and for $k = (k_1, \ldots, k_n) \in \mathbb{N}^n$, we define $J_k = J_k^o : \mathcal{C}^\infty(\Omega) \to \mathbb{C}$ by (1.8). We have seen in Theorem 3.1 that the averaging operators A_k, $k \in \mathbb{N}^n$, satisfy recurrence relations $A_k A_\ell = \sum_r c_{k,\ell;r} A_r$ for certain constants $c_{k,\ell;r}$, where $k, \ell, r \in \mathbb{N}^n$. We now show that the functionals $J_m : \mathcal{C}^\infty(\Omega) \to \mathbb{C}$ satisfy very similar relations.

Proposition 6.1. *Let $\varphi \in C^\infty(\Omega)$ be constant on each set $\Omega_o(v)$, $v \in S_{k_0}(o)$. Then*

$$\lambda_i J_k(\varphi) = \sum_r c_{k,e_i;r} J_r(\varphi)$$

for $i = 1, \ldots, n$ and for all $k \in \mathbb{N}^n$ such that $k \geq k_0$ (i.e., such that $k_j \geq (k_0)_j$ for $j = 1, \ldots, n$).

Proof. Let $k \geq k_0$. Each set $\Omega_o(v)$, $v \in S_k(o)$, is contained in a set $\Omega_o(u)$, $u \in S_{k_0}(o)$. Indeed, we can take u to be the unique element of $S_{k_0}(o)$ in the convex hull of o and v. Hence φ is a linear combination of indicator functions $\mathbf{1}_{\Omega_o(v)}$, $v \in S_k(o)$, and so we may assume that $\varphi = \mathbf{1}_{\Omega_o(v)}$ for such a v. Then

$$J_k(\varphi) = h(v)\nu_o(\Omega_o(v)) = h(v)/N_k\,.$$

Also,

$$\lambda_i N_i h(v) = \sum_{t \in S_{e_i}(v)} h(t) = \sum_r \Big(\sum_{t \in S_{e_i}(v) \cap S_r(o)} h(t) \Big)\,.$$

On the other hand, assuming still that $\varphi = \mathbf{1}_{\Omega_o(v)}$,

$$J_r(\varphi) = \sum_{t \in S_r(o)} h(t)\nu_o(\Omega_o(v) \cap \Omega_o(t))\,,$$

We now claim that

(a) if $S_{e_i}(v) \cap S_r(o) \neq \emptyset$, then for $t \in S_r(o)$, $\Omega_o(v) \cap \Omega_o(t) \neq \emptyset$ if and only if $t \in S_{e_i}(v) \cap S_r(o)$, and

(b)
$$\Omega_o(v) = \bigcup_{t \in S_{e_i}(v) \cap S_r(o)} \Omega_o(v) \cap \Omega_o(t)$$

is a partition of $\Omega_o(v)$ into $|S_{e_i}(v) \cap S_r(o)|$ sets of the same ν_o-measure.

Assuming these, if $S_{e_i}(v) \cap S_r(o) \neq \emptyset$ then (a) implies that

$$J_r(\varphi) = \sum_{t \in S_{e_i}(v) \cap S_r(o)} h(t)\nu_o(\Omega_o(v) \cap \Omega_o(t))$$

and so, by (b),

$$\begin{aligned}
|S_{e_i}(v) \cap S_r(o)| J_r(\varphi) &= |S_{e_i}(v) \cap S_r(o)| \sum_{t \in S_{e_i}(v) \cap S_r(o)} h(t)\nu_o(\Omega_o(v) \cap \Omega_o(t)) \\
&= \sum_{t \in S_{e_i}(v) \cap S_r(o)} h(t)\nu_o(\Omega_o(v)) \\
&= N_k^{-1} \sum_{t \in S_{e_i}(v) \cap S_r(o)} h(t)\,.
\end{aligned}$$

Thus

$$\lambda_i N_i h(v) = \sum_r \Big(\sum_{t \in S_{e_i}(v) \cap S_r(o)} h(t) \Big)$$
$$= N_k \sum_r \Big(|S_{e_i}(v) \cap S_r(o)| J_r(\varphi) \Big) .$$

But $|S_{e_i}(v) \cap S_r(o)|/N_i = c_{k,e_i;r}$ by (3.4), and we saw above that $J_k(\varphi) = h(v)/N_k$, and so the proposition follows.

Proof of (a): By hypothesis, there is a vertex $t^* \in S_{e_i}(v) \cap S_r(o)$. Thus $\{v, t^*\}$ forms a simplex, as does the vertex o. Hence there is an apartment A^* containing both these simplexes. As usual, there is an isomorphism $A^* \to \Sigma$ mapping o to 0. If v corresponds to the image of $j = (j_1, \dots, j_{n+1})$, then applying a suitable permutation, we may suppose that $j_1 \leqslant \cdots \leq j_{n+1}$. As $t^* \in S_{e_i}(v)$, it must correspond to the image of $j + \delta$ for some $\delta \in \mathcal{E}_i$. Applying a further permutation which fixes j, if necessary, we may suppose that also $j_1 + \delta_1 \leqslant \cdots \leq j_{n+1} + \delta_{n+1}$ (see Lemma 2.2 and the remark following it). Thus the isomorphism maps o, v and t^* into $\mathcal{S}_0$, and the preimage of $\mathcal{S}_0$ is therefore a sector S^* containing o, v and t^*, which represents an $\omega^* \in \Omega_o(v) \cap \Omega_o(t^*)$. This proves the "only if" part of (a). Now suppose that $t \in S_r(o)$ and that $\omega \in \Omega_o(v) \cap \Omega_o(t)$. Let A be an apartment containing the sector $S = S^o(\omega)$ (notation as at the beginning of §4). There is an isomorphism $A \to \Sigma$ mapping o to 0 and S to $\mathcal{S}_0$. We thus have an isomorphism $\psi : A \to A^*$ fixing o and mapping S to S^*. For each $m \in \mathbb{N}^n$, ψ maps the vertex $S^o_m(\omega)$ (the unique element in $S \cap S_m(o)$) to $S^o_m(\omega^*)$. Now $S^o_k(\omega) = v$ and $S^o_r(\omega) = t$. Since $S^o_k(\omega^*) = v$ and $S^o_r(\omega^*) = t^*$, and since $d(v, t^*) = 1$, we have that $d(v, t) = 1$ too. Moreover, since $t^* \in S_{e_i}(v)$, we know that $\tau(t^*) = \tau(v) + i \bmod n + 1$, and since ψ is type-preserving, we have $\tau(t) = \tau(v) + i \bmod n + 1$. Thus $t \in S_{e_i}(v)$, and so $t \in S_{e_i}(v) \cap S_r(o)$ and (a) is proved.

Proof of (b): Suppose that $t, t' \in S_{e_i}(v) \cap S_r(o)$. As in the proof of (a), there is an apartment A containing a sector S containing o, v and t, and an isomorphism $f : A \to \Sigma$ such that $f(S) = \mathcal{S}_0$. Let X be the convex hull of o, v and t in A. Then $f(X)$ equals

$$\{\ell_1 b_1 + \cdots + \ell_n b_n : \ell_j \in \mathbb{N} \text{ for each } j, \text{ and } \ell_i + \cdots + \ell_j \leq m_{i,j} \text{ for all } i \leq j\} \tag{6.1}$$

for certain integers $m_{i,j} \geq 0$. For such a set is the image in $\mathbb{R}^{n+1}/\mathbb{R}\mathbf{1}$ of the intersection of the half spaces $\{(x_1, \dots, x_{n+1}) : x_{j+1} - x_i \leq m_{i,j}\}$, for all $1 \leq i \leq j \leq n$, in $\mathbb{R}^{n+1}$. Here $m_{i,j}$ is the maximum of $k_i + \cdots + k_j$ and $r_i + \cdots + r_j$.

The set $\Omega_o(v) \cap \Omega_o(t)$ equals the set $\Omega_o(X)$ of all $\omega \in \Omega$ such that $S^o(\omega)$ contains X. We now show that for any set $X \subset A$ such that $f(X)$

is of the form (6.1), $\nu_o(\Omega_o(X))$ depends only on the numbers $m_{i,j}$ and not on the particular apartment containing X. Let $m_j = \max\{\ell_j : \ell \in f(X)\}$ ($\leq m_{j,j}$) for each j. Let X^* be the convex hull of o and the unique element u in $S \cap S_{m_1,\dots,m_n}(o)$. Then $f(X^*)$ equals

$$\{\ell_1 b_1 + \cdots + \ell_n b_n : 0 \leqslant \ell_j \leq m_j \text{ for all } j\}. \tag{6.2}$$

Now $X \subset X^*$ and $\nu_o(\Omega_o(X^*)) = \nu_o(\Omega_o(u)) = 1/N_{m_1,\dots,m_n}$ depends only on the numbers $m_{i,j}$. To show that $\nu_o(\Omega_o(X))$ depends only on these numbers, assume first that for each j there is an $\ell = \ell_1 b_1 + \cdots + \ell_n b_n \in f(X)$ such that $\ell_j > 0$. Then $m_{i,j} > 0$ for each $i \leq j$, and so $f(X)$ contains the vertices $b_1, \dots, b_n$ as well as o. Hence X contains the chamber C_0 in A for which $f(C_0)$ has vertices $o, b_1, \dots, b_n$. We show how to enlarge X to the set X^* defined above in a sequence of N steps, for some N which depends only on the numbers $m_{i,j}$, so that $\nu_o(\Omega_o(X))$ is reduced by a factor of q with each step. Thus the original $\nu_o(\Omega_o(X))$ equals $q^N/N_{m_1,\dots,m_n}$.

To see this, let $X \subsetneqq X^*$. Then there is an $\ell \in f(X^* \setminus X)$. Let $C_0, \dots, C_N$ be a minimal gallery from C_0 to a gallery C_N containing $f^{-1}(\ell)$. Then each C_i must be contained in X^*, because X^* is convex [18, Theorem 3.8 and Proposition 2.8]. However, there must be a $j < N$ such that $C_j \subset X$ and $C_{j+1} \not\subset X$. Let π be the panel common to C_j and C_{j+1}. There are exactly q chambers $D_m \neq C_j$, $m = 1, \dots, q$, in $\mathfrak{X}$ containing π, one of them C_{j+1}. For each m, $X \cup D_m$ is isometric to the subset $X \cup C_{j+1}$ of A, and so is contained in an apartment A_m [18, Theorem 3.6]. Hence we can write $\Omega_o(X) = \cup_{m=1}^{q} \Omega_o(X_m)$, a disjoint union, where X_m is the convex hull of $X \cup D_m$ in A_m, and where, if $\psi_m : A_m \to A$ is the isomorphism fixing X, $X \subsetneqq \psi_m(X_m) \subset X^*$ for each m. Each set $f(\psi_m(X_m))$ is of the form (6.1), with $m_{i,j}$'s different from the original ones, but independent of m. By induction on $|X^* \setminus X|$, we know that $\nu_o(\Omega_o(X_m))$ is independent of m, and equals $q^{N'}\nu_o(\Omega_o(X^*))$ for some N'. Hence $\nu_o(\Omega_o(X)) = q\nu_o(\Omega_o(X_m)) = q^{N'+1}\nu_o(\Omega_o(X^*))$ for each m.

Now suppose that there is a set $J \subsetneqq \{1, \dots, n\}$ such that for each $j \in J$ there is a vertex $\ell \in f(X)$ such that $\ell_j > 0$, but such that $\ell_j = 0$ for all $\ell \in f(X)$ if $j \notin J$. Note that $m_j = 0$ if $j \notin J$. As before, let u denote the unique element in $S \cap S_{m_1,\dots,m_n}(o)$. Let $v_j = f^{-1}(b_j)$ for $j \in J$. There is a certain number q_J of chambers C containing $\{v_j : j \in J\}$ and o. For each such C, there is an apartment A_C containing C and the vertex u. Then A_C contains o and u and therefore X^*, and so A_C contains the convex hull X_C of $X \cup C$. The set $\Omega_o(X)$ is the disjoint union of the q_J sets $\Omega_o(X_C)$. If $f_C : A_C \to \Sigma$ is the isomorphism mapping C to the chamber in Σ having vertices $o, b_1, \dots, b_n$, then each $f_C(X_C)$ is of the form (6.1), with the $m_{i,j}$'s different from the original ones, but dependent only on the original $m_{i,j}$'s, not on the particular C. By the case treated above, $\nu_o(\Omega_o(X_C)) = \alpha$, say, depends only on the original $m_{i,j}$'s, not on C, and so $\nu_o(\Omega_o(X)) = q_J\alpha$ depends only on the original $m_{i,j}$'s. This completes the proof of (b). $\square$

Lemma 6.2. *Suppose that we can find numbers $\{c_\ell : \ell \in F\}$, where $F \subset \mathbb{Z}^n$ is finite, so that for each vertex o and for each $\varphi \in C^\infty(\Omega)$ the expression $T^o_k(\varphi)$, as defined in (1.9), is independent of k, once $k_1, \ldots, k_n$ are sufficiently large. Write $T^o(\varphi)$ for the common value of $T^o_k(\varphi)$, k large. Then (1.10) holds.*

Proof. It is enough to prove (1.10) in the case when $d(o,x) = 1$ and $\varphi = \mathbf{1}_{\Omega_o(v)}$, where $v \in S_r(o)$, r large. Now $v \in S_t(x)$ for some $t \in \mathbb{N}^n$. If $r = (r_1, \ldots, r_n)$ is so large that $r_i \geq 2d(o,x)$ for all i, then arguing as in the proofs of Corollary 2.3 and Lemmas 2.4 and 2.5 of [6], we see that $\Omega_o(v) = \Omega_x(v)$ and that $m_i(o,x;\omega) = t_i - r_i$ for all $\omega \in \Omega_o(v)$. Hence

$$T^o_k(h_{o,x}\mathbf{1}_{\Omega_o(v)}) = \frac{s_1^{t_1-r_1+t_2-r_2+\cdots} \cdots s_n^{t_n-r_n}}{q^{\rho(t-r)}} T^o_k(\mathbf{1}_{\Omega_o(v)}),$$

and since

$$J^o_{k+\ell}(\mathbf{1}_{\Omega_o(v)}) = \frac{1}{N_{k+\ell}} \sum_{\substack{u \in S_{k+\ell}(o): \\ v \in [o,u]}} h(u),$$

$$T^o_k(\mathbf{1}_{\Omega_o(v)}) = C q^{\rho(k)} s_1^{k_1+\cdots+k_n} \cdots s_n^{k_n} \left(\sum_{\ell \in F} c_\ell q^{\rho(\ell)} \frac{1}{N_{k+\ell}} \sum_{\substack{u \in S_{k+\ell}(o): \\ v \in [o,u]}} h(u) \right).$$

On the other hand,

$$T^x_{k+t-r}(\mathbf{1}_{\Omega_o(v)}) = T^x_{k+t-r}(\mathbf{1}_{\Omega_x(v)}) = C q^{\rho(k+t-r)} \times$$

$$s_1^{k_1+t_1-r_1+\cdots} \cdots s_n^{k_n+t_n-r_n} \left(\sum_{\ell \in F} c_\ell q^{\rho(\ell)} \frac{1}{N_{k+t-r+\ell}} \sum_{\substack{u \in S_{k+t-r+\ell}(x): \\ v \in [x,u]}} h(u) \right).$$

So (1.10) will follow once we prove that, if $r_i \geq 2$ for all i, then

$$\{u \in S_{k+t-r}(x) : v \in [x,u]\} = \{u \in S_k(o) : v \in [o,u]\}. \tag{6.3}$$

Here we use the hypothesis made at the beginning of the proof that $d(o,x) = 1$. There is an apartment containing v and the simplex $\{o,x\}$, and it follows easily that $|t_i - r_i| \leq 1$ for all i. Hence $t_i \geq 1$ for all i. Suppose that $u \in S_k(o)$ and that $v \in [o,u]$. Choose a sector S with base vertex o containing u. As in the proof of Lemma 2.4, there is an apartment A containing S and x. There is an isomorphism $\psi : A \to \Sigma$ mapping o to 0, S to $\mathcal{S}_0$ and x to $\sum_{i=1}^n (\epsilon_{i+1} -$

$\epsilon_i)b_i$ for some $n+1$-tuple $(\epsilon_1,\dots,\epsilon_{n+1})$ of 0's and 1's. Then $v \in S_r(o)$ means that v is mapped to $\sum_{i=1}^n r_i b_i = \psi(x) + \sum_{i=1}^n (r_i + \epsilon_i - \epsilon_{i+1}) b_i$, and as $r_i + \epsilon_i - \epsilon_{i+1} \geq r_i - 1 \geq 0$ for all i, we have $t_i = r_i + \epsilon_i - \epsilon_{i+1}$. Also, u is mapped to $\sum_{i=1}^n k_i b_i = \psi(x) + \sum_{i=1}^n (k_i + \epsilon_i - \epsilon_{i+1}) b_i = \psi(x) + \sum_{i=1}^n (k_i + t_i - r_i) b_i$. So $u \in S_{k+t-r}(x)$, and as $0 \leq r_i + t_i - r_i \leq k_i + t_i - r_i$ for all i, we have $v \in [x,u]$. Conversely, if $u \in S_{k+t-r}(x)$ and $v \in [x,u]$, then as $t_i \geq 1$ for all i, interchanging the roles of o and x in the argument above, we find that $u \in S_k(o)$ and that $v \in [o,u]$. □

For $n=1$ and $n=2$ it was easy to find $\{c_\ell : \ell \in F\}$ satisfying the conditions of Lemma 6.2. Tim Steger pointed out to me that these satisfied

$$\sum_{\ell \in F} c_\ell s_{\sigma(1)}^{-(\ell_1+\cdots+\ell_n)} \cdots s_{\sigma(n)}^{-\ell_n} = 0 \tag{6.4}$$

for all permutations $\sigma \neq id$ of $\{1,\dots,n+1\}$. It is quite easy to construct families $\{c_\ell : \ell \in F\}$ satisfying this. First consider the product

$$\begin{aligned}\prod_{j=i+1}^{n+1} (s_j - s_i) &= s_i^{n-i} \sum_{\nu=0}^{n+1-i} (-1)^{n+1-i-\nu} \sigma_\nu(s_{i+1},\dots,s_{n+1}) s_i^{-(\nu-1)} \\ &= \sum_{\nu=0}^{n+1-i} a_\nu^{(i)} s_i^{-(\nu-1)}, \text{ say.}\end{aligned}$$

Clearly $\sum_{\nu=0}^{n+1-i} a_\nu^{(i)} s_{\sigma(i)}^{-(\nu-1)} = 0$ if σ is a permutation of $\{1,\dots,n+1\}$ such that $\sigma(i) \in \{i+1,\dots,n+1\}$, and the sum is $\prod_{j=i+1}^{n+1}(s_j - s_i)$ if $\sigma(i) = i$. Hence the n-fold sum

$$\sum_{\nu_1=0}^{n} \cdots \sum_{\nu_n=0}^{1} a_{\nu_1}^{(1)} \cdots a_{\nu_n}^{(n)} s_{\sigma(1)}^{-(\nu_1-1)} \cdots s_{\sigma(n)}^{-(\nu_n-1)}$$

equals $\prod_{i,j:1\leq i<j\leq n+1}(s_j - s_i)$ if σ is the identity permutation, and is 0 otherwise. We can write the above sum in the form (6.4), with $\ell_i = \nu_i - \nu_{i+1}$ for $i = 1\dots,n-1$ and $\ell_n = \nu_n - 1$.

For $k \in \mathbb{N}^n$ such that $k + \ell \geq 0$ for all $\ell \in F$, we define $T_k^o(\varphi)$ by (1.9), where c_ℓ are as above. It is easy to see how the constant C in (1.9) can be chosen so that (1.11) holds, at least when $s_1,\dots,s_{n+1}$ are distinct. For by Theorem 3.1 and Proposition 4.1, for each $\omega \in \Omega$ and each vertex o, $f(x) = h_{o,x}(\omega)$ defines an eigenfunction of each operator A_{e_r}, $r = 1,\dots,n$, and hence of A_k, for each $k \in \mathbb{N}^n$, and so we can write $A_k f = p_k f$, where p_k is a polynomial in $\lambda_1,\dots,\lambda_n$. These polynomials are the so called spherical functions, and it is well known that

$$p_{k_1,\dots,k_n} = C_{n+1}(q)^{-1} q^{-\rho(k)} \sum_\sigma s_{\sigma(1)}^{-(k_1+\cdots+k_n)} \cdots s_{\sigma(n)}^{-k_n} c(s_\sigma),$$

(see [14, p. 299]) where the sum is over all permutations σ of $\{1,\ldots,n+1\}$, and where

$$C_{n+1}(q)=\prod_{j=1}^{n+1}\left(\frac{q^j-1}{q-1}\right)\quad\text{and}\quad c(s_\sigma)=\prod_{1\le i<j\le n+1}\left(\frac{qs_{\sigma(j)}-s_{\sigma(i)}}{s_{\sigma(j)}-s_{\sigma(i)}}\right).$$

Using $J^o_{k+\ell}(\mathbf{1})=p_{k+\ell}\,h(o)$ and this formula for p_k, we see that

$$\begin{aligned}T^o_k(\mathbf{1})&=h(o)\frac{C}{C_{n+1}(q)}\sum_{\ell,\sigma}a_\ell s_{\sigma(1)}^{-(\ell_1+\cdots+\ell_n)}\cdots s_{\sigma(n)}^{-\ell_n}\,c(s_\sigma)\\&=h(o)\frac{C}{C_{n+1}(q)}\left(\sum_\ell a_\ell s_1^{-(\ell_1+\cdots+\ell_n)}\cdots s_n^{-\ell_n}\right)\prod_{1\le i<j\le n+1}\left(\frac{qs_j-s_i}{s_j-s_i}\right)\\&=h(o)\frac{C}{C_{n+1}(q)}\prod_{1\le i<j\le n+1}(qs_j-s_i)\end{aligned}$$

Hence if we set $C=C_{n+1}(q)/\prod_{1\le i<j\le n+1}(qs_j-s_i)$, then $T^o_k(\mathbf{1})=h(o)$, independent of $k\in\mathbb{N}^n$ (such that $\overline{k+\ell}\ge 0$ for $\ell\in F$).

To check that the c_ℓ defined above satisfy the hypotheses of Lemma 6.2, I need to solve the recurrence relations in Proposition 6.1. Using (2.2) and (3.4), we have $c_{k,e_i;r}=q^{N(\epsilon)}/M_{n+1,i}$ if $r=k+\Delta\epsilon$, $\epsilon\in\mathcal{E}_i$ and $k_\nu\ge 1$ for $\nu=1,\ldots,n$. Note that $N(\epsilon)=i(n+1-i)/2+\rho(\Delta\epsilon)$ for $\epsilon\in\mathcal{E}_i$. Hence $x_\ell=J_{k_0+\ell}(\varphi)/q^{\rho(\ell)}$ is defined if $\ell_\nu\ge 0$ for $\nu=1,\ldots,n$, and if $\ell_\nu\ge 1$ for all ν we have the simpler equations

$$\sigma_i x_\ell=\sum_{\epsilon\in\mathcal{E}_i}x_{\ell+\Delta\epsilon}\tag{6.5}$$

for $i=1,\ldots,n$, where $\sigma_i=\sigma_i(s_1^{-1},\ldots,s_{n+1}^{-1})$. For any permutation p of $\{1,\ldots,n+1\}$,

$$x_\ell^{(p)}=s_{p(1)}^{-(\ell_1+\cdots+\ell_n)}\cdots s_{p(n)}^{-\ell_n}$$

defines a solution of the recurrences (6.5).

Question. Assume for simplicity that $s_1,\ldots,s_{n+1}$ are distinct. Is any solution of the recurrences (6.5) a linear combination of the $(n+1)!$ solutions given above?

Assuming a positive answer to this question, it is not hard to show that the constants c_ℓ defined above satisfy the hypotheses of Lemma 6.2. The proof is similar to the proof that the above choice of C gives $T^o_k(\mathbf{1})=h(o)$.

7. Examples. The cases $n = 1$ and $n = 2$.

In this section, I give complete proofs of Theorem 1.1 in the cases $n = 1$ and $n = 2$. Suppose first that $n = 1$. Suppose that h is a λ-harmonic function on the vertex set of a homogeneous tree $\mathfrak{X}$ in which each vertex has $q+1$ neighbours (as noted above, this is a special case of an $\tilde{A}_1$ building). For any vertex o and any $k \in \mathbb{N}$, the functional $J_k = J_k^o : C^\infty(\Omega) \to \mathbb{C}$ is defined by (1.8). Here $N_0 = 1$ and $N_k = (q+1)q^{k-1}$, $k \geq 1$, and as is well-known and easy to verify, the averaging operators

$$(A_k f)(x) = \frac{1}{N_k} \sum_{y:d(x,y)=k} f(y)$$

satisfy $A_1 A_k = (qA_{k+1} + A_{k-1})/(q+1)$ for all $k \geq 1$, and so by Proposition 6.1, $\lambda J_k(\varphi) = (qJ_{k+1}(\varphi) + J_{k-1}(\varphi))/(q+1)$ provided that $k \geq k_0, 1$ if φ is constant on all sets $\Omega_o(v)$, where $d(o,v) = k_0$. It follows that writing

$$\lambda = \frac{\sqrt{q}}{q+1}(s + s^{-1}),$$

(here we write s in place of s_1, and s^{-1} in place of s_2), for any C,

$$T(\varphi) = T_k^o(\varphi) = Cq^{\frac{k}{2}} s^k \left(s^{-1} J_k(\varphi) - q^{-\frac{1}{2}} J_{k-1}(\varphi)\right) \tag{7.1}$$

satisfies $T_{k+1}^o(\varphi) = T_k^o(\varphi)$, and so is independent of k once $k \geq k_0, 1$. Now

$$T(\mathbf{1}) = T_1^o(\mathbf{1}) = C(\sqrt{q}J_1(\mathbf{1}) - sJ_0(\mathbf{1})) = C(\sqrt{q}\lambda - s)h(o) = C\frac{qs^{-1} - s}{q+1},$$

and so setting $C = (q+1)/(qs^{-1} - s)$, which we may if $s^2 \neq q$, we get $T(\mathbf{1}) = h(o)$. To show that the linear span V of the functions $h_{o,x}$, $x \in \mathfrak{X}^0$, equals $C^\infty(\Omega)$, we show that $\mathbf{1}_{\Omega_o(x)} \in V$ for any $x \in \mathfrak{X}^0$. If $d(o,x) = 0$, then $x = o$, and $\mathbf{1}_{\Omega_o(x)} = \mathbf{1} = h_{o,o}$. So suppose that $d(o,x) = d > 0$. Let $x_0 = o, \ldots, x_d = x$ be the vertices of the geodesic $[o,x]$ from o to x. Then $m_1(o,x;\omega) = d - 2j$ if $0 \leq j < d$ and $\omega \in \Omega_o(x_j) \setminus \Omega_o(x_{j+1})$. This is also true if $j = d$ if we interpret $\Omega_o(x_{d+1})$ as $\emptyset$. An easy calculation shows that

$$h_{o,x} = \left(\frac{s}{\sqrt{q}}\right)^d \mathbf{1} + \left(1 - \frac{s^2}{q}\right) \sum_{j=1}^{d} \left(\frac{s}{\sqrt{q}}\right)^{d-2j} \mathbf{1}_{\Omega_o(x_j)}. \tag{7.2}$$

Now replace x by $x^- = x_{d-1}$, the last vertex on $[o,x]$ before x. Multiply the corresponding equation for h_{o,x^-} by $s/\sqrt{q}$ and subtract it from (7.2) to obtain

$$\mathbf{1}_{\Omega_o(x)} = \left(1 - \frac{s^2}{q}\right)^{-1} \left(\frac{s}{\sqrt{q}}\right)^d \left(h_{o,x} - \frac{s}{\sqrt{q}} h_{o,x^-}\right), \tag{7.3}$$

assuming $s^2 \neq q$. This exhibits $\mathbf{1}_{\Omega_o(x)}$ as an element of V. This completes the proof of Theorem 1.1 in the case $n = 1$.

Remark. In (7.1), take $\varphi = \mathbf{1}_{\Omega_o(v)}$, where $v \in S_d(o)$ and $k = d \geq 1$. If μ is the finitely additive measure on Ω corresponding to T, then (7.1) implies that

$$\mu(\Omega_o(v)) = \frac{1}{qs^{-1} - s} \cdot \left(\frac{s}{\sqrt{q}}\right)^{d-1} (\sqrt{q}\, h(v) - sh(v^-)). \tag{7.4}$$

Also, from (1.7) we have $\mu(\Omega) = h(o)$. This shows how μ can be obtained explicitly from h. If we fix ω and let $h(x) = h_{o,x}(\omega)$, then μ is clearly the Dirac measure δ_ω concentrated at ω, and (7.4) specializes to (7.3).

Note also that we can use (7.4) to show that μ is positive if h is positive and λ-harmonic and if $\lambda \in [2\sqrt{q}/(q+1), \infty)$ (so that $s > 0$). For if say $0 < s \leq 1$, then we must show that $\sqrt{q}\, h(v) \geq sh(v^-)$. Let c_n denote the infimum of $f(x)/f(y)$ over all vertices x, y such that $d(x,y) = n$. It is enough to show that $c_1 \geq s/\sqrt{q}$. Clearly, $c_2 \geq c_1^2$. Also, the λ-harmonic property shows that $\lambda(q+1)c_1 \geq 1 + qc_2$. Combining these two inequalities, we see that $s\sqrt{q} \leq c_1 \leq \sqrt{q}/s$, which implies that $c_1 \geq s/\sqrt{q}$. The cases $1 < s < \sqrt{q}$ and $s > \sqrt{q}$ are similar.

The case $n = 2$.

Suppose that h is a (λ_1, λ_2)-harmonic function on the vertex set $\mathfrak{X}^0$ of a thick, locally finite $\tilde{A}_2$ building $\mathfrak{X}$. Fix a vertex o. For $m, n \in \mathbb{N}$, the functional $J_{m,n} = J^o_{m,n} : \mathcal{C}^\infty(\Omega) \to \mathbb{C}$ is defined by (1.8), where $k = (m, n)$. Then (see [5, Proposition 2.3]) the averaging operators $A_{m,n}$ satisfy

$$A_{m,n}A_{1,0} = (q^2+q+1)^{-1}(q^2 A_{m+1,n} + qA_{m-1,n+1} + A_{m,n-1}) \quad \text{and}$$
$$A_{m,n}A_{0,1} = (q^2+q+1)^{-1}(q^2 A_{m,n+1} + qA_{m+1,n-1} + A_{m-1,n})$$

for $m, n \geq 1$. Hence if $\varphi \in \mathcal{C}^\infty(\Omega)$ is constant on each set $\Omega_o(v)$, where $v \in S_{m_0,n_0}(o)$, and if $m \geq m_0, 1$ and $n \geq n_0, 1$ we have, by Proposition 6.1,

$$\lambda_1 J_{m,n}(\varphi) = (q^2+q+1)^{-1}(q^2 J_{m+1,n}(\varphi) + qJ_{m-1,n+1}(\varphi) + J_{m,n-1}(\varphi)) \tag{7.5}$$

and

$$\lambda_2 J_{m,n}(\varphi) = (q^2+q+1)^{-1}(q^2 J_{m,n+1}(\varphi) + qJ_{m+1,n-1}(\varphi) + J_{m-1,n}(\varphi)). \tag{7.6}$$

Proposition 7.1. *Let C be a constant. If $\varphi \in \mathcal{C}^\infty(\Omega)$ is constant on each set $\Omega_o(v)$, where $v \in S_{m_0,n_0}(o)$, then the value of the functional*

$$T^o_{m,n} = Cq^{m+n}s_1^{m+n}s_2^n\Big(-s_1q^{-1}J_{m,n-1} + s_1(s_2+s_3)J_{m+1,n-1} - qJ_{m+2,n-1} + s_1s_3q^{-1}J_{m-1,n} - s_1s_3(s_2+s_3)J_{m,n} + s_3qJ_{m+1,n}\Big) \tag{7.7}$$

at φ is independent of m and n once $m \geq m_0, 1$ and $n \geq n_0, 1$.

Proof. We must show that if $m \geq m_0, 1$ and $n \geq n_0, 1$, then, evaluated at φ,

$$\begin{aligned} qs_1\Big(&-s_1 q^{-1} J_{m+1,n-1} + s_1(s_2+s_3)J_{m+2,n-1} - qJ_{m+3,n-1} \\ &+ s_1 s_3 q^{-1} J_{m,n} - s_1 s_3(s_2+s_3)J_{m+1,n} + s_3 q J_{m+2,n}\Big) \\ = &-s_1 q^{-1} J_{m,n-1} + s_1(s_2+s_3)J_{m+1,n-1} - qJ_{m+2,n-1} \\ &+ s_1 s_3 q^{-1} J_{m-1,n} - s_1 s_3(s_2+s_3)J_{m,n} + s_3 q J_{m+1,n} \end{aligned} \tag{7.8}$$

and

$$\begin{aligned} qs_1 s_2\Big(&-s_1 q^{-1} J_{m,n} + s_1(s_2+s_3)J_{m+1,n} - qJ_{m+2,n} \\ &+ s_1 s_3 q^{-1} J_{m-1,n+1} - s_1 s_3(s_2+s_3)J_{m,n+1} + s_3 q J_{m+1,n+1}\Big) \\ = &-s_1 q^{-1} J_{m,n-1} + s_1(s_2+s_3)J_{m+1,n-1} - qJ_{m+2,n-1} \\ &+ s_1 s_3 q^{-1} J_{m-1,n} - s_1 s_3(s_2+s_3)J_{m,n} + s_3 q J_{m+1,n}\,. \end{aligned} \tag{7.9}$$

To obtain (7.8), we use (7.6) with (m,n) replaced by $(m+2,n)$, $(m+1,n)$, $(m,n+1)$ and as is, to eliminate $J_{m+3,n-1}$, $J_{m+1,n+1}$, $J_{m,n+2}$ and $J_{m,n+1}$, respectively, then (7.5), with (m,n) replaced by $(m+1,n+1)$, $(m+1,n)$ and as is, to eliminate $J_{m+1,n+1}$, $J_{m,n+1}$ and $J_{m-1,n+1}$, respectively. We find that everything cancels, and (7.8) holds.

To obtain (7.9), we use (7.5), with (m,n) replaced by $(m+1,n)$ and as is, to eliminate $J_{m+2,n}$ and $J_{m-1,n+1}$, respectively, then (7.6), with (m,n) replaced by $(m+1,n)$ and as is, to eliminate $J_{m+1,n+1}$ and $J_{m,n+1}$, respectively. Again, everything cancels, and (7.9) holds. □

Lemma 7.2. *Suppose that $qs_j \neq s_i$ for all $1 \leq i < j \leq 3$. In (7.7) above, take*

$$C = \frac{(q+1)(q^2+q+1)}{(qs_2-s_1)(qs_3-s_1)(qs_3-s_2)}\,.$$

Define $T^o : C^\infty(\Omega) \to \mathbb{C}$ by setting $T^o(\varphi)$ equal to $T^o_{m,n}(\varphi)$ for m and n sufficiently large. Then $T^o(\mathbf{1}) = h(o)$.

Proof. To calculate $T^o(\mathbf{1})$, we can take $m = n = 1$ in (7.7). We need to calculate $J_{1,0}(\mathbf{1})$, $J_{2,0}(\mathbf{1})$, $J_{3,0}(\mathbf{1})$, $J_{0,1}(\mathbf{1})$, $J_{1,1}(\mathbf{1})$ and $J_{2,1}(\mathbf{1})$. ¿From Proposition 6.1, and by Proposition 2.3 in [5], we have

$$\lambda_1 J_{m,0}(\varphi) = (q^2+q+1)^{-1}\big(q^2 J_{m+1,0}(\varphi) + (q+1)J_{m-1,1}(\varphi)\big)$$

if $\varphi \in C^\infty(\Omega)$ is constant on each set $\Omega_o(v)$, where $v \in S_{m_0,0}(o)$, and if $m \geq m_0, 1$. Also,

$$\lambda_2 J_{0,n}(\varphi) = (q^2+q+1)^{-1}(q^2 J_{0,n+1}(\varphi) + (q+1)J_{1,n-1}(\varphi))$$

if $\varphi \in C^\infty(\Omega)$ is constant on each set $\Omega_o(v)$, where $v \in S_{0,n_0}(o)$, and if $n \geq n_0, 1$, while

$$\lambda_1 J_{0,n}(\varphi) = (q^2+q+1)^{-1}(q(q+1)J_{1,n}(\varphi) + J_{0,n-1}(\varphi))$$

if $\varphi \in C^\infty(\Omega)$ is constant on each set $\Omega_o(v)$, where $v \in S_{0,n_0}(o)$, and if $n \geq n_0, 1$. Applying these formulas to $\varphi = \mathbf{1}$, and using the simple formulas $J_{0,0}(\mathbf{1}) = h(o)$, $J_{1,0}(\mathbf{1}) = \lambda_1 h(o)$ and $J_{0,1}(\mathbf{1}) = \lambda_2 h(o)$, routine but tedious calculations show that $T^o(\mathbf{1}) = h(o)$. □

Lemma 7.3. *Suppose that $qs_j - s_i \neq 0$ for $1 \leq i < j \leq 3$. Then for each fixed vertex o, the functions $h_{o,x}$, $x \in \mathfrak{X}^0$, span $C^\infty(\Omega)$.*

Proof. Let V^o denote the linear span of the functions $h_{o,x}$, $x \in \mathfrak{X}^0$. First note that $\mathbf{1}_{\Omega_o(o)} = \mathbf{1} = h_{o,o} \in V^o$. Next, if $x \in S_{1,0}(o)$, (with notation as in §4) let $A_x = \Omega_o(x) = \{\omega \in \Omega : S^o_{1,0}(\omega) = x\} = \Omega_o((0,1,1),x)$, $B_x = \Omega_o((1,0,1),x) = \{\omega \in \Omega : S^o_{1,0}(\omega) \neq x$ but $d(S^o_{0,1}(\omega),x) = 1\}$, and $C_x = \Omega_o((1,1,0),x) = \{\omega \in \Omega : d(S^o_{0,1}(\omega),x) = 2\}$; these sets are disjoint and have union Ω. Then it is easy to verify that $\sum_{y\in S_{0,1}(o):d(x,y)=1} h_{o,y}$ takes the constant values $q(s_2+s_3)$, s_1+qs_3 and s_1+s_2, respectively, on A_x, B_x and C_x, while $h_{o,x}$ takes the values q/s_1, $1/s_2$ and $1/(qs_3)$ on these sets. Hence

$$\begin{aligned}\mathbf{1}_{\Omega_o(x)} &= \frac{1}{(qs_3-s_1)(qs_2-s_1)}\Big\{qh_{o,x} + s_1^2 h_{o,o} - s_1 \sum_{\substack{y\in S_{0,1}(o):\\ d(x,y)=1}} h_{o,y}\Big\}\\ &= a_1 h_{o,x} + a_2 h_{o,o} + a_3 \sum_{\substack{y\in S_{0,1}(o):\\ d(x,y)=1}} h_{o,y}, \quad \text{say,}\end{aligned}$$

Similarly, for each $y \in S_{0,1}(o)$,

$$\begin{aligned}\mathbf{1}_{\Omega_o(y)} &= \frac{1}{(qs_3-s_2)(qs_3-s_1)}\Big\{qs_3 h_{o,y} + s_3^{-1} h_{o,o} - \sum_{\substack{x\in S_{1,0}(o):\\ d(x,y)=1}} h_{o,x}\Big\}\\ &= b_1 h_{o,y} + b_2 h_{o,o} + b_3 \sum_{\substack{x\in S_{1,0}(o):\\ d(x,y)=1}} h_{o,x}, \quad \text{say.}\end{aligned} \tag{7.10}$$

Suppose that $v \in S_{1,1}(o)$. Let x and y be the points in the convex hull of o and v which lie in $S_{1,0}(o)$ and $S_{0,1}(o)$, respectively. Then $v \in S_{0,1}(x)$, and so by (7.10), with x in place of o, we can write

$$\mathbf{1}_{\Omega_x(v)} = b_1 h_{x,v} + b_2 h_{x,x} + b_3 \sum_{\substack{u\in S_{1,0}(x):\\ d(u,v)=1}} h_{x,u}\,. \tag{7.11}$$

Multiply both sides of (7.11) by $h_{o,x} = (q/s_1)\mathbf{1}_{A_x} + (1/s_2)\mathbf{1}_{B_x} + (1/(qs_3)\mathbf{1}_{C_x}$. Now $\omega \in A_x \cap \Omega_x(v)$ if and only if $\omega \in \Omega_o(v)$. Thus $\mathbf{1}_{A_x}\cdot \mathbf{1}_{\Omega_x(v)} = \mathbf{1}_{\Omega_o(v)}$. It is also easy to see that $\mathbf{1}_{C_x}\cdot\mathbf{1}_{\Omega_x(v)} = 0$, and that

$$\mathbf{1}_{B_x}\cdot\mathbf{1}_{\Omega_x(v)} = \sum_{\substack{w\in S_{0,2}(o):\\ d(w,v)=1}} \mathbf{1}_{\Omega_o(w)} \quad - \mathbf{1}_{\Omega_o(v)}.$$

Hence

$$h_{o,x}\cdot\mathbf{1}_{\Omega_x(v)} = \frac{q}{s_1}\mathbf{1}_{\Omega_o(v)} + \frac{1}{s_2}\Big\{\sum_{\substack{w\in S_{0,2}(o):\\ d(w,v)=1}} \mathbf{1}_{\Omega_o(w)} \quad - \mathbf{1}_{\Omega_o(v)}\Big\}.$$

So using (4.2), we have

$$\begin{aligned}\Big(\frac{q}{s_1} - \frac{1}{s_2}\Big)\mathbf{1}_{\Omega_o(v)} + \frac{1}{s_2}\sum_{\substack{w\in S_{0,2}(o):\\ d(w,v)=1}} \mathbf{1}_{\Omega_o(w)} \\ = b_1 h_{o,v} + b_2 h_{o,x} + b_3 \sum_{\substack{u\in S_{1,0}(x):\\ d(u,v)=1}} h_{o,u} \quad \in V^o.\end{aligned} \tag{7.12}$$

If $w \in S_{0,2}(o)$ and $d(w,v) = 1$, then $w \in S_{0,1}(y)$, and so by (7.10),

$$\mathbf{1}_{\Omega_y(w)} = b_1 h_{y,w} + b_2 h_{y,y} + b_3 \sum_{\substack{\xi\in S_{1,0}(y):\\ d(\xi,w)=1}} h_{y,\xi}\,.$$

Multiply both sides by $h_{o,y}$. It is easy to convince oneself that $\Omega_y(w) = \Omega_o(w)$, and hence $\Omega_y(w) \subset \Omega_o(y)$. Now $h_{o,y}(\omega) = qs_3$ for all $\omega \in \Omega_o(y)$ by (4.3). So we obtain

$$\mathbf{1}_{\Omega_o(w)} = \frac{1}{qs_3}\Big\{b_1 h_{o,w} + b_2 h_{o,y} + b_3 \sum_{\substack{\xi\in S_{1,0}(y):\\ d(\xi,w)=1}} h_{o,\xi}\Big\} \in V^o.$$

Hence, using $qs_2 - s_1 \neq 0$, we see from (7.12) that $\mathbf{1}_{\Omega_o(v)} \in V^o$.

Suppose that $m, n \geq 0$ and that for all o we have shown that $\mathbf{1}_{\Omega_o(v)} \in V^o$ for every $v \in S_{m,n}(o)$. Let $v \in S_{m+1,n}(o)$. Let $x \in S_{1,0}(o)$ lie in the convex hull of o and v. Then $v \in S_{m,n}(x)$, and so by the induction hypothesis, we can write $\mathbf{1}_{\Omega_x(v)} = \sum_u c_{x,u} h_{x,u}$. By (4.2) again,

$$h_{o,x} \cdot \mathbf{1}_{\Omega_x(v)} = \sum_u c_{x,u} h_{o,u} \,. \tag{7.13}$$

It is easy to see that if $m \geq 1$, then $\Omega_x(v) = \Omega_o(v)$, and hence $\Omega_x(v) \subset \Omega_o(x)$. Now $h_{o,x}(\omega) = q/s_1$ for all $\omega \in \Omega_o(x)$. Hence $h_{o,x} \cdot \mathbf{1}_{\Omega_x(v)} = (q/s_1)\mathbf{1}_{\Omega_x(v)} = (q/s_1)\mathbf{1}_{\Omega_o(v)}$, and so (7.13) becomes:

$$\mathbf{1}_{\Omega_o(v)} = (s_1/q) \sum_u c_{x,u} h_{o,u} \in V^o \,.$$

Similarly, if $v \in S_{m,n+1}(o)$, let $y \in S_{0,1}(o)$ lie in the convex hull of o and v. Then $v \in S_{m,n}(y)$, and so by the induction hypothesis, we can write $\mathbf{1}_{\Omega_y(v)} = \sum_u c_{y,u} h_{y,u}$. It is easy to see that if $n \geq 1$, then $\Omega_y(v) = \Omega_o(v)$, and hence $\Omega_y(v) \subset \Omega_o(y)$. Now $h_{o,y}(\omega) = qs_3$ for all $\omega \in \Omega_o(y)$. Hence $h_{o,y} \cdot \mathbf{1}_{\Omega_y(v)} = qs_3 \mathbf{1}_{\Omega_y(v)} = qs_3 \mathbf{1}_{\Omega_o(v)}$. Hence $\mathbf{1}_{\Omega_o(v)} = (qs_3)^{-1} \sum_u c_{y,u} h_{o,u} \in V^o$. □

Remark. Evaluating the functional $T^o_{m,n}$ at $\varphi = \mathbf{1}_{\Omega_o(v)}$, where $v \in S_{m,n}(o)$, we find that (if μ corresponds to T^o), then $\mu(\Omega_o(v))$ equals

$$\frac{q}{q^{m+n}(qs_2 - s_1)(qs_3 - s_1)(qs_3 - s_2)} \times \Big(-qs_1 h(v^-) + qs_1(s_2 + s_3) \sum{}' h(u) \\ - \sum{}'' h(u) + qs_1 s_3 h({}^-v) - q^2 s_1 s_3 (s_2 + s_3) h(v) + qs_3 \sum{}''' h(u) \Big)$$

where v^- and ${}^-v$ are the unique elements in $S_{m,n-1}(o) \cap S_{1,0}(v)$ and in $S_{m-1,n}(o) \cap S_{0,1}(v)$, respectively, and where $\sum' h(u)$, $\sum'' h(u)$ and $\sum''' h(u)$ are sums over the sets $S_{m+1,n-1}(o) \cap S_{0,1}(v)$, $S_{m+2,n-1}(o) \cap S_{1,1}(v)$ and $S_{m+1,n}(o) \cap S_{1,0}(v)$, respectively (which have q, q^3 and q^2 elements, respectively). Taking in particular $h(x) = h_{o,x}(\omega)$, where we know that μ is δ_ω, we get an analogue of (7.3)

References.

[1] N. Bourbaki, *Groupes et algèbres de Lie, Chapitres 4, 5 et 6*, Hermann, Paris, 1968.

[2] Kenneth S. Brown, *Buildings*, Springer-Verlag, Berlin, New York, 1989.

[3] P. Cartier, Representations of $\mathfrak{p}$-adic groups: a survey, Proceedings of Symposia in Pure Mathematics, Vol. **33**, part 1, pp. 111-155.

[4] Donald I. Cartwright, A brief introduction to buildings, *Contemp. Math.* **206** (1997), 45–77.

[5] Donald I. Cartwright and W. Młotkowski, Harmonic analysis for groups acting on triangle buildings, *J. Austral. Math.* **56** (1994), 345–383.

[6] Donald I. Cartwright, W. Młotkowski and Tim Steger, Property (T) and $\tilde{A}_2$ groups, *Annales Inst. Fourier* **44** (1994), 213–248.

[7] W. Casselman, Introduction to the theory of admissible representations of $\mathfrak{p}$-adic reductive groups. Unpublished manuscript, 1974. Version worked over by the Séminaire Paul Sally, 1992–1993, available from //ftp.math.lsa.umich.edu/pub/moy.

[8] W. Casselman, The unramified principle series of $\mathfrak{p}$-adic groups I. The spherical function. *Comp. Math.* **40** (1980), 387–406.

[9] A. Figá-Talamanca and C. Nebbia, *Harmonic analysis and representation theory for groups acting on homogeneous tree*, London Mathematical Society Lecture Note Series **162**, Cambridge University Press 1991

[10] P. Gerardin, On harmonic functions on symmetric spaces and buildings, pp. 79–92 in "1980 Seminar on Harmonic Analysis", Proceedings of the Canadian Mathematical Society Seminar at McGill University, Montreal, Quebec, August 4–22, 1980. Edited by Carl Herz and Roger Rigelhof. CMS Conference Proceedings 1. Published by the American Mathematical Society, Providence RI, for the Canadian Mathematical Society, Ottawa, Ontario, 1981.

[11] Shin-ichi Kato, On eigenspaces of the Hecke algebra with respect to a good maximal compact subgroup of a p-adic reductive group. *Math. Annalen* **257** (1981), 1–7.

[12] Shin-ichi Kato, Irreducibility of principal series representations for Hecke algebras of affine type. *J. Fac. Sci. Univ. Tokyo Sect. 1A Math.* **28** (1981), 929–943.

[13] I.G. Macdonald, *Spherical functions on a group of p-adic type*, Ramanujan Inst. Publications **2** (University of Madras, 1971).

[14] I.G. Macdonald, *Symmetric functions and Hall polynomials, second edition*, Oxford University Press, 1995.

[15] A.M. Mantero and A. Zappa, Characterization of the eigenfunctions of the Laplace operators for an affine building of rank 2, *Contemp. Math.* **206**, 141–143.

[16] A.M. Mantero and A. Zappa, Eigenfunctions of the Laplace operators for a building of type $\tilde{A}_2$, preprint.

[17] A.M. Mantero and A. Zappa, Eigenfunctions of the Laplace operators for a building of type $\tilde{B}_2$ and $\tilde{G}_2$, preprint.

[18] Mark Ronan, *Lectures on buildings*, Academic Press, 1989.

[19] R. Stanley, *Enumerative Combinatorics I*, Wadsworth & Brooks/Cole, Monterey, California, 1986.

[20] Tim Steger, Local fields and buildings, *Contemp. Math.* **206** (1997), 79–107.

[21] Donald E. Taylor, *The geometry of the classical groups*, Helderman Verlag, Berlin, 1992.

Donald Cartwright
School of Math. and Statistics
University of Sidney
Sidney NSW 2006, Australia
email: donaldc@maths.usyd.edu.au

Spectre d'opérateurs différentiels sur le graphes

Yves Colin de Verdière

Abstract. Dans cet exposé de survol, nous commençons par présenter des ensembles naturels d'opérateurs de type Schrödinger associés à un graphe fini.

Nous étudions ensuite les limites singulières (au sens de la $\Gamma-$convergence) de tels opérateurs et montrons qu'elles sont associées à des relations naturelles entre graphes (mineurs, transformation étoile-triangle) ou à des limites d'un intérêt indépendant (processus de Markov (recuit simulé), réseaux électriques).

Cela conduit à introduire la notion de stabilité structurelle pour une valeur propre multiple d'un opérateur d'une familleen utilisant la transversalité dans les espaces d'opérateurs symétriques.

Les invariants numériques de graphes ainsi construits sont liés à des problèmes classiques de la combinatoire des graphes: planarité, genre, plongement non-noué dans $\mathbb{R}^3$, largeur d'arbre.

1 Introduction

La théorie spectrale des opérateurs différentiels dont le prototype est le laplacien sur une variété riemannienne compacte a connu un grand développement durant les 30 dernières années. En particulier, l'asymptotique des grandes valeurs propres, liée à la dynamique classique (géodésiques périodiques) a fait l'objet de nombreux travaux. Les propriétés purement spectrales (i.e. non asymptotiques, purement quantiques) présentent également un grand intérêt; il se trouve, ainsi que cette conférence le montre bien, qu'un cadre, a priori plus simple, se prête bien à l'étude spectrale des opérateurs différentiels: il s'agit de comprendre le spectre des opérateurs différentiels de type laplacien ou Schrödinger sur les graphes. En retour, on attend en particulier de nouveaux résultats pour le cas continu !!

Les *ensembles* d'opérateurs considérés dans la suite se rencontrent dans de nombreux domaines des mathématiques (discrétisation par éléments finis [8],[11], effet tunnel semi-classique [21], processus de Markov [12], limites de laplaciens de surfaces de Riemann à courbure -1 lorsque certaines géodésiques fermées simples ont des longueurs qui tendent vers 0 [9]) et de la physique (supraconducteurs [35], physique quantique [30], spectres moléculaires [24] ou même plus anciennement [29]).

A chaque graphe fini $G = (V, E)$ (on pose $n = |V|$), on associe des *ensembles* d'opérateurs différentiels auto-adjoints sur l'espace de Hilbert (de dimension finie) $H_G = \mathbb{C}^V$ muni du produit scalaire

$$(f|g) = \sum_{i \in V} f(i)\bar{g}(i) \, .$$

Un *opérateur différentiel* sur G est un endomorphisme linéaire A de H_G qui est *local*, c'est-à-dire que $Af(i) = \sum_j a_{i,j} f(j)$ où les $a_{i,j}$ sont nuls si $\{i,j\}$ n'est pas une arête et $i \neq j$ (on peut considérer qu'il y a des arêtes $\{i,i\}$ pour chaque $i \in V$) ; autrement dit $Af(i)$ ne dépend que de $f(j)$ pour j voisin de i.

Un opérateur différentiel A sur G est dit *elliptique* si $a_{i,j} \neq 0$ pour toute arête $\{i,j\}$, $i \neq j$.

A est dit *auto-adjoint* s'il l'est comme opérateur sur H_G.

M_G est l'ensemble de tous les opérateurs elliptiques auto-adjoints sur G ; on dira d'un élément de M_G que c'est un *opérateur de Schrödinger avec champ magnétique* sur G. M_G est une sous-variété de dimension $|V| + 2|E|$ de l'espace des matrices hermitiennes $V \times V$.

O_G est le sous-ensemble de M_G formé des matrices symétriques réelles dont les coefficients $a_{i,j}$ sont < 0 pour toute arête $\{i,j\}$. On dira d'un élément de O_G que c'est un *opérateur de Schrödinger* sur G. O_G est une sous-variété de dimension $|V| + |E|$ de $S(H_G)$, l'espace des endomorphismes symétriques de H_G.

L_G est le sous-ensemble des $A \in O_G$ tels que $A1 = 0$. Ce sont les *laplaciens*. L_G est de dimension $|E|$.

Les laplaciens canoniques habituellement considérés

$$\Delta_0 f(i) = \sum_{\{i,j\} \in E} (f(i) - f(j)), \quad \Delta_1 f(i) = - \sum_{\{i,j\} \in E} f(j) \, ,$$

sont des éléments de O_G, et pour Δ_0 de L_G. Le laplacien

$$\Delta_2 f(i) = \frac{1}{d_i} \sum_{\{i,j\} \in E} (f(i) - f(j)) \, ,$$

n'est pas auto-adjoint pour la métrique canonique sur H_G sauf si le degré d_i du sommet i est constant. Il est autoadjoint pour la métrique $\sum d_i x_i^2$ sur H_G. On peut le rendre auto-adjoint pour la métrique canonique par la transformation $(Ux)_i = \frac{1}{\sqrt{d_i}} x_i$; alors $U^{-1} \Delta_2 U$ est dans O_G.

Des généralisations possibles de ces ensembles sont les suivantes :

a) on attribue à chaque arête un signe $\pm$ et on impose le signe correspondant aux $a_{i,j}$. Le lien avec les diagrammes de noeuds et les invariants associés reste alors à éclaircir : le graphe médial d'un diagramme de noeud est de façon naturelle un graphe planaire avec des signes $\pm$ sur chaque arête qui indique quel brin passe dessus ou dessous.

b) On considère le cas vectoriel, i.e. celui où $H_G^N = \oplus_{i \in V} \mathbb{C}^N$. Cela intervient aussi comme outil lorsque l'on considère des produits de graphes.

c) Le cas des graphes infinis: on doit alors bien préciser les conditions de croissance des coefficients si possible de façon à avoir un spectre discret.

d) Dans le cas de M_G, il peut être utile de considérer des arêtes doubles et, si $\{i,j\}$ double, le coefficient $a_{i,j}$ de $A \in M_G$ peut être nul ($a_{i,j} = a+(-a)$, $a \neq 0$).

Soit $G = (V, E)$ donné. Le problème de départ est le suivant :

étant donné une suite $\lambda_1 \leqslant \cdots \leqslant \lambda_n$ et un des 3 ensembles précédents notés de façon générique Z_G, existe-t-il $A \in Z_G$ dont le spectre soit la suite précédente ?

Je vais commenter ce programme :

1) Tel quel le problème est mal posé (ou trop difficile), car non monotone par rapport à la complexité du graphe: la réponse est triviale pour un graphe sans arête, car $O_G = M_G$ est l'ensemble des matrices diagonales réelles, tout spectre est alors possible pour un opérateur convenable de O_G.

On sait par ailleurs que, si G est connexe et $A \in O_G$ la première valeur propre est de multiplicité 1 (Perron-Frobenius).

2) On va insister sur l'aspect ensemble d'opérateurs: ce n'est pas tant un opérateur isolé qui va nous intéresser, mais la position relative de Z_G et d'une variété W donnée par des conditions spectrales, par exemple $\lambda_2 = \lambda_3$. Cela permet d'introduire des idées de transversalité et donc de gagner une monotonie: tout ce qu'on peut faire avec un graphe G doit pouvoir être fait avec tout graphe G' plus complexe que G. On peut aussi rapprocher cela de la théorie des discriminants (Arnold, Vassiliev), qui consiste à regarder les objets singuliers de l'ensemble considéré, ici l'ensemble des opérateurs dont au moins une valeur propre est multiple.

3) Les arguments de transversalité permettent ainsi de construire des nouveaux invariants des graphes qui sont reliés à la théorie classique des graphes: plongements dans les surfaces, arbres, mineurs.

4) Un argument simple permet de montrer que toute suite $\lambda_1 < \cdots < \lambda_n$ de n nombres distincts est le spectre d'un $A \in O_G$:

on considère les applications

$$\Phi_\varepsilon : \{u_1 < u_2 < \cdots < u_n\} \to \lambda_1(\varepsilon) < \cdots < \lambda_n(\varepsilon)$$

où les $\lambda_i(\varepsilon)$ sont les valeurs propres (simples pour $\varepsilon > 0$ petit) de la forme quadratique

$$q_{\varepsilon,u} = \sum u_i x_i^2 + \varepsilon \sum_{\{i,j\} \in E} (x_i - x_j)^2 \ .$$

Pour $\varepsilon = 0$, $\Phi_\varepsilon = Id$ et donc l'application du théorème des fonctions implicites donne le résultat.

Le problème se concentre donc sur les multiplicités.

On notera $W_{l,k}$ la sous-variété de $S(H)$ formée des opérateurs A dont le spectre vérifie :

$$\cdots \leqslant \lambda_{k-1} < \lambda_k = \cdots = \lambda_{k+l-1} < \lambda_{k+l} \leqslant \cdots .$$

On note $W_l = \cup_k W_{l,k}$.

On note aussi $m_k^Z(G)$ le plus grand l tel que $Z_G \cap W_{l,k} \neq \emptyset$.

Par exemple $m_1^O(G)$ est le nombre de composantes connexes de G ; $m_2^O(G)$ est déjà un invariant moins trivial (voir plus loin).

Exemple: 1.1 les chemins.

Soit P_n le chemin à n sommets $\{1, \cdots, n\}$ et les arêtes $\{i, i+1\}$, $1 \leqslant i \leqslant n-1$. Toutes les valeurs propres d'un opérateur quelconque de M_{P_n} sont simples, car un vecteur propre x est entièrement déterminé par x_1.

Exemple: 1.2 les graphes cycliques.

Les sommets de ce graphe C_n sont les éléments de $\mathbb{Z}/n\mathbb{Z}$ et les arêtes joignent i à $i \pm 1$ (mod n). On alors, pour tout $A \in O_{C_n}$ ([21]) :

$$\lambda_1 < \lambda_2 \leqslant \lambda_3 < \lambda_4 \leqslant \lambda_5 < \cdots .$$

Listes de graphes vus plus loin:

$$P_n,\ C_n,\ K_n,\ E_{n+1},\ K_{3,3},\ S_{2n-1},\ T_{\frac{n(n+1)}{2}} ,$$

où l'indice désigne le nombre de sommets.

2 Stabilité structurelle

1 Perturbations des valeurs propres multiples

Soit H un espace euclidien de dimension n et $\varepsilon \to A_\varepsilon$ une application C^2 d'un voisinage de 0 à valeurs dans $S(H)$ l'espace des endomorphismes symétriques de H.

Soit λ_0 une valeur propre de A_0 et $E_0 \subset H$ l'espace propre associé de dimension l.

Soit I un intervalle fermé de centre λ_0 et tel que $\sigma(A_0) \cap I = \{\lambda_0\}$; alors pour ε assez petit, I contient exactement l valeurs propres de A_ε (comptées avec multiplicités) et notées :

$$\lambda_1(\varepsilon) \leqslant \cdots \leqslant \lambda_l(\varepsilon) .$$

Soit $B = \frac{d}{d\varepsilon}_{|\varepsilon=0} A_\varepsilon$ et q_B la forme quadratique sur E_0 définie par

$$q_B(x) = < x|B|x > \quad ,$$

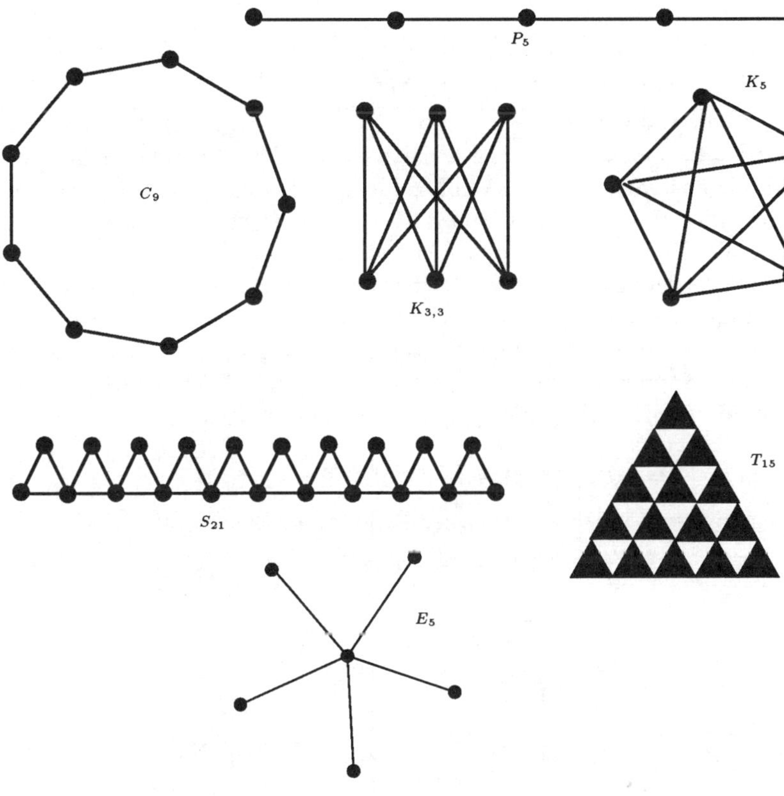

Figure 1: graphes

et $\mu_1 \leqslant \cdots \leqslant \mu_l$ le spectre de q_B.

Alors, on a [21]:

$$\forall i \text{ tel que } 1 \leqslant i \leqslant l,\ \lambda_i(\varepsilon) = \lambda_0 + \varepsilon\mu_i + O(\varepsilon^2)\ .$$

Soit $W_l^{\lambda_0} = \{A \in S(H) \mid \dim\ker(A - \lambda_0) = l\}$, alors $W_l^{\lambda_0}$ est une sous-variété de $S(H)$ dont l'espace tangent $T_{A_0}W_l^{\lambda_0}$ est formé des $B \in S(H)$ tels que $q_B = 0$. En particulier, $W_l^{\lambda_0}$ est de codimension $\frac{l(l+1)}{2}$.

Notations :

on note aussi $W_l^{\lambda_0} = \cup_k W_{l,k}^{\lambda_0}$, où $A \in W_{l,k}^{\lambda_0}$ si $\lambda_{k-1}(A) < \lambda_k(A) = \cdots = \lambda_{k+l-1}(A)\ (= \lambda_0) < \cdots$, et $W_{l,k} = \cup_{\lambda_0} W_{l,k}^{\lambda_0}$.

$W_{l,k}$ est de codimension $\frac{l(l+1)}{2} - 1$.

Pour $l = 2$, cette dernière codimension vaut 2: l'observation que la condition d'avoir une valeur propre double pour une matrice symétrique réelle est de codimension 2 remonte aux années 1930 (Wigner-Von Neumann [41], voir aussi [1], [3]). Donc, si $t \to A_t$ est un chemin générique dans $S(H)$, il n'y aura de valeurs propres multiples pour A_t pour aucun t. On aura typiquement une évolution des valeurs propres faisant apparaitre des *croisements évités*.

Soit maintenant $\Phi\ : (u,v) \to A_{u,v}$ une famille à 2 paramètres de matrices symétriques. Supposons que Φ coupe tranversalement $W_{2,k}$ en (u_0, v_0) ; on a donc, en notant $A_0 = A_{u_0,v_0}$, le spectre de A_0 qui vérifie

$$\cdots \leqslant \lambda_{k-1} < \lambda_k = \lambda_{k+1} < \cdots\ ,$$

et le graphe simultané de λ_k et λ_{k+1} a l'allure d'un cône.

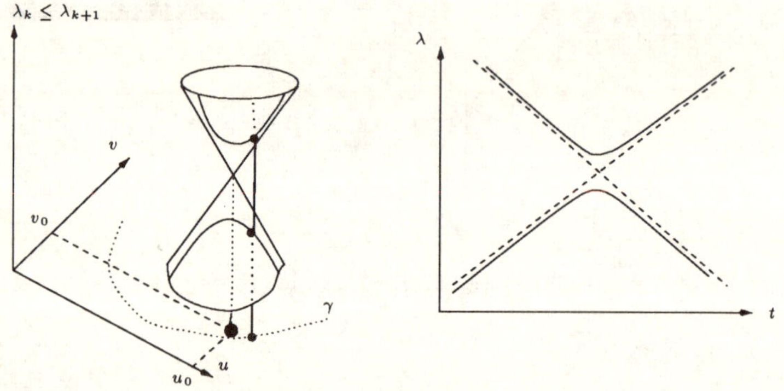

Figure 2: point diabolo

Soit γ un lacet du plan des (u,v) qui entoure (u_0, v_0) sans entourer d'autres points (u,v) avec $A_{u,v} \in W_{2,k}$, soit $\phi_k(t)$ un vecteur propre de $A_{\gamma(t)}$ correspondant à la valeur propre simple $\lambda_k(A_{\gamma(t)})$ supposé réel et de norme 1 et que l'on peut donc suivre par continuité autour de γ ; lorsqu'on fait un tour, $\phi_k(t)$ se transforme en $-\phi_k(t)$. Le fibré réel de dimension 1 donné par l'espace propre $\ker(A_{\gamma(t)} - \lambda_k(t))$ est donc non trivial (ruban de Möbius). Ce fait peut être mise à profit pour détecter des valeurs propres multiples dans une famille d'opérateurs dépendant de 2 paramètres. Voir [6]. La transversalité joue un grand rôle ici : sans elle, la monodromie pourrait être triviale !! C'est cette situation que nous allons généraliser maintenant.

Le cas hermitien complexe peut être traité de façon analogue. Il faut alors remplacer $\frac{l(l+1)}{2}$ par l^2, la dimension de l'espace des matrices hermitiennes $l \times l$. La variété $W_{2,k}$ est alors de codimension 3 et l'argument avec le lacet γ doit être remplacé par un argument avec une petite sphère plongée. Le fibré associé à une valeur propre simple est alors un fibré de rang 1 complexe dont la classe de Chern est 1 si la sphère entoure exactement un point de W_2. Voir [13] .

2 Stabilité structurelle des valeurs propres

On est ainsi conduit à donner une définition générale de la stabilité structurelle d'une valeur propre relativement à une variété Z d'opérateurs.

Définition 1 *Soit $A \in Z \subset S(H)$, où Z est une sous-variété. Supposons que $A \in Z_G \cap W_{l,k}$. On dit que la valeur propre λ_k de A est $Z-$structurellement stable si $W_{l,k}$ et Z se coupent transversalement en A. Autrement dit, l'apparition d'une valeur propre de multiplicité l dans Z est un phénomène stable par perturbation de Z.*

On peut tester de façon purement algébrique cette condition de stabilité si on connait l'espace propre :

si $\Phi : T_A Z \to S(E_0)$ (où E_0 est l'espace propre de A considéré) est donnée par

$$\Phi(B)(x) = < x|B|x > \quad ,$$

la condition de stabilité structurelle équivaut à la surjectivité de Φ.

3 Exemples

Exemple: 2.1 *Si $Z = O_G$, la première valeur propre λ_1 n'est stable que si elle est de multiplicité 1.*

Exemple: 2.2 *Soit E_N l'étoile à N branches, $A \in O_{E_N}$ où $-A$ est la matrice d'adjacence. La 2ème valeur propre de A est de multiplicité $N-1$ et n'est stable que si $N \leqslant 3$.*

Exemple: 2.3 graphes complets. *Il s'agit des graphes K_n dont toute paire de sommets est connectée par une arête. Si A est le laplacien canonique, la seconde valeur propre est de multiplicité $n-1$ et $O_{K_n}-$stable et même $L_{K_n}-$stable.*

Exemple: 2.4 graphe de Kuratowski. *Il s'agit du graphe noté $K_{3,3}$ dont l'ensemble des sommets est $V_1 \cup V_2$ (union disjointe), avec $|V_i| = 3$, et les 9 arêtes joignent les sommets de V_1 à ceux de V_2.*

Si $A = \Delta_0$ est le laplacien canonique, son spectre est

$$0 < 3 = 3 = \cdots = 3 < 6 \ ,$$

et la valeur propre 3 de multiplicité 4 est $O_{K_{3,3}}-$stable.

Exemple: 2.5 cas complexe. *Dans le cas hermitien complexe ($A \in M_G$), le théorème de Perron-Frobenius ne s'applique pas et on peut regarder la stabilité de λ_1.*

Soit d'abord S_{2n-1} le graphe planaire formé de $n-1$ triangles recollés en une chaine tel que chacun a un sommet commmun avec le suivant. Soit A tel que la forme quadratique associée est

$$q(x) = \sum |x_i + x_j + x_k|^2 \ ,$$

où la somme porte sur les triangles $\{i, j, k\}$. On voit facilement que $\ker A = q^{-1}(0)$ est de dimension n et n'est pas M_G-stable.

Soit ensuite $T_{n(n+1)/2}$ le graphe formé en subdivisant un triangle équilatéral en petits triangles équilatéraux grâce à une subdivision des côtés du triangle initial en $n-1$ intervalles égaux. Colorions en noir ou en blanc les petits triangles suivant qu'ils pointent vers le haut ou le bas. La même recette que précédemment, en sommant sur les triangles noirs, donne lieu a un fondamental de multiplicité n qui est M_G-stable.

4 Les invariants $\mu_k^Z(G)$

On définit maintenant une famille d'invariants numériques d'un graphe G grâce à une version stabilisée de $m_k^Z(G)$.

Définition 2 *Si Z_G est un des 3 ensembles d'opérateurs M_G, O_G, L_G définis plus haut, on définit*

$$\mu_k^Z(G)$$

comme le sup des entiers l tels qu'il existe $A \in Z_G$ avec $\cdots < \lambda_k = \cdots = \lambda_{k+l-1} < \cdots$ et $\lambda_k(A)$ est Z_G-stable.

Exemple: 2.6 *Pour tout G, $\mu_1^O(G) = 1$.*

Exemple: 2.7 $\mu_2^O(K_n) = n - 1$.

Exemple: 2.8 $\mu_2^M(T_{n(n+1)/2}) = n$.

Exemple: 2.9 $\mu_2^O(K_{3,3}) = 4$.

Exemple: 2.10 $\mu_2^O(E_3) = 2$ *et, pour* $N \geqslant 2$, $\mu_2^O(E_N) = 2$.

Exemple: 2.11 $\mu_k^O(C_n) = 1$ *si* k *est impair et* 2 *si* k *est pair.*

3 Mineurs des graphes et limites singulières d'opérateurs

1 Monotonie par mineurs

La propriété principale des invariants introduits précédemment est la monotonie par mineur.

Définition 3 *Un graphe* $G_1 = (V_1, E_1)$ *est dit mineur de* $G_2 = (V_2, E_2)$ *si* G_1 *peut être construit à partir de* G_2 *de la façon suivante : si*

$$V_2 = \cup_{\alpha \in B} W_\alpha$$

est une partition de V_2 *en sous-ensembles connexes,* V_1 *est un sous-ensemble de* B *et* E_1 *vérifie la condition suivante :*

$$\{\alpha, \beta\} \in E_1 \text{ implique qu'il existe } i \in W_\alpha,\ j \in W_\beta \text{ tels que } \{i, j\} \in E_2 .$$

On peut aussi dire que G_1 est un mineur de G_2 si on passe de G_2 à G_1 par une suite finie des opérations élémentaires suivantes : ôter une arête, contracter une arête, ôter un sommet isolé.

On a alors le :

Théorème 1 *Pour chaque ensemble* $Z = M, O$, *si* G_1 *est un mineur de* G_2 *:*

$$\mu_k^Z(G_1) \leqslant \mu_k^Z(G_2) .$$

Pour montrer ce théorème, il suffit de le montrer dans les cas particuliers où l'on ôte ou contracte une arête de G_1.

Le cas où l'on ôte une arête est simple, car $H_{G_1} = H_{G_2}$. On utilise alors la conservation d'une intersection transversale par perturbation C^1.

Donnons l'argument plus précis dans le cas de O_G. Soit H l'espace de Hilbert considéré et supposons que l'on a numéroté les sommets de G_2 de façon que G_1 s'obtienne en ôtant l'arête $\{1, 2\}$ de G_2. Soit $Z_\epsilon = O_{G_1} + \epsilon(x_1 - x_2)^2$ (où on a écrit les opérateurs comme des formes quadratiques), alors $Z_\epsilon \subset O_{G_2}$ pour $\epsilon > 0$.

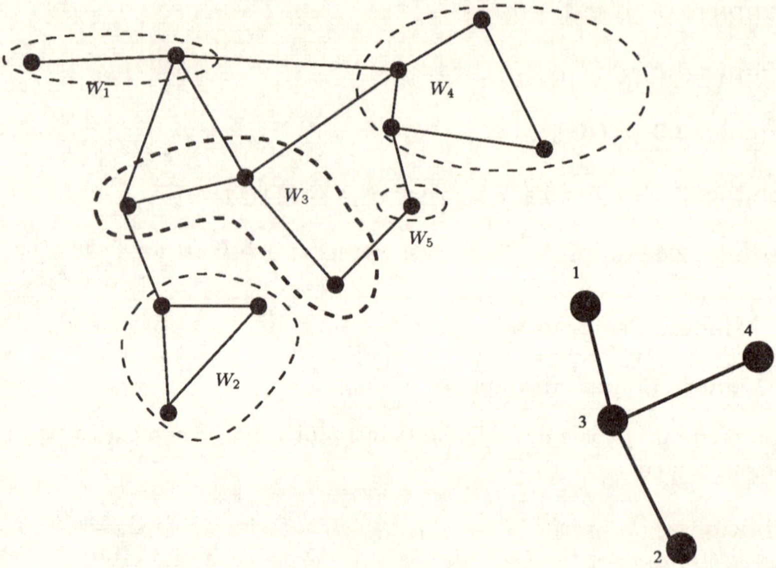

Figure 3: mineurs

On conclut ainsi: si $Z_0 = O_{G_1}$ et $W_{l,\lambda}$ se coupent transversalement en A, il en est de même pour Z_ϵ et $W_{l,\lambda}$ en A_ϵ et donc de O_{G_2} et $W_{l,\lambda}$ en A_ϵ. Il reste à vérifier que le numéro de la valeur propre λ ne change pas.

Le cas où l'on contracte l'arête $\{1, 2\}$ est plus délicat, car il n'est plus vrai que $H_{G_1} = H_{G_2}$; on a seulement une inclusion $H_{G_1} \subset H_{G_2}$ en identifiant H_{G_1} au sous-espace de $\mathbb{C}^{V_2}$ dont les éléments sont les vecteurs x tels que $x_1 = x_2$.

On doit pour faire marcher l'argument précédent mettre dans une même variété les opérateurs non partout définis. C'est le but du § suivant. Une autre difficulté est que la structure hilbertienne du sous-espace de H_{G_2} n'est pas la structure canonique de H_{G_1}.

Cela empêche de montrer le théorème pour $Z = L$.

Pour les autres ensembles, on se débrouille par changement de jauge ou plus simplement par translation par λId qui ramène à la valeur propre 0 indépendamment de la structure hilbertienne considérée.

2 Γ−convergence

Soit H un espace euclidien de dimension finie n. Soit A un opérateur symétrique sur H. On note q_A la forme quadratique associée.

Soit $\Gamma(A) = \{(x, Ax) \mid x \in H\} \subset H \oplus H$ le graphe de A.

Soit $Y \subset H$ un sous-espace euclidien de H et $B : Y \to Y$ un opérateur symétrique sur Y, on dira que le couple (Y, B) est un opérateur non borné sur H si $Y \neq H$. On étend la définition du graphe aux opérateurs non bornés de la façon suivante.

$$\Gamma(Y, B) = \{(y, z) \in Y \oplus H \mid \forall w \in Y, < z|w > = < By|w > \} .$$

Comme cas particulier, $\Gamma(A) = \Gamma(H, A)$. Il est clair que $\Gamma(Y, B)$ est toujours un sous-espace de dimension n de $H \oplus H$.

Soit ω la forme symplectique définie par

$$\omega((x, y), (x', y')) = < y'|x > - < x'|y > .$$

Un sous-espace de $H \oplus H$ de dimension n, qui est isotrope pour la forme symplectique ω est dit *lagrangien.*

On vérifie que tout sous-espace lagrangien est le graphe d'un unique opérateur symétrique, borné ou non, et que réciproquement tout graphe $\Gamma(Y, B)$ est lagrangien. L'ensemble des opérateurs symétriques avec domaine sur H est ainsi une compactification Λ_H de l'espace $S(H)$ des endomorphismes symétriques de H en une variété compacte de classe C^∞.

On dira que (Y, B) est *inversible* si $\sigma(\Gamma(Y, B))$, où $\sigma(x, y) = (y, x)$, est le graphe d'une application de H dans $Y \subset H$.

Pour $\lambda \in \mathbb{C}$, on définit

$$(Y, \lambda I - B)$$

par son graphe

$$\Gamma(Y, \lambda I - B) = \{(y, \lambda y - \zeta) \ , \ (y, \zeta) \in \Gamma(Y, B)\} .$$

Ceci définit du même coup les deux notions de résolvante (inverse de $\lambda I - B$) et de spectre (valeurs singulières de λ). Ainsi le spectre de (Y, B) est réunion du spectre de B et de la valeur propre ∞ répétée $(n - \dim Y)$ fois.

Remarque : seul un ordre cyclique sur les valeurs propres est pertinent ; on ne peut pas distinguer $-\infty$ de $+\infty$ sans perdre la continuité du spectre. Si A décrit un lacet de Λ_H les valeurs propres font collectivement un nombre de tours égal à l'indice de Maslov du chemin ([4]).

Définition 4 *On dira qu'une famille A_ε, $\varepsilon > 0$ d'opérateurs symétriques sur H $\Gamma-$converge vers (Y, B) si $\Gamma(A_\varepsilon)$ converge vers $\Gamma(Y, B)$ quand $\varepsilon \to 0$ au sens de la topologie naturelle de la grassmannienne des sous-espaces de dimension n de $H \oplus H$.*

On note

$$A_\varepsilon \xrightarrow{\Gamma} (Y, B) .$$

Le résultat suivant étend au cadre de la $\Gamma-$convergence les résultats classiques de Kato [32] sur la convergence des spectres de formes quadratiques.

Théorème 2 *Supposons* $A_\varepsilon \xrightarrow{\Gamma} (Y,B)$. *Soient*

$$\lambda_1(\varepsilon) \leqslant \cdots \leqslant \lambda_n(\varepsilon)$$

les valeurs propres de A_ε. *Alors le spectre de* A_ε *converge vers celui de* (Y,B) *au sens de la topologie de* $P^1(\mathbb{R}) = \mathbb{R} \cup \infty$.

L'existence de $\Gamma-$limites n'est pas exceptionnelle.

Proposition 1 *Soient* $A_i \ : X \to X$ *des applications linéaires symétriques,* $\alpha_1 < \cdots < \alpha_N$ *et, pour* $\varepsilon > 0$:

$$A_\varepsilon = \sum_{i=1}^{N} A_i \varepsilon^{\alpha_i} \ .$$

Soit Z *la grassmannienne des sous-espaces lagrangiens de dimension* n *de* $X \oplus X$ *et* $C_\varepsilon \in Z$ *le graphe de* A_ε.

Alors l'application $\varepsilon \to C_\varepsilon$ *de* $\mathbb{R}^+ \setminus 0$ *dans* Z *se prolonge par continuité en* $\varepsilon = 0$.

De plus, si les α_i *sont entiers, le prolongement est analytique réel.*

La $\Gamma-$convergence étant assurée dans de nombreux cas d'intérêt pratique, on cherchera à caractériser une $\Gamma-$limite (Y,B) en décrivant le domaine Y et les coefficients de la matrice B.

Dans toute la suite, les opérateurs symétriques sont toujours supposés positifs.

Définition 5 *On note* $x_\varepsilon \xrightarrow{\mathrm{s}} x_0$ *si* x_ε *converge vers* x_0 *avec* $Q_\varepsilon(x_\varepsilon) = O(1)$.

Lemme 1 *Si* $A_\varepsilon \xrightarrow{\Gamma} (Y,B)$, $x_0 \in Y$ *ssi il existe* x_ε *tel que* $x_\varepsilon \xrightarrow{\mathrm{s}} x_0$.

Lemme 2 *Soit* Y_ε *un sous-espace vectoriel de dimension* p *de* X *tel que* Y_ε *tend vers* Y *et que, pour* $\varepsilon \neq 0$, *la norme de la restriction de* A_ε *à* Y_ε *est bornée.*

Soit ϕ_ε *un isomorphisme de* Y *dans* Y_ε *tel que pour tout* $y \in Y$, *lorsque* $\varepsilon \to 0$, $\phi_\varepsilon(y)$ *tend vers* y *et*

$$Q_\varepsilon(y) = < \phi_\varepsilon(y) | A_\varepsilon | \phi_\varepsilon(y) > \ .$$

Alors Q_ε *admet une limite* Q_0 *comme forme quadratique sur* Y *et si* B *est l'opérateur associé, on a :*

$$A_\varepsilon \xrightarrow{\Gamma} (Y,B) \ .$$

Remarquons que l'hypothèse que la norme de la restriction de A_ε est bornée entraîne que $\phi_\varepsilon(y) \xrightarrow{s} y$. Une version pratique de ce lemme est la suivante : soit (y_i) une base de Y et $y_{i,\varepsilon}$ convergents vers y_i avec $A_\varepsilon y_{i,\varepsilon} = O(1)$, alors la matrice de $b_{i,j} =< y_i|B|y_j >$ de B est donnée par :

$$b_{i,j} = \lim < y_{i,\varepsilon}|A_\varepsilon|y_{j,\varepsilon} > \ .$$

Attention : il ne faut pas prendre pour définition de Q_ε

$$Q_\varepsilon(y) =< y|A_\varepsilon|y > ,$$

qui pourrait ne pas avoir de limite ou une limite incorrecte !

Exemple: 3.1 contraction d'une arête.

Soit $q_\varepsilon(x) = Q(x) + \frac{1}{\varepsilon}(x_1 - x_2)^2$. Lorsque $\varepsilon \to 0$, $q_\varepsilon \xrightarrow{\Gamma} (Y, B)$ où $Y = \{x_1 = x_2\}$ et q_B est la restriction de Q à Y.

Cet exemple permet de traiter la contraction d'une arête comme le cas où l'on ôte une arête.

Exemple: 3.2 transformations étoile-triangle.

Voir [5] pour ce qui suit.

Plaçons-nous sur $H = \mathbb{R}^4$, avec la forme quadratique

$$Q_\varepsilon(x) =< x|A_\varepsilon|x >= \sum_{i=1}^{3}(x_i - \frac{x_0}{\varepsilon})^2 \ .$$

Prenons $p = 3$ et $Y_\varepsilon = \{y_0 = (\varepsilon/3)(y_1+y_2+y_3)\}$. Alors $Q_0(y) = \dfrac{1}{3}\displaystyle\sum_{i=1}^{3}(y_i - y_{i+1})^2$ et non $\displaystyle\sum_{i=1}^{3} y_i^2$.

On dit que G' s'obtient de G par une transformation *étoile-triangle* si on remplace un sommet 0 de degré 3 de G et les arêtes $\{0, i\}$, $1 \leqslant i \leqslant 3$, qui s'y accrochent par le triangle $(1, 2, 3)$.

$O_{G'}$ est alors une strate de l'adhérence de O_G dans Λ_{H_G}.

Exemple: 3.3 recuit simulé.

Une présentation détaillée de cet exemple est donnée dans [12].

Soit $G = (V, E)$ un graphe connexe fini et $H : V \to \mathbb{N}$ une fonction (énergie). On suppose $V_0 \neq \emptyset$. On suppose aussi que, si $\{i, j\} \in E$, $|H(i) - H(j)| \leqslant 1$.

Soit $T > 0$ et $\varepsilon = e^{-\frac{1}{2T}}$. On considère la mesure de probabilité de Gibbs π_T sur V, à température T donnée par $p_i = \frac{1}{Z}e^{-\frac{H(i)}{T}}$. Lorsque $T \to 0^+$ cette

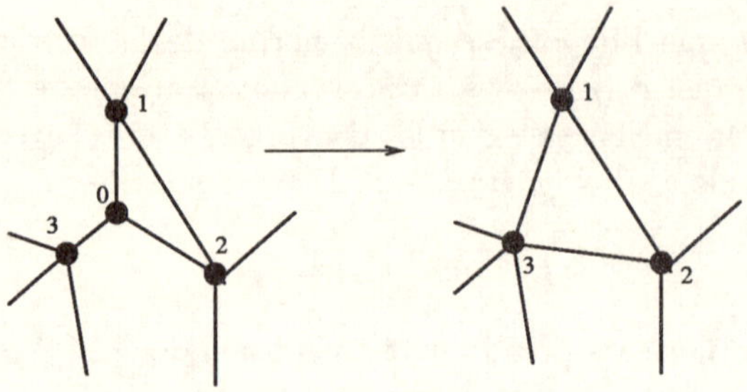

Figure 4: transformation étoile-triangle

mesure de probabilité converge vers la probabilité uniforme sur l'ensemble V_0 des minimas absolus de H.

On peut donc espérer trouver ces minimas en simulant un processus de Markov ayant π_T comme mesure d'équilibre et en faisant tendre T vers 0 à une vitesse judicieusement choisie: c'est le principe de base de l'algorithme dit de *recuit simulé*.

On définit le processus de Markov par un opérateur Λ_T défini par

$$(\Lambda_T x)_i = \sum_{j\sim i} \lambda_{i,j}(x_i - x_j) ,$$

avec $\lambda_{i,j} = 0$ si $\{i,j\} \notin E$ et :

$$\lambda_{i,j} = \begin{cases} 1, & \text{si } H(j) \geqslant H(i) \ ; \\ e^{-\frac{H(j)-H(i)}{T}}, & \text{si } H(j) > H(i) \ . \end{cases} \tag{1}$$

L'opérateur Λ_T est symétrique sur $l^2(V, \pi_T)$. Posant $y_i = \sqrt{p_i}x_i$, on obtient un opérateur de O_G noté A_ε, unitairement équivalent à Λ_T, associé à la forme quadratique

$$Q_\epsilon(x) = \sum_{\{i,j\}\in E,\ H(i)<H(j)} (\epsilon^{H(j)-H(i)}x_i - x_j)^2 + \sum_{\{i,j\}\in E,\ H(i)=H(j)} (x_i - x_j)^2 \ .$$

On note

$$\lambda_1(\epsilon) = 0 < \lambda_2(\epsilon) \leqslant \cdots \leqslant \lambda_n(\epsilon) \ ,$$

les valeurs propres de Λ_ε (et donc de A_ε).

On souhaite donner une description précise du comportement asymptotique des $\lambda_i(\epsilon)$ lorsque ε tend vers 0, et en particulier du gap $g(\varepsilon) = \lambda_2(\varepsilon)$.

On a une filtration

$$\cdots \subset \mathcal{H}_l \subset \cdots \subset \mathcal{H}_0 = \mathbb{R}^V$$

où $\mathcal{H}_l$ est l'espace engendré par les fonctions propres de valeurs propres $O(\varepsilon^{2l})$.

Le but est de décrire cette filtration et l'asymptotique des valeurs propres associées. L'existence de la filtration résulte de Kato et du fait que $Q_\varepsilon \geqslant 0$ est isospectral à $Q_{-\varepsilon}$: en effet si $U : \mathcal{H}_0 \to \mathcal{H}_0$ est l'opérateur unitaire qui consiste à changer x_i en $-x_i$ lorsque $H(i)$ est impair, on a: $Q_\varepsilon(Ux) = Q_{-\varepsilon}(x)$.

$\varepsilon^{-2l}Q_\varepsilon$ admet une Γ−limite $R_l = (\mathcal{H}_l, S_l)$ qui s'identifie à un élément de O_{G_l} que l'on va décrire.

Décrivons maintenant les graphes G_l et les espaces $\mathcal{H}_l$ (les domaines) par récurrence sur l :

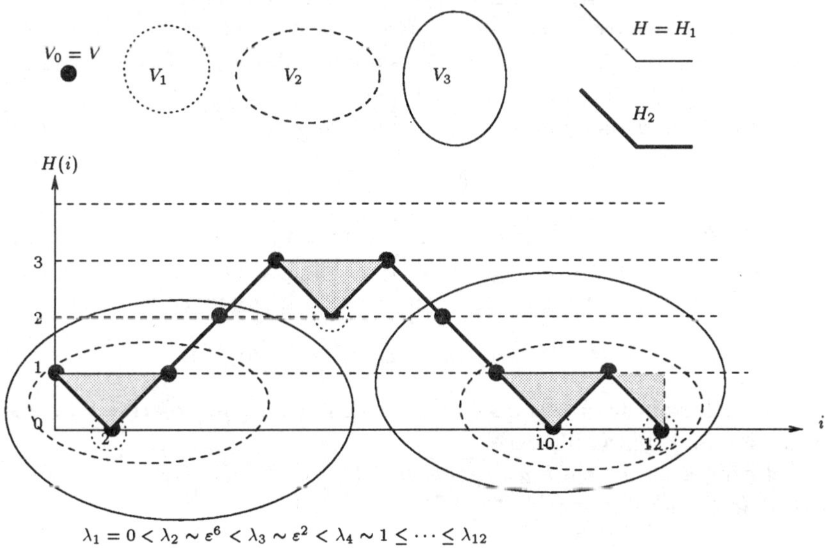

Figure 5: les graphes G_l

(i) $G_1 = (V_1, E_1)$ est ainsi défini: $i \in V$ sera dit minimum local de H s'il existe $A \subset V$ connexe, $i \in A$ tel que H est constante sur A et vaut $H(i) + 1$ sur les voisins d'un sommet de A qui ne sont pas dans A. L'ensemble des ces parties connexes A est l'ensemble V_1. Si $\alpha \in V_1$, on note $A_\alpha \subset V$ la composante connexe du minimum local correspondante: les A s'appelle puits d'ordre 1.

$\mathcal{H}_1$ est l'espace des fonctions constantes sur les $A \in V_1$ et nulles ailleurs. Sa dimension est bien sûr $\#V_1$.

(ii) On remplace alors $H = H_1$ par $H + 1$ sur chacune de ces parties de V_1, on obtient une énergie H_2 et on construit ainsi un ensemble de parties V_2 en considérant de même les minimas locaux de H_2; les sommets d'une composante d'un minima de H_2 s'appellent puits d'ordre 2; 2 éléments A, B

de V_1 (éventuellement confondues) sont connectés par une arête de G_1 s'ils sont contenus dans un même puits C de V_2: le graphe G_l est donc un graphe avec des boucles.

Et on itère la construction:

(ii) on remplace H_2 par $H_2 + 1$ sur les ensembles de V_2, etc...

Soit $Q_l = \lim_{\varepsilon \to 0} \varepsilon^{-2l} Q_\varepsilon$, alors Q_l s'identifie naturellement à un opérateur de type Schrödinger sur G_l avec un potentiel $\geqslant 0$ nul seulement aux sommets qui ont des boucles.

De ceci, on peut calculer le nombre de valeurs propres qui sont $O(\varepsilon^{2l})$ qui est donné par le nombre de sommets de G_l, ie le nombre de puits d'ordre l.

En fait les formes Q_l sont déterminées entièrement en restreignant Q_ϵ à un sous-graphe de G: celui formé des sommets des $A \in V_l$ et des sommets inondés si on augmente H_l de η, $\eta > 0$ petit sur les $A \in V_l$.

Tout ceci détermine l'ordre de grandeur du gap:

$$g(\epsilon) = C\epsilon^{2k}(1 + O(\epsilon^2)) ,$$

où k est la *hauteur du col le plus haut* ; un col est un chemin γ entre 2 minimas locaux a et b dont l'altitude maximale $h = \sup_{x \in \gamma} H(x)$ est minimale parmi tous les chemins qui joignent ces 2 minimas. La hauteur de ce col est

$$H(a, b) = \inf(h - H(a), h - H(b)) .$$

k est aussi le sup des l tels qu'il existe au moins 2 puits d'ordre l (= 0 s'il n'existe qu'un minimum local).

Cette description n'est pas nouvelle. On la déduit classiquement de la théorie de Freidlin et Wentzell ([27]).

Exemple: 3.4 limites croissantes.

Dans le cas particulier d'une suite croissante de formes quadratiques positives, la notion de Γ−limite se ramène à celle de limite simple.

Proposition 2 *Soit A_n une suite croissante de formes quadratiques positives définies sur X et $q_n(x) =< x|A_n|x >$ la suite croissante des formes quadratiques associées. Soit*

$$q_\infty : X \to \mathbb{R} \cup +\infty$$

la limite simple des q_n, et

$$Y = \{x \in X \mid q_\infty(x) < \infty\} .$$

Soit (Y, B) la forme quadratique avec domaine associée (si $y \in Y$, $q_\infty(y) =< y|B|y >$). Alors

$$A_n \xrightarrow{\Gamma} (Y, B) .$$

On en déduit en particulier le :

Théorème 3 *L'adhérence dans Λ_{H_G} de l'ensemble L_G est la réunion des $L_{G'}$ pour G' mineur de G (attention la structure euclidienne sur les $\mathbb{R}^{V'}$ n'est pas la structure canonique).*

Cette proposition est en particulier utilisée dans [19] et [10] pour montrer que la réponse d'un réseau électrique varie continument lorsque le réseau dégénère.

Exemple: 3.5 réseaux électriques.

Un *réseau électrique* est un quadruplet $R = (V_0, V_1, E, \rho)$ où V_0 et V_1 sont 2 ensembles finis disjoints, $G = (V = V_0 \cup V_1, E)$ est un graphe fini et $\rho = (\rho_{i,j}) \in (\mathbb{R}^+ \setminus 0)^E$ est l'ensemble des conductances des arêtes. V_0 est l'ensemble des sommets terminaux, V_1 l'ensemble des sommets intérieurs. L'énergie électrique d'un potentiel $u \in \mathbb{R}^V$ est donnée par

$$e(u) = \frac{1}{2}(\sum_{\{i,j\}\in E} \rho_{i,j}(u_i - u_j)^2) .$$

Un *potentiel d'équilibre* de R est un élément de $u = (u_i) \in \mathbb{R}^V$ tel que, pour tout $i \in V_1$,

$$\sum_{\{i,j\}\in E} \rho_{i,j}(u_i - u_j) = 0 .$$

Les réponses de R sont les couples $(u_0, w_0) \subset \mathbb{R}^{V_0} \oplus (\mathbb{R}^{V_0})^\star$, où u_0 est la restriction d'un potentiel d'équilibre u à V_0 et $(w_0)_i = \sum_{\{i,j\}\in E} \rho_{i,j}(u_i - u_j)$. On note $\Phi(R)$ l'ensemble des réponses de R.

Si $L_e \subset \mathbb{R}^V \oplus (\mathbb{R}^V)^\star$ est le graphe de la différentielle de e, l'ensemble des réponses est obtenue de L_e par réduction symplectique et donc $\Phi(R)$ est un sous-espace lagrangien de $\mathbb{R}^{V_0} \oplus (\mathbb{R}^{V_0})^\star$.

Soit maintenant $\rho \in [0, +\infty]$. A ce ρ, on associe un nouveau réseau électrique (G', ρ') dont le graphe G' sous-jacent est le mineur de G obtenu en contractant les arêtes telles que $\rho_{i,j} = \infty$ et en ôtant celles telles que $\rho_{i,j} = 0$. ρ' est défini à partir de e par restriction à $\mathbb{R}^{V'} \subset \mathbb{R}^V$.

On montre ([19]) que l'application Φ s'étend continument comme application de $[0, \infty]^V$ dans la grassmannienne lagrangienne de $\mathbb{R}^{V_0} \oplus (\mathbb{R}^{V_0})^\star$.

Exemple: 3.6 limites d'éléments de O_G.

Le théorème suivant montre que toute $\Gamma-$ limite d'éléments de O_G est de façon naturelle élément d'un $O_{G'}$, où G' est un mineur généralisé de G. On notera $G' << G$ cette relation d'ordre.

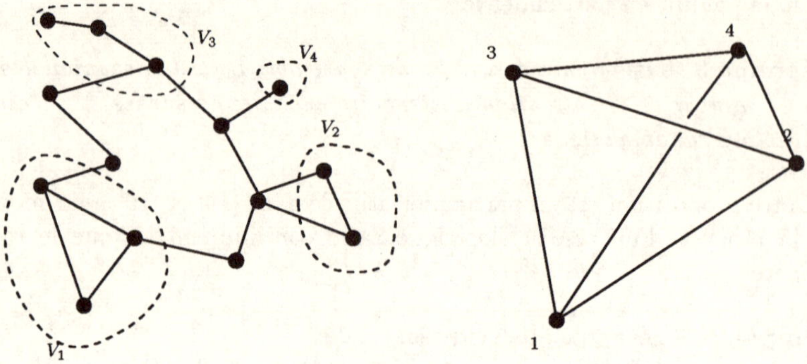

Figure 6: G et G'

Théorème 4 *Soit $A_\varepsilon \in O_G$. Supposons que $A_\varepsilon \xrightarrow{\Gamma} (Y, B)$ avec $p = \dim Y$.*
Description de Y :
il existe une partition (définie par A_ε)

$$V = W_0 \cup V_1 \cup \cdots \cup V_p \ ,$$

où les V_i, $i \geqslant 1$ sont connexes, telle que si $F_i = \{y \in Y t.q.\ Supp(y) \subset V_i\}$, les F_i, $i \geqslant 1$ sont de dimension 1 et $Y = \oplus_{i=1}^{p} F_i$.

Chaque F_i est engendré par une fonction $\varphi_i \geqslant 0$ de norme 1. Ces ϕ_i forment donc une base orthonormée de Y.

Description de B :

la matrice $(b_{i,j})$ de B dans la base des ϕ_i est un élément de O_{G_0} où $V(G_0) = \{1, \cdots, p\}$ et, si $\{i,j\} \in E(G_0)$, il existe un chemin γ de G qui joint V_i à V_j sans rencontrer les autres V_l.

On a donc les inclusions :

$$\cup_{G'<G} O_{G'} \subset \bar{O}_G \subset \cup_{G"<<G} O_{G"} \ .$$

3 Stabilité structurelle pour les opérateurs non bornés

On se ramène au cas où $\lambda_0 = 0$. L'adhérence dans Λ_H de W_l^0 est une sous-variété lisse. En effet, si $L = H \oplus 0$, $A \in W_l^0$ si et seulement si $\Gamma(A) \cap L$ est de dimension l. L'adhérence de W_l^0 est donc l'ensemble des espaces lagrangiens définis par cette condition de dimension. C'est une sous-variété lisse de Λ_H.

Pour l'étude de ces variétés et de la stratification correpondante, voir [2].

On peut ainsi étendre aux opérateurs non bornés la théorie de Kato et la notion de stabilité structurelle.

4 Plongements de graphes dans les surfaces

1 Le théorème de Cheng et sa réciproque

Dans [7], Cheng a prouvé le résultat suivant :

Théorème 5 *Si $A = \Delta_g + V$ est un opérateur de Schrödinger sur S^2 munis d'une métrique riemannienne g, la multiplicité de $\lambda_2(A)$ est inférieure ou égale à 3 ; cette borne est optimale (cas $V = 0$ et $g = g_{can}$).*

Les ingrédients de la preuve sont le théorème de Jordan et le théorème de Courant sur les domaines nodaux. La preuve s'appuie sur la structure locale des lignes nodales des fonctions propres.

Voici une esquisse de la preuve du théorème de Cheng: soit A un opérateur de Schrödinger sur S^2 tel que, si $F = \ker(A - \lambda_2(A))$, $\dim F \geqslant 4$.

Soit $x_0 \in S^2$, par un argument d'algèbre linéaire, il existe $\varphi \in F \setminus 0$ telle que

$$\varphi(x_0) = 0, \ \nabla\varphi(x_0) = 0 \ .$$

Soit $C = \varphi^{-1}(0)$. On montre que C est un graphe plongé dans S^2 dont tous les sommets sont de degré pair (les branches issues d'un sommet s forment un système équiangulaire dont le degré est $2p$ si p est l'ordre d'annulation de φ en s). La formule d'Euler montre alors que le nombre de composantes connexes de $S^2 \setminus C$ est $\geqslant 3$ dès que C a aumoins un sommet de degré $\geqslant 4$ ce qui est le cas pour le φ choisi.

Il reste à appliquer le théorème de Courant sur les ensembles nodaux qui affirme que, si $\varphi \in \ker(A - \lambda_k(A))$ où A est un opérateur de Schrödinger sur une variété quelconque X, alors le nombre de composantes connexes de $X \setminus \varphi^{-1}(0)$ est $\leqslant k$, ici 2.

L'extension du théorème de Courant au cas des graphes ne marche pas en toute généralité, mais H. van der Holst a remarqué qu'il s'applique pour $\varphi \in \ker(A - \lambda_2(A)) = F$, avec $A \in O_G$, pourvu que le support de φ soit minimal dans $F \setminus 0$.

Ce théorème a été généralisé aux graphes par Hein van der Holst ([38] et [39]) en le :

Théorème 6 *Si G est le 1-squelette d'une triangulation de S^2 et $A \in O_G$, la multiplicité de $\lambda_2(A)$ est au plus 3. Cette borne est optimale et atteinte pour la laplacien canonique de K_4.*

Utilisant le fait que tout graphe planaire est mineur du 1-squelette d'une triangulation de S^2, on en déduit que, si G est planaire, $\mu_2^O(G) \leqslant 3$.

En fait, en utilisant la théorème de Kuratowski qui caractérise les graphes non planaires comme ceux dont K_5 ou $K_{3,3}$ est mineur, on montre ([18]) le :

Théorème 7 *Un graphe G est planaire si et seulement si :*

$$\mu_2^O(G) \leqslant 3 .$$

2 Extension aux autres surfaces: Cheng, Besson, Nadirashvili, Sévennec

Des majorations de la multiplicité de la seconde valeur propre d'opérateurs de Schrödinger sur les surfaces ont été obtenues ensuite (voir par exemple [37]) ; par exemple, si S est orientable, la borne optimale est 6 pour le tore ; la borne prouvée est $4g$ en général, mais on ne sait pas si elle est optimale.

3 Les graphes ayant un plongement non-noué et le résultat de Lovász-Schrijver

Un généralisation intéressante de la classe des graphes planaires est la classe des graphes G admettant un plongement non noué dans $\mathbb{R}^3$, i.e. tel que 2 circuits disjoints quelconques de G ne soient pas entrelacés comme lacets de $\mathbb{R}^3$. L'exemple le plus simple est K_5 qui n'est pas planaire, mais n'admet aucune paire de circuits disjoints.

L'analogue du théorème de Kuratowski est que ces graphes admettent une caractérisation par mineurs exclus: les graphes exclus étant les transformés par étoile-triangle de K_6. On montre dans [5] que $\mu_2^O(G) = 5$ pour ces graphes et Lovász et Schrijver [31] ont réussi à montrer le joli :

Théorème 8 $\mu_2^O(G) \leqslant 4$ *si et seulement si G admet un plongement non-noué dans $\mathbb{R}^3$.*

5 Largeur d'arbre

Pour tout ce §, voir [23].

Les invariants $\mu_k^M(G)$ ne sont pas reliés aux propriétés topologiques des graphes. On peut s'en douter par analogie avec le cas continu: l'opérateur de Schrödinger avec champ magnétique constant $Bdx \wedge dy$, $B > 0$, dans le plan $\mathbb{R}^2$,

$$H = (\frac{1}{i}\partial_x - By)^2 + (\frac{1}{i}\partial_y)^2$$

admet un spectre discret

$$B < 3B < 5B < \cdots$$

où chaque valeur propre est de multiplicité infinie (ces valeurs propres sont appelées *niveaux de Landau* en physique). Les graphes importants à considérer ici sont les arbres !

1 Transformations de jauge

Soit $G = (V, E)$ un graphe et $D = (\varepsilon_i)_{i \in V}$ une matrice diagonale unitaire. Si $A = (a_{i,j}) \in M_G$, il en est de même pour $A_D = D^{-1}AD$ et $\sigma(A_D) = \sigma(A)$. On a :

$$(a_D)_{i,j} = \varepsilon_i^{-1}\varepsilon_j a_{i,j} \ .$$

Si $\gamma = (i_1, i_2, \cdots, i_k)$ ($\forall m$ *tel que* $1 \leqslant m \leqslant k$, $\{i_m, i_{m+1}\} \in E$), le produit

$$h(A, \gamma) = (-1)^k a_{i_1,i_2} \cdots a_{i_k,i_1}$$

vérifie :

$$h(A_D, \gamma) = h(A, \gamma) \ .$$

De plus, on a, bien sûr, $h(A, \gamma) > 0$ si $A \in O_G$ et la réciproque est vraie quitte à faire un changement de jauge.

Si $T = (V, E)$ est un arbre, il n'admet pas de circuit non triviaux et, pour tout $A \in M_T$, on peut trouver D tel que $A_D \in O_T$.

On en déduit, par application du théorème de Perron-Frobenius, que

$$m_1^M(T) = 1 \ .$$

2 Les $\mu_k^M(T)$

Les arguments suivants sont détaillés dans [23].

On a besoin des lemmes suivants :

Lemme 3 *Si* $G = (V, E)$ *est connexe et* $\{1, 2\} \in E$ *est telle que* $G' = (V, E \setminus \{1, 2\})$ *n'est pas connexe, si* $A \in M_G$ *et* $F = \ker A$, *si* $r \ : \ F \to \mathbb{C}^2$ *est définie par* $r(x) = (x_1, x_2)$, *l'image de* r *est de dimension* $\leqslant 1$.

Preuve.

Soit V_1 la composante connexe dans G' du sommet 1 et pour tout $x \in F$ soit x' la restriction à V_1 étendue par 0. On calcule explicitement le membred e droite de l'égalité :

$$0 =< Ax'|y' > - < x'|Ay' > \ ,$$

en fonctions des valeurs de x et y en 1 et 2.

On trouve :

$$0 = a_{1,2}(x_2\bar{y}_1 - x_1\bar{y}_2) \ ,$$

d'où la conclusion.

□

Lemme 4 *Soit T un arbre dont les sommets sont de degré $\leqslant 3$ et pour $\{i,j\} \in E(T)$, T' et T" les 2 arbres obtenus en ôtant l'arête $\{i,j\}$. On note F' et F" les restrictions à V' et V" de $\ker r$. Si $\dim F > 2$, il existe $\{i,j\} \in E(T)$ telle que $F' \neq 0$ et $F" \neq 0$.*

On en déduit la:

Proposition 3 *Si T est un arbre dont tous les sommets sont de degré leq3,*

$$\forall k,\ \mu_k^M(T) \leqslant 2\ .$$

Preuve.

Soit $x' \in F' \setminus 0$ et $x" \in F" \setminus 0$, si on prend ces 2 vecteurs normalisés comme premiers vecteurs d'une base de F, on contredit la transversalité: $\forall B \in T_A M_G$, $q_B(x', x") = 0$.

□

Du fait que tout arbre est mineur d'un arbre dont les sommets sont de degré $\leqslant 3$, on déduit le

Théorème 9 *Pour tout arbre T,*

$$\forall k,\ \mu_k^M(T) \leqslant 2\ .$$

3 Largeur d'arbre

Définition 6 *Soit G un graphe fini, on définit la largeur d'arbre de G, notée $la(G)$ comme le inf des entiers k tel que G est mineur d'un graphe $T \times K$ avec $|V(K)| = k$.*

On peut alors montrer, à l'aide d'une version vectorielle des arguments précédents, le :

Théorème 10 *Pour tout graphe G, on a :*

$$\mu_1^M(G) \leqslant la(G)\ ,$$

et, $\forall k$, :

$$\mu_k^M(G) \leqslant 2la(G)\ .$$

On peut aussi montrer que $\mu_1^M(G) = 1$ si et seulement si G est une forêt. H. van der Holst ([40]) a trouvé une caractérisation des graphes G tels que $\mu_1^M(G) \leqslant 2$ (ce sont exactement ceux tels que $la(G) \leqslant 2$) et de ceux tels que $\mu_1^M(G) \leqslant 3$.

4 Fonctions holomorphes discrètes

Les majorations précédentes sont optimales, même dans le cadre des des graphes planaires: on a

$$\mu_1^M(T_{\frac{n(n+1)}{2}}) = n = la(T_{\frac{n(n+1)}{2}}) .$$

En fait, la majoration de la largeur d'arbre par n est simple et la minoration résulte du théorème précédent.

On peut donner une version géométrique d'un opérateur A dont le fondamental est de mutiplicité n et M−stable au moyen des fonctions holomorphes discrètes.

On prend comme forme hermitienne sur $\mathbb{C}^V$

$$q_A(x) = \sum |(x_k - x_i) - \omega(x_j - x_i)|^2 ,$$

où la somme porte sur les triangles noirs (i, j, k) orientés et $\omega = e^{i\frac{\pi}{3}}$.

Le noyau de A corespond à une notion de fonction holomorphe de G dans $\mathbb{C}$. L'image de chaque triangle noir par un $x \in \ker A$ prolongé affinement est un triangle équilatéral direct.

6 Problèmes

1 Nombres chromatiques

Le nombre chromatique $\chi(G)$ du graphe G est le nombre minimum de couleurs nécessaires pour colorier les sommets d'un graphe de façon que 2 sommets voisins quelconques soient de couleurs différentes.

Le nombre chromatique $\chi(S)$ d'une surface compacte S est le sup des nombres chromatiques des graphes plongés dans S. Lorsque S n'est pas une sphère, ce nombre est donné par le sup des n tels que K_n se plonge dans S (cf [34]). Pour la sphère, le théorème des 4 couleurs affirme que $\chi(S^2) = 4$.

Un des problèmes les plus excitants est de relier nos invariants $\mu_k^Z(G)$ aux nombres chromatiques. Une conjecture possible est la suivante :

Conjecture *([17]): pour tout graphe G, on a:*

$$\chi(G) \leqslant \mu_2^O(G) + 1 .$$

Cette conjecture qui implique le théorème des 4 couleurs est plus faible que la conjecture classique de Hadwiger qui dit que, si $\chi(G) = l$, K_l est un mineur de G.

2 Majoration optimale en terme du genre

Il résulte des travaux de Cheng, Besson, Nadirashvili et Sévennec que, pour une surface orientable compacte de genre g, on a pour $g \geqslant 2$:

$$\chi(S) - 1 = E[\frac{5}{2} + \sqrt{12g + \frac{1}{4}}] \leqslant \mu_2^O(S) \leqslant m_2^O(S) \leqslant 4g ,$$

alors que pour $g = 0$ et $g = 1$, on a $\mu_2^O(S) = m_2^O(S) = \chi(S) - 1$.

Peut-on améliorer ces inégalités ?

3 Stabilisation de A. Schrijver

Soit $\nu_k^Z(G) = \inf_{G \leqslant G'} m_k^Z(G')$. On a bien sûr $\nu_k^Z(G) \geqslant \mu_k^Z(G)$.

A-t-on égalité ?

4 Perron-Frobenius quantitatif

Peut-on majorer la multiplicité de la première valeur propre $\lambda_1(A)$ pour $A \in M_G$ en fonction des arguments des holonomies $h(A, \gamma)$ pour les cycles de G ?

On peut trouver un analogue continu, le théorème de Aharonov-Casher, voir [25] p. 125-129.

References

[1] V. Arnold. Modes and quasi-modes. *Functional analysis and its applications*, 6:94-101, 1972.

[2] V. Arnold. The Sturm theorems and symplectic geometry. *Functional analysis and its applications*, 19:251-259, 1995.

[3] V. Arnold. Remarks on eigenvalues and eigenvectors of Hermitian matrices, Berry phase, adiabatic connections and quantum Hall effect. *Sel. Math., New Ser.*, 1:1-19, 1995.

[4] V. Arnold. On a characteristic class intervening in quantization conditions. *Functional analysis and its applications*, 1:1-14, 1967.

[5] R. Bacher et Y. Colin de Verdière. Multiplicités de valeurs propres et transformations étoile-triangle des graphes *Bull. Soc. Math. Fr.*, 123:101-117, 1995.

[6] M. Berry, M. Wilkinson. Spectra of triangles. *Proc. Royal Soc. London*, A392:15-43, 1984.

[7] S.Y. Cheng. Eigenfunctions and nodal sets. *Comment. Math. Helv.*, 51:432-55, 1976.

[8] Ph. Ciarlet. The finite element method for elliptic problems. *North Holland*, 1978.

[9] B. Colbois et Y. Colin de Verdière. Sur la multiplicité de la première valeur propre d'une surface de Riemann à courbure constante. *Comment. Math. Helv.*, 63:194-208, 1988.

[10] Y. Colin de Verdière, I. Gitler, D. Vertigan. Réseaux électriques planaires II. *Comment. Math. Helvetici*, 71:144-167, 1996.

[11] Y. Colin de Verdière et A. Marin. Triangulations presque equilatérales des surfaces. *Journal of Diff. Geometry*, 32:199-207, 1990.

[12] Y. Colin de Verdière, Y. Pan, B. Ycart. Singular Limits of Schrödinger operators and Markov Processes, à paraitre dans *Journal of Operator Theory*, 41, 1999.

[13] Y. Colin de Verdière, N. Torki. Opérateur de Schrödinger avec champs magnétiques. *Séminaire de théorie spectrale et géométrie de Grenoble*, 1992–1993.

[14] Y. Colin de Verdière. Sur la multiplicité de la première valeur propre non nulle du Laplacien. *Comment. Math. Helv.*, 61:254-270, 1986.

[15] Y. Colin de Verdière. Spectre de variétés Riemanniennes et spectre de graphes. *Proc. Intern. Cong. Math. (Berkeley)*, 522:530, 1986.

[16] Y. Colin de Verdière. Sur une hypothèse de transversalité d'Arnold. *Comment. Math. Helv.*, 63:184:193, 1988.

[17] Y. Colin de Verdière. Construction de laplaciens dont une partie finie du spectre est donnée. *Ann. scient. E.N.S.*, 20:599–615, 1987.

[18] Y. Colin de Verdière. Un nouvel invariant des graphes finis et application à un critère de planarité. *Journal of Combinatorial theory B*, 50:11-21, 1990.

[19] Y. Colin de Verdière. Réseaux électriques planaires I. *Commentarii Math. Helv.*, 69:351-374, 1994.

[20] Y. Colin de Verdière. Multiplicités de valeurs propres: laplaciens discrets et continus. *Rendiconti di Matematica*, VII,13:433-460, 1993.

[21] Y. Colin de Verdière. Spectres de graphes. Cours Spécialisés no. 4 *Soc. Math. de France*, 1998.

[22] Y. Colin de Verdière. Théorèmes de Courant et de Cheng combinatoires. *Séminaire de théorie spectrale et géométrie de Grenoble*, 13:9-13, 1994–1995.

[23] Y. Colin de Verdière. Multiplicities of eigenvalues and Tree-width of Graphs, *J. Comb. Th.*, serie B, vol. 74:121-146, 1998.

[24] F. Chung, S. Sternberg. Mathematics and the Buckyball. *American scientist*, 81:56-71, 1993.

[25] H. L. Cycon, R.G. Froese, W. Kirsch, B. Simon. Schroedinger operators, with application to quantum mechanics and global geometry. *Springer Study, Texts and Monographs in Physics*, 1987.

[26] A. Dal Maso. An introduction to Γ-convergence. *Birkhäuser*, 1993.

[27] M. Freidlin, A. Wentzell. Random perturbations of dynamical systems. *Springer (Berlin)*, 1984.

[28] K. Kuratowski. Sur le problème des courbes gauches en topologie. *Fund. Math.*, 15:271-283, 1930.

[29] J.L. de Lagrange. Recherches sur la propagation du son. *Oeuvres (Gauthier-Villars)*, 1:72-79, 1867.

[30] E. Lieb. Fluxes and Dimers in the Hubbard model. *Proc. Internat. Congress of Math. (Zürich)*, 2:1279-1280, 1994.

[31] L. Lovász, A. Schrijver. A Borsuk theorem for antipodal links and a spectral characterization of linklessy embeddable graphs. *Preprint (à paraitre dans Proc. AMS.)*, 1-11, 1996.

[32] T. Kato. Perturbation theory for linear operators. *Springer (Berlin)*, 1976.

[33] N. Nadirashvili. Multiple eigenvalues of Laplace operators. *Math. USSR Sbornik*, 61:225-238, 1988.

[34] G. Ringel. Map color theorem. *Springer Grundlehren*, 1974.

[35] J. Rubinstein, M. Schatzman. Asymptotic for thin superconducting rings. *J. Maths. Pures et Appliquées (à paraitre)*.

[36] J.P. Serre. Répartition asymptotique des valeurs propres de l'opérateur de Hecke T_p. *J. of the AMS*, 10:75-102, 1997.

[37] B. Sévennec. Multiplicité du spectre des surfaces: une approche topologique. *Prépublication ENS Lyon*, 1-16, 1994. *(http://umpa.ens-lyon.fr/UMPA/)*

[38] H. van der Holst. A short proof of the planarity characterization of Colin de Verdière. *J. Comb. Th B*, 65:269-272, 1995.

[39] H. van der Holst. Topological and Spectral Graph Characterizations. *Thèse (Amsterdam)*, 1-135, 1996.

[40] H. van der Holst. A Characterization of the class of graph G with $\nu(G) \leqslant 3$. *Preprint (Princeton)*, 1-15, 1997.

[41] E. Wigner, J. von Neumann. Über das Verhalten von Eigenwerten bei adiabatischen Prozessen. *Phys. Zeit.*, 30:467-470, 1929.

[42] A. Zuk. La propriété (T) de Kazhdan pour les groupes agissant sur les polyèdres. *C. R. Acad. Sci. Paris*, 323:453-458, 1996.

Y. Colin de Verdière
Institut Fourier,BP 74,
38402-Saint Martin d'Hères Cedex (France)
email: `yves.colin-de-verdiere@ujf-grenoble.fr`

Analysis on infinite graphs with regular volume growth

Thierry Coulhon

Abstract. We give an overview of recent results that characterise the validity of upper and lower Gaussian estimates for kernels of random walks on graphs with regular volume growth. We also review some analytic consequences of these estimates, such as boundedness of Riesz transforms and $H^1 - BMO$ duality. We finally state on-diagonal upper and lower estimates for graphs with general volume growth.

1 Introduction

Let Γ be an infinite connected graph, endowed with a reversible Markov chain, whose k-th iterate has kernel $p_k(x,y)$, $x,y \in \Gamma$.

We shall report on recent progress made by a group of people, including A. Grigor'yan, T. Delmotte, E. Russ and the author ([10], [13], [14], [21], [22], [33], [34], [35]), in answering the following two questions:

- what is the connection between the asymptotics of $p_k(x,x)$, for fixed $x \in \Gamma$, or $\sup_{x\in\Gamma} p_k(x,x)$, as k goes to infinity, on the one hand, and the large scale geometry of the graph Γ, on the other hand?

- assume that Γ satisfies such geometric conditions that $p_k(x,x)$ behaves more or less as it does in a Euclidean graph; to what extent can one carry on in this setting the analogue of what is usually called real analysis, e.g. boundedness of natural singular integrals such as the Riesz transforms, or $H^1 - BMO$ duality?

The same questions can be asked for the heat kernel on a non-compact Riemannian manifold. In this setting, part or the first question, say for manifolds satisfying the doubling property (or regular volume growth) and for more general volume growth as far as upper bounds are concerned, was answered several years ago, but the discrete case resisted. So that one of the aims of the above program was to reduce the gap between analysis of heat kernels on manifolds, as developed for instance in [26], [27], [36], [37], [39], and the corresponding analysis in discrete time and discrete space. Other aspects of the first question, concerning lower bounds, as well as the second one, were answered more recently, essentially simultaneously for graphs and for manifolds, with however always a more or less important additional difficulty in the discrete setting.

2 Definitions and notation

Let Γ be an infinite graph. Write $x \sim y$ if $x, y \in \Gamma$ are neighbours. Assume that Γ is locally uniformly finite, i.e. there exists $N \in \mathbb{N}^*$ such that any vertex of Γ has at most N neighbours. A path of length n between x and y in Γ is a sequence x_i, $i = 1, ..., n$ such that $x_0 = x$, $x_n = y$ and $x_i \sim x_{i+1}$, $i = 0, ..., n-1$. We shall assume that Γ is connected, i.e. there exists a path between any two points of Γ. Let d be the natural metric on Γ: $d(x, y)$ is the minimal length of a path between x and y. Denote by $B(x, r)$ the closed ball of center $x \in \Gamma$ and radius $r > 0$.

Let m be a positive function on Γ, and $p : \Gamma \times \Gamma \to \mathbb{R}_+$ a Markov kernel with respect to m, i.e.

$$p(x, y)m(x) = p(y, x)m(y), \ \forall x, y \in \Gamma,$$

$$\sum_y p(x, y) = 1, \ \forall x \in \Gamma.$$

Assume that p has range one (in other terms we shall consider a nearest neighbour random walk), i.e.

$$p(x, y) = 0 \text{ if } d(x, y) \geqslant 2,$$

but the following can be easily adapted to the case of a bounded range.

It is convenient to consider

$$\mu_{xy} = p(x, y)m(x) = \mu_{yx}.$$

Once the graph structure is given, all the remaining information is contained in the function $\mu : \Gamma \times \Gamma \to \mathbb{R}_+$, since $m(x) = \sum_{y \sim x} \mu_{xy}$ and $p(x, y) = \frac{\mu_{xy}}{m(x)}$. The object under consideration in the sequel will therefore be the weighted graph (Γ, μ). Note that a posteriori one might consider as neighbours in Γ only the vertices x, y for which $\mu_{xy} > 0$, (which would modify the distance function accordingly), but another point of view is to keep the graph structure and the distance fixed, and to consider various reversible Markov chains thereon.

Define

$$p_k(x, y) = \sum_{z \in \Gamma} p_{k-1}(x, z)p(z, y).$$

Note that the correct analogue of the heat kernel on Riemannian manifolds is $\frac{p_k(x,y)}{m(y)}$ rather than $p_k(x, y)$. We shall denote by P the Markov operator associated with p:

$$Pf(x) = \sum_y p(x, y)f(y).$$

The volume $|\Omega|$ of a subset Ω of Γ will be defined by

$$|\Omega| = m(\Omega) = \sum_{x\in\Omega} m(x).$$

Denote by $V(x,r)$ the volume $|B(x,r)|$ of the ball $B(x,r)$. In what follows, the ℓ^p norms of functions in $\mathbb{R}^\Gamma$ will be taken with respect to the measure m. We shall denote by $(.,.)$ the scalar product on $\ell^2(\Gamma, m)$; the operator P is self-adjoint on the latter space.

We shall say that (Γ,μ) has regular volume growth, or satisfies the doubling property, if there exists C such that

$$V(x,2r) \leqslant C\, V(x,r),\ \forall x\in\Gamma,\ r>0. \tag{D}$$

For $f\in\mathbb{R}^\Gamma$, define the length of its gradient by

$$|\nabla f|(x) = \left(\frac{1}{2}\sum_{y\in\Gamma, x\sim y} |f(x)-f(y)|^2 p(x,y)\right)^{1/2}.$$

One has

$$\||\nabla f|\|_2^2 = \frac{1}{2}\sum_{x,y\in\Gamma} |f(x)-f(y)|^2 \mu_{xy},$$

and an easy computation shows that the gradient and the operator P are connected via

$$\||\nabla f|\|_2^2 = \|f\|_2^2 - (Pf,f) = \|f\|_2^2 - \|P^{1/2}f\|_2^2.$$

The latter quantity is called the Dirichlet form associated with P.

We shall denote by $c_0(\Gamma)$ the set of finitely supported functions on Γ and if Ω is a finite subset of Γ, by $c_0(\Omega)$ the set of functions of $\mathbb{R}^\Gamma$ supported in Ω.

If Ω is a subset of Γ, define its boundary $\partial\Omega$ by

$$\partial\Omega = \{x\in\Omega;\ \exists y\in\Gamma\setminus\Omega,\ y\sim x\},$$

and the size of its boundary by

$$|\partial\Omega|_s = \sum_{x\in\Omega, y\notin\Omega} \mu_{xy}.$$

Now let Γ be an infinite locally uniformly finite connected graph. Fix $\alpha, c, C>0$, and consider varying Markov kernels such that

$$p(x,y)\geqslant\alpha,\ \forall x,y\in\Gamma,\ x\sim y,$$

that are reversible with respect to varying measures m such that

$$c \leqslant m(x) \leqslant C, \ \forall x \in \Gamma,$$

(or, equivalently, fix $c', C' > 0$ and consider a class of weights μ on Γ such that

$$c' \leqslant \mu_{xy} \leqslant C', \ \forall x, y \in \Gamma, \ x \sim y).$$

Then the geometric properties of the graph encoded in the volume, the gradient, the size of the boundary, are, up to multiplicative constants, independent of the Markov kernel under consideration.

3 Upper estimates for random walks: uniformly polynomial volume growth

The subject started in the mid-eighties when Varopoulos proved in [41] that, for $D > 2$, the estimate

$$p_k(x,x) \leqslant Cm(x)k^{-D/2}, \ \forall k \in \mathbb{N}^*, \ x \in \Gamma, \tag{1}$$

is equivalent to the following Sobolev inequality

$$\|f\|_{\frac{2D}{D-2}} \leqslant C\|\nabla f\|_2, \ \forall f \in c_0(\Gamma).$$

Note that this equivalence holds for any Markov chain on a denumerable state space, regardless of the graph structure, therefore regardless of a finite range assumption. The graph structure really comes in when one has to consider the distance function, e.g. to give a proper estimate of $p_k(x,y)$, $x \neq y$ (see below), or to estimate $p_k(x,x)$ in terms of the volume of balls in Γ.

Around the same time, Carlen, Kusuoka and Stroock proved in [2] that (1) is also equivalent to the Nash type inequality

$$\|f\|_2^{1+\frac{2}{D}} \leqslant C\|f\|_1^{2/D}\|\nabla f\|_2, \ \forall f \in c_0(\Gamma)$$

(this equivalence holds for $D > 0$).

It is only recently that it was understood directly why the Sobolev and Nash inequalities were equivalent when $D > 2$ (see [1]). Both of them are also equivalent to the Faber-Krahn type inequality

$$\lambda_1(\Omega) \geqslant c|\Omega|^{-D/2}, \ \forall \Omega \text{ finite subset of } \Gamma,$$

where

$$\lambda_1(\Omega) = \inf\left\{\frac{\||\nabla f|\|_2^2}{\|f\|_2^2}; f \in c_0(\Omega)\right\}$$

(see [4], [9]). We shall see in §7 that the Nash (or Faber-Krahn) inequalities are well adapted to the treatment of decays with respect to k that are more general than $k^{-D/2}$. The potential usefulness of Faber-Krahn inequalities in the subject first appeared in [27].

The proof of the equivalence between (1) and one or another of the above inequalities has become very simple. The point is that what one has to estimate is

$$\sup_{x\in\Gamma} \frac{p_k(x,x)}{m(x)} = \|P^k\|_{1\to\infty}$$

as a function of k, which can be done by functional analytic methods (see [2], [17], [18], and the survey [8]).

To give a flavour of those methods, let us prove the following basic result, that is to be found in [2], Theorem 5.14. The articles [2], [17] and [18] consist essentially in a quantitative (one has to introduce the parameter D!) refinement of the ideas in the proof below.

Proposition 3.1 *Let (Γ,μ) be any infinite connected weighted graph such that $\mu_{xy} \geqslant c > 0$, $\forall\, x,y \in \Gamma$, $x \sim y$. Then there exists C such that*

$$p_k(x,x) \leqslant Cm(x)k^{-1/2}, \ \forall k \in \mathbb{N}^*.$$

Proof: Let $f \in \mathbb{R}^\Gamma$, with finite support. Let x be the point where $|f|$ reaches its maximum and y any vertex outside the support of f (this is the point where the assumption on Γ is used!). One may write

$$\|f\|_\infty = |f(x) - f(y)|.$$

By the triangle inequality one has

$$c|f(x) - f(y)| \leqslant \|\nabla f\|_1$$

(here we use the assumption on μ), therefore

$$\|f\|_\infty \leqslant C\|\nabla f\|_1, \ \forall f \in c_0(\Gamma)$$

(this is the analytic translation of the fact that the boundary of every finite subset of Γ contains at least one point). Applying this to f^2 and using Hölder yields

$$\|f\|_\infty^2 \leqslant C\|f\|_2\|\nabla f\|_2.$$

Fix $f \in c_0(\Gamma)$ with $\|f\|_2 = 1$. Consider the operator $Q = P^{1/2}$, defined by spectral theory, which is contractive on $\ell^2(\Gamma,m)$. Taking squares and replacing f by $Q^k f$ in the above inequality, one gets, because of the relation between the gradient and the operator P,

$$\|Q^k f\|_\infty^4 \leqslant C\left(\|Q^k f\|_2^2 - \|Q^{k+1} f\|_2^2\right).$$

Since Q is Markovian (expand it in a series of powers of $I - P$) and therefore contractive on $\ell^\infty(\Gamma)$, adding up these inequalities gives

$$(k+1)\|Q^k f\|_\infty^4 \leq \sum_{\ell=0}^{k} \|Q^\ell f\|_\infty^4 \leqslant C\left(\|f\|_2^2 - \|Q^{k+1} f\|_2^2\right) \leqslant C,$$

which implies

$$\|Q^k\|_{2\to\infty} = O(k^{-1/4}),$$

therefore

$$\|P^k\|_{2\to\infty} = O(k^{-1/4}).$$

The estimate

$$\|P^k\|_{1\to\infty} = O(k^{-1/2})$$

follows by duality and is equivalent to the claim. □

What is the connection between the Sobolev inequality

$$\|f\|_{\frac{2D}{D-2}} \leqslant C\|\nabla f\|_2, \ \forall f \in c_0(\Gamma),$$

therefore the estimate (1), and the geometry of (Γ, μ)?

The above Sobolev inequality implies (see [4])

$$V(x,r) \geqslant cr^D, \ \forall r > 0,$$

for some $c > 0$; in particular, m has to be bounded below. Conversely, the isoperimetric inequality

$$|\Omega|^{\frac{D-1}{D}} \leqslant C|\partial\Omega|_s$$

easily implies (S_D), therefore (1), but it is strictly stronger (for the latter fact, see [16]).

However, for Cayley graphs (i.e. graphs associated with finitely generated groups), all three properties are equivalent. This was discovered by Varopoulos. For a straightforward approach, see [19], and for more information on random walks on Cayley graphs, the survey [30]. To give an idea of what happens, let us say that, on groups, some inequalities called pseudo-Poincaré inequalities hold for free, and that they ensure very easily that the volume growth controls the isoperimetry (see also the survey [11] for a wider application of these ideas).

On an arbitrary graph, there are combinatorial criteria that ensure the validity of such pseudo-Poincaré inequalities, but they are not very tractable, because they involve the existence of certain well-behaved families of paths, which is hard to predict (see [19], §5). However, these techniques proved quite useful in the context where they first appeared ([24], [23]).

Let us consider now an arbitrary infinite connected graph and suppose that our only information is its volume growth (in particular, we have no idea of its isoperimetry). Suppose for example that the volume growth is uniformly polynomial from above and below. Which behaviour can one expect for $\sup_{x\in\Gamma}\frac{p_k(x,x)}{m(x)}$ (which in this case is comparable to $\sup_{x\in\Gamma}p_k(x,x)$)?

This question can be answered completely.

Theorem 3.2 *Assume that* Γ *satifies*

$$cr^D \leq V(x,r) \leqslant Cr^D,\ \forall x\in\Gamma,\ r>0.$$

Then there exists $c', C' > 0$ *such that*

$$c'k^{-D/2} \leqslant \sup_{x\in\Gamma} p_{2k}(x,x) \leqslant C'k^{-\frac{D}{D+1}},\ \forall k\in\mathbb{N}^*,$$

and both bounds are optimal.

The lower bound is a consequence of [13], Theorem 4.6. The proof consists in noticing that since from the volume assumption Γ satisfies (D) (in fact, regular volume growth around a fixed point x_0 is enough), the λ_1 of any ball of radius r is bounded by $\frac{C}{r^2}$. This is enough to ensure

$$\sup_{x\in\Gamma} p_{2k}(x,x) \geqslant \frac{c}{V(x_0,\sqrt{k})}$$

because

$$\sup_{x\in\Gamma} p_{2k}(x,x) \geqslant \sup_{\Omega}\left\{\frac{\exp\left(-C\lambda_1(\Omega)k\right)}{|\Omega|}\right\}$$

(see [13], §4 for details).

The upper bound comes from [19], Théorème 7. Its proof consists in showing that the volume upper bound implies a weak pseudo-Poincaré inequality, which together with the volume lower bound implies the Nash inequality

$$\|f\|_2^{1+\frac{D+1}{D}} \leqslant C\|f\|_1^{\frac{D+1}{D}}\|\nabla f\|_2,\ \forall f\in c_0(\Gamma),$$

which in turn implies the decay by the above theory. For further results concerning the upper bound, with more general and different upper and lower volume estimates, see [10], Proposition V.5. This upper bound goes over by discretization to the heat kernel on a Riemannian manifold; this is an example of a situation where the discrete theory comes to the rescue of the continuous one, essentially because counting geodesics is easier on a graph!

As we have seen, all graphs with polynomial growth and isoperimetric inequality of exponent D satisfy

$$\sup_k p_k(x,x) \leqslant Ck^{-D/2}.$$

Finally, A. Grigor'yan and myself have recently been able ([15]) to build for any $D \geqslant 1$ a graph satisfying

$$C^{-1}n^D \leqslant V(x,n) \leqslant Cn^D$$

and such that

$$\sup_k p_{2k}(x,x) \geqslant ck^{-\frac{D}{D+1}}$$

(in particular such a graph is recurrent).

Let us come back to the estimate (1) itself. It immediately raises the question: what about $p_k(x,y)$ for x, y distinct?

Cauchy-Schwarz and reversibility immediately yield

$$p_{2k}(x,y) \leqslant \sqrt{\frac{m(y)}{m(x)}}\sqrt{p_{2k}(x,x)p_{2k}(y,y)},$$

but one would like to have an upper bound that becomes stronger for x, y far apart (however, at distance less than k, otherwise $p_k(x,y)$ is zero!).

Indeed, Varopoulos proved in [41] (see also [3] for a simpler proof) that

$$p_k(x,y) \leqslant C\sqrt{\frac{m(y)}{m(x)}} \exp\left(-c\frac{d^2(x,y)}{k}\right).$$

This estimate can be useful (Lemma 4.3 below is nothing but an integrated form of it), but, since the term $k^{-D/2}$ has disappeared, it does not carry any more a geometric information on (Γ, μ), and does not give (1) back again.

Later, Hebisch and Saloff-Coste filled these gaps and proved in [28] that (1) self-improves into the following off-diagonal estimate

$$p_k(x,y) \leqslant Cm(y)k^{-D/2} \exp\left(-c\frac{d^2(x,y)}{k}\right)$$

(note that this was substantially harder to prove than the continuous time analogue). We shall encounter the same self-improvement phenomenon in a more general context in §4.

If in addition

$$V(x,r) \leqslant Cr^D,$$

the estimate (1) can be reformulated as

$$p_k(x,x) \leqslant \frac{C\,m(x)}{V(x,\sqrt{k})}, \ \forall x \in \Gamma, k \in \mathbb{N}^*. \tag{2}$$

It is now tempting to look for such a direct relationship between $p_k(x,x)$ and the volume growth of balls centered at x, independently of any a priori prescription on the volume growth. The above functional analysis can no longer help on such a non-uniform ground. We shall present alternative methods in §§4 and 5.

Another natural generalisation of the above result is to consider general uniform decays of the form

$$p_k(x,x) \leqslant C\frac{m(x)}{\gamma(k)}, \ \forall k \in \mathbb{N}^*, \ x \in \Gamma, \tag{3}$$

and to connect them with Sobolev or isoperimetric type inequalities. This will be our concern in §7.

Finally another question was left open by the above uniform and polynomial theory: the question of finding proper lower bounds for $p_k(x,x)$ and also for $p_k(x,y)$ for x,y far apart. Till recently, the only satisfactory answer to this question was for Cayley graphs with polynomial growth (see [28]) and for graphs with volume growth less than quadratic ([20]).

In §§4 and 5 we shall give an essentially complete answer to the question of lower bounds for $p_k(x,x)$ (on-diagonal lower bounds) and $p_k(x,y)$ (off diagonal lower bounds) for graphs having regular volume growth. The situation is quite different for more rapid volume growths. There, as we shall see in §7, on-diagonal lower bounds are more or less understood, but criteria to ensure off-diagonal lower bounds, and even the form they should take, remain unknown.

4 Upper estimates for random walks: regular volume growth

We shall now consider a weighted graph (Γ,μ) with no explicit or implicit (like in Sobolev) prescribed volume growth assumption. For a graph satisfying (D), we shall give a geometric characterization of the estimate (2), that we shall now call (DUE), in terms of an isoperimetric type inequality.

From §3, one can expect that an L^2 inequality, e.g. a Faber-Krahn inequality, is relevant, but it turns out that, to capture the non-uniform estimate (DUE), one has to consider a whole family of Faber-Krahn inequalities indexed by balls. We shall call such a family a relative Faber-Krahn inequality and denote it by (FK). The results presented in this section come from [14]. They are the discrete analogues of some results in [26]. However, the discrete setting creates new difficulties so that additional technical tools have to be brought into the game.

Let us call (UE) the following off-diagonal version of (DUE)

$$p_k(x,y) \leqslant \frac{Cm(y)}{V(x,\sqrt{k})}\exp\left(-c\frac{d^2(x,y)}{k}\right) \quad \forall x,y \in \Gamma, \ k \in \mathbb{N}^*.$$

The corresponding on-diagonal lower bound

$$p_{2k}(x,x) \geqslant \frac{cm(x)}{V(x,\sqrt{k})}, \ \forall x \in \Gamma, k \in \mathbb{N}^*$$

will be called (DLE) (one considers only even times since otherwise $p_k(x,x)$ may be zero).

One says that Γ satisfies a relative Faber-Krahn inequality (see [26]) if there exists $a > 0$, $\nu > 0$ such that, for every $x \in \Gamma$, $r \geqslant 1/2$, and for every non-empty finite subset Ω of Γ contained in $B(x,r)$,

$$(FK) \qquad \lambda_1(\Omega) \geqslant \frac{a}{r^2}\left(\frac{V(x,r)}{|\Omega|}\right)^{\nu}.$$

Note that if (FK) holds with $\nu = 2/D$ and if $V(x,r) \geqslant cr^D$, then

$$\lambda_1(\Omega) \geqslant \frac{c'}{|\Omega|^{2/D}},$$

in other terms the flat Sobolev inequality holds:

$$\|f\|_{\frac{2D}{D-2}} \leqslant C\|\nabla f\|_2, \ \forall f \in c_0(\Gamma).$$

Conversely, the latter inequality together with a volume upper bound $V(x,r) \leqslant Cr^D$ implies (FK), but one can easily imagine situations where Sobolev holds but not (FK) (trees, for instance, but one can also stay inside the regular volume growth realm). Therefore, though the following result has many advantages upon the uniform theory of §3, it does not make it totally obsolete. We shall come back to this in §7.

Theorem 4.1 *The following properties of* (Γ, μ) *are equivalent:*

1. *The relative Faber-Krahn inequality* (FK).
2. *The full upper estimate* (UE) *in conjunction with the doubling property* (D).
3. *The on-diagonal upper estimate* (DUE) *in conjunction with the doubling property* (D).

Moreover, each of them implies the on-diagonal lower estimate (DLE).

Note that in particular (DUE) together with (D) implies (UE). It would be nice to have a direct proof of this fact. In contrast, the proof that (UE) together with (D) implies (DLE) is quite simple.

Let us give the main steps of the proof that (FK) implies (UE).

The first thing is to realize that one can as well assume in addition the following condition (α): there exists $\alpha > 0$ such that $p(x, y) \geqslant \alpha$ if $x \sim y$ or $x = y$. If the claim holds for kernels with (α), one shows that it holds for any kernel.

Then assuming (α) (the technical importance of this assumption appeared in [17], and even more in [22]), one proves a discrete parabolic Caccioppoli inequality, which may be of interest in itself.

Let

$$\nabla_{xy} f = f(y) - f(x)$$

and

$$\partial_k f(x) = f(k+1, x) - f(k, x) .$$

A function u on $\mathbb{N} \times \Gamma$ is called *a subsolution* of the heat equation if it satisfies the inequality

$$\partial_k u \leqslant (P - I) u,$$

which can be rewritten as

$$u_{k+1}(x) \leqslant \sum_y p(x, y) u_k(y) = (P u_k)(x) . \tag{4.1}$$

Proposition 4.2 *There exist $c, A > 0$ depending on α such that, for every non-negative subsolution f and any function φ with finite support*

$$\sum \partial_k(f^2)\varphi^2 m + c \sum_{x,y} |\nabla_{xy}(f\varphi)|^2 \mu_{xy} \leqslant A \sum_{x,y} |\nabla_{xy}\varphi|^2 f^2 \mu_{xy}. \tag{4.2}$$

This inequality together with (FK) enables one to compare the L^2 norms of solutions of the heat equation in smaller and smaller time-space cylinders. An iterative process leads from there to an L^2 mean value inequality, and another iterative process to an L^1 mean value inequality of the form

$$u(T, z) \leqslant \frac{\text{const}}{m(\Psi)} \sum_{k,x \in \Psi} u(k, x) m(x) \tag{4.3}$$

where Ψ is the cylinder $[0, 2T] \times B(z, R)$.

The other input is a discrete version of Gaffney's lemma, which is an extension of a result in [28] (note that unfortunately, this step of the proof is the only one that is not yet entirely discrete, because it uses a continuous-time semigroup).

Lemma 4.3 *There exist universal constants $C, c > 0$ such that, for all finite subsets E, F of Γ,*

$$\sum_{x\in E}\sum_{y\in F} p_k(x,y)m(x) \leqslant Ce^{-c\frac{d^2(E,F)}{k}}|E|^{1/2}|F|^{1/2}, \quad \forall k \in \mathbb{N}^*,$$

where $d(E,F) = \min_{x\in E, y\in F} d(x,y)$.

Then, if $x, y \in \Gamma$ are at distance r, one takes $E = B(x, r/3)$, $F = B(y, r/3)$, and applying Lemma 4.3, then the mean value inequality twice, one gets the claimed estimate for $p_k(x,y)$.

As far as the connection with the geometry of (Γ, μ) is concerned, Theorem 4.1 says in particular that (FK) implies (D) (the proof is relatively straightforward). Conversely, (FK) is implied by a relative isoperimetric inequality of the type

$$\frac{|\partial\Omega|_s}{|\Omega|} \geq \frac{a'}{r}\left(\frac{V(x,r)}{|\Omega|}\right)^{\nu/2},$$

for some $a', \nu > 0$, and every $x \in \Gamma$, $r \geqslant 1/2$, and every non-empty finite subset Ω of Γ contained in $B(x,r)$ (see [14], §2.1 for details).

5 Off-diagonal lower estimates and parabolic Harnack inequality

The result in this section is due to T. Delmotte.

Since (D) and (FK) give the Gaussian upper estimate and the on-diagonal lower estimate, what is the additional price to pay in order to get the full Gaussian lower estimate?

In the continuous world, it is well-known that the conjunction of upper and lower off-diagonal estimates is equivalent to a parabolic Harnack inequality. We have no room here to explain what a parabolic Harnack inequality is, except that it enables one to compare the supremum and the infimum of any non-negative solution of the heat equation in time-space cylinders of large size but fixed proportions. We refer the reader to [39] for a survey in the continuous setting, and directly to [22] for the discrete version. To get Harnack inequalities, the most efficient tool remains the Moser iteration scheme. It was performed in the setting of manifolds in [36],[37]; see also [39] for a complete exposition and [26] for a more geometric approach to Harnack inequalities. However, running the Moser iteration scheme in the discrete setting brings in again new difficulties that are very well analysed in [22]. In fact, if one is interested in the elliptic version of Harnack inequality, a relatively easy, though non-trivial, adaptation goes through; it was performed independently by Delmotte, by Holopainen-Soardi, and by Rigoli-Salvatori-Vignati (see [21] and the references therein). It is even more difficult to work with discrete time that with discrete space. This is why the parabolic Harnack inequality, i.e. the analogue of the result of [37], requires more work, in particular a careful analysis of the interplay between a discrete-time and a continuous-time process on a graph. For details see [22]. Note that the latter work was completed before [13], and in fact, it stimulated the search for a completely discrete approach to the whole thing, which is not yet at hand, and of which [13] is the first step.

One says that Γ satisfies the Poincaré inequality if there exists $C > 0$ and $C' \geqslant 1$ such that

(P)

$$\sum_{y \in B(x,r)} |f(y) - f_r(x)|^2 m(y) \leqslant Cr^2 \sum_{B(x,C'r)} |\nabla f|^2(y) m(y), \forall f \in \mathbb{R}^\Gamma, \forall r > 0,$$

where

$$f_r(x) = \frac{1}{V(x,r)} \sum_{y \in B(x,r)} f(y) m(y).$$

See [19], §5 for combinatorial conditions that ensure (P).

Theorem 5.1 *Let (Γ, μ) be as above. Assume in addition that there exists $\alpha > 0$ such that $p(x,y) \geq \alpha$ if $x \sim y$ or $x = y$. Then the conjunction of the doubling property (D) and of the Poincaré inequality (P) is equivalent to*

$$\frac{c\,m(y)}{V(x,\sqrt{k})} \exp\left(-C\frac{d^2(x,y)}{k}\right) \leq p_k(x,y) \leqslant \frac{C\,m(y)}{V(x,\sqrt{k})} \exp\left(-c\frac{d^2(x,y)}{k}\right),$$

$\forall x, y \in \Gamma, k \in \mathbb{N}^*$ *such that* $d(x,y) \leqslant k$.

The conclusion of the two preceding sections is that, among the graphs satisfying (D), two distinguished subclasses appear:

-those who satisfy (DUE), and consequently (UE) and (DLE); in geometric terms, they are characterised by (FK).

-those who satisfy the full upper and lower Gaussian estimate for $p_k(x,y)$; they are characterised by (P).

Of course, one can see directly that (P) and (D) imply (FK) ([14], §2). On the other hand, it is easy to build graphs with (D) and (FK) but without (P): glue along a single edge two graphs X_1, X_2 with the same polynomial volume growth. If the two graphs satisfy Sobolev, the resulting graph $X_1 \coprod X_2$ will also satisfy Sobolev, therefore (FK), but it is easy to violate (P) just by taking $f = 0$ on X_1 and $f = 1$ on X_2. We will encounter again this type of example (which was first introduced in [38]) in §6.

6 Riesz transforms and $H^1 - BMO$ duality

The results in this section are due to E. Russ ([33], [34], [35]).

Trying to generalize to Euclidean graphs $\mathbb{Z}^D$ and then to more general graphs as much a possible of the theorems that hold in $\mathbb{R}^D$ is an aim in itself. It is well-known that certain tools, like the Calderon-Zygmund decomposition, are available as soon as the graph, seen as a metric measured space, has regular volume growth ([6]). However, this is far from enough to ensure, for example, the boundedness of the Riesz transforms or the $H^1 - BMO$ duality (less us recall that till recently, on manifolds, non-negative curvature was essentially required!). What we are going to see in this section is that one can get such results for the two classes of graphs that appeared in §4 and 5, and also that the difference between the two classes can be felt at this level. The conclusion is that under (FK), and even more under (D) and (P), there is a very tight connection between the behaviour of the random walk and the geometry (say, between spaces of functions defined on a pure geometric basis and spaces defined with the help of the operator P).

The first result is the graph version of the main theorem of [12], where Riesz transforms on Riemannian manifolds are considered. Both results rely on a new approach to boundedness of singular integral operators introduced by X.T. Duong and A. McIntosh ([25]). Here, what is more difficult to obtain than in the discrete setting is the control of the time differences $p_k(x,y) - p_{k+1}(x,y)$ of the random walk. This can be obtained by using an intricate work by M. Christ ([5]), that gives the proper Gaussian pointwise estimate of such differences, but the following relies on a cheaper estimate that is obtained directly in [33].

Theorem 6.1 *Assume that (Γ,μ) satisfies (DUE). Then the Riesz transform $T = \nabla(I-P)^{-\frac{1}{2}}$ is weak $(1,1)$ and bounded on L^p if $1 < p \leqslant 2$, which means that for all $p \in [1,2]$, there exists $C_p > 0$ such that*

$$\|\nabla f\|_p \leqslant C_p \left\|(I-P)^{\frac{1}{2}} f\right\|_p, \ \forall f \in c_0(\Gamma), \quad 1 < p \leqslant 2,$$

and

$$\|\nabla f\|_{1,\infty} \leqslant C_1 \left\|(I-P)^{\frac{1}{2}} f\right\|_1, \ \forall f \in c_0(\Gamma).$$

The above statement cannot hold for $p > 2$: consider the graph $\Gamma = \mathbb{Z}^2 \coprod \mathbb{Z}^2$ made of two copies of $\mathbb{Z}^2$ glued by a single edge, and p the kernel of the standard random walk. First notice that on Γ, $V(x,r)$ is uniformly quadratic, and that the optimal upper estimate for $p_k(x,y)$ holds (thanks to the theory exposed in §3). Now, if for some $p > 2$ one had

$$\|\nabla f\|_p \leqslant C_p \left\|(I-P)^{\frac{1}{2}} f\right\|_p,$$

one shows that this would yield the oscillation estimate

$$|p_k(x,x) - p_k(x,y)| \leqslant C \left(\frac{d(x,y)}{\sqrt{k}} \right)^{1-\frac{2}{p}} p_k(x,x),$$

therefore

$$p_k(x,y) \geqslant c p_k(x,x),$$

for $d(x,y) \leqslant a\sqrt{k}$, with a small enough. From there, the off-diagonal lower estimate

$$p_k(x,y) \geqslant \frac{c}{V(x,\sqrt{k})} \exp\left(-\frac{d^2(x,y)}{Ck} \right)$$

would easily follow.

But this is impossible, because of Theorem 4.1 and of the fact that the Poincaré inequality (P) is false on Γ.

The question whether for graphs satisfying (P) and (D) the estimate

$$\|\nabla f\|_p \leqslant C_p \left\| (I-P)^{\frac{1}{2}} f \right\|_p,$$

for $p > 2$ could be true remains open.

As far as the case $p = 1$ is concerned, let us recall that in the singular integral theory $L^1 - L^1$ boundedness is usually excluded. There are classically two substitutes: the $L^1 - L^{1,\infty}$ boundedness, and the $H^1 - L^1$, and even the $H^1 - H^1$, boundedness. The first property holds under the assumptions of Theorem 5.1 (in fact this is what one proves first, via a Calderon-Zygmund decomposition, and the $L^p - L^p$ boundedness follows by interpolation). If in addition Γ satisfies (P), one can obtain the $H^1 - L^1$ boundedness. This is the main result in [34]. Again, $\mathbb{Z}^2 \coprod \mathbb{Z}^2$ is a counterexample, therefore (D) and (P) do not suffice.

We shall be brief on the subject of $H^1 - BMO$ duality on graphs, which is somewhat alien to the mainstream of this survey; it is however striking that the key to the analogue of what is considered as a deep result in Euclidean harmonic analysis is given by the full Gaussian estimate for $p_k(x,y)$, therefore by the geometric conditions (D) and (P).

For $f \in \mathbb{R}^\Gamma$, say that $f \in BMO(\Gamma)$ if and only if

$$\sup_B \frac{1}{V(B)} \int_B |f(x) - f_B| \, d\mu(x) < +\infty$$

where the supremum is taken over all the balls B of Γ and

$$f_B = \frac{1}{V(B)} \int_B f(x) d\mu(x).$$

If this is the case, define

$$\|f\|_{BMO} = \sup_B \int_B |f(x) - f_B|\, d\mu(x).$$

We now turn to the definition of $H^1_{at}(\Gamma)$. A function a on X is said to be an atom if it is supported in a ball $B = B(x_0, r_0)$ with

$$\|a\|_\infty \leqslant \frac{1}{V(B)} \text{ and } \int a(x)d\mu(x) = 0.$$

A function of $H^1_{at}(\Gamma)$ is any sum

$$f = \sum_{k\in\mathbb{N}} \lambda_k a_k,$$

where the a_k's are atoms and $\sum_k |\lambda_k| < +\infty$. Of course $\|f\|_{H^1_{at}}$ is then defined as the infimum of $\sum_k |\lambda_k|$ on all possible such decompositions of f.

It is well-known ([7]) that on any metric measured space with regular volume growth the dual of H^1_{at} is BMO.

Finally, if $f \in \mathbb{R}^\Gamma$, one says that $f \in H^1_{max}(\Gamma)$ if and only if $f^+ \in L^1(\Gamma, m)$, where

$$f^+(x) = \sup_{k\in\mathbb{N}} \left|P^k f(x)\right|,$$

and one sets $\|f\|_{H^1_{max}} = \|f^+\|_{L^1}$.

One has then the following result.

Theorem 6.2 *Let (Γ, μ) satisfy (D) and (P). Then the spaces H^1_{at} and H^1_{max} coincide. As a consequence, the dual of H^1_{max} is BMO.*

The identification between atomic spaces and spaces defined through the random walk, and also the identification of the dual space, can also be performed for H^p spaces, for $p < 1$ sufficiently close to one (see [35]).

7 Beyond regular volume growth

Let us now drop the condition of regular volume growth on (Γ, μ). We are going to give isoperimetric type conditions on Γ that ensure upper or lower on-diagonal bounds for $\sup_{x\in\Gamma} \frac{p_k(x,x)}{m(x)}$ not in terms of the volume growth function, but in terms of a function of k, whose rate of decay in essentially arbitrary.

The following result is taken from [10], §V. It is the discrete version of the main result in [27], where upper bounds for the heat kernel on Riemannian manifolds are shown to be equivalent to Faber-Krahn type inequalities.

Theorem 7.1 *If γ is a C^1 function from $\mathbb{R}_+$ to itself satisfying (δ), one has*

$$p_k(x,x) \leqslant \frac{m(x)}{\gamma(k)}, \; \forall x \in \Gamma, \tag{3}$$

if and only if

$$\lambda_1(\Omega) \geqslant \frac{1}{\psi^2(|\Omega|)}, \tag{4}$$

for every Ω finite subset of Γ, where ψ and γ are related by

$$t = \int_0^{\gamma(t)} \psi^2(x) \frac{dx}{x}$$

or

$$\psi(x) = \sqrt{\frac{x}{\gamma' \circ \gamma^{-1}(x)}}.$$

We won't explain here what condition (δ) really is (see [10], p.524); let us say that it is a technical condition on an increasing differentiable bijection of $\mathbb{R}_+$ that ensures a certain regularity of growth, but does not affect its ratc of growth, as long as it is subexponential. Therefore all reasonable rates of decay for a random walk can be considered. For instance, $\gamma(t) = e^{ct^\alpha}$, for $0 < \alpha < 1$, satisfies (δ).

The above one-to-one correspondance between functions γ and ψ is to be understood up to multiplicative constants: we identify $\gamma(\cdot)$ and $C\gamma(c\,\cdot)$, $\psi(\cdot)$ and $C\psi(c\,\cdot)$.

What is the connection between the Faber-Krahn type inequality (4) and the geometry of (Γ, μ)? First (4) can be reformulated into a Sobolev type inequality

$$\|f\|_2 \leqslant \psi(|\Omega|)\||\nabla f|\|_2,$$

for every finite subset Ω of Γ and every function f supported in Ω, or into a Nash type inequality

$$\|f\|_2 \leqslant \psi\left(\frac{\|f\|_1^2}{\|f\|_2^2}\right) \||\nabla f|\|_2, \; \forall f \in c_0(\Gamma)$$

(for more on these reformulations, see [1], and [9] for a simple account in the polynomial case). Also, (4), which should be considered as an L^2 isoperimetric inequality, is implied by the L^1 isoperimetric inequality

$$\frac{|\Omega|}{\psi(|\Omega|)} \leqslant C|\partial\Omega|.$$

If the geometry of the graph is sufficiently regular, in the sense that pseudo-Poincaré inequalities hold, then a volume lower bound

$$V(x,r) \geqslant \psi^{-1}(r)$$

is enough to imply (4) (see [10], proposition V.3). This is the case for Cayley graphs.

Conversely, the fact that (4) implies the volume lower bound breaks down outside the polynomial scale: there are random walks that decay faster that what one could tell from the volume growth. More precisely, for any $n \in \mathbb{N}^*$, there are graphs with exponential volume growth such that

$$p_k(x,x) \simeq \exp(-ck^{\frac{n}{n+2}}),$$

whereas the rate of decay corresponding to exponential growth, i.e. $\psi(x) = \log x$ is only $\exp(-k^{1/3})$, i.e. $n = 1$ (see [31]).

If the only information one has is a uniform upper and lower estimate for the volume growth (possibly different), one can nevertheless get a weaker upper bound for $p_k(x,x)$, that generalizes the upper bound in Theorem 3.2 ([10], Proposition V.5).

Finally recall from the discussion in §4 about the relationship between (FK) and Sobolev (or more generally flat, or non relative, Faber-Krahn inequalities of the type just considered), that, even if the function γ is expected to have slow growth (e.g. $\gamma(2t) \leqslant C\gamma(t)$), and depending on the informations available on Γ, Theorem 7.1 may provide a sharper information than Theorem 4.1.

To finish here with the subject of on-diagonal upper bounds, let us point out that once one controls from above the decay of a reversible Markov chain, one also controls the decay of the iterates of any submarkovian kernel that is more spread out and satisfies a weak form of reversibility (see [10], Proposition IV.4). This uses the fact that the two implications in Theorem 7.1 have in fact a different set of assumptions: from the Faber-Krahn inequality to the decay, only uses a weak form of symmetry of the operator P, namely, that P^* is Markov. Then the proof goes like this: let P be the original Markov operator; from its decay, one gets a Faber-Krahn inequality. If Q is the new operator, the assumptions of the kernels ensure that the Dirichlet forms associated with P and Q are uniformly comparable. Since one assumes Q^* is submarkovian, the decay for Q^k follows.

Proposition 7.2 *Let* (Γ, μ) *be as above and let be* $q(x,y)$ *be any non-negative kernel on* Γ *such that*

$$\sum_y q(x,y) \leqslant 1$$

$$\sum_y \frac{q(y,x)m(y)}{m(x)} \leqslant 1$$

and

$$p(x,y) \leqslant Cq(x,y),\ \forall x,y \in \Gamma.$$

Let γ be a C^1 function satisfying (δ). Then

$$\sup_{x\in\Gamma} \frac{p_k(x,x)}{m(x)} = O(1/\gamma(k))$$

implies

$$\sup_{x\in\Gamma} \frac{q_k(x,x)}{m(x)} = O(1/\gamma(ck))$$

for some $c > 0$.

In the case of two symmetric and group invariant random walks on a Cayley graph, one can even compare directly the probabilities of return to the origin, see [32]. An early version of Proposition 7.2 (for a power function γ) already appears in [18].

Let us turn to lower bounds. Upper bounds of the random walks are governed by Faber-Krahn inequalities: in heuristic terms, $p_k(x,x)$ decays quickly enough if and only if every finite subset of Γ has a large enough L^2 boundary (through which the random particle can escape) with respect to its size. What is proved in [13] is that $p_k(x,x)$ decays slowly enough if at all scales there are subsets of Γ whose λ_1 is not bigger than what is prescribed by a Faber-Krahn inequality. In other terms, the upper bound

$$\sup_{x\in\Gamma} \frac{p_{2k}(x,x)}{m(x)} \geqslant \frac{1}{\gamma(k)}, \forall k > 0$$

is true as soon as the Faber-Krahn inequality which is equivalent to the corresponding upper bound is optimal at all scales. We call the latter property an anti-Faber-Krahn inequality.

Theorem 7.3 *Suppose that γ and ψ are as in Theorem 7.1, and that for every $\xi \in \mathbb{R}_+^*$, there exists Ω_ξ such that $|\Omega_\xi| \leqslant \xi$ and $\lambda_1(\Omega_\xi) \leqslant \frac{1}{\psi(\xi)}$. Then there exists $C, c > 0$ such that*

$$\sup_{x\in\Gamma} \frac{p_{2k}(x,x)}{m(x)} \geqslant \frac{C}{\gamma(ck)}, \forall k > 0.$$

For more results (including a discussion on situations where anti-isoperimetric inequalities imply anti-Faber-Krahn inequalities, therefore a lower bound on p_k), see [13], §4, and also the fore-runner paper [29].

As we already said, a theory of off-diagonal lower bounds (but also, a good theory of off-diagonal upper bounds) remains to be built in the case of non-polynomial decays.

Let us call non-amenable a graph that satisfies a linear isoperimetric inequality

$$|\Omega| \leqslant C|\partial\Omega|_s$$

or equivalently

$$\|P\|_{2\to 2} = a < 1.$$

Another limit of the theory is that we are still unable to decide on which non-amenable graphs one can have a better estimate than the trivial one

$$p_k(x,x) \leqslant m(x)a^k,$$

for example

$$p_k(x,x) \leqslant m(x)a^k k^{-3/2},$$

as this is the case on free groups.

See however [13], §5 for some preliminary remarks.

I thank Sasha Grigory'an for helpful conversations while I was preparing this survey, as well as Thierry Delmotte, Emmanuel Russ and Laurent Saloff-Coste for remarks on the manuscript.

Note added in proof: a purely discrete approach to Theorem 5.1 has recently been given in P. Auscher, T. Coulhon, Gaussian lower bounds for random walks from elliptic regularity, (to appear in *Ann. I.H.P., proba. stats.*).

References

[1] Bakry D., Coulhon T., Ledoux M., Saloff-Coste L., Sobolev inequalities in disguise, *Indiana Univ. Math. J.*, 124, 4, 1033-1074, 1995.

[2] Carlen E., Kusuoka S., Stroock D., Upper bounds for symmetric Markov transition functions, *Ann. Inst. H. Poincaré, proba. et stat.*, suppl. au no 2, 245-287, 1987.

[3] Carne K., A transmutation formula for Markov chains, *Bull. Sc. Math.*, 2ème série, 109, 399-403, 1985.

[4] Carron G., Inégalités isopérimétriques de Faber-Krahn et conséquences, in *Actes de la table ronde de géométrie différentielle en l'honneur de Marcel Berger*, Collection SMF Séminaires et congrès, no 1, 1994.

[5] Christ M., Temporal regularity for random walk on discrete nilpotent groups, *Journal of Fourier Analysis and Applications*, Kahane special issue, 141-151, 1995.

[6] Coifman R., Weiss G., *Analyse harmonique non-commutative sur certains espaces homogènes*, Springer L. N. 242, 1971.

[7] Coifman R., Weiss G., Extension of Hardy spaces and their use in analysis,*Bull. Amer. Math. Soc.*, 83, 569-645, 1977.

[8] Coulhon T., Dimensions of continuous and discrete semigroups on the L^p spaces, in *Semigroup theory and evolution equations*, Clément, Mitidieri, De Pagter éd., Marcel Dekker L. N. in Pure and Appl. Math., vol. 135, 93-99, 1991.

[9] Coulhon T., Espaces de Lipschitz et inégalités de Poincaré, *J. Funct. Anal.*, 136, 1, 81-113, 1996.

[10] Coulhon T., Ultracontractivity and Nash type inequalities, *J. Funct. Anal.*, 141, 2, 510-539, 1996.

[11] Coulhon T., Dimensions at infinity for Riemannian manifolds, *Pot. Anal.*, 4, 4, 335-344, 1995.

[12] Coulhon T., Duong X.T., Riesz transforms for $1 \leq p \leqslant 2$, *T.A.M.S.*, 351, 1151-1169, 1999.

[13] Coulhon T., Grigor'yan A., On-diagonal lower bounds for heat kernels and Markov chains, *Duke Math. J.*, 89, 1, 133-199, 1997.

[14] Coulhon T., Grigor'yan A., Random walks on graphs with regular volume growth, *G.A.F.A.*, 8, 656-701, 1998.

[15] Coulhon T., Grigor'yan A., Manifolds with big heat kernels, preprint.

[16] Coulhon T., Ledoux M., Isopérimétrie, décroissance du noyau de la chaleur et transformations de Riesz: un contre-exemple, *Ark. för Mat.*, 32, 63-77, 1994.

[17] Coulhon T., Saloff-Coste L., Puissances d'un opérateur régularisant, *Ann. Inst. Henri Poincaré, Prob. et Stat.*, 26, 3, 419-436, 1990.

[18] Coulhon T., Saloff-Coste L., Marches aléatoires non symétriques sur les groupes unimodulaires, *C.R.A.S. Paris,* t. 310, série I, 627-630, 1990.

[19] Coulhon T., Saloff-Coste L., Isopérimétrie pour les groupes et les variétés, *Rev. Mat. Iberoamericana*, 9, 2, 293-314, 1993.

[20] Coulhon T., Saloff-Coste L., Minorations pour les chaînes de Markov unidimensionnelles, *Prob. Th. Rel. F.* 97, 423-431, 1993.

[21] Delmotte T., Inégalité de Harnack elliptique sur les graphes, *Colloq. Math.*, 72, 1, 19-37, 1997.

[22] Delmotte T., Parabolic Harnack inequality and estimates of Markov chains on graphs, in *Rev. Mat. Iberoamericana*, 15, 1, 181-232, 1999.

[23] Diaconis P., Saloff-Coste L., Nash's inequality for finite Markov chains, *J. Th. Prob.*, 9, 459-510, 1996.

[24] Diaconis P., Stroock D., Geometric bounds for eigenvalues of Markov chains, *Ann. Appl. Prob.*, 1, 36-61, 1991.

[25] Duong X., McIntosh A., Singular integral operators with non-smooth kernels on irregular domains, to appear in *Rev. Mat. Iberoamericana.*

[26] Grigor'yan A., The heat equation on non-compact Riemannian manifolds, in Russian: *Matem. Sbornik*, 182, 1, 55-87, 1991; English translation: *Math. USSR Sb.*, 72, 1, 47-77, 1992.

[27] Grigor'yan A., Heat kernel upper bounds on a complete non-compact manifold, *Rev. Mat. Iberoamericana*, 10, 2, 395-452, 1994.

[28] Hebisch W., Saloff-Coste L., Gaussian estimates for Markov chains and random walks on groups, *Ann. Probab.*, 21, 673-709, 1993.

[29] Lust-Piquard F., Lower bounds on $||K^n||_{1\to\infty}$ for some contractions K of $L^2(\mu)$, with some applications to Markov operators, *Math. Ann.*, 303, 699-712, 1995.

[30] Pittet C., Saloff-Coste L., A survey on the relationship between volume growth, isoperimetry, and the behaviour on simple random walk on Cayley graphs, with examples, preprint.

[31] Pittet C., Saloff-Coste L., Amenable groups, isoperimetric profiles and random walks, in *Geo,etric group theory down under. Proceedings of a special year in geometric group theory, Canberra, Australia, 1996*, J. Cossey, C.F. Miller III, W.D. Neumann and M. Shapiro eds, De Gruyter, 1999.

[32] Pittet C., Saloff-Coste L., On the stability of the behavior of random walks on groups, to appear in *J. Geom. Analysis*.

[33] Russ E., Riesz transforms on graphs for $1 \leqslant p \leqslant 2$, to appear in *Math. Scand.*.

[34] Russ E., $H^1 - BMO$ duality on graphs, preprint.

[35] Russ E., $H^1 - H^1$ boundedness for Riesz transforms on Riemannian manifolds and on graphs, to appear in *Pot. Anal.*.

[36] Saloff-Coste L., Uniformly elliptic operators on Riemannian manifolds. *J. Diff. Geom.* 36, 417-450, 1992.

[37] Saloff-Coste L., A note on Poincaré, Sobolev, and Harnack inequalities, *Duke Math. J., I.M.R.N.*, 2, 27-38, 1992.

[38] Saloff-Coste L., On global Sobolev inequalities, *Forum Mat.*, 6, 271-286, 1994.

[39] Saloff-Coste L., Parabolic Harnack inequality for divergence form second order differential operators, *Pot. Anal.*, 4, 4, 429-467, 1995.

[40] Varopoulos N., Long range estimates for Markov chains, *Bull. Sc. Math.*, 2ème série, 109, 113-119, 1985.

[41] Varopoulos N., Isoperimetric inequalities and Markov chains, *J. Funct. Anal.*, 63, 215-239, 1985.

Thierry Coulhon
Université de Cergy-Pontoise
email: `coulhon@u-cergy.fr`

On the asymptotic spectrum of random walks on infinite families of graphs

Rostislav I. Grigorchuk, Andrzej Żuk

Abstract. We observe that the spectral measure of the Markov operator depends continuously on the graph in the space of graphs with uniformly bounded degree. We investigate the behaviour of the largest eigenvalue and the density of eigenvalues for infinite families of finite graphs. The relations to the theorems of Alon-Boppana and Greenberg are indicated.

1 Introduction

The theory of the spectrum of the Markov (or Laplace) operator on graphs and groups is an important part of the theory of Markov processes. It is related to many topics in the theory of random walks on groups and graphs, abstract harmonic analysis, the theory of operator algebras, K-theory etc.

The spectral radius, spectrum and the spectral measure of the Markov operator on the homogeneous tree T_k of degree k was computed by Kesten in [12]. The number $\frac{2\sqrt{k-1}}{k}$, the value of the spectral radius for this graph, plays an important role in the theory.

First of all there is a large class of infinite regular graphs of degree k with the same value of the spectral radius as was first observed in [6]. These are graphs with cogrowth beween 1 and $\sqrt{k-1}$. (In [12] and [6] the case when k is even is considered. But the same arguments work for odd k if we keep in mind the spherical homogeneity of the simple random walk on a group $\mathbb{Z}_2 * \ldots * \mathbb{Z}_2$).

The second place where these numbers appear is the theory of Ramanujan graphs (see [14], [16], [20] and [23]).

The third place is the theory of the largest eigenvalue λ_1, not equal to 1, of the Markov operator on finite graphs. In particular, the theorem of Alon-Boppana claims:

Theorem 1 (Alon-Boppana) *Let $\{X_{n,k}\}_{n=1}^{\infty}$ be a sequence of k-regular, finite, connected graphs where k is fixed and the number of vertices of $X_{n,k}$ tends to infinity. Then*

$$\liminf_{n\to\infty} \lambda_1(X_{n,k}) \geqslant \frac{2\sqrt{k-1}}{k}.$$

In this paper we will clarify and simplify some considerations related to this kind of problem. Another goal is to consider the problem of approximation of the spectrum of the Markov operator on infinite graphs by the spectra of operators on the sequence of finite graphs which approximate the given graph. For this purpose we introduce the compact topology on the space of graphs with uniformly bounded degree (which generalizes the topology introduced in [8]) and observe that the spectral measure (a measure on $[-1,1]$) depends continuously on the graph in this space. More precisely in Section 3 we will show:

Theorem 2 *The spectral measure is a continuous (measure valued) function on the space of marked graphs.*

The most regular case to which we can apply our results is the case of Cayley graphs; Let F_k be the free group on k generators. Let H_n be a sequence of normal subgroups in F_k such that $H_{n+1} \subset H_n$ for every $n \in \mathbb{N}$. Let $H = \bigcap H_n$. Then the Cayley graphs of F_k/H_n converge to the Cayley graph F_k/H (for generators we choose the images of generators of F_k).

Some versions of convergence of graphs and corresponding limit theorems for spectral measures were considered in [4] and [18] (the latter paper is also a good survey on spectra of infinite graphs). The idea of approximation of an infinite graph by finite graphs is presented also in [14] and other books and articles. It comes from the group theory where the corresponding property is called residual finiteness. But in contrast to the group case where not every finitely generated group is residually finite every infinite regular graph can be approximated by finite regular graphs (see Proposition 1 below).

We hope that the approximation method may help to compute the spectrum (and even the spectral measure) of the Markov operator for some interesting infinite residually finite groups (for instance for the groups of intermediate growth constructed in [7] and [8]).

2 The space of graphs with uniformly bounded degree

In this paper we consider only locally finite, connected, non-oriented graphs which we identify with their geometric realization, i.e. simplicial complexes. This makes the concept of covering of the graph clear. Moreover, our graphs will be supplied with a natural distance metric.

Theorems 2 and 3 (see below) also have analogues in the cases of oriented graphs and coloured graphs when the number of colours is finite. This is important for constructions with Cayley graphs of finitely generated groups.

So let $X = (V_X, E_X)$ be a connected graph and V_X, E_X its set of vertices and edges respectively. Let the distance between two vertices $v_a \neq v_b \in V_X$

be the minimal number of edges needed to connect them, i.e.

$$\begin{aligned}\mathrm{dist}(v_a, v_b) &= \min\{n; \exists v_0, \ldots, v_n \in V_X, v_0 = v_a, v_n = v_b, \\ &\quad (v_i, v_{i+1}) \in E_X \text{ for } 0 \leqslant i \leqslant n-1\}.\end{aligned}$$

Let us consider a family $\{(X_n, v_n)\}$ of marked graphs, i.e. graphs with chosen vertices $v_n \in X_n$.

On the space of marked graphs there is a metric *Dist* defined as follows

$$Dist((X_1, v_1), (X_2, v_2)) = \inf\left\{\frac{1}{n+1}; B_{X_1}(v_1, n) \text{ is isometric to } B_{X_2}(v_2, n)\right\}$$

where $B_X(v, n)$ is the ball of radius n in X centered on v.

For a sequence of marked graphs (X_n, v_n) we say that (X, v) is the limit graph if

$$\lim_{n\to\infty} Dist((X, v), (X_n, v_n)) = 0.$$

The limit graph is unique up to the isometry.

We will consider locally finite graphs, i.e. the degree $deg(v)$ of each vertex v is finite and we will always assume that the graphs are connected. Now we prove:

Lemma 1 *Let $\{(X_n, v_n)\}_{n=1}^{\infty}$ be a sequence of marked graphs whose degree is uniformly bounded, i.e. there exists $k > 0$, such that $\deg(X_n) \leqslant k$ for all $n \in \mathbb{N}$. Then there exists a subsequence $\{(X_{n_i}, v_{n_i})\}_{i=1}^{\infty}$ which converges to some marked graph (X, v).*

Proof Because the degrees of the graphs are uniformly bounded, we can use the diagonal argument. □

Lemma 1 has as a corollary the following theorem:

Theorem 3 *The space of marked graphs of uniformly bounded degree is compact.*

Remark The analogous statement for the case of the space of Cayley graphs of groups with a fixed set of generators was mentioned in [8].

Proposition 1 *For any regular marked graph (X, v) there exists a sequence of finite marked regular graphs (X_n, v_n) converging to (X, v).*

Proof First of all let us suppose that the degree of X is even and equal to $2n$. Then X can be represented as the Schreier graph of F_n/H where F_n is a free group on n generators, H some subgroup of F_n and as generators for F_n/H we take the images of standard generators of F_n. We can suppose that the vertex v in X is the image of the identity element e in F_n. Now

$H = \bigcup_{i=1}^{\infty} H_i$, where H_i is a sequence of subgroups of F_n such that for every i we have $H_i \subset H_{i+1}$ and H_i is finitely generated. By a theorem of M. Hall [9] every finitely generated subgroup of H_i can be represented as the intersection $\bigcap_{j=1}^{\infty} H_{ij}$ where H_{ij} are subgroups of F_n of finite index. By a diagonal process we can choose a sequence $\{H_{i_k j_k}\}_{k=1}^{\infty}$ such that the finite marked Schreier's graphs $\{F_n/H_{i_k j_k}, e\}_{k=1}^{\infty}$ converge to (X, v).

In the case when the degree of X is odd, the proof is similar but we have to use the version of Hall's theorem for $\mathbb{Z}_2 * \ldots * \mathbb{Z}_2$. □

Remark There are other more direct proofs of Proposition 1. One of them was indicated to us by L. Bartholdi.

3 Continuity of the spectral measure of the Markov operator on graphs

On the locally finite, connected graph $X = (V_X, E_X)$ we can consider a random walk operator M acting on functions $f \in l^2(X, deg)$ as follows:

$$Mf(v) = \frac{1}{deg(v)} \sum_{w \sim v} f(w),$$

where $w \sim v$ means that w is a neighbor of v. This is a self-adjoint operator on $l^2(X, deg)$.

This random walk operator is related to the following simple random walk on X: starting at vertex v, we choose uniformly at random one of its neighbors w and then go to w.

By the random walk operator on marked graphs we will mean the random walk operator on the corresponding non-marked graph.

Let $\rho(M)$ be the spectral radius of M. Then it is known and easy to see that

$$\lim_{n \to \infty} \sqrt[2n]{p^{2n}(v, v)} = \rho(M) = ||M||_{l^2(X, deg)}$$

for any vertex $v \in X$, where $p^{2n}(v, v)$ is the probability of going from v to v in $2n$ steps. In particular $||M|| = \rho(M) \leqslant 1$ and $\rho(M)$ is in the spectrum of M. If X is finite then $||M|| = 1$ and 1 is the largest eigenvalue of M. For a finite connected graph X let $\lambda_1(X)$ denote the second eigenvalue after 1 in a natural decreasing ordering of points in the spectrum of M.

Lemma 2 *Let $f : X_1 \to X_2$ be a covering between two graphs X_1 and X_2. Then*

$$\rho(X_2) \geqslant \rho(X_1).$$

Proof By definition of the cover, different loops in X_1 are projected onto different loops in X_2. The loops in X_1 which start and finish in v are projected onto loops in X_2 which start and finish in $f(v)$. Thus

$$p^n(v,v) \leqslant p^n(f(v), f(v)),$$

which implies that $\rho(X_1) \leqslant \rho(X_2)$. □

Since M is a bounded ($||M|| \leqslant 1$) and self-adjoint operator, it has the spectral decomposition

$$M = \int_{-1}^{1} \lambda E(\lambda),$$

where E is the spectral measure. This spectral measure is defined on Borel subsets of the interval $[-1,1]$ and takes the values in projections on the Hilbert space $l^2(X, deg)$. The matrix μ^X of measures μ^X_{xy} for vertices $x, y \in X$ can be associated with E as follows:

$$\mu^X_{xy}(B) = \langle E(B)\delta_x, \delta_y \rangle,$$

where B is a Borel subset of $[-1,1]$ and δ_x is the function which equals 1 at x and 0 elsewhere.

Now, in general, $\lambda \in Sp(M)$ if and only if for every $\varepsilon > 0$ there exists μ^X_{xy} such that $|\mu^X_{xy}((\lambda - \varepsilon, \lambda + \varepsilon))| > 0$. But we also have the following result (see [12]):

Lemma 3 *$\lambda \in Sp(M)$ if and only if for every $\varepsilon > 0$ there exists $x \in X$ such that $\mu^X_{xx}((\lambda - \varepsilon, \lambda + \varepsilon)) > 0$.*

Proof We need only show that if, for $B = (\lambda - \varepsilon, \lambda + \varepsilon)$, we have $|\mu^X_{xy}(B)| > 0$ then $\mu^X_{xx}(B) > 0$. As $E(B)$ is a projection, one has

$$\begin{aligned} 0 &< (\mu^X_{xy}(B))^2 = \langle E(B)\delta_x, \delta_y \rangle^2 \leqslant \langle E(B)\delta_x, E(B)\delta_x \rangle \langle \delta_y, \delta_y \rangle \\ &= \langle E(B)\delta_x, \delta_x \rangle deg(y) = \mu^X_{xx}(B) deg(y), \end{aligned}$$

which ends the proof. □

Our main tool will be the weak convergence of the measures $\mu^{X_n}_{v_n v_n}$ to the measure μ^X_{vv} (provided that the sequence of marked graphs (X_n, v_n) converges to the marked graph (X, v)), i.e.

$$\lim_{n \to \infty} \int_{-1}^{1} f \mu^{X_n}_{v_n v_n} = \int_{-1}^{1} f \mu^X_{vv}$$

for any $f \in C[-1,1]$. The weak convergence implies (see for instance [3]) that for any open interval $B \subset [-1,1]$:

$$\liminf_{n \to \infty} \mu^{X_n}_{v_n v_n}(B) \geqslant \mu^X_{vv}(B).$$

Lemma 4 *Let us suppose that the sequence of marked graphs (X_n, v_n) converges to the marked graph (X, v). Then the measures $\mu^{X_n}_{v_n v_n}$ converge weakly to the measure μ^X_{vv}.*

Proof We are going to prove that the moments of the measures $\mu^{X_n}_{v_n v_n}$ converge to the moments of the measure μ^X_{vv}. As it is easy to see, in our situation this implies weak convergence of corresponding measures (see for example [3]). The l-th moment of the measure μ^Y_{yy} for a graph Y and $y \in Y$ is given by

$$(\mu^Y_{yy})^{(l)} = \int_{-1}^{1} \lambda^l \mu^Y_{yy}(\lambda) = \int_{-1}^{1} \lambda^l \langle E(\lambda)\delta_y, \delta_y\rangle = \langle M^l \delta_y, \delta_y\rangle.$$

Thus the l-th moment of the measure μ^Y_{yy} is equal to the probability of going from y to y in l steps. But for n sufficiently large, the balls $B_{X_n}(v_n, l)$ and $B_X(v, l)$ are isometric and the l-th moment of the measure $\mu^{X_n}_{v_n v_n}$ is the same as the l-th moment of the measure μ^X_{vv}.

This ends the proof of Lemma 4. □

Proof of Theorem 2 This is a consequence of Lemma 4 and the fact that the space of graphs that we are considering is a metric space. □

4 Some applications

Proposition 2 *Let $\{(X_n, v_n)\}_{n=1}^{\infty}$ be a sequence of marked graphs which converges to the marked graph (X, v). Then for every α in the spectrum of the random walk operator M_X on X, i.e. $\alpha \in Sp(M_X)$, and for every $\varepsilon > 0$, there exists $N \in \mathbb{N}$ such that for $n \geqslant N$ there is always a spectral value of M_{X_n} which is in the interval $(\alpha - \varepsilon, \alpha + \varepsilon)$.*

Proof Let x be any vertex in X, and let d be its distance from $v \in X$, i.e. $d = \text{dist}_X(v, x)$. For n sufficiently large the balls $B_{X_n}(v_n, d)$ and $B_X(v, d)$ are isometric. Let x_n be the image of x in X_n. Now, for any $\lambda \in Sp(M_X)$ and for every $\varepsilon > 0$, there exists $x \in X$ such that

$$\mu^X_{xx}((\alpha - \varepsilon, \alpha + \varepsilon)) = c > 0.$$

This means that for any $\varepsilon' > 0$ and n sufficiently large, by Lemma 4 we have

$$\mu^{X_n}_{x_n x_n}((\alpha - \varepsilon, \alpha + \varepsilon)) \geqslant c - \varepsilon',$$

which ends the proof of Proposition 2. □

We can prove the following proposition:

Proposition 3 *Let $\{(X_n, v_n)\}_{n=1}^{\infty}$ be a sequence of finite marked graphs convergent to the marked graph (X, v). If the number of vertices of the graph X_n tends to infinity, then*

$$\liminf_{n\to\infty} \lambda_1(X_n) \geqslant \rho(X).$$

Proof As $\rho(X) \in Sp(M_X)$, Proposition 3 follows immediately from Proposition 2 in the case where $\rho(X) < 1$. In the case $\rho(X) = 1$ we need the following:

Lemma 5 *Let X be an infinite, connected graph such that $\rho(X) = 1$. Then 1 is not an isolated eigenvalue in $Sp(M_X)$.*

Proof If 1 were an isolated eigenvalue then there would be an eigenfunction in $l^2(X, deg)$ with the eigenvalue 1. Because the l^2 norm of this function is finite, it has to attain either a maximum or minimum. But as it corresponds to the eigenvalue 1, it attains the extremum also on the neighbours of the vertex on which it attains the extremum. And as the graph is connected, this implies that this eigenfunction is constant. However, as the graph is infinite this means that l^2 norm is infinite and we obtain a contradiction. □

Thus if $\rho(X) = 1$, Proposition 2 and Lemma 5 imply that for n sufficiently large there are eigenvalues of X_n different from 1 but arbitrarily close to 1. This ends the proof of Proposition 3. □

As a corollary we obtain the following generalization of Alon-Boppana's theorem:

Theorem 4 *Let $\{X_n\}_{n=1}^{\infty}$ be a sequence of finite connected graphs. If the degree of each graph (which does not have to be regular) is bounded by k and the number of vertices of X_n tends to infinity, then*

$$\liminf_{n\to\infty} \lambda_1(X_n) \geqslant \frac{2\sqrt{k-1}}{k}.$$

Proof Suppose this is not true, i.e.

$$\liminf_{n\to\infty} \lambda_1(X_n) < \frac{2\sqrt{k-1}}{k}.$$

Let v_n be any vertex in the graph X_n. By Lemma 2, there exists a subsequence of marked graphs $\{(X_{n_i}, v_{n_i})\}_{i=1}^{\infty}$ convergent to the marked graph (X, v). The degree of the graph X is bounded by k. So one knows (see [6] for the case when X is a k-regular graph and see [24] for the general case) that

$$\rho(X) \geqslant \frac{2\sqrt{k-1}}{k}.$$

Thus by Proposition 3 we have

$$\liminf_{i\to\infty} \lambda_1(X_{n_i}) \geqslant \frac{2\sqrt{k-1}}{k}$$

which gives a contradiction. □

If $\alpha \in Sp(M_X)$ is not an isolated eigenvalue in $Sp(M_X)$ then Proposition 2 has as a corollary the following proposition:

Proposition 4 *Let $\{(X_n, v_n)\}_{n=1}^{\infty}$ be a sequence of marked graphs which converges to the marked graph (X, v). If $\alpha \in Sp(M_X)$ is not an isolated point in $Sp(M_X)$ then for every $\varepsilon > 0$:*

$$\#\{\lambda \in Sp(M_{X_n}); \lambda \in (\alpha-\varepsilon, \alpha+\varepsilon)\} \to_{n\to\infty} \infty.$$

Below is one more example (besides the one given in Lemma 5) of infinite graphs for which the spectral radius is not an isolated point in the spectrum:

Proposition 5 *If there exists a vertex in X whose orbit under the group $Aut(X)$ of automorphisms of X is infinite then $\rho(X)$ is not an isolated point in the spectrum $Sp(M_X)$.*

Proof If $\rho(X)$ were an isolated eigenvalue then there would be an eigenfunction f in $l^2(X, deg)$ corresponding to $\rho(X)$. As the norm of M is attained on f and because for any $g \in l^2(X, deg)$ one has $\|Mg\| \leqslant \|M|g|\|$, we can suppose that f is positive. The function f must be strictly positive, because if it were zero somewhere, it would be zero at its neighbours, and as X is connected this gives a contradiction. Now, using the group $Aut(X)$ we can translate the function f and obtain a new eigenfunction f'. As there is a vertex whose orbit under $Aut(X)$ is infinite, we can obtain the eigenfunction f' which is different from f. Now, let us consider the function F on X:

$$F(x) = \max\{f(x), f'(x)\}.$$

For F one has

$$MF(x) \geqslant \rho(X)F(x)$$

for each $x \in X$ and F belongs to $l^2(X, deg)$. As f and f' have the same l^2 norms, there exist v and w in X such that $f(v) > f'(v)$ and $f(w) < f'(w)$. There is a vertex $z \in X$ such that the strict inequality

$$MF(z) > \rho(X)F(z), \tag{1}$$

holds. Indeed, if $MF(v) = \rho(X)F(v)$ then $f(u) \geqslant f'(u)$ for all vertices u at distance 1 from v. If $MF(u) = \rho(X)F(u)$ for every such vertex then

$f(t) \geqslant f'(t)$ for all vertices t at distance 2 from v, etc. As $f(w) < f'(w)$ and X is connected, there must be a vertex z such that the inequality (1) holds.

But this means that $||M|| > \rho(X)$ which gives a contradiction. □

Now we will consider graphs with a finite number of orbits, i.e. graphs X such that $\#(X/Aut(X))$ is finite.

Theorem 5 *Let $\{(X_n, v_n)\}_{n=1}^{\infty}$ be a sequence of finite marked graphs with a degree uniformly bounded by k and such that $\#(X_n/Aut(X_n)) \leqslant K$ for each graph X_n. If (X, v) is the limit graph for the sequence $\{(X_n, v_n)\}_{n=1}^{\infty}$ then for any open interval $B \subset [-1, 1]$ and any vertex $v \in X$ we have:*

$$\liminf_{n\to\infty} \frac{\#\{\lambda \in Sp(M_{X_n}); \lambda \in B\}}{\#X_n} \geqslant \frac{(\mu_{vv}^X(B))^2}{K \cdot k^2},$$

where the eigenvalues are counted with multiplicities.

Proof We will prove Theorem 5 in the case where $K = 1$. The proof in the general case is similar. Thus we will consider homogenous graphs, i.e. graphs X such that $\#(X/Aut(X)) = 1$. For a homogenous graph, the spectral measure μ_{xx}^X does not depend on $x \in X$ and we will write for simplicity μ^X. Also, if we consider a sequence of marked homogenous graphs $\{(X_n, v_n)\}_{n=1}^{\infty}$ convergent to the marked graph (X, v), the limit space does not depend on the sequence v_n and the limit graph is homogenous. We can suppose that the degree of each graph X_n is k.

Let $d(n) = \#\{\lambda \in Sp(M_{X_n}); \lambda \in B\}$.

For $\lambda_i \in Sp(M_{X_n})$, let E_{λ_i} be the eigenspace in $l^2(X_n, deg)$ corresponding to λ_i. The $l^2(X_n, deg)$ admits the following decomposition

$$l^2(X_n, deg) = \bigoplus_{i=1}^{\#X_n} E_{\lambda_i},$$

where we suppose E_{λ_i} to be one-dimensional and thus we take as many one-dimensional eigenspaces as multiplicity indicates. So

$$E_{\lambda_i} = \{\lambda f_i; \lambda \in \mathbb{R}, ||f_i||_{l^2(X_n,deg)} = 1\}.$$

Let us choose a vertex $x \in X_n$. Then we have

$$\delta_x = \sum_{i=1}^{\#X_n} a_i f_i,$$

where a_i are real numbers such that $\sum_{i=1}^{\#X_n} a_i^2 = ||\delta_x||_2^2 = k$. And by definition of the projection $E(B)$

$$E(B)\delta_x = \sum_{j=1}^{d(n)} a_{i_j} f_{i_j}.$$

Because of the weak convergence of the measures μ^{X_n} to the measure μ^X (Lemma 4), for any $\varepsilon > 0$ there exists $N \in \mathbb{N}$, such that for $n > N$ we have

$$\mu^X(B) - \varepsilon \leqslant \mu^{X_n}(B).$$

By definition we have

$$\mu^{X_n}(B) = \mu^{X_n}_{xx}(B),$$

where x is any vertex in X_n. So

$$\begin{aligned}
\mu^X(B) - \varepsilon &\leqslant \mu^{X_n}(B) = \mu^{X_n}_{xx}(B) = \langle E(B)\delta_x, \delta_x\rangle \\
&= \left\langle \sum_{j=1}^{d(n)} a_{i_j} f_{i_j}, \delta_x \right\rangle = \sum_{j=1}^{d(n)} a_{i_j} f_{i_j}(x) k \\
&\leqslant \sqrt{\sum_{j=1}^{d(n)} a_{i_j}^2} \sqrt{\sum_{j=1}^{d(n)} f_{i_j}^2(x) k} \leqslant ||\delta_x||_2 \sqrt{\sum_{j=1}^{d(n)} f_{i_j}^2(x) k} \\
&= \sqrt{\sum_{j=1}^{d(n)} f_{i_j}^2(x) k}\sqrt{k}.
\end{aligned}$$

Thus

$$(\mu^X(B) - \varepsilon)^2 \leqslant \sum_{i=1}^{d(n)} f_{i_j}^2(x) k^3.$$

This gives us

$$\begin{aligned}
\# X_n (\mu^X(B) - \varepsilon)^2 &= \sum_{x \in X_n} (\mu^X(B) - \varepsilon)^2 \\
&\leqslant \sum_{x \in X_n} \sum_{j=1}^{d(n)} f_{i_j}^2(x) k^3 \\
&= \sum_{j=1}^{d(n)} \left(\sum_{x \in X_n} f_{i_j}^2(x) k \right) k^2 \\
&= d(n) k^2,
\end{aligned}$$

which ends the proof. □

For future reference, we notice that the above proof shows:

$$\sum_{x \in X_n} \left(\mu^{X_n}_{xx}(B)\right)^2 \leqslant d(n) k^2. \tag{2}$$

We are going to prove two statements analogous to Proposition 3 and Theorem 5 but in a slightly different situation.

Proposition 6 *Let $\{X_n\}_{n=1}^{\infty}$ be a sequence of finite graphs such that*

1. *each graph X_n is covered by a graph X and*

2. *the number of vertices of X_n tends to infinity, i.e. $\#X_n \to_{n\to\infty} \infty$.*

Then

$$\liminf_{n\to\infty} \lambda_1(X_n) \geqslant \rho(X).$$

Proof Suppose that this is not true, i.e. there exists $\varepsilon > 0$ and a subsequence $\{X_{n_i}\}_{i=1}^{\infty}$ such that

$$\lambda_1(X_{n_i}) \leqslant \rho(X) - \varepsilon.$$

Because X covers a finite graph, the degree of graphs $\{X_n\}_{n=1}^{\infty}$ is uniformly bounded. Let us choose a vertex v in X, and let v_n be its image in X_n. By Lemma 1 there exists a subsequence $\{X_{n_{i_j}}, v_{n_{i_j}}\}_{j=1}^{\infty}$ of marked graphs which is convergent to a marked graph (X', v'). Now we prove:

Lemma 6 *The limit graph X' is covered by the graph X.*

Proof We will construct a covering of X' by X such that v is mapped onto v'. For $m > 0$, let $\delta(m)$ be a constant such that the ball $B_{X_{n_{i_{\delta(m)}}}}(v_{n_{i_{\delta(m)}}}, m)$ is isometric to the ball $B_{X'}(v', m)$. Let $p_{n_{i_{\delta(m)}}} : X \to X_{n_{i_{\delta(m)}}}$ be a cover mapping between the graph X and the graph $X_{n_{i_{\delta(m)}}}$ such that $p_{n_{i_{\delta(m)}}}(v) = v_{n_{i_{\delta(m)}}}$. By definition of the covering, it follows that the ball $B_X(v, m)$ covers the ball $B_{X_{n_{i_{\delta(m)}}}}(v_{n_{i_{\delta(m)}}}, m)$. As the latter is isometric to $B_{X'}(v', m)$ let

$$f_m : B_X(v, m) \to B_{X'}(v', m)$$

be a covering of $B_{X'}(v', m)$ by $B_X(v, m)$ such that $f_m(v) = v'$.

By the diagonal argument, there exists a subsequence $\{f_{m_j}\}_{j=1}^{\infty}$ of coverings such that f_{m_j}, restricted to $B_X(v, m_{j'})$ for $m_{j'} \leqslant m_j$, coincides with $f_{m_{j'}}$, i.e.

$$f_{m_j}|_{B_{X(v,m_{j'})}} = f_{m_{j'}}.$$

Thus there exists a covering

$$f : X \to X',$$

such that $f|_{B_{X(v,m_j)}} = f_{m_j}$, which ends the proof of Lemma 6. □

Thus by Lemma 2

$$\rho(X') \geqslant \rho(X).$$

Hence Proposition 3 gives us the desired contradiction. This ends the proof of Proposition 6. □

Theorem 6 *Let $\{X_n\}_{n=1}^{\infty}$ be a sequence of finite connected graphs with uniformly bounded degree, such that the number of vertices of X_n tends to infinity.*

1. *Then there exists a subsequence $\{X_{n_i}\}_{i=1}^{\infty}$ and a graph X' such that for any open interval $B \subset [-1,1]$ and any vertex $x' \in X'$ there exists a constant $c_{x'} > 0$ such that*

$$\liminf_{n\to\infty} \frac{\#\{\lambda \in Sp(M_{X_{n_i}}); \lambda \in B\}}{\#X_n} \geqslant c_{x'} \cdot (\mu^{X'}_{x'x'}(B))^2.$$

2. *If the graphs X_n are covered by a graph X such that there exists a group Γ acting cocompactly on X then X' can be chosen in such a way that X covers X'.*

Proof As the graphs X_n have uniformly bounded degree, there exist positive integers $N^1, \dots, N^R, \dots$ with the following property:

Property (P) *Let us consider any graph X_n. Then up to the isometry there are at most N^R different balls of radius R in X_n.*

Let $\{B_R^{X_n}\}_{R=1}^{\infty}$ be any sequence of sets of balls in X_n of radius R with the following properties:

1. For fixed n and R the balls in $B_R^{X_n}$ are isometric to each other;
2. If $B(v,R) \in B_R^{X_n}$ then $B(v,R-1) \in B_{R-1}^{X_n}$.

Now we prove:

Lemma 7 *It is possible to find a sequence $\{B_R^{X_n}\}_{R=1}^{\infty}$ in such a way that*

$$\#B_R^{X_n} \geqslant \#X_n \cdot c_R,$$

where $c_R = (N^1 \cdot \ldots \cdot N^R)^{-1}$.

Proof We construct the sequence $\{B_R^{X_n}\}_{R=1}^{\infty}$ by induction:

1. There are $\#X_n$ balls of radius 1 in X_n. According to Property (P) there are at least $\#X_n \cdot (N^1)^{-1}$ among these balls which are isometric to each other. So let $B_1^{X_n}$ be any set of cardinality at least $\#X_n \cdot (N^1)^{-1}$ consisting of balls of radius 1 which are isometric to each other.

2. Suppose that we constructed the sets $B_1^{X_n}, \dots, B_{R-1}^{X_n}$ with the desired properties. Let us consider the balls of radius R with the same centers as the balls in $B_{R-1}^{X_n}$. According to Property (P) there are at least $\#B_{R-1}^{X_n} \cdot (N^R)^{-1}$ among these balls which are isometric to each other and let $B_R^{X_n}$ be any set of them. Thus by induction

$$\#B_R^{X_n} \geqslant \#B_{R-1}^{X_n} \cdot (N^R)^{-1} \geqslant \#X_n \cdot c_{R-1} \cdot (N^R)^{-1} = \#X_n \cdot c_R,$$

which ends the proof of Lemma 7. □

Because the graphs X_n are finite, for each graph X_n there exists an integer $R(n)$ such that for $R > R(n)$ one has

$$B_R^{X_n} = B_{R(n)}^{X_n}.$$

So for each X_n there exists a vertex $v_n \in X_n$ such that for every R there is at least one ball in $B_R^{X_n}$ centered on v_n. By Lemma 1 there exists a subsequence of marked graphs $\{(X_{n_j}, v_{n_j})\}_{j=1}^{\infty}$ which is convergent to some marked graph (X', v). We will show that the sequence of graphs $\{X_{n_j}\}_{j=1}^{\infty}$ and the graph X' have the properties stated in Theorem 6.

Let $C\left(B_R^{X_n}\right)$ be the set of centers of the balls in $B_R^{X_n}$, i.e.

$$C\left(B_R^{X_n}\right) = \left\{v \in X_n; B(v,R) \in B_R^{X_n}\right\}.$$

We need the following:

Lemma 8 *For any interval $B \subset [-1,1]$ and $\varepsilon > 0$ there exist integers $N > 0$ and $R > 0$ such that for any $n_j > N$ and for any vertex $x \in C(B_R^{X_{n_j}})$, we have*

$$\mu_{xx}^{X_{n_j}}(B) \geqslant \mu_{vv}^{X'}(B) - \varepsilon.$$

Proof Suppose this is not true, i.e. there exist an interval $B \subset [-1,1]$, $\varepsilon > 0$, a subsequence of graphs $\{X_{n_{j_i}}\}_{i=1}^{\infty}$, a sequence of integers $\{R_i\}_{i=1}^{\infty}$ ($\lim_{i\to\infty} R_i = \infty$) and a sequence of vertices $x_i \in X_{n_{j_i}}$ ($x_i \in C\left(B_{R_i}^{X_{n_{j_i}}}\right)$) such that

$$\mu_{x_i x_i}^{X_{n_{j_i}}}(B) < \mu_{vv}^{X'}(B) - \varepsilon. \tag{3}$$

By construction, the sequence of marked graphs $\{(X_{n_{j_i}}, x_i)\}_{i=1}^{\infty}$ is convergent to the marked graph (X', v). So by Lemma 4 there exists an integer i such that

$$\mu_{x_i x_i}^{X_{n_{j_i}}}(B) \geqslant \mu_{vv}^{X'}(B) - \varepsilon,$$

which contradicts (3). □

We are now in a position to prove the first statement of Theorem 6. We will prove this for $x' = v$ and later remark that the same proof works for other vertices. First of all if $\mu_{vv}^{X'}(B) = 0$ there is nothing to prove.

Let us consider an open interval $B \subset [-1,1]$ such that $\mu_{vv}^{X'}(B) > 0$. Let $\varepsilon = \frac{1}{2}\mu_{vv}^{X'}(B)$. By Lemma 8 there exist positive integers N and R such that for $n_j > N$ and for any vertex $x \in C\left(B_R^{X_{n_j}}\right)$ we have:

$$\mu_{xx}^{X_{n_j}}(B) \geqslant \frac{1}{2}\mu_{vv}^{X'}(B).$$

Let $k = deg(x)$ (there exists a positive integer N' such that for $n_j > N'$ the degree of vertices in $C\left(B_R^{X_{n_j}}\right)$ is the same for all X_{n_j}). Then for $n_j > max\{N, N'\}$ by Lemma 7 and (2) we have:

$$\begin{aligned}
\#X_{n_j}\left(\frac{1}{2}\mu_{vv}^{X'}(B)\right)^2 &\leqslant (c_R)^{-1}\#B_R^{X_{n_j}}\cdot\left(\frac{1}{2}\mu_{vv}^{X'}(B)\right)^2 \\
&= (c_R)^{-1}\sum_{x\in C\ B_R^{X_{n_j}}}\left(\frac{1}{2}\mu_{vv}^{X'}(B)\right)^2 \\
&\leqslant (c_R)^{-1}\sum_{x\in C\ B_R^{X_{n_j}}}\left(\mu_{xx}^{X_{n_j}}(B)\right)^2 \\
&\leqslant (c_R)^{-1}d(n_j)k^2.
\end{aligned}$$

As c_R and k are constants which do not depend on X_{n_j}, this ends the proof of Theorem 6 for $x' = v$. In the case when $x' \neq v$ the proof is almost the same, because for R sufficently large, x' is in the ball $B(v, R)$ and therefore in our considerations we can replace v by x'.

Now we prove the second statement in Theorem 6. Because $\Gamma \setminus X$ is finite, there exists a vertex $w \in X$ and a subsequence of marked graphs $\{(X_{n_{j_i}}, v_{n_{j_i}})\}_{i=1}^{\infty}$ such that a covering of $X_{n_{j_i}}$ by X maps w onto $v_{n_{j_i}}$. As the sequence $\{(X_{n_{j_i}}, v_{n_{j_i}})\}_{i=1}^{\infty}$ is convergent to the marked graph (X', v), by Lemma 6 we have that X' is covered by X. □

Theorem 6 has the following corollary:

Corollary 1 *Let X be an infinite connected graph and X_n the family of finite connected graphs covered by X. (Where here we mean that for each X_n there exists a group Γ_n acting on X such that $\Gamma_n \setminus X = X_n$.) Then for every $\varepsilon > 0$ there exists $c = c(X, \varepsilon)$, $0 < c < 1$, such that at least $c\#X_n$ eigenvalues λ of X_n satisfy $\lambda \geqslant \rho(X) - \varepsilon$. In particular for any $r \in \mathbb{N}$ we have*

$$\liminf_{n\to\infty} \lambda_r(X_n) \geqslant \rho(X),$$

where $\lambda_1(X_n) \geqslant \lambda_2(X_n) \geqslant \ldots$ are the eigenvalues of M_{X_n} listed in ordered way.

The first statement of Corollary 1 was proved by Greenberg [5] (see also [14]) and the second statement was proved by Burger [2]. In the case when X is a regular tree Corollary 1 was observed by Serre [21]. The exposition of all statements contained in Corollary 1 can be also found in [13] (Theorem 13 in Chapter 9).

Proof of Corollary 1 Suppose this is not true, i.e. there exists $\varepsilon > 0$ and a subsequence of graphs $\{X_{n_i}\}_{i=1}^{\infty}$ such that

$$\lim_{n\to\infty} \frac{\#\{\lambda \in Sp(M_{X_{n_i}}); \lambda \geqslant \rho(X) - \varepsilon\}}{\#X_{n_i}} = 0. \tag{4}$$

Clearly $\#X_n$ tends to infinity in this case. By Theorem 6 there exist a subsequence $\{X_{n_{i_j}}\}_{j=1}^{\infty}$ and a graph X' which is covered by the graph X, such that for any vertex $x' \in X'$ there is a constant $c_{x'} > 0$ such that

$$\liminf_{j\to\infty} \frac{\#\{\lambda \in Sp(M_{X_{n_{i_j}}}); \lambda \geqslant \rho(X') - \varepsilon\}}{\#X_{n_{i_j}}} \geqslant c_{x'}\mu_{x'x'}^{X'}\left((\rho(X') - \varepsilon, 1]\right).$$

As $\rho(X')$ belongs to $Sp(M_{X'})$, by Lemma 3 there exists $x' \in X'$ such that $\mu_{x'x'}^{X'}((\rho(X') - \varepsilon, 1]) = c > 0$. By Lemma 2, $\rho(X') \geqslant \rho(X)$. Thus

$$\liminf_{j\to\infty} \frac{\#\{\lambda \in Sp(M_{X_{n_{i_j}}}); \lambda \geqslant \rho(X) - \varepsilon\}}{\#X_{n_{i_j}}} > 0,$$

which contradicts (4). □

5 Conclusions

Let $\{X_n\}_{n=1}^{\infty}$ be a sequence of finite graphs covered by a graph X or converging to X. According to our results there is an extremal case when $\lambda_1(X_n) \leqslant \rho(X)$ for all n and this leads to the following definitions, the first of which generalizes the ones given in [5] and [14]:

Definition 1 *Let $\{X_n\}$ be an infinite family of finite graphs, covered by an infinite graph X. The family $\{X_n\}$ is called X-**Ramanujan** if every eigenvalue not equal to 1 of the simple random walk on X_n is less than the spectral radius $\rho(X)$ of the simple random walk on X, i.e.*

$$\lambda_1(X_n) \leqslant \rho(X),$$

for each graph X_n.

Definition 2 *Let $\{(X_n, v_n)\}_{n=1}^{\infty}$ be a sequence of finite marked graphs, convergent to the infinite marked graph (X, v). The sequence $\{(X_n, v_n)\}_{n=1}^{\infty}$ is called (X, v)-**Ramanujan** if every eigenvalue not equal to 1 of the simple random walk on X_n is less than the spectral radius $\rho(X)$ of the simple random walk on X, i.e.*

$$\lambda_1(X_n) \leqslant \rho(X),$$

for each graph X_n.

Ramanujan graphs were defined in [16], [17] and implicitly in [11] as finite k-regular graphs with $\lambda_1 \leqslant \frac{2\sqrt{k-1}}{k}$. For a given k of the form $k = q + 1$ where q is a power of a prime number, the constructions of the infinite families of Ramanujan graphs covered by a regular tree (namely a Cayley graph of a free group of corresponding rank) were constructed by Lubotzky, Phillips and Sarnak in [16] and by Margulis in [17].

It would be interesting to find other examples of a graph X (not a tree) with the spectral radius $\frac{2\sqrt{k-1}}{k}$ and a sequence X_n which is X-Ramanujan or at least (X, v)-Ramanujan for some $v \in X$.

In [15] an example of an infinite non-regular tree X was constructed such that X admits infinitely many finite quotients and none of its quotients is X-Ramanujan.

Acknowledgments This work was done during our stay at the Université de Genève. We would like thank Pierre de la Harpe for providing a very stimulating mathematical environment and for his useful remarks. We would also like to thank Alexander Lubotzky for several enlightening discussions.

References

[1] N. Alon, *Eigenvalues and expanders*, Combinatorica, 6 (1986), pp. 83-96.

[2] M. Burger, *Cheng's inequality for graphs*, preprint 1987.

[3] W. Feller, *An introduction to Probability Theory and its Applications*, vol. II, Wiley, 1966.

[4] C. D. Godsil, B. Mohar, *Walk generating functions and spectral measures of infinite graphs*, Linear Algebra Appl. 107 (1988), 191-206.

[5] Y. Greenberg, *Ph.D. thesis*, Hebrew University, Jerusalem, 1995.

[6] R. I. Grigorchuk, *Symmetrical random walks on discrete groups*, in Multicomponent random systems, ed. R. L. Dobrushin, Ya. G. Sinai, p. 132-152, Nauka, Moscow 1978 (English translation: Advances in Probability and Related Topics, vol. 6, p. 285-325, Marcel Dekker, New York, 1980).

[7] R. I. Grigorchuk, *On Burnside's problem on periodic groups*, Funktsional. Anal. i Prilozhen. 14 (1980), no. 1, 53-54.

[8] R. I. Grigorchuk, *Degrees of growth of finitely generated groups and the theory of invariant means*, Izv. Akad. Nauk SSSR Ser. Mat. 48 (1984), no. 5, 939-985.

[9] M. Hall, *Coset representations in free groups*, Trans. Amer. Math. Soc., 67, (1949), 421-432.

[10] P. R. Halmos, *Introduction to Hilbert space and the theory of spectral multiplicity*, New York, Chelsea Publishing Company, 1951.

[11] Y. Ihara, *Discrete subgroups of* $PL(2, k_p)$, in Algebraic groups and discontinuous subgroups, Proc. Symp. Pure Math IX, Amer. Math. Soc., pp. 272-278, 1966.

[12] H. Kesten, *Symmetric random walks on groups*, Trans. Amer. Math. Soc., Vol. 92, 1959, 336-354.

[13] W. Li, *Number theory with applications*, World Scientific, 1996.

[14] A. Lubotzky, *Discrete Groups, Expanding graphs and Invariant Measures*, Progress in Mathematics, Birkhäuser 1994.

[15] A. Lubotzky, T. Nagnibeda, *Not every uniform tree covers Ramanujan graphs*, to appear in Journal of Combinatorial Theory, Series B.

[16] A. Lubotzky, R. Phillips, P. Sarnak, *Ramanujan graphs*, Combinatorica, 8 (1988), p. 261-277.

[17] G. Margulis, *Explicit group-theoretical constructions of combinatorial schemes and their application to the design of expanders and concentrators*, J. Problems of Information Transmission, 24 (1988), p. 39-46.

[18] B. Mohar, W. Woess, *A survey on spectra of infinite graphs*, Bull. London Math. Soc. 21 (1989), no. 3, 209-234.

[19] A. Nili, *On the second eigenvalue of a graph*, Discrete Math., 91 (1991), p. 207-210.

[20] P. Sarnak, *Some Applications of Modular Forms*, Cambridge Tracts in Mathematics 99, Cambridge University Press 1990.

[21] J.-P. Serre, *Lettre à Winnie Li*, 8 octobre 1990.

[22] J.-P. Serre, *Répartition asymptotique des valeurs propres de l'opérateur de Hecke T_p*, J. Amer. Math. Soc., 10, no. 1, p. 75-102, 1997.

[23] A. Valette, *Graphes de Ramanujan et applications*, Séminaire Bourbaki, Exposé 829, mars 1997.

[24] A. Żuk, *A generalized Følner condition and the norms of random walk operators*, preprint, 1996.

Rostislav I. Grigorchuk
Steklov Mathematical Institute
Vavilova Str. 42 Moscow, 117966, Russia
email: grigorch@alesia.ips.ras.ru

Andrzej Żuk
CNRS, Ecole Normale Supérieure de Lyon
Unité de Mathématiques Pures et Appliquées
46, Allée d'Italie, F-69364 Lyon cedex 07 France
email: `azuk@umpa.ens-lyon.fr`

Stochastic pin-ball

Geoffrey R. Grimmett

Abstract. A ball is propelled through a random environment of obstacles off which it rebounds with perfect reflection. What is the behaviour of the trajectory of the ball? We summarise known mathematical results concerning this model, which we call 'stochastic pin-ball', and which is known also as a 'Lorentz lattice gas' and a version of which is sometimes termed the 'Ehrenfest wind–tree model'. The rigorous theory is more extensively developed if the environment is allowed to include a positive density of space in which the ball behaves in the manner of a random walk. For a lattice model of this type, one may employ arguments of percolation theory in order to prove theorems concerning non-localisation, transience, and asymptotic normality, under certain assumptions on the environment.

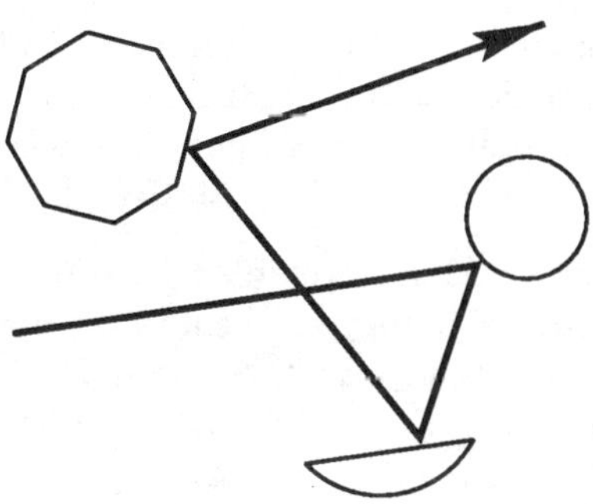

1. The origins of stochastic pin-ball

There is a modern version of the game of bagatelle involving a ball which is propelled about an inclined plane and which suffers deflections as a result of collisions with protruding nails. Modern pin-ball is an electrified version of this game, with a variety of obstacles and with interaction with the player. In a simplified stochastic model for the motion of the ball, we position smooth obstacles about $\mathbb{R}^2$ at random, and we then project a ball through the ensuing environment, requiring that the ball be reflected at the obstacles on which it impacts. What can be said about the trajectory of the ball?

Such a mathematical model is often named after Hendrik Lorentz, following his 1905 papers [21] concerning the motion of an electron through a field of massive particles. The 'Lorentz lattice gas' has generated considerable interest amongst physicists (see [8, 9, 10, 24, 26, 27]), but very little is known about the rigorous mathematical theory. The apparent difficulty of the problem is due to the fact that the model postulates a dynamical system within a random environment; the asymptotic behaviour of the dynamical system can be rather sensitive to small variations in the environment.

Lorentz's exposition was developed by Ehrenfest [14], and a simple version of the lattice gas model has become known as the 'Ehrenfest wind–tree model', a title with a natural interpretation. Another modern interpretation of such a system is that of a ray of light shining through a medium of mirrors: reflecting bodies are placed randomly in $\mathbb{R}^d$, and the trajectory of a ray of light through the subsequent environment is studied. These interpretations have as common requirement the definition of a probability measure governing the dispositions and shapes of the obstacles (massive particles, mirrors, trees, etc.). Once this is prescribed, then one seeks to categorise the trajectory of the ball (electron, light ray, wind, etc.) using words of the type 'recurrent, transient, ergodic, (non-)localised, diffusive'.

Only fragmentary progress has been made with the required mathematics, and we summarise some of this in the following paragraphs.

1. *Periodic pin-ball in* $\mathbb{R}^2$. Circular obstacles are distributed about a finite box of $\mathbb{R}^2$ in the manner of a Poisson process, and the contents of the finite box are copied periodically in a tiling of the plane $\mathbb{R}^2$. Subject to certain assumptions, Bunimovich and Sinai [3] have established a central limit theorem for the trajectory of the ball. See [25] for related material.

2. *Plane mirrors in* $\mathbb{R}^2$. Plane two-sided mirrors of unit length are distributed as follows about $\mathbb{R}^2$: they are centred at points of a Poisson process with intensity λ, and their orientations are chosen independently and randomly from a given countable set S having a certain property. We now project light from the origin. Harris [19] has proved that the light is a.s. localised (i.e., confined to a bounded region of $\mathbb{R}^2$) if $\lambda < \lambda_c$, and is non-localised with strictly positive probability when $\lambda > \lambda_c$. Here, λ_c is the critical density of the continuum percolation system of mirrors, viewed as unit rods. (Related results concerning continuum percolation may be found in [22].)

3. *Diagonal mirrors on* $\mathbb{Z}^2$. Each vertex of $\mathbb{Z}^2$ is designated a *mirror* with probability p and a *crossing* otherwise. Given that a vertex is a mirror, it is designated a north-west (nw) mirror with probability $\frac{1}{2}$ and a north-east (ne) mirror otherwise. We now place small plane two-sided mirrors at those vertices which have been designated mirrors, in the directions of the designations (see Figure 1). Light is shone northwards (say) from the origin, and we ask for properties of the ensuing trajectory.

It is trivial that light is non-localised if $p = 0$, and it is known but non-trivial that light is a.s. localised if $p = 1$ (see [4, 15, 16]). The latter statement is proved by a simple but beautiful argument from percolation

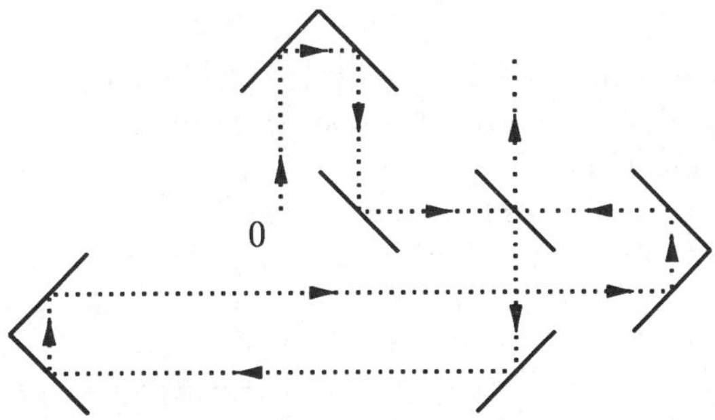

Fig. 1. A labyrinth of mirrors on the square lattice. The ray of light is reflected by the mirrors, and it is a problem is to determine, for a given density of mirrors, whether or not the light is a.s. restricted to a finite region.

theory. Some partial progress for the case of general p has been made recently by Quas [23], but mathematicians have no proof of the physicists' conjecture ([10, 27]) that light is a.s. localised whenever $p > 0$.

4. *Pin-ball with scatterers.* Menshikov and Volkov [18] have proposed a model in which the obstacles are augmented by zones in which the pin-ball behaves in the manner of a random walk. In a lattice model, for example, a positive density of vertices are designated random walk (rw) points. When the ball arrives at a rw point, it chooses its exit direction uniformly at random from those available. Not surprisingly, this additional randomness provides a degree of flexibility in the environment which may be harnessed by mathematicians. The subsequent theory for lattice pin-ball has been developed in [2, 16], and is summarised in Sections 2–4 of this survey.

5. *Rotator pin-ball.* Ruijgrok and Cohen [24] have proposed a general study of mirror and 'rotator' models. In a rotator model in $\mathbb{R}^2$, the direction of the ball is rotated clockwise through an angle $\theta(x)$ when it arrives at a vertex x; here, the $\theta(x)$ are independent, identically distributed random variables. One may also allow a stochastic variation in the environment, in the sense that the $\theta(x)$ may be allowed to vary randomly as time passes. See [4, 5] for partial results.

2. Generalised pin-ball

Our model for generalised pin-ball involves a random environment of reflecting bodies distributed around the vertices of the d-dimensional cubic lattice $\mathbb{L}^d$. Each vertex is designated either a reflector (of a randomly chosen type) or a 'random walk point'. The interpretation of the term 'random walk point' is as follows: when a ball impacts on such a point, then it departs in a direction chosen randomly from the $2d$ available directions, this exit being chosen independently of everything else. Such models were introduced in

[18] and have been studied further in [2, 16]. A similar model was proposed in [7] in the context of train sets, together with an application to the design of a computing machine.

There are many types of reflector, especially in three and more dimensions. The defining properties of a reflector ρ are that (i) to each incoming direction u there is assigned a unique outgoing direction $\rho(u)$, and (ii) the ball will retrace its path if the path's direction is reversed. Let $I = \{u_1, u_2, \dots, u_d\}$ be the set of positive unit vectors of $\mathbb{Z}^d$, and let $I^{\pm} = \{\alpha u_j : \alpha = \pm, 1 \leqslant j \leqslant d\}$. A *reflector* is defined to be a map $\rho : I^{\pm} \to I^{\pm}$ with the property that $\rho(-\rho(u)) = -u$ for all $u \in I^{\pm}$ (this condition is in response to the reversibility of light paths). We write $\mathcal{R}$ for the set of all reflectors. One particular reflector is special, namely the identity map satisfying $\rho(u) = u$ for all $u \in I^{\pm}$; we call this the *crossing*, and we denote it by $+$. Crossings do not deflect the ball.

A *random labyrinth* is defined as follows. Let p_{rw} and p_+ be non-negative reals such that $p_{\mathrm{rw}} + p_+ \leqslant 1$, and let π be a probability mass function on the set $\mathcal{R}\backslash\{+\}$ of 'non-trivial' reflectors (that is, $\pi(\rho) \geqslant 0$ for $\rho \in \mathcal{R}\backslash\{+\}$ and $\sum_{\rho \in \mathcal{R}\backslash\{+\}} \pi(\rho) = 1$). Let $Z = (Z_x : x \in \mathbb{Z}^d)$ be a family of independent random variables, taking values in $\mathcal{R} \cup \{\varnothing\}$, with probabilities

$$\mathbb{P}(Z_x = \beta) = \begin{cases} p_{\mathrm{rw}} & \text{if } \beta = \varnothing, \\ p_+ & \text{if } \beta = +, \\ (1 - p_{\mathrm{rw}} - p_+)\pi(\rho) & \text{if } \beta = \rho \in \mathcal{R}\backslash\{+\}. \end{cases}$$

A vertex x is called a *crossing* if $Z_x = +$, and a *random walk (rw) point* if $Z_x = \varnothing$.

We now study admissible paths in the labyrinth Z. Consider a path in $\mathbb{L}^d$ which visits (in order) the vertices $x_0, x_1, \dots, x_n$; we allow the path to revisit a given vertex more than once, and to traverse a given edge more than once. This path is *admissible* if it conforms to the reflectors which it meets, which is to say that

$$x_{j+1} - x_j = Z_{x_j}(x_j - x_{j-1}) \quad \text{for all } j \text{ such that } Z_{x_j} \neq \varnothing.$$

If $p_{\mathrm{rw}} = 0$, then very little is known about such systems except that which has been summarised in Section 1. Henceforth, we assume that $p_{\mathrm{rw}} > 0$, and we define a 'random walk in the labyrinth Z'. Let x be a rw point. A walker starts at x, and flips a fair $2d$-sided coin in order to determine the direction of its first step. Henceforth, it is required to traverse admissible paths only, and it flips the coin to determine its exit direction from any rw point encountered. We write P_x^Z for the law of the random walk in the labyrinth Z, starting from a rw point x.

There is a natural equivalence relation on the set N of rw points of $\mathbb{Z}^d$, namely $x \leftrightarrow y$ if there exists an admissible path with endpoints x and y. Let C_x be the equivalence class containing the rw point x. We may follow the progress of a random walk starting at x by writing down (in order) the rw points which it visits, say X_0 $(= x), X_1, X_2, \ldots$. Now, given Z, $X = (X_n)$ is an irreducible Markov chain on the countable state space C_x; furthermore it is reversible with respect to the measure μ given by $\mu(y) = 1$ for $y \in C_x$. We say that x is *Z-localised* if $|C_x| < \infty$, and *Z-non-localised* otherwise. We call Z *localised* if all rw points are Z-localised, and we call Z *non-localised* otherwise. By a zero–one law, we have that $\mathbb{P}(Z \text{ is localised})$ equals either 0 or 1.

We say that the rw point x is *Z-recurrent* if

$$P_x^Z(X_N = x \text{ for some } N \geqslant 1) = 1,$$

and *Z-transient* otherwise. The labyrinth Z is called *recurrent* if all rw points are Z-recurrent, and *transient* otherwise. By an appropriate zero–one law, we have that $\mathbb{P}(Z \text{ is recurrent})$ equals either 0 or 1.

We now state four problems concerning the random labyrinth Z. Only partial information about these problems is known.

1. Decide when it is the case that

$$\mathbb{P}(Z \text{ is localised}) = 1.$$

2. If $\mathbb{P}(Z \text{ is non-localised}) = 1$, decide when it is the case that

$$\mathbb{P}\Big(0 \text{ is } Z\text{-recurrent} \,\Big|\, 0 \text{ is a rw point}\Big) = 1.$$

3. Decide when it is the case that

$$\mathbb{P}\Big(P_0^Z(|X_n|^2) \,\Big|\, 0 \text{ is a rw point}\Big) \sim cn \qquad \text{as } n \to \infty$$

   for some $c > 0$. (Here, $m(Y)$ denotes the mean of Y under the measure m, and $|\cdot|$ denotes Euclidean distance.) There is also a 'pointwise', or 'quenched', version of this question.

4. If $\mathbb{P}(Z \text{ is non-localised}) = 1$ and $X_0 = 0$, $|C_0| = \infty$, decide when it is the case that (X_n) satisfies a central limit theorem, in the limit as $n \to \infty$.

Problems 3 and 4 are versions of the 'diffusivity' problem discussed in the physics literature (see, for example, [8, 9, 10, 24, 26, 27]). We note that the mean-square displacement of X_n could (in principle) grow linearly with n even when the walk is localised. In contrast, one cannot have a full central limit theorem without non-localisation.

3. Non-localisation and recurrence

We concentrate in this section on the property of non-localisation for generalised pinball. Let $p_c = p_c(\mathbb{L}^d)$ denote the critical probability of site percolation on $\mathbb{L}^d$; see [15] for an account of percolation theory.

Theorem 3.1. *Let $d \geqslant 2$ and $p_{rw} > 0$.*

(a) *The number M of infinite equivalence classes of $(N, \leftrightarrow)$ satisfies*

$$\text{either } \mathbb{P}(M = 0) = 1 \text{ or } \mathbb{P}(M = 1) = 1.$$

(b) *There exists a strictly positive constant $A = A(p_{rw}, d)$ such that*

$$\mathbb{P}(Z \text{ is non-localised}) = 1 \tag{3.1}$$

if either $1 - p_{rw} - p_+ < A$ or $p_{rw} > p_c$.

Part (a) is proved by adapting the scheme of Burton and Keane [6] who proved the uniqueness of infinite clusters in percolation-type models. The details may be found in [2]. As for part (b), two related but distinct proofs have appeared in [16, 18]. The major difficulty is to prove non-localisation under the assumption that the density $1 - p_{rw} - p_+$ is small. Of greatest value in the proofs of part (b) is the 'block method' of [18], which provides a powerful tool for controlling the geometry of the labyrinth, and which is useful for other problems too.

One may find cases of labyrinths which are localised, and also non-localised labyrinths which fall outside the conditions of part (b) of Theorem 3.1. See [16, 18] for the latter.

We turn now to the question of determining whether a labyrinth is transient or recurrent. For this problem, the most useful arguments appear to be those related to certain corresponding electrical networks; see [13, 17]. One may use block arguments, referred to above, in order to compare a random walk in a random labyrinth with a random walk on the infinite cluster of a certain related percolation model. Another feasible approach might be to employ the results of [1].

Theorem 3.2. *Let $p_{rw} > 0$.*

(a) *If $d = 2$, the labyrinth Z is $\mathbb{P}$-a.s. recurrent.*

(b) *Let $d \geqslant 3$. There exists a strictly positive constant $A = A(p_{rw}, d)$ such that: (3.1) holds, and in addition*

$$\mathbb{P}(Z \text{ is transient}) = 1,$$

whenever either $1 - p_{rw} - p_+ < A$ or $p_{rw} > p_c$.

Part (a) is a minor extension of Theorem 3 of [18], proved similarly. The conclusion of part (b) has appeared in [16, 18] under the condition that $1 - p_{rw} - p_+ < A$; when $p_{rw} > p_c$, the claim follows from the results and arguments of [17, 18].

4. Central limit theorem

When $p_{rw} > 0$, the ball follows a type of random walk in a random environment, the environment being the rather rigid one provided by the pin-ball table. Whenever the walk is non-localised, it is natural to seek a central limit theorem (CLT) for its displacement after n units of time have elapsed. The basic methodology of the theorem which follows is the CLT of Kipnis and Varadhan [20], together with its application to percolation by DeMasi, Ferrari, Goldstein, and Wick [11, 12]. Numerous complications arise in applying such techniques in the present setting.

Suppose that the origin 0 is a rw point. As before, we consider the sequence $X_0 \ (= 0), X_1, X_2, \ldots$ of rw points visited in sequence by a random walk in Z beginning at the origin 0. For $\epsilon > 0$, we let

$$X^\epsilon(t) = \epsilon X_{\lfloor \epsilon^{-2} t \rfloor} \qquad \text{for } t \geqslant 0,$$

and we are interested in the behaviour of the process $X^\epsilon(\cdot)$ in the limit as $\epsilon \downarrow 0$. We shall study X^ϵ under the probability measure $\mathbb{P}_0$, defined as the measure $\mathbb{P}$ conditional on the event $\{0 \text{ is a rw point, and } |C_0| = \infty\}$.

Theorem 4.1. *Let $d \geqslant 2$ and $p_{rw} > 0$. There exists a strictly positive constant $A = A(p_{rw}, d)$ such that the following holds whenever either $1 - p_{rw} - p_+ < A$ or $p_{rw} > p_c$:*

(a) *$\mathbb{P}(0$ is a rw point, and $|C_0| = \infty) > 0$, and*

(b) *as $\epsilon \downarrow 0$, the re-scaled process $X^\epsilon(\cdot)$ converges $\mathbb{P}_0$-dp to $\sqrt{\delta} W$, where W is a standard Brownian motion in $\mathbb{R}^d$ and δ is a strictly positive constant.*

The convergence '$\mathbb{P}_0$ dp' means that

$$P_0^Z(f(X^\epsilon)) \to E(f(\sqrt{\delta} W)) \quad \text{in } \mathbb{P}_0\text{-probability}, \tag{4.1}$$

for all bounded continuous functions on the appropriate Skorohod path-space $D([0,\infty), \mathbb{R}^d)$. Here, E stands for the canonical expectation operator.

We finish with some remarks concerning Theorem 4.1 (full details of the proof of which may be found in [2, 16, 18]). In applying the CLT of [12, 20], one needs certain information about the geometry of the pin-ball table. Non-trivial reflectors (i.e., reflectors other than crossings) may have complicated geometries, and so one works as much as possible on volumes of space which contain only rw points and crossings. The geometry of such regions may be controlled as follows. First, one states an appropriate property of a large block, and then one utilises arguments from percolation theory (see [15]) to describe the set of 'good' blocks. More precisely, when the density of good blocks exceeds the critical percolation probability p_c, then there exists a.s. an infinite cluster of good blocks, all of whose rw points lie in a single inter-communicating class within the pin-ball table. The validity of a CLT then follows in a fairly straightforward manner from the results of [12, 20]. It is however a substantial problem to prove that the diffusion constant δ is *strictly* positive, and this may be achieved using arguments relating random walks to electrical networks.

Acknowledgements. This paper was written partly in Ste Lucie-de-Porto-Vecchio. The work was done with partial financial support from the European Union under contracts CHRX-CT93-0411 and FMRX-CT96-0075A, and from the Engineering and Physical Sciences Research Council of the UK under grant GR/L15426. The author congratulates the organisers of the 1997 Cortona Conference on 'Probability and Discrete Potential Theory' for the stimulating scientific environment of the event.

References.

1. Benjamini, I., Pemantle, R., and Peres, Y., *Unpredictable paths and percolation*, Annals of Probability **26** (1998), 1198–1211.
2. Bezuidenhout, C. E. and Grimmett, G. R., *A central limit theorem for random walks in random labyrinths*, Annales de l'Institut Henri Poincaré, Probabilités et Statistiques (1997) (to appear).
3. Bunimovich, L. A. and Sinai, Ya. G., *Statistical properties of Lorentz gas with periodic configuration of scatterers*, Communications in Mathematical Physics **78** (1981), 479–497.
4. Bunimovich, L. A. and Troubetzkoy, S. E., *Recurrence properties of Lorentz lattice gas cellular automata*, Journal of Statistical Physics **67** (1992), 289–302.
5. Bunimovich, L. A. and Troubetzkoy, S. E., *Rotators, periodicity and absence of diffusion in cyclic cellular automata*, Journal of Statistical Physics **74** (1994), 1–10.
6. Burton, R. M. and Keane, M., *Density and uniqueness in percolation*, Communications in Mathematical Physics **121** (1989), 501–505.
7. Chalcraft, A. and Greene, M., *Train sets*, Eureka (1994), 5–12, Eureka is the Journal of the Archimedeans, Cambridge University.
8. Cohen, E. G. D., *New types of diffusions in lattice gas cellular automata*, Microscopic Simulations of Complex Hydrodynamical Phenomena (M. Mareschal and B. L. Holian, eds.), Plenum Press, New York, 1991, pp. 137–152.
9. Cohen, E. G. D. and Wang, F., *New results for diffusion in Lorentz lattice gas cellular automata*, Journal of Statistical Physics **81** (1995), 445–466.
10. Cohen, E. G. D. and Wang, F., *Novel phenomena in Lorentz lattice gases*, Physica A **219** (1995), 56–87.
11. DeMasi, A., Ferrari, P. A., Goldstein, S., and Wick, W. D., *Invariance principle for reversible Markov processes with application to diffusion in the percolation regime*, Particle Systems, Random Media and Large Deviations (R. T. Durrett, ed.), Contemporary Mathematics no. 41, American Mathematical Society, Providence, R. I., 1985, pp. 71–85.
12. DeMasi, A., Ferrari, P. A., Goldstein, S., and Wick, W. D., *An invariance principle for reversible Markov processes. Applications to random motions in random environments*, Journal of Statistical Physics **55** (1989), 787–855.

13. Doyle, P. G. and Snell, E. L., *Random Walks and Electric Networks*, Carus Mathematical Monograph no. 22, AMA, Washington, D. C., 1984.
14. Ehrenfest, P., *Collected Scientific Papers* (M. J. Klein, ed.), North-Holland, Amsterdam, 1959.
15. Grimmett, G. R., *Percolation*, 2nd edition, Springer-Verlag, Berlin, 1999.
16. Grimmett, G. R., *Percolation and disordered systems*, Ecole d'Eté de Probabilités de Saint Flour XXVI–1996 (P. Bernard, ed.), Lecture Notes in Mathematics no.1665, Springer-Verlag, Berlin, 1997, pp. 153–300.
17. Grimmett, G. R., Kesten, H., and Zhang, Y., *Random walk on the infinite cluster of the percolation model*, Probability Theory and Related Fields **96** (1993), 33–44.
18. Grimmett, G. R., Menshikov, M. V., and Volkov, S. E., *Random walks in random labyrinths*, Markov Processes and Related Fields **2** (1996), 69–86.
19. Harris, M., *Nontrivial phase transition in a continuum mirror model* (1996) (to appear).
20. Kipnis, C. and Varadhan, S. R. S., *Central limit theorem for additive functionals of reversible Markov processes and applications to simple exclusion*, Communications in Mathematical Physics **104** (1986), 1–19.
21. Lorentz, H. A., *The motion of electrons in metallic bodies, I, II, and III*, Koninklijke Akademie van Wetenschappen te Amsterdam, Section of Sciences **7** (1905), 438–453, 585–593, 684–691.
22. Meester, R. and Roy, R., *Continuum Percolation*, Cambridge University Press, Cambridge, 1996.
23. Quas, A., *Some properties of Lorentz lattice gas models*, Probability Theory and Related Fields (1996) (to appear).
24. Ruijgrok, T. W. and Cohen, E. G. D., *Deterministic lattice gas models*, Physics Letters A **133** (1988), 415–418.
25. Spohn, H., *Large Scale Dynamics of Interacting Particles*, Springer-Verlag, Berlin, 1991.
26. Wang, F. and Cohen, E. G. D., *Diffusion in Lorentz lattice gas cellular automata: the honeycomb and quasi-lattices compared with the square and triangular lattices*, Journal of Statistical Physics **81** (1995), 467–495.
27. Ziff, R. M., Kong, X. P., and Cohen, E. G. D., *Lorentz lattice-gas and kinetic-walk model*, Physical Review A **44** (1991), 2410–2428.

Geoffrey R. Grimmett
Statistical Laboratory, University of Cambridge,
16 Mill Lane, Cambridge CB2 1SB, UK
email: g.r.grimmett@statslab.cam.ac.uk
http://www.statslab.cam.ac.uk/~grg/

A discrete time Harnack inequality and its applications

Vadim A. Kaimanovich

Furstenberg in his paper [Fu71] considered the Brownian motion on a Riemannian symmetric space S and proved that if G is a lattice in the corresponding semi-simple Lie group, then for any point $o \in S$ there exists a probability measure μ on G such that the harmonic measure ν_o of the point o on the Furstenberg boundary of the space S is μ-stationary, i.e., $\nu_o = \sum_g \mu(g) g\nu_o$. It implies that for any bounded harmonic function on S its restriction to the G-orbit of o is a μ-harmonic function. In fact, his argument required just recurrence of the Brownian motion on the quotient space S/G. Lyons and Sullivan [LS84] noticed that Furstenberg's construction is applicable to the Brownian motion on an arbitrary covering Riemannian manifold $\widetilde{M}$ provided the quotient manifold M is recurrent (actually, they showed how it can be done even without any homogeneity assumptions) and proved that the resulting random walk on the deck transformations group G well approximates the original Brownian motion in various senses. As it was later shown by the author in [Ka92a], this random walk also has the same bounded harmonic functions as the original Brownian motion: if f is a bounded harmonic function of the Laplace-Beltrami operator on $\widetilde{M}$, then its restriction to the orbit $Go \cong G$ is μ-harmonic, every bounded μ-harmonic function can be uniquely obtained in this way, and this correspondence is an isometry between the spaces of bounded harmonic functions on $\widetilde{M}$ and G. On a measure-theoretical level it just means that the corresponding Poisson boundaries are isomorphic as measure spaces [Ka92b].

The main idea of this discretization procedure is to use the *Harnack inequality*. In the Riemannian case it almost automatically follows from boundedness of geometry of $\widetilde{M}$ and continuity of sample paths of the Brownian motion. In order to obtain a Harnack inequality for discrete time Markov operators one has to impose on the operator some locality and "bounded geometry" assumptions. We establish a Harnack inequality (Theorem 1) for discrete time Markov operators on a general metric space (X, d) under several additional conditions (P_1) – (P_5), the most important of which are absolute continuity of transition probabilities with respect to a given smooth reference measure on X and the "bounded range condition" (P_3), i.e., existence of a number R such that for any $x \in X$ the transition probability π_x

is concentrated on the ball of radius R around x.

Then we apply a discretization procedure similar to the one used in the Riemannian case to obtain a Markov chain on a recurrent net $\Gamma \subset X$ with the same Poisson boundary (i.e., with the same space of bounded harmonic functions) as the original Markov operator P on X (Theorem 2). If the operator P and the net Γ are both G-invariant with respect to an action of a group G, then our construction is G-equivariant, and the resulting Markov chain on Γ has G-invariant transition probabilities. In particular, for a corecurrent covering operator with the deck group G (cf. [Ka95]) we obtain a random walk on the group G with the same Poisson boundary as that of the operator P. Note that the Harnack inequality from Theorem 1 can be also used in other discretization procedures, cf. [An90], [BL96].

In the particular case of certain Markov operators on Teichmüller space the discrete time Harnack inequality was first considered in [KM95], and our Theorems 1 and 2 are inspired by the corresponding results from [KM95]. In the present paper we also consider yet another particular case, that of geodesic random walks on Riemannian manifolds. Their transition probabilities are images of the uniform measures on spheres in the tangent space under the exponential map of the Riemannian metric. In other words, the "Markov jump" from any given point $x \in X$ consists in choosing a random direction on the unit sphere in the tangent space at the point x (with respect to the normalized Lebesgue measure on the sphere) and then issuing from x a geodesic segment of fixed length in this direction. Using Theorem 2 we prove coincidence of the Poisson boundary of geodesic random walks with the Poisson boundary of an appropriate Markov chain on a uniform net (Theorem 3).

1. Background definitions and notations

Consider a metric space (X, d) endowed with a σ-finite reference Radon measure m on it (i.e., such that $m(K) < \infty$ for all compact sets $K \subset X$). Assigning in a measurable way a probability measure π_x to any point $x \in X$ determines a *Markov chain* on X with *transition probabilities* π_x. For a point $x \in X$ denote by $\mathbf{P}_x$ the probability measure in the space $X^{\mathbb{Z}_+}$ of *sample paths* $\boldsymbol{x} = \{x_0, x_1, \dots\}$ of this chain corresponding to the initial distribution δ_x (i.e., the measure $\mathbf{P}_x$ is concentrated on sample paths which start from the point $x_0 = x$ at time 0). For an arbitrary σ-finite initial distribution θ (not necessarily a probability one!) put $\mathbf{P}_\theta = \int \mathbf{P}_x \, d\theta(x)$.

Below we shall always assume (unless otherwise specified) that the transition probabilities π_x are absolutely continuous with respect to the reference measure m with densities $p(x, y) = d\pi_x / dm(y)$. Then the transition operator of the chain acts in the space $L^\infty(X,m)$ and takes the form

$$Pf(x) = \langle f, \pi_x \rangle = \int f(y)\, p(x,y)\, dm(y)\,.$$

Denote by Q the *adjoint operator* of P acting in the space of measures on X. In probabilistic terms,

$$Q\theta(E) = \mathbf{P}_\theta[x_1 \in E]\,, \qquad E \subset X\,,$$

i.e., Q assigns to an initial distribution θ the distribution of the position of the Markov chain after one step. Since P has absolutely continuous transition probabilities, the operator Q preserves the type of the measure m (moreover, $Q\theta \prec m$ for *any* measure θ on X), and it acts on the space of densities $\varphi = d\theta/dm$ by the formula

$$Q\varphi(y) = \frac{dQ\theta}{dm}(y) = \int p(x,y)d\theta(x) = \int \varphi(x)p(x,y)dm(x)\,. \qquad (1)$$

A function f on X is called *P-harmonic* if $Pf = f$. Denote by ∂P the *Poisson boundary* of the operator P, i.e., the space of *ergodic components* of the shift in the unilateral path space $(X^{\mathbb{Z}_+}, \mathbf{P}_m)$, and by **bnd** the corresponding projection $X^{\mathbb{Z}_+} \to \partial P$. Then (by the definition of the space of ergodic components) **bnd** is an isomorphism (mod 0) between the σ-algebra of shift invariant measurable subsets of the path space and the σ-algebra of all measurable subsets of the Poisson boundary. By $[\nu]$ denote the *harmonic measure class* on ∂P, i.e., the class of measures $\nu_\theta = \mathbf{bnd}(\mathbf{P}_\theta)$, where θ is a probability measure equivalent to m. For any point $x \in X$ the *harmonic measure* $\nu_x = \mathbf{bnd}(\mathbf{P}_x)$ is absolutely continuous with respect to the type $[\nu]$, and the *Poisson formula*

$$f(x) = \langle \widehat{f}, \nu_x \rangle$$

is an isometric isomorphism between the space of bounded measurable P-harmonic functions $H^\infty(P) = \{f \in L^\infty(X,m) : Pf = f\}$ and the space $L^\infty(\partial P, [\nu])$ [Ka92b].

2. Discrete time balayage operators

Let V be a measurable subset of the state space X such that both V and its complement $\complement V = X \setminus V$ are non-negligible with respect to the measure m. Denote by $\Lambda = \Lambda_{V,P}$ the *balayage operator* determined by the set V and the Markov operator P. The operator Λ assigns to an initial distribution θ on

X the distribution of the first exit point from V of the associated Markov chain, i.e.,

$$\Lambda\theta(E) = \mathbf{P}_\theta[x_\tau \in E] ,$$

where

$$\tau(\boldsymbol{x}) = \tau_{\complement V}(\boldsymbol{x}) = \min\{n \geq 0 : x_n \in \complement V\}$$

is the time of the first exit from V. The measure $\Lambda\theta$ is called the *balayage* of the measure θ. Note that we define balayage for *all* measures on X, not only for those supported on V; if $\theta(V) = 0$, then by definition $\Lambda\theta = \theta$. In general, the total mass $\|\Lambda\theta\|$ of the measure $\Lambda\theta$ can be less than the total mass of θ (if the measure $\mathbf{P}_\theta$ of those sample paths which never leave V is non-zero). However, if the set $\complement V$ is *recurrent* in the sense that $\mathbf{P}_m$-a.e. sample path eventually hits $\complement V$, then $\|\Lambda\theta\| = \|\theta\|$ for any measure θ on X. Denote by P_V the *sub-Markov operator* with the state space V obtained by restricting P to V, so that it has the transition densities

$$p_V(x,y) = \begin{cases} p(x,y) , & \text{if } x, y \in V ; \\ 0 , & \text{otherwise} . \end{cases}$$

Then $P - P_V$ is also a sub-Markov operator. In terms of the adjoint operators Q and Q_V the result of applying the balayage operator Λ to a measure θ living on V can be expressed as

$$\Lambda\theta = \sum_{n=0}^{\infty} (Q - Q_V) Q_V^n \theta = (Q - Q_V) \sum_{n=0}^{\infty} Q_V^n \theta$$

(each term $(Q - Q_V)Q_V^n\theta$ in this sum corresponds to staying in V for the first n steps and exiting to $\complement V$ at the time $n+1$). Hence, we have

Lemma 1. *If there is a constant H such that*

$$\sum_{n=0}^{\infty} Q_V^n \theta_1 \leq H \sum_{n=0}^{\infty} Q_V^n \theta_2$$

for two measures θ_1, θ_2 on V, then

$$\Lambda\theta_1 \leq H \Lambda\theta_2 .$$

If the set $\complement V$ is recurrent, then the harmonic measures ν_θ and $\nu_{\Lambda\theta}$ on the Poisson boundary coincide. This is so because $\Lambda\theta$ is the distribution of the first exit point $x(\tau_{\complement V})$ determined by the *Markov stopping time* $\tau_{\complement V}$, see [Ka92b]. In particular, in this case the values on V of any bounded P-harmonic function can be recovered from its values on $\complement V$ by the formula

$$f(x) = \langle \widehat{f}, \nu_x \rangle = \langle \widehat{f}, \nu_{\Lambda\delta_x} \rangle = \langle f, \Lambda\delta_x \rangle . \tag{2}$$

Let

$$\tau_1(\boldsymbol{x}) = \min\{n > 0 : x_n \in \complement V\} ,$$
$$\tau_{k+1}(\boldsymbol{x}) = \min\{n > \tau_k : x_n \in \complement V\}$$

be the times when a sample path $\boldsymbol{x} = \{x_n\}$ hits the set $\complement V$, then $\{x_{\tau_k}\}$ are sample paths of the *induced chain* on $\complement V$ corresponding to the operator $P^{\complement V}$ with transition probabilities $\Lambda\pi_x$, $x \in \complement V$. The Poisson boundaries of the operators P and $P^{\complement V}$ are isomorphic; for any bounded P-harmonic function its restriction to $\complement V$ is $P^{\complement V}$-harmonic, and, conversely, any bounded $P^{\complement V}$-harmonic function uniquely extends to a P-harmonic function on X by formula (2) [Ka92b].

3. Bounded geometry conditions

We shall always assume that the metric space X is complete, separable, non-compact with compact balls. Below we shall impose on the measure m and the transition densities $p(\cdot,\cdot)$ the following additional conditions. Here $B(x,\varepsilon)$ denotes the ball in X of radius ε centered at a point $x \in X$.

(P_1) There exist $\varepsilon, \delta > 0$ such that $mB(x,\varepsilon) > \delta$ for all $x \in X$.

(P_2) There exists $C > 0$ such that $p(x,y) \le C$ for all $x, y \in X$.

(P_3) There exists $R > 0$ such that $p(x,y) = 0$ whenever $d(x,y) > R$.

(P_4) There exist $c > 0$ and r_1, r_2 with $0 \le r_1 < r_2 \le R$, $r_2 - r_1 \ge 4\varepsilon$ such that $p(x,y) \ge c$ whenever $r_1 \le d(x,y) \le r_2$.

Here (P_1) and (P_2) can be considered as *"bounded geometry"* conditions, (P_3) is a *bounded range* condition, and (P_4) is an *irreducibility* condition. Note that condition (P_4) implies that $mB(x,\varepsilon) \le 1/c$ for all $x \in X$. An analysis of the arguments below shows that condition (P_4) could be significantly relaxed. The constants $\varepsilon, \delta, c, C, r_1, r_2, R$ from conditions (P_1) – (P_4) will be used through the rest of the paper without further notice. We shall also impose on the space X the following "chain connectivity" condition

(P_5) For any number $M > 0$ there exists an integer N such that for any $o \in X$ and any two points $z, \bar{z} \in B(o, M + R)$ there is a chain of points $z = z_0, z_1, \ldots, z_N = \bar{z}$ in X with the property that

$$d(z_i, z_{i+1}) = (r_1 + r_2)/2 = r \qquad \forall i = 0, 1, \ldots, N-1$$

and

$$d(z_i, o) \leq M + R - \varepsilon \qquad \forall i = 1, 2, \ldots, N-1 ,$$

where r_1, r_2, R, ε are the constants from conditions (P_1), (P_3), (P_4).

The following Lemma provides us with a large class of metric spaces on which condition (P_5) is satisfied.

Lemma 2. *Let X be a geodesic metric space, i.e., for any two points $x, y \in X$ there exists an isometric embedding φ of the interval $[0, d(x,y)]$ into X such that $\varphi(0) = x$ and $\varphi(d(x,y)) = y$. If all spheres in X are non-empty and connected, then condition (P_5) is satisfied.*

Proof. It takes at most $[(M+R)/r]$ steps of length r to attain a point $z' \in B(o,r)$ by moving from z to o along the geodesic segment $[z, 0]$ (here $[\cdot]$ is the integer part). Then by continuity of the function $x \mapsto d(z', x)$ on the sphere $S(o,r)$ it takes at most 2 steps to attain o from z'. Concatenating the chains joining z and $\bar{z}$ with o we obtain that one can get from z to $\bar{z}$ in not more than $2[(M+R)/r] + 4$ steps. As we want all chains to have the same length, we can further add to any such chain 2 segments $[o,x], [x,o]$ or 3 segments $[o,x], [x,x'], [x',o]$ with $x, x' \in S(o,r)$, $d(x,x') = r$ several times until we get a chain of length $N = 2[(M+R)/r] + 6$. The chain obtained in this way satisfies condition (P_5), because $\varepsilon < r < R - \varepsilon$ by (P_4). □

4. Discrete time Harnack inequality

To prove our Harnack inequality we shall need the following auxiliary statement:

Lemma 3. *If the distance between two points $x_0, y_0 \in X$ equals r, and if θ is a measure on X such that $d\theta/dm \geq 1$ on $B(x_0, \varepsilon)$, then $dQ\theta/dm \geq c\delta$ on $B(y_0, \varepsilon)$.*

Proof. Since

$$|d(x,y) - r| \leq 2\varepsilon \qquad \forall\, x \in B(x_0, \varepsilon),\ y \in B(y_0, \varepsilon) ,$$

by formula (1) and by conditions (P_1), (P_4) for any $y \in B(y_0, \varepsilon)$

$$\frac{dQ\theta}{dm}(y) = \int p(x,y) \frac{d\theta}{dm}(x) dm(x) \geq c\, mB(x_0, \varepsilon) \geq c\delta .$$

□

Theorem 1 (Harnack inequality). *Let P be a Markov operator on X satisfying conditions (P_1) – (P_5). For a point $o \in X$ and a number $M > 0$ denote by Λ the balayage operator of the ball $V = B(o, M+R)$. Then there exists a constant $H > 0$ depending only on the constants from conditions (P_1) – (P_5) and M such that*

$$\Lambda\delta_x \le H\Lambda\delta_y \qquad \forall x, y \in B(o, M)\,.$$

Proof. First notice that by condition (P_3) the measure $Q\delta_x$ is supported on V for any $x \in B(o, M)$, so that $\Lambda\delta_x = \Lambda Q\delta_x$. Now, by Lemma 1 it is sufficient to prove that for any $x \in B(o, M)$ the density of the measure $(\sum Q_V^n)Q\delta_x$ with respect to the restriction m_V of the measure m to V is uniformly bounded from below and from above.

By condition (P_4) there exists an ε-ball in V such that the density of the measure $Q\delta_x$ is at least c on this ball. Take the number N from condition (P_5), then Lemma 3 applied to the operator Q_V implies that $c(c\delta)^N$ is a lower bound of the density of the measure $Q_V^N Q\delta_x$ with respect to the measure m_V. In particular, since the total mass of the measure $Q_V^N Q\delta_x$ does not exceed 1, we have the uniform estimate $\|m_V\| \le 1/c(c\delta)^N$.

To obtain an upper bound note that the operator Q_V (the adjoint operator of the sub-Markov operator P_V introduced in Section 2) also acts on the space $L^\infty(m_V)$, and

$$\|Q_V\|_\infty \le \operatorname{ess\,sup}_{x,y} p_V(x,y) m(V) \le Cm(V)\,. \tag{3}$$

The transition densities p_V^n of the operators P_V^n satisfy the relation

$$p_V^{n+k}(x,y) = \int p_V^k(x,z) p_V^n(z,y)\,dm(z) \qquad \forall n, k \ge 1\,,$$

so that

$$\operatorname{ess\,sup} p_V^{n+k}(x,y) \le \operatorname{ess\,sup} p_V^n(x,y) \qquad \forall n, k \ge 1\,. \tag{4}$$

Moreover, there exists a constant k_0 and a number $\alpha > 0$ such that

$$\int p_V^{k_0}(x,z)\,dm(z) \le 1 - \alpha \qquad \forall x \in V$$

(cf. condition (P_5) and the proof of Lemma 3). Hence,

$$\operatorname{ess\,sup} p_V^{n+k_0}(x,y) \le (1-\alpha) \operatorname{ess\,sup} p_V^n(x,y) \qquad \forall n \ge 1\,. \tag{5}$$

Formulas (4) and (5) imply that $\operatorname{ess\,sup} p_V^n(x,y)$ decays exponentially on n. Thus, by (3) $\|Q_V^n\|_\infty$ also decays exponentially, hence

$$\left\|\frac{d(\sum Q_V^n)Q\delta_x}{dm}\right\|_\infty \le \sum \|Q_V^n\|_\infty \left\|\frac{dQ\delta_x}{dm}\right\|_\infty \le C \sum \|Q_V^n\|_\infty < \infty\,.$$

Note that the convergence $\|Q_V^n\|_\infty \to 0$ implies that the set $\complement V$ is recurrent. □

5. Discretization of corecurrent Markov operators

The following Theorem uses a discretization procedure first suggested by Furstenberg [Fu71] for diffusion processes (see also [LS84]). The fact that this procedure leads to a discrete space Markov chain with the same Poisson boundary as the original process was proved in [Ka92a]. Here we have to consider a discrete time Markov operator on X, so that the constructions and results from [Fu71], [LS84], [Ka92a] are not applicable directly. However, the Harnack inequality from Theorem 1 (used instead of the usual elliptic Harnack inequality) allows us to proceed with the discretization (see also [Ka95]).

Theorem 2. *Let P be a Markov operator on a metric state space X satisfying conditions* (P_1) – (P_5). *Let Γ be a subset of X such that there is a number $M > 0$ with the property that the balls $B(\gamma, M)$, $x \in \Gamma$ are all pairwise disjoint, and both the union $\bigcup_{\gamma\in\Gamma} B(\gamma, M)$ and its complement are recurrent for the operator P. Then there exists a family of probability measures μ_γ, $\gamma \in \Gamma$ concentrated on Γ such that the Poisson boundary of the operator P is isomorphic to the Poisson boundary of the Markov operator P_Γ on Γ with the transition probabilities μ_γ: the restriction of any bounded P-harmonic function from X to Γ is P_Γ-harmonic, and, conversely, any bounded P_Γ-harmonic function can be uniquely extended from Γ to a bounded P-harmonic function on X.*

The proof will consist of several steps. First we describe a construction of the measures μ_γ (Section 6), and then show coincidence of the Poisson boundaries ∂P and ∂P_Γ (Sections 7, 8).

Remark. The assumptions of mutual disjointness of the balls $B(\gamma, M)$, $x \in \Gamma$ and of recurrence of the complement $\complement\left[\bigcup_{\gamma\in\Gamma} B(\gamma, M)\right]$ (i.e., that there is "enough" free space left after removing from X all the balls $B(\gamma, M)$) are made here for the sake of technical convenience only and could be avoided along the same lines as in [Ka92a] where an analogous result for the Brownian motion on Riemannian manifolds is proved without any assumptions of this kind.

6. Proof of Theorem 2. Construction of the measures μ_x

For every point $x \in X$ we shall construct a probability measure μ_x on the set Γ in such a way that the harmonic measure ν_x on ∂P satisfies the relation

$$\nu_x = \sum_{\gamma\in\Gamma} \mu_x(\gamma)\nu_\gamma \,. \tag{6}$$

We do it by an iterative construction described below. Namely, we construct a sequence $\theta_k = \theta_k^x$ of measures on X and a sequence $\varkappa_k = \varkappa_k^x$ of measures on Γ such that

(i) $\nu_{\theta_0} = \nu_x$;
(ii) $\nu_{\theta_k} = \nu_{\theta_{k+1}} + \nu_{\varkappa_{k+1}}$ for all $k \geq 0$;
(iii) $\|\theta_k\| \to 0$.

Thus, we begin with the harmonic measure $\nu_x = \nu_{\theta_0}$, and at each step we single out a part of it which can be replaced with the harmonic measure of a distribution (denoted $\varkappa_{k+1}$) concentrated on Γ. Condition (iii) says that finally the whole measure ν_x will be exhausted. The resulting measure

$$\mu_x = \sum_{k \geq 1} \varkappa_k^x$$

then clearly has the property (6).
Denote by Λ_γ the balayage operator of the set $V_\gamma = B(\gamma, M + R)$, and let $\omega_\gamma = \Lambda_\gamma \delta_\gamma$. Put

$$\theta_0 = \begin{cases} \delta_x, & x \notin \Gamma ; \\ \omega_\gamma, & x = \gamma \in \Gamma . \end{cases}$$

Clearly, this choice of θ_0 satisfies condition (i) above.
Now we define the iterative procedure. By the assumption, the set $Z = \bigcup_{\gamma\in\Gamma} B(\gamma, M)$ is recurrent, and we can apply the corresponding balayage operator to the measure θ_k. Denote the resulting measure on Z by ξ_{k+1}, and denote by ξ_{k+1}^γ, $\gamma \in \Gamma$ the restrictions of ξ_{k+1} to the balls $B(\gamma, M)$, so that $\xi_{k+1} = \sum_\gamma \xi_{k+1}^\gamma$, and

$$\nu_{\theta_k} = \nu_{\xi_{k+1}} = \sum_{\gamma\in\Gamma} \nu_{\xi_{k+1}^\gamma} .$$

Let $\zeta_{k+1}^\gamma = \Lambda_\gamma \xi_{k+1}^\gamma$ be the balayage of the measure ξ_{k+1}^γ to the set $\complement V_\gamma$, then

$$\nu_{\zeta_{k+1}^\gamma} = \nu_{\xi_{k+1}^\gamma} \qquad \forall \gamma \in \Gamma ,$$

and

$$\nu_{\theta_k} = \sum_{\gamma\in\Gamma} \nu_{\zeta_{k+1}^\gamma} .$$

As it follows from Theorem 1, for any $\gamma \in \Gamma$

$$\frac{\zeta_{k+1}^\gamma}{\|\zeta_{k+1}^\gamma\|} \geq \frac{1}{H} \omega_\gamma ,$$

thus if we put

$$\varkappa_{k+1}(\gamma) = \frac{\|\zeta_{k+1}^\gamma\|}{H} = \frac{\|\xi_{k+1}^\gamma\|}{H} ,$$

then all measures

$$\theta_{k+1}^\gamma = \zeta_{k+1}^\gamma - \varkappa_{k+1}(\gamma)\omega_\gamma$$

are non-negative. Let

$$\theta_{k+1} = \sum_{\gamma\in\Gamma} \theta_{k+1}^\gamma ,$$

then, since $\nu_{\omega_\gamma} = \nu_\gamma$ for any $\gamma \in \Gamma$ by definition of the measures ω_γ, we have

$$\begin{aligned} \nu_{\theta_{k+1}} &= \sum_\gamma \nu_{\zeta_{k+1}^\gamma} - \sum_\gamma \varkappa_{k+1}(\gamma)\nu_{\omega_\gamma} \\ &= \sum_\gamma \nu_{\xi_{k+1}^\gamma} - \sum_\gamma \varkappa_{k+1}(\gamma)\nu_\gamma \\ &= \nu_{\theta_k} - \nu_{\varkappa_{k+1}} , \end{aligned}$$

and condition (ii) is satisfied. Finally, by the construction

$$\|\theta_{k+1}\| = (1 - 1/H)\|\theta_k\| ,$$

so that the total masses $\|\theta_k\|$ decay exponentially, and condition (iii) is also satisfied.

7. Proof of Theorem 2. Markov stopping times

The fact that the restriction of any bounded P-harmonic function from X to Γ is a P_Γ-harmonic function follows from formula (6). Indeed, if f is a bounded P-harmonic function, then by the Poisson formula there exists a bounded measurable function $\widehat{f}$ on the Poisson boundary ∂P such that

$$f(x) = \langle \widehat{f}, \nu_x \rangle \qquad \forall x \in X .$$

Then by (6) for any $\gamma \in \Gamma$

$$f(\gamma) = \langle \widehat{f}, \nu_\gamma \rangle = \sum_{\gamma'} \mu_\gamma(\gamma')\langle \widehat{f}, \nu_{\gamma'} \rangle = \sum_{\gamma'} \mu_\gamma(\gamma') f(\gamma') .$$

Now we want to show that, conversely, for any bounded P_Γ-harmonic function f on Γ its extension to X by the formula

$$f(x) = \sum_\gamma \mu_x(\gamma) f(\gamma)$$

is P-harmonic. We shall prove this statement by constructing a sequence of Markov operators connecting the operator P with the operator P_Γ, and such that the Poisson boundaries of any two consecutive operators in this sequence coincide (this argument will also give another proof of the fact that the restriction of any bounded P-harmonic function to Γ is P_Γ-harmonic).
First we have to reformulate the iterative process from Section 6 in terms of *Markov stopping times.* This process includes balayages, for which such reformulation is straightforward, and subtracting from the measures ζ_{k+1}^γ the measures $(\|\zeta_{k+1}^\gamma\|/H)\omega_\gamma$. The latter operation can be realized by introducing a new random variable α uniformly distributed on $[0,1]$ and independent of all the rest, and stopping the process when α does not exceed the Radon–Nikodym derivative of the measure $(\|\zeta_{k+1}^\gamma\|/H)\omega_\gamma$ with respect to the measure ζ_{k+1}^γ.
For a sample path $\boldsymbol{x} = \{x_n\}$ let

$$S_0(\boldsymbol{x}) = \begin{cases} 0, & x \notin \Gamma\,; \\ \min\{n > 0 : x_n \in \complement V_\gamma\}, & x = \gamma \in \Gamma\,, \end{cases}$$

and we define inductively

$$\begin{aligned} R_{k+1}(\boldsymbol{x}) &= \min\{n \geq S_k(\boldsymbol{x}) : x_n \in Z\}\,, \\ \gamma_{k+1}(\boldsymbol{x}) &= \gamma(x_{R_{k+1}}) \in \Gamma\,, \\ S_{k+1}(\boldsymbol{x}) &= \min\{n > R_{k+1}(\boldsymbol{x}) : x_n \in \complement V_{\gamma_{k+1}}\}\,, \end{aligned}$$

where $\gamma(x)$, $x \in Z$ is the uniquely determined point $\gamma \in \Gamma$ such that $x \in B(\gamma, M)$.
Take a sequence of i.i.d. random variables $\boldsymbol{\alpha} = \{\alpha_n\}_{n \geq 0}$ which are independent of the chain $\{x_n\}$ and have the Lebesgue measure ρ on the interval $[0,1]$ as their common distribution. Formally, it means that from now on we pass from the original path space $(X^{\mathbb{Z}_+}, \mathbf{P}_x)$ to its product with the measure space $([0,1]^{\mathbb{Z}_+}, \rho^{\mathbb{Z}_+})$, where $\rho^{\mathbb{Z}_+}$ is the infinite product of measures ρ indexed by the set $\mathbb{Z}_+$. Denote the product measure $\mathbf{P}_x \otimes \rho^{\mathbb{Z}_+}$ on $(X \times [0,1])^{\mathbb{Z}_+}$ by $\widetilde{\mathbf{P}}_x$. The measure $\widetilde{\mathbf{P}}_x$ corresponds to the initial distribution $\delta_x \otimes \rho$ in the space of the sample paths $(\boldsymbol{x}, \boldsymbol{\alpha})$ of the Markov operator

$$\widetilde{P} f(x, \alpha) = \iint f(y, \underline{\ }) \, d\pi_x(y) \, d\rho(\underline{\ })$$

whose transition probabilities $\pi_x \otimes \rho$ do not depend on the $[0,1]$-component.
For given $\gamma \in \Gamma$ and points $y \in B(\gamma, M)$, $z \in \complement V_\gamma$ let

$$\Delta(\gamma; y, z) = \frac{d\omega_\gamma}{d\Lambda_\gamma \delta_y}(z)$$

be the Radon–Nikodym derivative of the measure $\omega_\gamma = \Lambda_\gamma\delta_\gamma$ with respect to the measure $\Lambda_\gamma\delta_y$ evaluated at the point z. Now define

$$T_0(\boldsymbol{x}, \boldsymbol{\alpha}) = 0\,,$$

and for $m \geq 0$ by induction

$$T_{m+1}(\boldsymbol{x}, \boldsymbol{\alpha}) = \min\{k > T_m : \alpha_{S_k} < \frac{1}{H}\Delta(\gamma_k; x_{R_k}, x_{S_k})\}$$

(as $\Delta(\gamma_k; x_{R_k}, x_{S_k}) \geq 1/H$ by Theorem 1, the times T_m are a.e. finite). We claim that the measures μ_x constructed in the first part of the proof can be presented as

$$\mu_x(\gamma) = \widetilde{\mathbf{P}}_x[\gamma_{T_1} = \gamma]\,. \tag{7}$$

Indeed, by definition of the stopping time S_0 the distribution of x_{S_0} coincides with the measure θ_0. Now we shall prove by induction that θ_k (in notations from Section 6) is the distribution of x_{S_k} restricted to the set $[T_1 > k]$ (by "restricted" we mean here that θ_k is the image under the map $(\boldsymbol{x}, \boldsymbol{\alpha}) \mapsto x_{S_k}$ of the restriction of the measure $\widetilde{\mathbf{P}}_x$ to the set $[T_1 > k]$), and $\varkappa_k$ is the distribution of γ_k restricted to the set $[T_1 = k]$. As $\mu_x = \sum \varkappa_k$, the latter will imply (7).

Suppose we have already proved this assertion for θ_k and $\varkappa_k$. Then the stopping time R_{k+1} corresponds to the balayage of the measure θ_k to the set Z, so that the distribution of $x_{R_{k+1}}$ restricted to the set $[T_1 > k]$ is the measure $\xi_{k+1} = \sum_\gamma \xi^\gamma_{k+1}$. After that the stopping time S_{k+1} corresponds to the balayage of each of the measures ξ^γ_{k+1} from the corresponding set $B(\gamma, M)$ to the set $\complement V_\gamma$, so that the distribution of $x_{S_{k+1}}$ restricted to the set $[T_1 > k]$ is $\sum_\gamma \zeta^\gamma_{k+1}$. Now the definition of the stopping times T_m means that given $x_{R_{k+1}}$ and $x_{S_{k+1}}$, we have $T_1 = k+1$ with the conditional probability $\Delta(\gamma_{k+1}; x_{R_{k+1}}, x_{S_{k+1}})/H$ which is the Radon–Nikodym derivative of the measure $\omega_{\gamma_{k+1}}/H$ with respect to the measure $\Lambda_{\gamma_{k+1}}\delta_{x_{R_{k+1}}}$ evaluated at $x_{S_{k+1}}$.

In order to find the unconditional probability of the event $[T_1 = k+1]$ we have to integrate these conditional probabilities with respect to the conditions. Here we condition by $x_{R_{k+1}}$ and $x_{S_{k+1}}$, so that we have to integrate first with respect to the conditional distribution of $x_{S_{k+1}}$ conditioned by $x_{R_{k+1}}$, and then with respect to the unconditional distribution of $x_{R_{k+1}}$. As the measure $\Lambda_{\gamma_{k+1}}\delta_{x_{R_{k+1}}}$ is precisely the conditional distribution of $x_{S_{k+1}}$ provided $x_{R_{k+1}}$ is fixed, the result of the first integration is the measure $\omega_{\gamma_{k+1}}/H$ for any $x_{R_{k+1}}$ (i.e., it depends only on $\gamma_{k+1} = \gamma(x_{R_{k+1}})$). Thus, the second integration with respect to $x_{R_{k+1}}$ just reduces to multiplying the measures ω_γ/H by the probability that γ_{k+1} takes a given value γ, the latter being exactly $\|\xi^\gamma_{k+1}\|$. So, we obtain that the distribution of γ_{k+1} (resp.,

$x_{S_{k+1}}$) restricted to the set $[T_1 = k+1]$ is $\varkappa_{k+1}$ (resp., $\sum \varkappa_{k+1}(\gamma)\omega_\gamma$), and that the remaining measure θ_{k+1} is the distribution of S_{k+1} restricted to the set $[T_1 > k+1]$.

8. Proof of Theorem 2. Extended Markov chains

In defining the measures μ_x we had to use two stopping times R_k and S_k, both of which are Markov. However, the definition of S_k includes the position x_{R_k} of the Markov chain $\{x_n\}$ at the time R_k. If we want to use formula (2) for proving coincidence of the Poisson boundaries, we have to extend the original Markov chain $\{x_n\}$ by adding a component x'_n which keeps track of the positions x_{R_k} until the moment S_k is attained. The second component x_n of the *extended chain* $\{(x'_n, x_n)\}$ on $X \times X$ coincides with the original chain on X, whereas the first component x'_n once in Z remains unchanged until the second component leaves the set $V_{\gamma(x'_n)}$; otherwise $x'_n = x_n$. In other words, paths $\{x_n\}$ of the original chain determine paths $\{(x'_n, x_n)\}$ of the extended chain by the formula

$$x'_n = \begin{cases} x_{R_k} , & R_k < n \le S_k ; \\ x_n , & \text{otherwise} . \end{cases} \tag{8}$$

The transition probabilities of the extended chain are

$$\begin{cases} \delta_{x'} \otimes \pi_x , & \text{if } x' \in Z, x \in V_{\gamma(x')} ; \\ \delta_x \otimes \pi_x , & \text{if } x' \in Z, x \in \complement V_{\gamma(x')} \cap Z ; \\ \operatorname{diag} \pi_x , & \text{otherwise} . \end{cases}$$

We claim that all sequences of random variables in the succession

$$\begin{aligned} &\{x_n\} \xrightarrow{1} \{(x'_n, x_n)\} \xrightarrow{2} \{(x'_n, x_n, \alpha_n)\} \\ &\xrightarrow{3} \{(x_{R_{T_m}}, x_{S_{T_m}}, \alpha_{S_{T_m}})\} \xrightarrow{4} \{x_{S_{T_m}}\} \xrightarrow{5} \{\gamma_{T_m}\} . \end{aligned} \tag{9}$$

are Markov chains, and all these Markov chains have the same (in a natural sense to be specified in each case) bounded harmonic functions, hence, the same Poisson boundary. We shall consider the transformations in (9) step by step.

1. As we have just seen, the chain $\{(x'_n, x_n)\}$ is Markov. The set $\complement Z$ is a recurrent set for the chain $\{x_n\}$ (cf. the arguments in Section 3 and 4), and $\operatorname{diag} \complement Z$ is a recurrent set for the chain $\{(x'_n, x_n)\}$. By definition of $\{(x'_n, x_n)\}$, the corresponding induced chains on $\complement Z$ and $\operatorname{diag} \complement Z$ are isomorphic, so that by formula (2) the chains $\{x_n\}$ and $\{(x'_n, x_n)\}$ also have the same Poisson boundary. In particular, all harmonic functions of the chain $\{(x'_n, x_n)\}$ depend on the second component only.

Another explanation of why $\{x_n\}$ and $\{(x'_n, x_n)\}$ have the same Poisson boundary can be obtained by using directly the definition of the Poisson boundary as the space of ergodic components of the time shift in the path space. Indeed, formula (8) states an isomorphism of the measure spaces of sample paths of the chains $\{x_n\}$ and $\{(x'_n, x_n)\}$. As the chain $\{x_n\}$ is a quotient of the chain $\{(x'_n, x_n)\}$, its Poisson boundary must be a quotient of the Poisson boundary of $\{(x'_n, x_n)\}$. On the other hand, if two paths $\{x_n\}$ and $\{y_n\}$ are trajectory equivalent for the shift in the path space, i.e., if there exist integers n_1, n_2 such that $x_{n_1+n} = y_{n_2+n}$ $\forall n \geq 0$, then the corresponding paths $\{(x'_n, x_n)\}$ and $\{(y'_n, y_n)\}$ are also equivalent (for, with probability 1, there exists an arbitrarily large N such that $x_N \in \complement Z$, and for any such N all stopping times $R_k, S_k > N$ are determined by the positions x_n, $n \geq N$ only).

2. This transformation consists in adding a sequence of i.i.d. random variables $\{\alpha_n\}$ independent of $\{x_n\}$ (hence, of $\{(x'_n, x_n)\}$). As the transition probabilities of the chain $\{(x'_n, x_n, \alpha_n)\}$ do not depend on the α-component, the chain $\{(x'_n, x_n, \alpha_n)\}$ has the same harmonic functions and the same Poisson boundary as the chain $\{(x'_n, x_n)\}$.

3. This step consists in passing to the induced chain on the recurrent subset

$$A = \{(x', x, \alpha) : x' \in Z,\ x \in \complement V_{\gamma(x')},\ \alpha < \frac{1}{H}\Delta(\gamma(x'); x', x)\}$$

which again does not change the Poisson boundary by (2).

4. The transition probabilities $\pi'_{x',x,\alpha}$ of the chain obtained on step 3 depend only on the component x of the triple (x', x, α); denote their projections to the x-component by π'_x. Then clearly $\{x_{S_{T_m}}\}$ is a Markov chain with transition probabilities π'_x, and all $(\pi'_{x',x,\alpha})$-harmonic functions F have the form $F(x', x, \alpha) = f(x)$, where f is a (π'_x)-harmonic function on X.

5. As we have shown in Section 7, the transition probabilities π'_x are convex combinations of the measures ω_γ:

$$\pi'_x = \sum_\gamma \mu_x(\gamma)\omega_\gamma \,. \tag{10}$$

On the other hand,

$$\mu_\gamma = \int \mu_x \, d\omega_\gamma(x) \,,$$

because μ_γ is obtained from the balayage of ω_γ, whereas μ_x, $x \in X \setminus \Gamma$ is obtained from the balayage of δ_x, and the balayage of ω_γ is the integral with respect to ω_γ of the balayée measures of δ_x. Then

$$\int \pi'_x \, d\omega_\gamma(x) = \sum_{\gamma'} \mu_\gamma(\gamma')\omega_{\gamma'} \,, \tag{11}$$

i.e., the result of applying the transition probabilities π'_x to the measure ω_γ is a sum of measures $\omega_{\gamma'}$ with weights $\mu_\gamma(\gamma')$. It implies that averaging with respect to the measures ω_γ determines a natural isomorphism between the spaces of bounded (π'_x)-harmonic functions and of bounded P_Γ-harmonic functions on Γ. Indeed, if f is a (π'_x)-harmonic function, then the function $\overline{f}(\gamma) = \langle f, \omega_\gamma \rangle$ on Γ is P_Γ-harmonic by formula (11). Conversely, if $\overline{f}$ is a P_Γ-harmonic functions on Γ, then by (10) and (11) the function $f(x) = \langle \overline{f}, \pi'_x \rangle$ is (π'_x)-harmonic, and $\overline{f}(\gamma) = \langle f, \omega_\gamma \rangle$. Going backwards along the sequence (9) we see that a bounded function f on X is P-harmonic if and only if it is (π'_x)-harmonic. Finally, since ω_γ is the balayage of the measure δ_γ to $\complement V_\gamma$, we have

$$f(\gamma) = \langle f, \omega_\gamma \rangle = \overline{f}(\gamma) \qquad \forall \gamma \in \Gamma \,.$$

Remark. Group invariant case. If the set X is endowed with an action of a group G, such that the set Γ and the operator P are both G-invariant, then our construction is G-equivariant, and the resulting Markov chain on Γ has G-invariant transition probabilities. In particular, if P is a corecurrent *covering operator* with the deck group G (see [Ka95]), then we obtain a random walk on the group G with the same Poisson boundary as that of the operator P.

9. Geodesic random walks

An example of Markov operators which satisfy conditions $(\mathsf{P}_1) - (\mathsf{P}_5)$, and which can be "discretized" by using Theorem 2 is provided by so called geodesic random walks (cf. [Su81], [Ma95], [KM95]).

Let X be a complete Riemannian manifold with bounded geometry (with uniformly bounded sectional curvatures and bounded away from 0 injectivity radii). Fix a number $D > 0$ smaller than the lower bound of the injectivity radii, and suppose that the "Markov jump" from any given point $x \in X$ consists in choosing a random direction on the unit sphere in the tangent space at the point x (with respect to the normalized Lebesgue measure on the sphere) and then issuing from x a geodesic segment of length D in this direction. Thus, the transition probabilities of the resulting Markov chain (the *geodesic random walk* of length D) are images of the uniform measures on D-spheres in the tangent space under the exponential map of the Riemannian metric. Denote by P the corresponding Markov operator. The transition probabilities of the operator P itself are singular with respect to the Riemannian volume m on X. However, the 2-step transition probabilities (the transition probabilities of the operator P^2) can be easily seen to satisfy conditions $(\mathsf{P}_1) - (\mathsf{P}_5)$ with an appropriate choice of the corresponding constants (depending on the bounds on the sectional curvature and the injectivity radius on X). Using the 0–2 laws [Ka92b], one shows

that the spaces of bounded harmonic functions of the operators P and P^2 coincide.
Take for Γ a uniform net in X (i.e., pairwise distances between the points from Γ are bounded away from 0 and the function $d(x,\Gamma)$ on X is uniformly bounded). Then boundedness of geometry of X implies that for any $M > 0$ the union $\bigcup_{\gamma\in\Gamma} B(\gamma, M)$ is a recurrent set for the Brownian motion on X. Thus, by Theorem 2 we obtain

Theorem 3. *Let X be a bounded geometry Riemannian manifold, and Γ – a uniform net in X. Then there exists a Markov operator P_Γ on Γ such that its space of bounded harmonic functions is isomorphic to the space of bounded harmonic functions of the geodesic random walk on X: the restriction of any such function to Γ is P_Γ-harmonic, and, conversely, any bounded P_Γ-harmonic function can be uniquely extended to a bounded harmonic function of the geodesic random walk.*

The analogous result for the space of usual bounded harmonic functions on a Riemannian manifold X was proved in [Ka92a] (also see [An90]). Note that conditions of Theorems 1 and 2 leave a lot of freedom for modifying the definition of geodesic random walks (one can make the length dependent on starting point, consider not necessarily uniform measures on spheres, etc.).

References.

[An90] A. Ancona, *Théorie du potentiel sur les graphes et les variétés*, Springer Lecture Notes in Math. **1427** (1990), 4–112.

[BL96] W. Ballmann, F. Ledrappier, *Discretization of positive harmonic functions on Riemannian manifolds and Martin boundary*, Sémin. Congr. (Soc. Math. France) **1** (1996), 77–92.

[Fu71] H. Furstenberg, *Random walks and discrete subgroups of Lie groups*, Adv. Probab. Related Topics, vol. 1, Dekker, New York, 1971, pp. 3–63.

[Ka92a] V. A. Kaimanovich, *Discretization of bounded harmonic functions on Riemannian manifolds and entropy*, Proceedings of the International Conference on Potential Theory (Nagoya, 1990) (M. Kishi, ed.), de Gruyter, Berlin, 1992, pp. 213–223.

[Ka92b] ———, *Measure-theoretic boundaries of Markov chains, 0–2 laws and entropy*, Proceedings of the Conference on Harmonic Analysis and Discrete Potential Theory (Frascati, 1991) (M. A. Picardello, ed.), Plenum, New York, 1992, pp. 145–180.

[Ka95] ———, *The Poisson boundary of covering Markov operators*, Israel J. Math **89** (1995), 77–134.

[Ka96] ———, *Boundaries of invariant Markov operators : the identification problem*, Proceedings of the Conference on Algebraic and Number Theoretic Aspects of Ergodic Theory (Warwick, 1994) (. M. Pollicott, K. Schmidt, eds.), Cambridge Univ. Press, Cambridge, 1996, pp. 127–176.

[KM95] V. A. Kaimanovich, H. Masur, *The Poisson boundary of Teichmüller space*, preprint (1995).

[LS84] T. Lyons, D. Sullivan, *Function theory, random paths, and covering spaces*, J. Diff. Geom. **19** (1984), 299–323.

[Ma95] H. Masur, *Random walks on Teichmüller space and the mapping class group*, J. Anal. Math. **67** (1995), 117–164.

[Su81] T. Sunada, *Spherical means and geodesic chains on a Riemannian manifold*, Trans. Amer. Math. Soc. **267** (1981), 483–501.

Vadim A. Kaimanovich
IRMAR, Université Rennes-1
Campus Beaulieu, Rennes 35042, France
email:kaimanov@univ-rennes1.fr

Multifractal Nature of Two Dimensional Simple Random Walk Paths

Gregory F. Lawler

Abstract. The multifractal spectrum of discrete harmonic measure of a two dimensional simple random walk path is considered. It is shown that the spectrum is the same as for Brownian motion, is nontrivial, and can be given in terms of a quantity known as the intersection exponent.

1 Introduction

Let S_n be a simple random walk in $\mathbb{Z}^2$ starting at the origin and let

$$A_n = S[0, 2n^2] = \{S_j : 0 \leqslant j \leqslant 2n^2\}.$$

The set A_n is a random set with "fractal dimension" two. In this paper we will study F_n, the frontier or outer boundary of A_n, i.e., the set of x in A_n such that there exists a nearest neighbor path starting at x going to infinity that never reenters the set A_n. We will study the "multifractal" nature of F_n. More precisely, we consider the multifractal spectrum of (discrete) harmonic measure on A_n, which is the same as harmonic measure on F_n, and show how it can be given in terms of a certain exponent called the intersection exponent. We will summarize the main results in a heuristic way in this introduction and will be more precise in the remainder of the paper. Our definition of multifractal spectrum is based on the definitions found in the physics literature, see, e.g., [1, 20].

Assume A is a large finite, connected subset of $\mathbb{Z}^2$ containing the origin. Let r denote the radius of the set A. (Discrete) harmonic measure H_A on A is the hitting measure of A from infinity, which can be defined by taking an appropriate limit. It can also be defined, up to a multiplicative constant, as the sum over all edges adjacent to the vertex of the gradient of the function on $\mathbb{Z}^2$ that is zero on A, is harmonic on $\mathbb{Z}^2 \setminus A$, and grows logarithmically at infinity. Harmonic measure is only positive at those points from which one can reach infinity; we will call these points the frontier or outer boundary of A, and denote the set by $F = F_A$. The multifractal spectrum of H_A is a measure of the "dimension" of the set of points where the measure H_A looks like a given dimensional measure. We will consider dimension here in a very naive sense: a subset $A_1 \subset A$ has dimension D if the cardinality of A_1 is of order r^D. Also H_A looks D dimensional at the point x if $H_A(x) \approx r^{-D}$. The multifractal spectrum is the function $f(a)$ defined by saying that the set of points where the measure H_A looks like an a or smaller dimensional measure has dimension $f(a)$, i.e.,

$$\#\{x : H_A(x) \geqslant r^{-a}\} \approx r^{f(a)}.$$

Of course we are being very cavalier about "large" and "$\approx$" here. The natural way to make precise definitions is to take a sequence of finite sets A_n whose cardinality increases to infinity and then define $f(a)$ in terms of a limit. We do this in the later sections, but for the time being we will stay with this informal view.

The multifractal spectrum $f(a)$ is closely related to a function $\tau(\lambda)$ that measures the moments of harmonic measure. For $\lambda > 0$, $\tau(\lambda)$ is defined by

$$\sum_{x \in A} H_A(x)^\lambda = \sum_{x \in F} H_A(x)^\lambda \approx r^{\tau(\lambda)},$$

and $\tau(0)$ is given by

$$\#\{H_A(x) > 0\} = \#(F) \approx r^{\tau(0)}.$$

The functions f and τ (in nice cases) are related by the Legendre transform relations

$$\tau(\lambda) = \sup_{a>0}[f(a) - a\lambda],$$

$$f(a) = \inf_\lambda[\tau(\lambda) + \lambda a].$$

The interesting, multifractal case is when the function τ is *strictly* convex so that the infimum in the second relation is taken on at a unique point. Let us assume that the somewhat stronger condition holds: τ is C^2 with $\tau'' > 0$. Then the infimum in the second expression is taken on at λ_a, where λ_a is defined by

$$\tau'(\lambda_a) = -a.$$

Since τ'' is positive, the relationship $a \leftrightarrow \lambda_a$ is one-to-one and we can equally well write $\lambda \leftrightarrow a_\lambda$. The supremum in the first expression is then taken on at a_λ and we have

$$\tau(\lambda) = f(a_\lambda) - a_\lambda \lambda,$$

In particular, the multifractal spectrum $f(a)$ can be determined from τ. Roughly speaking, the measure H_A^λ concentrates on points x with $H_A^\lambda(x) \approx r^{-a_\lambda}$ and the set of such points has dimension $f(a_\lambda)$.

The above analysis is fairly straightforward if we know that $\tau(\lambda)$ is C^2 with a *strictly positive* second derivative (we do this analysis in Section 3). However, it is not so easy to produce interesting examples for which one can show this condition holds. However, recent results on the intersection exponent for Brownian motion and simple random walk allow us to prove that this condition *does* hold for a random walk path in two dimensions. Suppose $A = A_n$ is a random walk path as above of radius $r \approx n$ and of order n^2 points (we will ignore logarithmic factors in this introduction). The intersection exponent $\xi(\lambda)$ is defined in a way so that

$$\mathbf{E}[H_A(S_{n^2})^\lambda] \approx n^{-\xi(\lambda)}, \tag{1}$$

Here, S_{n^2} is a typical point away from the "endpoints" of A. It has recently been shown that: the intersection exponent $\xi(\lambda)$ for random walk is the same as that for Brownian motion for all $\lambda > 0$ [15], and the Brownian motion intersection exponent $\xi(\lambda)$ is C^2 with $\xi''(\lambda) < 0$ [13]. From (1), one gets essentially

$$\mathbf{E}[\sum_{x\in A} H_A(x)^\lambda] \approx r^{2-\xi(\lambda)}.$$

(Actually this is not quite true for large λ because the endpoints start to dominate the sum; however, it will turn out that we will not need to include the endpoints in the sum and so for practical purposes we can regard this as true.) It does not necessarily follow from the above that for a typical realization of A, $\tau(\lambda) = 2 - \xi(\lambda)$. In fact, this will not be true for large λ. We will show that there is a $\bar{\lambda} > 0$ such that

$$\tau(\lambda) = 2 - \xi(\lambda), \quad \lambda \leqslant \bar{\lambda}. \tag{2}$$

In particular, for $0 \leqslant \lambda \leqslant \bar{\lambda}$, $\tau(\lambda)$ satisfies the conditions of the previous paragraph.

The Brownian motion analogue of this result has been proved in both two and three dimensions [14]. Let $B_t, 0 \leqslant t \leqslant 1$ denote a standard Brownian motion in $\mathbb{R}^d$, $d = 2, 3$, and consider the set

$$A = B[0, 1].$$

Harmonic measure $H = H_A$ is the hitting measure of A by another Brownian motion started at infinity. Let G_a be the set of all $x \in A$ such that

$$\lim_{r\to 0} \frac{\log H(\mathcal{B}(x, r))}{\log r} = a,$$

where $\mathcal{B}(x, r)$ is the closed ball of radius r about x. Then with probability one

$$\dim_h(G_{a+d-2}) = \min\{2 - \xi(\lambda), 0\},$$

where $\lambda = \lambda_a$ and $\dim_h$ denotes Hausdorff dimension.

This paper will make the above ideas precise for random walk. We start with a quick review of harmonic measure and then give a precise definition of the multifractal spectrum of harmonic measure of a sequence of finite subsets. The intersection exponent is described next and then it is used to analyze the function $\tau(\lambda)$ on random walk paths. The next two sections sketch the proof of (2), first concentrating on $\lambda = 0$ (the dimension of the frontier) and then on $\lambda > 0$. The final section discusses the situation in higher dimensions.

2 Harmonic Measure in Two Dimensions

In this section we review some standard facts about discrete harmonic measure (from infinity) in $\mathbb{Z}^2$. See [6] for more details. Let A be a finite subset of $\mathbb{Z}^2$ containing the origin. Let

$$\mathrm{rad}(A) = \sup\{|x| : x \in A\},$$

and let C_n be the discrete ball of radius n,

$$C_n = \{x : |x| < n\},$$

with boundary

$$\partial C_n = \{x \notin C_n : |x - y| = 1 \text{ for some } y \in C_n\}.$$

Let $S(t)$ denote a simple random walk in $\mathbb{Z}^2$ with integer time t and let

$$\tau = \tau_A = \inf\{t \geqslant 1 : S(t) \in A\},$$

and

$$T_n = \inf\{t \geqslant 1 : S(t) \in \partial C_n\}.$$

Harmonic measure starting at y is the measure $H_A(\cdot, y)$ given by

$$H_A(x, y) = \mathbf{P}^y\{S(\tau) = x\}.$$

Harmonic measure at infinity is the measure $H_A(\cdot)$ given by

$$H_A(x) = \lim_{|y| \to \infty} H(x, y).$$

It is well known that the limit exists; moreover, if $A \subset C_n$ and $|y| \geqslant 2n$,

$$H_A(\cdot) \asymp H_A(\cdot, y),$$

i.e., there exist constants c_1, c_2, such that if $A \subset C_n$ and $|y| \geqslant 2n$,

$$c_1 H_A(x) \leqslant H_A(x,y) \leqslant c_2 H_A(x).$$

If $\phi(x) = \phi_A(x)$ denotes the unique function on $\mathbb{Z}^2$ with

$$\phi(x) = 0, \quad x \in A,$$

$$\Delta\phi(x) = 0, \quad x \in A^c,$$

$$\phi(x) \sim \frac{2}{\pi} \log|x|, \quad |x| \to \infty,$$

then for $x \in A$,

$$H_A(x) = \frac{1}{4} \sum_{|y-x|=1} \phi(y).$$

The relationship between escape probabilities and harmonic measure is particularly nice in the case of connected subsets of $\mathbb{Z}^2$. If A is a connected subset containing the origin and $r/2 \leqslant \mathrm{rad}(A) \leqslant r$, then

$$H_A(x) \asymp \mathbf{P}^x\{T_{2r} < \tau\}. \tag{3}$$

3 Multifractal Spectrum

Suppose now we have a sequence of finite subsets A_n of $\mathbb{Z}^2$ with $\#(A_n) \to \infty$. For ease we will assume that $0 \in A_n$ for each n. In the example of interest in this paper, A_n will be an increasing sequence, but we do not need to assume that here. We set

$$r_n = \mathrm{rad}(A_n) = \sup\{|x| : x \in A_n\}.$$

Let

$$Q(\lambda, A_n) = \sum_{x \in A_n} H_{A_n}(x)^\lambda.$$

We will assume that there exists a $\tau = \tau(\lambda)$ such that

$$Q(\lambda, A_n) \approx r_n^{\tau(\lambda)}, \quad n \to \infty,$$

where $\approx$ means that the logarithms of boths sides are asymptotic. (We will write $\sim$ for asymptotic and $\asymp$ to mean both sides of an expression are bounded by a constant times the other side.) Note that

$$Q(0, A_n) \leqslant c r_n^2,$$

so $\tau(0) \leqslant 2$. Let

$$N(a, A_n) = \#\{x : H_n(x) \geqslant r_n^{-a}\}.$$

We will sometimes write just $Q(\lambda, n)$ and $N(a, n)$ for $Q(\lambda, A_n)$ and $N(a, A_n)$. The multifractal spectrum is the function $f(a)$ defined as the right continuous version of the function $\tilde{f}$ defined by

$$N(a, n) \approx r_n^{\tilde{f}(a)}, \quad n \to \infty.$$

We assume that such a function $\tilde{f}$ exists, and hence that f exists. Note that f and $\tilde{f}$ are nondecreasing functions and $\tilde{f} \leqslant f$. Intuitively, the set of points at which $H_n(x) \geqslant n^{-a}$ has "dimension" $f(a)$. (We have defined the multifractal spectrum in terms of the dimension of the set of points for which H_n is greater than n^{-a}. One could also define it to be the dimension of the set of points at which H_n is about n^{-a}. The definitions will be equivalent in the interesting cases where f is strictly increasing.)

We consider first two easy examples. First if $A_n = C_n$, then $r_n \sim n$. It is easy to see that

$$Q(0, C_n) \asymp n, \quad n \to \infty,$$

and it can be shown that for every x with $H_{C_n}(x) > 0$,

$$H_{C_n}(x) \asymp n^{-1}.$$

From this we can check easily that

$$\tau(\lambda) = 1 - \lambda,$$

$$f(a) = \begin{cases} -\infty, & a < 1, \\ 1, & a \geqslant 1. \end{cases}$$

As a second example consider a line segment,

$$A_n = \{(k, 0) \in \mathbb{Z}^2 : 0 \leqslant k \leqslant n\}.$$

It can be shown that (see [6, Section 2.4]),

$$H_{A_n}(k, 0) \asymp (\min\{k+1, n-k+1\})^{-1/2} n^{-1/2},$$

where $\asymp$ means that both sides are bounded by a constant (independent of k and n) times the other side. A quick calculation shows that

$$\tau(\lambda) = \begin{cases} 1 - \lambda, & \lambda \leqslant 2, \\ -\lambda/2, & \lambda \geqslant 2. \end{cases}$$

Also,

$$f(a) = \begin{cases} -\infty, & \lambda < 1/2 \\ 2(\lambda - 1/2), & 1/2 \leqslant \lambda < 1 \\ 1, & \lambda \geqslant 1. \end{cases}$$

In this case, if $\lambda < 2$, the measure H^λ spreads itself over the set A_n, while for $\lambda > 2$, the measure is concentrated near the endpoints.

The next two lemmas give the relationships between $f(a)$ and $\tau(\lambda)$. These relationships sometimes go under the name Legendre transforms.

Lemma 3.1

$$\tau(\lambda) = \sup_{a>0}[\tilde{f}(a) - a\lambda] = \sup_{a>0}[f(a) - a\lambda].$$

Proof. The second equality is immediate so we will only consider the first. For every $a > 0$,

$$Q(\lambda, n) \geqslant N(a, n) r_n^{-a\lambda}.$$

Therefore,

$$\tau(\lambda) \geqslant \tilde{f}(a) - a\lambda.$$

Let $K = K_\lambda = (\tau(0) - \tau(\lambda) + 2)/\lambda$, so that

$$\sum_{H_n(x) \leqslant r_n^{-K}} H_n(x)^\lambda \leqslant \#\{x : H_n(x) > 0\} r_n^{-K\lambda} < r_n^{\tau(\lambda)-1}, \tag{4}$$

for n sufficiently large, and hence

$$\sum_{H_n(x) > r_n^{-K}} H_n(x)^\lambda \approx r_n^{\tau(\lambda)}, \quad n \to \infty.$$

Let

$$s = s(\lambda) = \sup_a[\tilde{f}(a) - a\lambda].$$

Let $\epsilon > 0$ and choose $0 = a_0 < a_1 < \ldots < a_m = K$ with $a_i - a_{i-1} < \epsilon$. Then

$$\begin{aligned} \sum_{H_n(x) > r_n^{-K}} H_n(x)^\lambda &= \sum_{i=1}^{m} \sum_{r_n^{-a_i} < H_n(x) \leqslant r_n^{-a_{i-1}}} H_n(x)^\lambda \\ &\leqslant \sum_{i=1}^{m} N(a_i, n) r_n^{-(a_i - \epsilon)\lambda}. \end{aligned}$$

Since $N(a_i, n) \approx r_n^{\tilde{f}(a_i)}$, for sufficiently large n, this is bounded above by

$$r_n^{s+2\lambda\epsilon},$$

and hence

$$\tau(\lambda) \leqslant s + 2\lambda\epsilon.$$

Letting $\epsilon \to 0$, we get the result. □

It follows from the lemma (or directly from the definition) that τ is a convex function, i.e., if $0 < t < 1$,

$$\tau(t\lambda_1 + (1-t)\lambda_2) \leqslant t\tau(\lambda_1) + (1-t)\tau(\lambda_2).$$

In many examples of interest, such as the random walk paths we consider in this paper, the function τ is strictly convex. We will call the following property the *strong multifractal assumption*: $\tau(\lambda)$ is a C^2 function for $\lambda > 0$ with

$$\tau''(\lambda) > 0, \quad \lambda > 0.$$

Choose $a > 0$. By the argument in the proof of the lemma above, for every $\lambda > 0$,

$$f(a) \leqslant \tau(\lambda) + \lambda a.$$

Assume that a is in the image of $-\tau'$, i.e. that there exists a $\lambda_a \in (0, \infty)$ with

$$\tau'(\lambda_a) = -a.$$

Note that the function

$$a \mapsto \lambda_a,$$

is a strictly decreasing function on its domain.

Proposition 3.2 *Under the strong multifractal assumption, if a is in the range of τ',*

$$f(a) = \inf_{\lambda}[\tau(\lambda) + \lambda a] = \tau(\lambda_a) + \lambda_a a. \tag{5}$$

Proof. The second equality can be seen easily by differentiating, and we have already established that

$$f(a) \leqslant \tau(\lambda_a) + \lambda_a a.$$

To prove the other direction, it suffices to show that for every $\epsilon > 0$,

$$\sum_{r_n^{-a-\epsilon} \leqslant H_n(x) \leqslant r_n^{-a+\epsilon}} H_n(x)^{\lambda_a} \approx r_n^{\tau(\lambda_a)}. \tag{6}$$

To see that (6) suffices, note that

$$\#\{x : H_n(x) \geqslant r_n^{-a-\epsilon}\} r_n^{(-a+\epsilon)\lambda_a} \geqslant \sum_{r_n^{-a-\epsilon} \leqslant H_n(x) \leqslant r_n^{-a+\epsilon}} H_n(x)^{\lambda_a},$$

and hence (6) implies for all $\epsilon > 0$,

$$f(a+\epsilon) \geqslant a\lambda_a - \lambda_a \epsilon + \tau(\lambda_a).$$

Replacing a and $a+\epsilon$ with $a-\epsilon$ and a we see that for all $\epsilon > 0$,

$$f(a) \geqslant (a-\epsilon)\lambda_{a-\epsilon} - \lambda_{a-\epsilon}\epsilon + \tau(\lambda_{a-\epsilon}),$$

and hence by right-continuity,

$$f(a) \geqslant a\lambda_a + \tau(\lambda_a).$$

To prove (6), let

$$\hat{s} = \hat{s}(a) = \inf_{\lambda}[\tau(\lambda) + a\lambda],$$

and let $K > 0$ be sufficiently large that

$$\sum_{H_n(x) \geqslant r_n^{-K}} H_n(x)^{\lambda_a} \approx r_n^{\tau(\lambda_a)-1},$$

for n sufficiently large (see (4)). Choose $\epsilon > 0$. By strict concavity, there exists a $\delta > 0$ such that for all $t \notin (a-\epsilon, a+\epsilon)$,

$$\hat{s}(t) \leqslant \tau(\lambda_a) + t\lambda_a - 3\delta\lambda_a,$$

and hence

$$f(t) \leqslant \tau(\lambda_a) + t\lambda_a - 3\delta\lambda_a.$$

Let $s_1, \ldots, s_m$ be a finite set of points in $[0, K] \setminus (a-\epsilon, a+\epsilon)$ such that every point in $[0, K] \setminus (a-\epsilon, a+\epsilon)$ is contained within δ of some s_i. For a fixed i,

$$\sum_{r_n^{-s_i} \leqslant H_n(x) \leqslant r_n^{-s_i+\delta}} H_n(x)^{\lambda_a} \leqslant N(s_i, n) r_n^{-\lambda_a s_i} r_n^{\lambda_a \delta}.$$

The term on the right is bounded above by $r_n^{\tau(\lambda_a)-\lambda_a\delta}$ for n sufficiently large. Since there are only a finite number of s_i, the same argument shows that the sum over all the s_i is bounded above by $cr_n^{\tau(\lambda_a)-\lambda_a\delta}$, for large n. This proves (6). $\square$

Note that in the last proof we showed (under the strong multifractal assumption) that for every $\lambda > 0$ and every $\epsilon > 0$, there is a $\delta > 0$ such that if

$$A_{n,a,\epsilon} = \{x : r_n^{-a-\epsilon} < H_n(x) \leqslant r_n^{-a+\epsilon}\},$$

then

$$\lim_{n\to\infty} \frac{\log \sum_{x\notin A_{n,a,\epsilon}} H_n(x)^{\lambda_a}}{\log r_n} \leqslant \tau(\lambda_a) - \delta.$$

In particular, if we let $\mu_{n,\lambda}$ be the probability measure

$$\mu_{n,\lambda}(x) = \frac{H_n(x)^\lambda}{\sum_y H_n(y)^\lambda},$$

we see that

$$\lim_{n\to\infty} \frac{-\sum_x \mu_{n,\lambda}(x) \log H_n(x)}{\log r_n} = -\tau'(\lambda).$$

If we let Ent denote the entropy of a probability measure,

$$\mathrm{Ent}(P) = -\sum P(x) \log P(x),$$

then we can rephrase this as

$$\mathrm{Ent}(\mu_{n,\lambda}) \sim [\tau(\lambda) - \tau'(\lambda)] \log r_n, \quad n \to \infty.$$

Note by (5),

$$\frac{d(f(a))}{da} = \tau'(\lambda_a)\frac{d(\lambda_a)}{da} + a\frac{d(\lambda_a)}{da} + \lambda_a = \lambda_a > 0.$$

For compact subsets of $\mathbb{Z}^2$, Makarov [17] showed that (continuous) harmonic measure is carried on a set of Hausdorff dimension one. In [8], it was shown how to use this result to prove an analogous result for discrete harmonic measure. There exists a function $g(n)$ going to infinity slower than any power of n such that if A is any connected subset of $\mathbb{Z}^2$ containing the origin of radius n,

$$ng(n)^{-1} \leqslant \#\{x : n^{-1}g(n)^{-1} \leqslant H_A(x) \leqslant n^{-1}g(n)\} \leqslant ng(n). \qquad (7)$$

In particular, this implies that if the A_n are connected subsets of $\mathbb{Z}^2$,

$$f(1) = 1.$$

4 Intersection Exponent

In this section, we introduce the intersection exponent for Brownian motion and random walk. We restrict ourselves to the exponent for nonintersection of two random walks with λ random walks in $\mathbb{Z}^2$, for this is the exponent relevant to the multifractal spectrum of random walk paths. In general, there is an intersection exponent for k walks avoiding λ walks in $\mathbb{Z}^d$, where k is any positive integer, $\lambda > 0$, and $d < 4$.

Let S^1, S^2 be independent simple random walks starting at the origin in $\mathbb{Z}^2$, defined on the probability space $(\Omega, \mathbf{P})$ with stopping times

$$T_n^j = \inf\{t > 0 : S^j(t) \in \partial C_n\}.$$

Let S be another simple random walk starting at the origin defined on a different space $(\Omega_1, \mathbf{P}_1)$ with corresponding stopping times T_n. For any $A \subset Z^2$, $x \in \mathbb{Z}^2$, $1 \leqslant m \leqslant n$, let

$$\mathrm{Es}_A(x, m, n) = \mathbf{P}_1^x\{S[T_m, T_n] \cap A = \emptyset\},$$

$$\mathrm{Es}_A(m, n) = \mathrm{Es}_A(0, m, n).$$

Let

$$\Gamma_n = S^1[0, T^1_{e^n}] \cup S^2[0, T^1_{e^n}],$$

and for $m < n$,

$$\Gamma(m, n) = S^1[T^1_{e^m}, T^1_{e^n}] \cup S^2[T^2_{e^m}, T^2_{e^n}].$$

For $0 \leqslant m < n < \infty$, let

$$X(m, n) = \mathrm{Es}_{\Gamma(m,n)}(e^m, e^n).$$

We write X_n for $X(0, n)$ and

$$\Phi_n = -\log X_n, \quad \Phi(m, n) = -\log X(m, n).$$

Note that $S[T_0, T_n]$ is the same as $S(0, T_n]$. We let $\mathcal{G}_n$ denote the σ-algebra generated by

$$\{S^1(t) : 0 \leqslant t \leqslant T^1_{e^n}\} \cup \{S^2(t) : 0 \leqslant t \leqslant T^2_{e^n}\}.$$

By the discrete Harnack inequality [6, Theorem 1.7.2], for any event $V \in \mathcal{G}_{n-1}$, $x, y \in \partial C_{e^n}$,

$$\mathbf{P}\{S^1(T^1_{e^n}) = x, S^2(T^2_{e^n}) = y; V\} \asymp \mathbf{P}\{S^1(T^1_{e^n}) = x, S^2(T^2_{e^n}) = y\}\mathbf{P}(V).$$

If we apply the discrete Harnack inequality to the random walk S in the definition of $\mathrm{Es}_{\Gamma(m,n)}$ we see that there exists a constant c such that for $0 \leqslant m \leqslant l \leqslant n$,

$$X(m,n) \leqslant cX(m,l)X(l+1,n).$$

If we apply the discrete Harnack inequality to S_1 and S_2, we see that the conditional distribution of $\Gamma(l+1,n)$ conditioned on any event in $\mathcal{G}_l$ is within a multiplicative factor c of the unconditional distribution.

The intersection exponent $\xi(\lambda)$ is defined for $\lambda > 0$ by the relation

$$\mathbf{E}[X_n^\lambda] \approx e^{-n\xi(\lambda)}.$$

The intersection exponent was defined for Brownian motion in [3, 4, 6, 13], and it can shown [3, 5, 11, 15] that random walk has the same intersection exponent. More precisely, we can prove the following.

Theorem 4.1 *[15] For every $0 < \lambda_1 < \lambda_2 < \infty$, there exist $0 < c_1 < c_2 < \infty$ such that for all $0 \leqslant m, n < \infty$ and all $\lambda_1 \leqslant \lambda \leqslant \lambda_2$,*

$$c_1 e^{-\xi(\lambda)n} \leqslant \mathbf{E}[X(m,m+n)^\lambda] \leqslant c_2 e^{-\xi(\lambda)n}.$$

The exponent $\xi(0)$, which is sometimes called the disconnection exponent, is defined by the relation

$$\mathbf{P}\{X_n > 0\} \approx e^{-\xi(0)n}.$$

Theorem 4.2 *[15] There exist $0 < c_1 < c_2 < \infty$ such that for all $0 \leqslant m, n < \infty$,*

$$c_1 e^{-\xi(0)n} \leqslant \mathbf{P}\{X(m,m+n) > 0\} \leqslant c_2 e^{-\xi(0)n}.$$

A slightly stronger version of these theorems that is useful for doing estimates is also given in [15]. Let

$$X^*(m,n) = \sup_{|x| \leqslant e^m} \mathbf{P}_1^x\{S(0,T_n] \cap \Gamma_n = \emptyset\}.$$

Then if $|y|, |z| \leqslant e^m$,

$$\mathbf{E}^{y,z}[X^*(m,m+n)^\lambda] \leqslant ce^{-\xi(\lambda)n}, \tag{8}$$

$$\mathbf{P}^{y,z}\{X^*(m,m+n) > 0\} \leqslant ce^{-\xi(0)n}. \tag{9}$$

The only value of the intersection exponent known rigorously is $\xi(1) = 2$. The disconnection exponent $\xi(0)$ is related to the dimension of the outer boundary of Brownian motion; in fact, it has been shown [12] that the dimension of this boundary is $2 - \xi(0)$. Mandelbrot [16], making an analogy with self-avoiding walks, conjectured that this dimension is 4/3 and hence we have the conjecture $\xi(0) = 2/3$. Rigorously, it is known [2, 4, 21, 22] that

$$.52 < \xi(0) < .985.$$

It is not difficult to show that $\xi(\lambda)$ is a concave function of λ. In fact, it has recently been shown that ξ is *strictly* concave.

Theorem 4.3 *[13] For $\lambda > 0$, $\xi(\lambda)$ is a C^2 function with*

$$\xi''(\lambda) < 0.$$

Moreover,

$$\xi(0) = \lim_{\lambda \downarrow 0} \xi(\lambda).$$

Let $a(\lambda) = \xi'(\lambda)$. The measure

$$\frac{X(m, m+n)^\lambda}{\mathbf{E}[X(m, m+n)^\lambda]}$$

concentrates on paths with $\Phi(m, m+n)$ about equal to $a(\lambda)n$. The next two propositions use the strict concavity of $\xi(\lambda)$ to quantify this.

Proposition 4.4 *For every $0 < \lambda_1 < \lambda_2 < \infty$, there exists a constant c such that for all $0 \leqslant m, n < \infty$ and all $\lambda_1 < \lambda < \lambda_2$,*

$$\mathbf{E}[X(m, m+n)^\lambda; |\Phi(m, m+n) - a(\lambda)n| \geq n^{2/3}\} \leqslant ce^{-\xi(\lambda)n}e^{-n^{1/6}}.$$

Proof. For ease we will do the proof for $X_n = X(0, n)$; the general proof is essentially the same. It suffices to prove the result for n sufficiently large. We will allow constants to depend on λ but one can check that the constants can be chosen uniformly for $\lambda \in [\lambda_1, \lambda_2]$. Note that since

$$\xi'(\lambda) = a(\lambda),$$

and ξ is C^2, there is c such that for all n sufficiently large

$$|\xi(\lambda + n^{-1/2}) - \xi(\lambda) - a(\lambda)n^{-1/2}| \leqslant cn^{-1},$$

$$|\xi(\lambda - n^{-1/2}) - \xi(\lambda) + a(\lambda)n^{-1/2}| \leqslant cn^{-1}.$$

Hence, if $a = a(\lambda)$,

$$\begin{aligned}
\mathbf{E}[e^{-\lambda\Phi_n}; \Phi_n \leqslant an - n^{2/3}] &\leqslant \exp\{an^{1/2} - n^{1/6}\}\mathbf{E}[\exp\{-(\lambda + n^{-1/2})\Phi_n\}] \\
&\leqslant c\exp\{an^{1/2} - n^{1/6} - n\xi(\lambda + n^{-1/2})\} \\
&\leqslant ce^{-\xi(\lambda)n}e^{-n^{1/6}},
\end{aligned}$$

and

$$\begin{aligned}
\mathbf{E}[e^{-\lambda\Phi_n}; \Phi_n \geqslant an + n^{2/3}] &\leqslant \exp\{-an^{1/2} - n^{1/6}\}\mathbf{E}[\exp\{-(\lambda - n^{-1/2})\Phi_n\}] \\
&\leqslant c\exp\{-an^{1/2} - n^{1/6} - n\xi(\lambda - n^{-1/2})\} \\
&\leqslant ce^{-\xi(\lambda)n}e^{-n^{1/6}}.
\end{aligned}$$

□

A similar proof can be used to show the following.

Proposition 4.5 *For every $0 < \lambda_1 < \lambda_2 < \infty$, there exist constants c_1, c_2 such that for all $0 \leqslant m, n < \infty$ and all $\lambda_1 < \lambda < \lambda_2$,*

$$\mathbf{E}[X(m, m+n)^\lambda; |\Phi(m, m+n) - a(\lambda)n| \leqslant c_2 n^{1/2}] \geqslant c_1 e^{-\xi(\lambda)n}.$$

We will need the following corollary.

Corollary 4.6 *For every $s > 0$ and every $0 < \lambda_1 < \lambda_2 < \infty$, there exists an $r > 0$ such that for every $\lambda_1 < \lambda < \lambda_2$ and every $0 < m < n < \infty$,*

$$\mathbf{E}[e^{-\lambda\Phi_n}; \Phi(m, n) \leqslant a(\lambda)(n - m) - r(n - m)^{2/3}$$

$$\textit{for some } 0 \leqslant m \leqslant n] \leqslant se^{-\xi(\lambda)n},$$

$$\mathbf{E}[e^{-\lambda\Phi_n}; \Phi(0, m) \leqslant a(\lambda)m - rm^{2/3} \textit{ for some } 0 \leqslant m \leqslant n] \leqslant se^{-\xi(\lambda)n}.$$

Proof. We will prove the first inequality; the second is proved similarly. Let $a = a(\lambda), \xi = \xi(\lambda)$. We note that it suffices to prove that there is an $M = M_s$ such that

$$\mathbf{E}[e^{-\lambda\Phi_n}; \Phi(m, n) \leqslant a(n - m) - (n - m)^{2/3}$$

$$\text{for some } 0 \leqslant m \leqslant n - M] \leqslant se^{-\xi n},$$

for if we can prove this, the more general fact follows easily by choosing an appropriate value of r. For any fixed m, using the Harnack inequality and Proposition 4.4,

$$\begin{aligned}
&\mathbf{E}[e^{-\lambda\Phi_n}; \Phi(m,n) \leqslant a(n-m) - (n-m)^{2/3}] \\
&\quad\leqslant c\mathbf{E}[e^{-\lambda\Phi_{m-1}}e^{-\lambda\Phi(m,n)}; \Phi(m,n) \leqslant a(n-m) - (n-m)^{2/3}] \\
&\quad\leqslant ce^{-\xi m}\mathbf{E}[e^{-\lambda\Phi(m,n)}; \Phi(m,n) \leqslant a(n-m) - (n-m)^{2/3}] \\
&\quad\leqslant ce^{-\xi n}e^{-(n-m)^{1/6}}.
\end{aligned}$$

Hence by choosing M sufficiently large we can make

$$e^{\xi n}\sum_{m=0}^{n-M}\mathbf{E}[e^{-\lambda\Phi_n}; \Phi(m,n) \leqslant a(n-m) - (n-m)^{2/3}]$$

as small as we like. □

We will write a_λ for $a(\lambda)$ and a_0, a_∞ are defined as the appropriate limits. The function $\lambda \to a(\lambda)$ is continuous and strictly decreasing. Hence there is a well-defined inverse

$$a \to \lambda_a,$$

defined for $a \in (a_\infty, a_0)$. Choose $a \in (a_\infty, a_0)$ and let

$$b(a) = \xi(\lambda_a) - a\lambda_a.$$

Note that

$$\mathbf{P}\{X(m,m+n) \geqslant e^{-an}\} \leqslant e^{\lambda_a an}\mathbf{E}[X(m,m+n)^{\lambda_a}] \leqslant ce^{-b(a)n}.$$

However, Proposition 4.5 implies that

$$\begin{aligned}
&\mathbf{P}\{\Phi(m,m+n) \geqslant an - c_2n^{1/2}\} \\
&\quad\geqslant \mathbf{E}[X(m,m+n)^{\lambda_a}; |\Phi(m,m+n) - a(\lambda)n| \leqslant c_2n^{1/2}]e^{\lambda_a an - c_2\lambda_a n^{1/2}} \\
&\quad\geqslant ce^{-b(a)n}e^{-c_2\lambda_a n^{1/2}}.
\end{aligned}$$

Noting that

$$\mathbf{P}\{\Phi(m,m+n) \geqslant an\} = \mathbf{P}\{\Phi(m,m+n) \geqslant (a + c_2n^{-1/2})n - c_2n^{1/2}\},$$

we can then use the continuity of $b(a)$ and λ_a to conclude

$$\mathbf{P}\{X(m,m+n) \geqslant e^{-an}\} \approx e^{-b(a)n}.$$

We define $b(a)$ for other values of a by the above relation.

The (discrete) Beurling projection theorem [6, Theorem 2.5.2] states that for $d = 2$ there is a constant c such that for any connected subset A containing the origin of radius at least n,

$$\mathrm{Es}_A(0, 2n) \leqslant cn^{-1/2}.$$

This implies that $b(a) = \infty$ for all $a < 1/2$. In fact, it is not too difficult to show that (see [4] for the Brownian motion version)

$$\lim_{a \downarrow 1/2} b(a) = \infty.$$

From this one can easily verify that $a_\infty = 1/2$. In [13] it is shown that $a_0 < \infty$. This is done as part of the proof that the intersection exponent $\xi(\lambda)$ is continuous from the right at 0. This can be used [15] to show that for some $\alpha < \infty$,

$$\mathbf{P}\{X(m, m+n) \geqslant e^{-\alpha n}\} \asymp \mathbf{P}\{X(m, m+n) > 0\} \asymp e^{-\xi(0)n}. \tag{10}$$

By differentiating both sides of

$$b(a) = \xi(\lambda_a) - \lambda_a a$$

with respect to a and using the fact that $\xi'(\lambda_a) = a$, we see that $b'(a) = -\lambda_a$. The discrete version of Makarov's theorem (see (7)) can be used to show that $a_1 = 1, b(a_1) = 1$. No other values of b are known.

We finish this section by stating some technical improvements to the main estimate, Theorem 4.1 (see [15]). The first allows us to replace the hitting times T_n with fixed times. Let

$$\underline{T}_n = \min\{n^2, T_n\}, \quad \overline{T}_n = \max\{n^2, T_n\},$$

and define $\underline{T}^i_n, \overline{T}^i_n$ similarly. Let

$$\underline{\Gamma}_n = S^1[0, \underline{T}^1_{e^n}] \cup S^2[0, \underline{T}^2_{e^n}],$$

$$\overline{\Gamma}_n = S^1[0, \overline{T}^1_{e^n}] \cup S^2[0, \overline{T}^2_{e^n}],$$

and let

$$\underline{X}_n = \mathbf{P}_1\{S(0, \underline{T}_{e^n}] \cap \underline{\Gamma}_n = \emptyset\},$$

$$\overline{X}_n = \mathbf{P}_1\{S(0, \overline{T}_{e^n}] \cap \overline{\Gamma}_n = \emptyset\}.$$

Proposition 4.7 *For every $0 < \lambda_1 < \lambda_2 < \infty$, there exist $0 < c_1 < c_2 < \infty$ such that for all $\lambda_1 \leqslant \lambda \leqslant \lambda_2$,*

$$c_1 e^{-\xi(\lambda)n} \leqslant \mathbf{E}[\overline{X}_n^\lambda] \leqslant \mathbf{E}[\underline{X}_n^\lambda] \leqslant c_2 e^{-\xi(\lambda)n}.$$

Proposition 4.8 *There exist $0 < c_1 < c_2 < \infty$ such that*

$$c_1 e^{-\xi(0)n} \leqslant \mathbf{P}\{\overline{X}_n > 0\} \leqslant \mathbf{P}\{\underline{X}_n > 0\} \leqslant c_1 e^{-\xi(0)n}.$$

The other improvement deals with specifying certain beginning and ending configurations on the random walk. This improvement makes precise the "obvious" notion that random walks conditioned not to intersect have a reasonable probability of being a reasonable distance apart. We will state one result of this type; others that we need can easily be deduced from this one by appropriately adding on configurations. Let $u = (1, 0), u_n = [e^n]u$. Let $\mathcal{Q}(r)$ be the half-space

$$\mathcal{Q}(r) = \{(x^1, x^2) : x^1 \geqslant r\}.$$

Let $J(m,n)$ be the event

$$J(m,n) = \left\{\Gamma_n \subset \mathcal{B}(0, e^{n-1}) \cup \mathcal{Q}(\frac{e^n}{4});\ \Gamma_n \cap \mathcal{B}(0, e^{m+1}) \cap -\mathcal{Q}(\frac{e^m}{4}) = \emptyset\right\}.$$

Let

$$\tilde{X}(m,n) = \mathbf{P}^{-u_m}\Big\{S[0,T_n] \cap \Gamma_n = \emptyset;$$

$$S[0,T_n] \cap \mathcal{B}(0, e^{m+1}) \subset -\mathcal{Q}(\frac{e^m}{2});$$

$$S[0,T_n] \subset \mathcal{B}(0, e^{n-1}) \cup -\mathcal{Q}(\frac{e^n}{4});\ |S(T_n) + u_n| \leqslant \frac{e^n}{4}\Big\}.$$

Proposition 4.9 *For every $0 < \lambda_1 < \lambda_2 < \infty$, there exist c such that for all $\lambda_1 \leqslant \lambda \leqslant \lambda_2$, $0 \leqslant m \leqslant n$, $|x - u_m|$, $|y - u_m| \leqslant e^m/4$,*

$$\mathbf{E}^{x,y}[\tilde{X}(m,n)^\lambda; J(m,n)] \geqslant c e^{-\xi(\lambda)(n-m)}.$$

Proposition 4.10 *There exists a c and an α such that for $0 \leqslant m \leqslant n$, $|x - u_m|$, $|y - u_m| \leqslant e^m/4$,*

$$\mathbf{P}^{x,y}[\tilde{X}(m,n) > c e^{-\alpha(n-m)}; J(m,n)] \geqslant c e^{-\xi(0)(n-m)}.$$

5 Frontier of Random Walk Paths

Let S^1, S^2 be independent simple random walks starting at the origin in $\mathbb{Z}^2$ defined on $(\Omega, \mathbf{P})$ and let $W_t = W(t)$ denote the two-sided walk

$$W(t) = \begin{cases} S^1(-t), & t \leqslant 0 \\ S^2(t), & t \geqslant 0. \end{cases}$$

Let $A_n = W[-n^2, n^2]$. We have chosen two-sided walks for convenience so that the origin is a "typical" point of the path rather than an "endpoint". Note that with probability one

$$\text{rad}(A_n) \approx n.$$

Let S be another simple random walk in $\mathbb{Z}^2$ defined on a different probability space $(\Omega_1, \mathbf{P}_1)$ with stopping times

$$T_n = \inf\{t : |S(t)| \geqslant n\},$$

and let

$$\text{Es}_A(x, n) = \mathbf{P}_1^x\{S(0, T_n] \cap A = \emptyset\}.$$

The frontier or outer boundary of A_n is the set

$$F_n = \{x \in A_n : \text{Es}_{A_n}(x, m) > 0 \text{ for all } m\}.$$

We let $A^n = A_{2^n}, F^n = F_{2^n}$. Since random walk paths are connected it follows from (3) that for any $x \in F_n$,

$$H_{A_n}(x) \asymp \mathbf{P}_1^x\{S(0, T_{2r}] \cap A_n = \emptyset\}, \tag{11}$$

where

$$r = r_n = \text{rad}(A_n).$$

The following is essentially a corollary of the previous section.

Proposition 5.1 *For every* $0 < \lambda_1 < \lambda_2 < \infty$ *there exist* $0 < c_1 < c_2 < \infty$ *such that for all* n*, and all* $\lambda_1 < \lambda < \lambda_2$*,*

$$c_1 n^{-\xi(\lambda)} \leqslant \mathbf{E}[H_{A_n}(0)^\lambda] \leqslant c_2 n^{-\xi(\lambda)}, \tag{12}$$

$$c_1 n^{-\xi(0)} \leqslant \mathbf{P}\{H_{A_n}(0) > 0\} \leqslant c_2 n^{-\xi(0)}. \tag{13}$$

Proof. The lower bounds follow almost immediately from (11) and Propositions 4.9 and 4.10 (with $m = 0$) by appropriately extending paths by events of positive probability. We will do the upper bound for (12); the upper bound for (13) is proved similarly. Let $A = A_n$ and $V_s = V_{s,n} = A_{2^{-s}n}$. By (11),

$$\mathbf{E}[H_A(0)^\lambda] \leqslant c\mathbf{E}[\mathrm{Es}_A(0,n)^\lambda] + c\sum_{s=1}^{\infty} \mathbf{E}[\mathrm{Es}_A(0,2^{-s})^\lambda; \mathrm{rad}(A) \leqslant 2^{-s}n].$$

If $Y_s = Y_{s,n} = \underline{X}_{\log(2^{-s}n)}$ as defined in Proposition 4.7,

$$\begin{aligned}\mathbf{E}[&\mathrm{Es}_A(0,2^{-s})^\lambda; \mathrm{rad}(A) \leqslant 2^{-s}n] \\ &\leqslant \mathbf{E}[\mathrm{Es}_{V_s}(0,2^{-s})^\lambda; \mathrm{rad}(A) \leqslant 2^{-s}n] \\ &\leqslant \mathbf{E}[Y_s^\lambda] \sup_x \mathbf{P}^x\{|S^1(t)| \leqslant 2^{-s}n,\ 0 \leqslant t \leqslant [1-2^{-2s}]n^2\} \\ &\leqslant c[2^{-s}n]^{-\xi(\lambda)} \exp\{-\beta 2^{2s}\},\end{aligned}$$

for an appropriate choice of β. The last inequality uses Proposition 4.7 and the invariance principle. By summing over s we get the upper bound.

It is the goal of this section to show that with probability one

$$\#(F^n) \approx (2^n)^{2-\xi(0)},$$

and hence for the sequence A^n,

$$\tau(0) = 2 - \xi(0).$$

The upper bound follows almost immediately from Proposition 5.1. Note that for $0 \leqslant j \leqslant n^2$,

$$\mathbf{P}\{H_{A_n}(W_j) > 0\} \leqslant \mathbf{P}\{H_{W[2j-n^2,n^2]}(W_j) > 0\} \leqslant c[n^2 - j]^{-\xi(0)/2}.$$

A similar bound holds for $-n^2 \leqslant j \leqslant 0$. By summing over j and using the fact that $0 < \xi(0) < 2$, we see that

$$\mathbf{E}[\#(F_n)] \leqslant \sum_{j=-n^2}^{n^2} \mathbf{P}\{H_{A_n}(W_j) > 0\} \leqslant cn^{2-\xi(0)}.$$

A routine argument using Markov's inequality and the Borel-Cantelli Lemma then allow us to conclude that with probability one, for every $\epsilon > 0$ and all n sufficiently large,

$$\#(F^n) \leqslant 2^{n(2-\xi(0)+\epsilon)}.$$

Let $l(n,x)$ be the "local time"

$$l(n,x) = \#\{t : -n^2 \leqslant t \leqslant n^2, W_t = x\}.$$

It is standard that by choosing c sufficiently large we have

$$\mathbf{P}\{l(n,x) \geqslant c(\log n)^2\} \leqslant cn^{-4},$$

and hence by a standard Borel-Cantelli argument, with probability one for all n sufficiently large and all x,

$$l(n,x) \leqslant c(\log n)^2.$$

In particular if f is any nonnegative function on $\mathbb{Z}^2$, for all n sufficiently large,

$$\sum_{x \in A_n} f(x) \leqslant \sum_{j=-n^2}^{n^2} f(W_j) \leqslant c(\log n)^2 \sum_{x \in A_n} f(x). \tag{14}$$

(One can improve this and show that for any f supported on the frontier F_n,

$$\sum_{j=-n^2}^{n^2} f(W_j) \leqslant c(\log n) \sum_{x \in F_n} f(x),$$

but we will not need this stronger version.)

Let

$$Y_{j,n} = \mathbf{P}_1^{W(j)}\{S(0,T_n] \cap A_n = \emptyset;$$

$$S(0,T_n] \subset \mathcal{B}(0,\frac{n}{2}) \cup \mathcal{B}(nu,\frac{2n}{3}); |S(T_n) - nu| \leqslant \frac{n}{10}\},$$

and let U_n be the event

$$U_n = \{A_n \subset \mathcal{B}(0,\frac{n}{2}) \cup \mathcal{B}(-nu,\frac{2n}{3});\ \mathrm{rad}(A_{n/2}) \leqslant \frac{n}{4};$$

$$|W(n^2) + nu| \leqslant \frac{n}{5},\ |W(-n^2) + nu| \leqslant \frac{n}{5}\}.$$

Here, as before, $u = (1,0)$ and $\mathcal{B}(x,r)$ is the closed ball of radius r about x. From Proposition 4.10 (attaching an appropriate end configuration if necessary) we can deduce that for every $0 < \lambda_1 < \lambda_2 < \infty$ there exist c_1, c_2 such that for $-n^2/4 \leqslant j \leqslant n^2/4$,

$$\mathbf{P}\{Y_{j,n} > 0; U_n\} \geqslant cn^{-\xi(0)}. \tag{15}$$

Note that on the event U_n,

$$Y_{j,n} \geqslant cH_{A_n}(W_j), \quad -\frac{n^2}{4} \leqslant j \leqslant \frac{n^2}{4}.$$

Let

$$Z_n = \sum_{j=-n^2/4}^{n^2/4} I\{Y_{j,n} > 0\}.$$

From (15) we know that

$$\mathbf{E}[Z_n; U_n] \geqslant cn^{2-\xi(0)}.$$

To get a bound on the second moment, we note that

$$\mathbf{P}\{Y_{j,n}Y_{k,n} > 0\} \leqslant cn^{-\xi(0)}|k-j|^{-\xi(0)/2}.$$

The basic argument is as follows. Assume for ease $k = 0, j > 0$. The probability that $W[-j/4, j/4]$ does not disconnect the origin from infinity is bounded above by $cj^{-\xi(0)/2}$. Similarly, the probability that $W[3j/4, 5j/4]$ does not disconnect W_j from infinity is bounded above by $cj^{-\xi(0)/2}$. Finally, given that neither of these disconnections occurs, the random walk has to go from distance of order $\sqrt{j}$ to distance of order n without producing a disconnection. This contributes a term bounded above by $c(n/\sqrt{j})^{-\xi(0)}$. Filling in the details of this argument is not too difficult, using Propositions 4.9 and 4.10. This estimate can then be used to show that

$$\mathbf{E}[Z_n^2] \leqslant cn^{2(2-\xi(0))}.$$

Hence a standard second moment argument shows that there exists a constant c such that

$$\mathbf{P}\{Z_n \geqslant cn^{2-\xi(0)}; U_n\} \geqslant c. \tag{16}$$

This shows that the dimension of the frontier is at least $2 - \xi(0)$ with positive probability. We need another argument to show that it is true with probability one. We will sketch an argument here that will show that for every $\epsilon > 0$ there is a $\delta > 0$ such that for all n sufficiently large

$$\mathbf{P}\{\#(F_n) \leqslant n^{2-\xi(0)-\epsilon}\} \leqslant n^{-\delta}.$$

From this and the upper bound already established, another application of Borel-Cantelli will tell us that with probability one

$$\#(F^n) \approx (2^n)^{2-\xi(0)}.$$

The basic idea is to look at the point of $W[-n^2, n^2]$ at which the first component is a minimum. We write $W_t = ([W_t]_1, [W_t]_2)$. Let

$$\eta = \eta_n = \inf\{t : -n^2 \leqslant t \leqslant n^2 : [W_t]_1 \leqslant [W_s]_1, \ -n^2 \leqslant s \leqslant n^2\},$$

i.e., η is chosen so that

$$[W_s]_1 > [W_\eta]_1, \quad -n^2 \leqslant s < \eta,$$

$$[W_s]_1 \geqslant [W_\eta]_1, \quad \eta \leqslant s \leqslant n^2.$$

A standard estimate (gambler's ruin estimate for one-dimensional random walks) tells us that for $1 \leqslant r \leqslant n^2$,

$$\mathbf{P}\{|\eta| = n^2 + 1 - r\} \asymp n^{-1} r^{-1/2}.$$

In particular, there is a constant c such that for all ϵ,

$$\mathbf{P}\{-n^2 + n^{2-\epsilon} \leqslant \eta \leqslant n^2 - n^{2-\epsilon}\} \geqslant 1 - cn^{-\epsilon/2}.$$

Fix $\epsilon > 0$ small and consider only the event where

$$-n^2 + n^{2-\epsilon} \leqslant \eta \leqslant n^2 - n^{2-\epsilon}. \tag{17}$$

Given η, the distribution of $W[-n^2, n^2]$ is the same, up to a translation, as the distribution of $W[-n^2 - \eta, n^2 - \eta]$ conditioned by

$$[W(t)]_1 \geqslant 0, \quad 0 \leqslant t \leqslant n^2 - \eta,$$

$$[W(t)]_1 > 0, \quad -n^2 - \eta \leqslant t < 0.$$

We will assume (17) and this conditioning. For any m, let

$$\sigma_m^+ = \inf\{t \geqslant 0 : [W_t]_1 = 2^m\},$$

$$\sigma_m^- = \inf\{t \geqslant 0 : [W_{-t}]_1 = 2^m\},$$

$$\rho_m^+ = \sup\{t \leqslant n^2 - \eta : [W_t]_1 = 2^m\},$$

$$\rho_m^- = \inf\{t \geqslant -n^2 - \eta : [W_t]_1 = 2^m\}.$$

Another standard estimate gives that for $2^m \leqslant n^{1-2\epsilon}$ and all n sufficiently large,

$$\mathbf{P}\{\sigma_m^+ > \rho_{m-1}^+\} \geqslant 1/3,$$

$$\mathbf{P}\{\sigma_m^- > \rho_{m-1}^-\} \geqslant 1/3.$$

We call an m good if

$$\sigma_m^+ > \rho_{m-1}^+, \ \sigma_m^- > \rho_{m-1}^-.$$

Then one can use standard large deviation estimates to show that with probability greater than $1 - O(n^{-\delta_\epsilon})$ there exist at least $c_\epsilon \log n$ good values of m with

$$n^{1-3\epsilon} \leqslant 2^m \leqslant n^{1-2\epsilon}.$$

Given a good m, we can use the estimate (16), extending paths appropriately, to show that the probability that the frontier of the set with first coordinate between 2^{m-1} and 2^m has at least $n^{(1-3\epsilon)(2-\xi(0))}$ points is bounded below by some constant c_1. We then have $c_\epsilon \log n$ events each with probability greater than c_1 of occurring. The probability that none of these events occurs is then bounded by $n^{-\delta_\epsilon}$ for some δ_ϵ. This is only a sketch, but it is not too difficult to make this argument precise.

6 Multifractal Spectrum for Random Walk

Let $\lambda > 0$, A_n as in the previous section, and let $H_n = H_{A_n}$. Let $a = \xi'(\lambda)$ and $b = \xi - \lambda a$. As before let

$$Q(\lambda, A_n) = \sum_{x \in A_n} H_n(x)^\lambda.$$

Let

$$\tilde{Q}(\lambda, A_n) = \sum_{j=-n^2}^{n^2} H_n(W_j)^\lambda.$$

From (14) we know that for large n

$$Q(\lambda, A_n) \leqslant \tilde{Q}(\lambda, A_n) \leqslant c(\log n)^2 Q(\lambda, A_n).$$

For the remainder of this section we will assume that $2 - b > 0$. It is our goal to show with probability one,

$$Q(\lambda, A^n) \approx \tilde{Q}(\lambda, A^n) \approx (2^n)^{2-\xi(\lambda)}.$$

The upper bound is not too difficult, although we do need one lemma to take care of "endpoint" effects.

Lemma 6.1 *There exists a $c > 0$ such that for all n sufficiently large*

$$\mathbf{P}\{H_{W[0,n^2]}(W[0,n^2(\log n)^{-2c}]) \neq 0\} \leqslant (\log n)^{-2}.$$

Proof. (sketch) Assuming $c > 4$, in order for

$$H_{W[0,n^2]}(W[0,n^2(\log n)^{-2c}]) \neq 0$$

one of the following must hold:

$$1)\ W[0,n^2(\log n)^{-2c}] \not\subset \mathcal{B}(0,n(\log n)^{-c+2});$$

$$2)\ W[0,n^2] \subset \mathcal{B}(0,n(\log n)^{-2});$$

or 3) the path $W[T_{n(\log n)^{-c+2}}, T_{n(\log n)^{-2}}]$ fails to disconnect $\mathcal{B}(0,n(\log n)^{-c+2})$ from infinity. Standard estimates show that the probability of 1) and 2) are bounded by terms decaying faster than any power of the logarithm and the probability of 3) decays like $(\log n)^{-c\alpha}$ for some α. By choosing c sufficiently large we can make sure that the probability is less than $(\log n)^{-2}$.

Choose some c so that the above lemma holds and let

$$\hat{Q}(\lambda, A_n) = \sum_{j=-n^2+n^2(\log n)^{-2c}}^{n^2-n^2(\log n)^{-2c}} H_n(W_j)^\lambda.$$

By Lemma 6.1 and the Borel-Cantelli lemma, with probability one, for all n sufficiently large,

$$\tilde{Q}(\lambda, A^n) = \hat{Q}(\lambda, A^n).$$

Using Proposition 5.1, we can see that

$$\mathbf{E}[\hat{Q}(\lambda, A_n)] \approx n^{2-\xi(\lambda)}.$$

Hence it follows from an easy application of Markov's inequality and the Borel-Cantelli lemma that with probability one

$$\limsup_{n\to\infty} \frac{\log \tilde{Q}(\lambda, A^n)}{\log 2^n} = \limsup_{n\to\infty} \frac{\log \hat{Q}(\lambda, A^n)}{\log 2^n} \leqslant 2 - \xi(\lambda).$$

Note that this estimate needs only that $2-\xi(\lambda) > 0$, which is clearly implied by $2 - b > 0$.

Let

$$N(a,n) = \#\{x : H_n(x) \geqslant n^{-a}\}.$$

This is slightly different than the definition of $N(a, n)$ in Section 3, but since $r_n \approx n$, this will not be important. Also, let

$$\tilde{N}(a, n) = \#\{j : -n^2 \leqslant j \leqslant n^2, H_n(W_j) \geqslant n^{-a}\}.$$

Note from (14) , for all n sufficiently large,

$$\tilde{N}(a, n) \leqslant c(\log n)^2 N(a, n).$$

Since

$$N(a, n) \leqslant Q(\lambda, A_n) n^{\lambda a}, \tag{18}$$

and $b = b(\lambda) = \xi - \lambda a$, we can conclude that

$$\limsup_{n \to \infty} \frac{\log N(a, 2^n)}{\log 2^n} \leqslant 2 - b,$$

and similarly for $\tilde{N}(a, 2^n)$.

In the remainder of this section we will describe how to show

$$\lim_{n \to \infty} \frac{\log N(a, 2^n)}{\log 2^n} = 2 - b. \tag{19}$$

Before proceeding with the argument, we will show what (18) implies for $b \geqslant 2$. If $b > 2$, we can see from (18) that for n large and some $\epsilon = \epsilon_b$,

$$\mathbf{E}[N(a, n)] \leqslant n^{\ \epsilon}.$$

Hence by an easy Borel-Cantelli argument, with probability one for all n sufficiently large,

$$N(a, 2^n) = 0.$$

Let λ^* satisfy $b(\lambda^*) = 2$, and let $a^* = \xi'(\lambda^*)$. Since for all $a > a^*$,

$$N(a, 2^n) \approx (2^n)^{2 - b(a)},$$

we see from (18) that

$$\liminf_{n \to \infty} \frac{\log Q(\lambda, A^n)}{\log 2^n} \geqslant 2 - b(a^*) - \lambda a^* = -\lambda a^*.$$

Hence, if we consider the sequence $\{A^n\}$, with probability one,

$$\tau(\lambda) \approx \begin{cases} 2 - \xi(\lambda), & b \leqslant 2 \\ -a^* \lambda, & b > 2. \end{cases}$$

For the remainder of this section we fix $a > 0$ such that $b = b(a) < 2$ and sketch the argument to derive (19). We have already established the upper bound so we need only show that with probability one

$$\liminf_{n\to\infty} \frac{\log N(a, 2^n)}{\log 2^n} \geq 2 - b.$$

This is done by proving three estimates. We will use $q(n), q_i(n)$ to denote nondecreasing sequences (in n) of positive integers that grow slower than any power, i.e.,

$$\lim_{n\to\infty} \frac{\log q_i(n)}{\log n} = 0.$$

The sequence $q(n)$ may differ from line to line, but sequences $q_i(n)$ will remain fixed.

Lemma 6.2 *There exists a sequence $q_1(n)$ such that*

$$\mathbf{P}\{N(a,n) \geqslant n^{2-b} q_1(n)^{-1}\} \geqslant q_1(n)^{-1}.$$

Lemma 6.3 *There exists a sequence $q_2(n)$ and a constant $c_3 > 0$ such that*

$$\mathbf{P}\{N(a,n) \geqslant n^{2-b} q_2(n)^{-1}\} \geqslant c_3.$$

Lemma 6.4 *For every $\epsilon > 0$ there exists a $\delta > 0$ such that for all n sufficiently large,*

$$\mathbf{P}\{N(a,n) \leqslant n^{2-b-\epsilon}\} \leqslant n^{-\delta}.$$

In other words, we first show that $N(a,n) \approx n^{2-b}$ with probability going to zero slower than any power law; then we show that it is true with positive probability, independent of n; and then we finally show that it is true with probability near 1. It is not difficult to see that Lemma 6.4 implies (19). The argument to get Lemma 6.4 from Lemma 6.3 is very similar to the argument at the end of the previous section so we will not discuss it here.

The proof of Lemma 6.2 uses a second moment argument. If Y is any nonnegative random variable, it is easy to show that

$$\mathbf{P}\{Y \geqslant \frac{1}{2}\mathbf{E}(Y)\} \geq \frac{[\mathbf{E}(Y)]^2}{4\mathbf{E}(Y^2)}.$$

We will find a sequence $q(n)$ and a (random) subset D_n of

$$\{x : H_{A_n}(x) \geqslant n^{-a} q(n)^{-1}\}$$

such that

$$\mathbf{E}[\#(D_n)] \geqslant n^{2-b} q(n)^{-1}, \tag{20}$$

$$\mathbf{E}[(\#(D_n))^2] \leqslant n^{2(2-b)} q(n). \tag{21}$$

This will imply that there exists a sequence $q(n)$ with

$$\mathbf{P}\{N(a, nq(n)^{-1}) \geqslant n^{2-b} q(n)^{-1}\} \geqslant q(n)^{-1},$$

and the continuity of $b(a)$ then gives Lemma 6.2.

Consider the two-sided walk W. As before, let $\mathcal{Q}(r)$ denote the half space

$$\mathcal{Q}(r) = \{(x^1, x^2) : x^1 \geqslant r\}.$$

and let $u = (1,0)$. For a given sequence $q(n)$ and a given n we define some random variables and events below; these will depend on n and $q(n)$ but we have suppressed that dependence in the notation. Let

$$\tilde{G} = \mathbf{P}_1\{S(0, n^2] \cap W[-\frac{1}{2}n^2, \frac{1}{2}n^2] = \emptyset;\ \ S(\frac{1}{2}n^2, n^2) \subset \mathcal{B}(-nu, \frac{1}{4}n);$$

$$S(0, n^2] \subset [\mathcal{B}(0, n) \setminus \mathcal{Q}(\frac{1}{4}n)]\},$$

$$G(m) = \mathbf{P}_1\{S(0, m^2] \cap W[-\frac{1}{2}m^2, \frac{1}{2}m^2] = \emptyset\},$$

$$G'(m) = \mathbf{P}_1\{S[m^2, n^2] \cap W[-\frac{1}{2}n^2, \frac{1}{2}n^2] = \emptyset\}.$$

We call the origin good (more precisely, good for $a, n, q(n)$) if

$$W[-\frac{1}{2}n^2, \frac{1}{2}n^2] \cap -\mathcal{Q}(\frac{1}{4}n) = \emptyset,$$

$$W[-\frac{1}{2}n^2, -\frac{1}{4}n^2] \cup W[\frac{1}{4}n^2, \frac{1}{2}n^2] \subset \mathcal{B}(nu, \frac{1}{4}n),$$

$$W[-\frac{1}{100}n^2, \frac{1}{100}n^2] \subset \mathcal{B}(0, \frac{1}{100}n),$$

$$G(m) \geqslant m^{-a} q(n)^{-1}, \ \ 0 < m \leqslant n,$$

$$G'(m) \geqslant (n-m)^{-a} q(n)^{-1}, \quad 0 < m < n.$$

We call the time t good if the origin is good for the translated walk

$$\tilde{W}(s) = W(s-t) - W(t).$$

It follows from Proposition 4.4, Corollary 4.6, and Proposition 4.9 that for an appropriate choice of $q(n)$,

$$\mathbf{E}[\tilde{G}^{\lambda}; 0 \text{ good}; n^{-a} q(n)^{-1} \leqslant \tilde{G} \leqslant n^{-a} q(n)] \geqslant c n^{-\xi},$$

and hence

$$\mathbf{P}\{0 \text{ good}\} \geqslant n^{-b} q(n)^{-1}.$$

for an appropriately chosen $q(n)$. If we set

$$D_n = \{0 \leqslant t \leqslant n^2/200 : t \text{ good}\},$$

we see that (20) holds. Let $G_t(m), G_t'(m)$ be the analogues of $G(m), G'(m)$ for the translated walks. Then, as in the argument for the frontier of the path we can see that for $t > 0$,

$$\mathbf{P}\{0, t^2 \text{ good}\} \leqslant \mathbf{P}\{G(t) \geqslant t^{-a} q(n)^{-1}, G_{t^2}(t) \geqslant t^{-a} q(n)^{-1}\} \cdot$$

$$\cdot \mathbf{P}\{G(n) \geqslant n^{-a} q(n)^{-1} \mid G(t) \geqslant t^{-a} q(n)^{-1}, G_t(t) \geqslant t^{-a} q(n)^{-1}\}$$

$$\leqslant q(n) t^{-b} t^{-b} (n/t)^{-b} = q(n) n^{-b} t^{-b},$$

for perhaps a larger value of $q(n)$. With this estimate we can conclude (21).

We have actually shown a little more since we also have some control of the final configuration of the paths. It will be convenient if we state this stronger result. Let T, T^1, T^2, Γ_n be as before, and let $\tilde{Y}_n(x)$ be the random variable

$$\tilde{Y}_n(x) = \mathbf{P}_1^x\{S(0, T_n] \cap \Gamma_n = \emptyset; \ \ |S(T_n) + un| \leqslant \frac{1}{4} n;$$

$$S(0, T_n] \subset [\mathcal{B}(0, \frac{1}{10} n) \cup -\mathcal{Q}(0)]\ \};$$

let J_n be the event

$$J_n = \{\Gamma_n \cap \mathcal{B}(-un, \frac{1}{2} n) = \emptyset; \ \ n^2 \leqslant T_n^1, T_n^2 \leqslant 2n^2;$$

$$|S^i(T^i_n) - un| \leqslant \frac{1}{4}n,\ i = 1, 2\},$$

and if $q(n)$ is a given sequence, let $K_n = K_n(a, q(n))$ be the event

$$K_n = \{\ \#\{x \in \Gamma_n, |x| \leqslant \frac{1}{4}n; \tilde{Y}_n(x) \geqslant n^{-a}q(n)^{-1}\}\ \geqslant\ n^{2-b}q(n)^{-1}\ \}.$$

Then we can conclude the following.

Proposition 6.5 *There exists a sequence $q_1(n)$ so that*

$$\mathbf{P}(J_n \cap K_n) \geqslant q_1(n)^{-1}.$$

We now fix such a sequence $q_1(n)$ and we let

$$\alpha(n) = \mathbf{P}(J_n \cap K_n)$$

as in the proposition. To derive Lemma 6.3 we will choose a sequence $q_2(n)$ that grows faster than $q_1(n)$ and divide the time into $q_2(n)$ disjoint pieces, each of $8n^2$ steps. (We will prove the lemma for $A_{32nq_2(n)}$ rather than A_n.) We will use Proposition 6.5 to give the probability of a "locally nice" configuration in each of the intervals ("locally nice" essentially meaning that the local equivalent of the event in Proposition 6.5 holds) and then using the fact that there are $q_2(n)$ pieces try to conclude that with some positive probability (independent of n) a good number of the pieces are locally nice. We have to worry about possible global effects ruining the "niceness" so we have to be somewhat careful. Let $q_2(n)$ satisfy

$$\frac{q_2(n)^{2-\xi(0)}}{q_1(n)} \to \infty.$$

For a fixed n, let

$$j_i = j_{i,n} = 8in^2, \quad i = 0, 1, 2, \dots, q_2(n).$$

We define the following random variables and events; all of these depend on n, but we have suppressed the n in the notation. These quantities are identical to those in Proposition 6.5, except that we have translated the origin to $W(j_i)$.

$$T_i^+ = \inf\{t \geqslant j_i : |W(t) - W(j_i)| \geqslant n\};$$

$$T_i^- = \sup\{t \leqslant j_i : |W(t) - W(j_i)| \geqslant n\};$$

$$\rho_i = \inf\{t : |S(t) - W(j_i)| \geqslant n\};$$

$$\tilde{Y}_i(x) = \mathbf{P}_1^x\{S(0,\rho_i] \cap W[T_i^-, T_i^+] = \emptyset; |S(\rho_i) - W(j_i) + nu| \leqslant \frac{1}{4}n\,;$$

$$S(0,\rho_i] \subset \mathcal{B}(W(j_i), \frac{n}{10}) \cup [-\mathcal{Q}(0) + W(j_i)]\ \}\,;$$

$$\tilde{J}_i = \{W[T_i^-, T_i^+] \cap \mathcal{B}(-nu + W(j_i), \frac{1}{2}n) = \emptyset; n^2 \leqslant T_i^-, T_i^+ \leqslant 2n^2\,;$$

$$|W(T_i^+) - nu - W(j_i)|, |W(T_i^-) - nu - W(j_i)| \leqslant \frac{1}{4}n\ \}\,;$$

$$\tilde{K}_i = \tilde{K}_i(q_1, a) =$$

$$\{\ \#\{x \in W[T_i^-, T_i^+] : |x - W(j_i)| \leqslant \frac{1}{4}n;$$

$$\tilde{Y}_i(x) \geq n^{-a} q_1(n)^{-1}\}\ \geqslant n^{2-b} q_1(n)^{-1}\ \}.$$

By Proposition 6.5, for each i,

$$\mathbf{P}(\tilde{J}_i \cap \tilde{K}_i) = \alpha(n) \geqslant q_1(n)^{-1}.$$

We now define another event and another random variable to handle the global aspects. Again, these depend on n although that has been suppressed in our notation. Let L_i be the event

$$L_i = \{\Lambda_{32nq_2(n)} \setminus \Lambda_{16nq_2(n)} \subset \mathcal{Q}(8nq_{2n});$$

$$(\Lambda_{32nq_2(n)} \setminus W[T_i^-, T_i^+]) \cap \mathcal{B}(W(j_i), 2n) \subset \mathcal{Q}(\frac{1}{4}n) + W(j_i)\},$$

and let g_n be the random variable

$$g_n = \mathbf{P}_1^{-nu}\{S(0, T_{32nq(n)}] \cap (\Lambda_{32nq_2(n)} \setminus W[T_i^-, T_i^+]) = \emptyset;$$

$$S[T_{16nq_2(n)}, T_{32nq_2(n)}] \subset -\mathcal{Q}(8nq_2(n))\}.$$

We will suppose another sequence $q_3(n)$ is given (we will be specific about this in a moment), and set

$$\tilde{g}_n = g_n I\{g_n \geqslant q_3(n)^{-1}\}.$$

We let $Y_0 = Y_{0,n}$ be the random variable

$$Y_0 = J_n K_n L_n \tilde{g}_n,$$

where we allow events to denote also the indicator function of the events. The strong Markov property and our estimates (see (10)) tell us that for an appropriate choice of q_3,

$$\mathbf{E}[Y_0] \geqslant c\alpha(n) q_2(n)^{-\xi}.$$

We fix such a $q_3(n)$. Also, for $0 < i \leqslant q_2(n)$,

$$\mathbf{E}[Y_0 Y_i] \leqslant c\alpha(n)^2 q_2(n)^{-\xi} i^{-\xi}.$$

Hence if we let

$$h_n = \sum_{i=0}^{q_2(n)} Y_i,$$

we get

$$\mathbf{E}[h_n^2] \leqslant c[\mathbf{E}(h_n^2)]^2,$$

and hence by second moment arguments there is a c such that

$$\mathbf{P}\{h_n > 0\} \geqslant c.$$

We can see we have made the following improvement on Proposition 6.5.

Proposition 6.6 *There exists a sequence $q_4(n)$ and an constant c_4 such that if $K_n = K_n(q_4)$,*

$$\mathbf{P}(J_n \cap K_n) \geqslant c.$$

This gives Lemma 6.3, and also allows us to control the end configuration of the walk. With this control we can then use the idea at the end of Section 5 to derive Lemma 6.4.

7 Higher Dimensions

In this section we discuss higher dimensions. The other interesting dimension for multifractal spectrum is $d = 3$. In fact, most of what is shown in this paper can be proved for $d = 3$; there is one technical fact that we have not been able to prove, however. The results for multifractal spectrum for Brownian motion [14] do hold for $d = 3$. The intersection exponent $\xi(\lambda) = \xi_3(\lambda, 2)$ is well-defined, is strictly concave, and it can be shown

$$\mathbf{E}[H_{A_n}(0)^\lambda] \asymp n^{-\xi(\lambda)-\lambda}.$$

Note the extra λ which comes from the fact that in d dimensions ($d \geqslant 2$), $H_{A_n}(0)$ is analogous to $n^{-(d-2)}\mathrm{Es}_{A_n}(0, n)$. This implies that at least for λ not too large

$$\mathbf{E}[Q(\lambda, A_n)] \asymp n^{2-\xi(\lambda)-\lambda}.$$

(When λ gets very large the "endpoints" will dominate the sum.) However, we have yet to overcome a necessary hurdle to prove the equivalent of Lemma 6.3 so we only conjecture at this point that with probability one for λ small

$$\tau(\lambda) = 2 - \xi(\lambda) - \lambda.$$

If $d \geqslant 4$, the path is not multifractal. In fact

$$\mathbf{E}[H_{A_n}(0)^\lambda] \asymp n^{-2\lambda} \asymp \#(F_n)^{-\lambda},$$

and

$$\tau(\lambda) = 2 - 2\lambda.$$

For $d \geqslant 5$ this follows easily from the fact that

$$\mathbf{E}[\mathrm{Es}_{A_\infty}(0)] > 0.$$

For $d = 4$, more work is needed. In this case

$$\mathbf{E}[\mathrm{cap}(A_n)] \approx n^2(\log n)^{-1},$$

and

$$\mathbf{E}[\mathrm{Es}_{A_n}(0)^\lambda] \asymp \mathbf{E}[\mathrm{Es}_{A_n}(0)]^\lambda \asymp (\log n)^{-\lambda}.$$

The first relationship is nontrivial but can be justified using the idea of slowly recurrent sets [7].

References

[1] Aharony, A. (1991). Fractal growth. In *Fractals and Disordered Systems* (A. Bunde and S. Havlin, Ed.), Springer-Verlag, Berlin, 151-174.

[2] Bishop, C., Jones, P., Pemantle, R., and Peres, Y. (1997). The dimension of the Brownian frontier is greater than 1. J. Funct. Anal. **143**, 309-336.

[3] Burdzy, K. and Lawler, G. (1990). Non-intersection exponents for random walk and Brownian motion. Part I: Existence and an invariance principle. Prob. Theor. and Rel. Fields **84** 393-410.

[4] Burdzy, K. and Lawler, G. (1990). Non-intersection exponents for random walk and Brownian motion. Part II; Estimates and applications to a random fractal. Ann Prob **18** 981-1009.

[5] Cranston, M. and Mountford, T. (1991). An extension of a result of Burdzy and Lawler. Probab. Th. and Rel. Fields **89**, 487-502.

[6] Lawler, G. (1991). *Intersections of Random Walks*. Birkhäuser-Boston.

[7] Lawler, G.(1992). Escape probabilities for slowly recurrent sets, Probab. Theory Relat. Fields **94**, 91 -117.

[8] Lawler, G. (1993). A discrete analogue of a theorem of Makarov, Combin. Probab. and Comp. **2**, 181-200.

[9] Lawler, G. (1995). Nonintersecting planar Brownian motions. Mathematical Physics Electronic Journal **1**, paper no. 4.

[10] Lawler, G. (1996). Hausdorff dimension of cut points for Brownian motion. Electronic Journal of Probability **1**, paper no. 2 .

[11] Lawler, G. (1996). Cut points for simple random walk, Electronic Journal of Probability **1**, #13.

[12] Lawler, G. (1996). The dimension of the frontier of planar Brownian motion, Electronic Communications in Probability **1**, paper no. 5, 29-47.

[13] Lawler, G. (1997). Strict concavity of the intersection exponent for Brownian motion in two and three dimensions, preprint.

[14] Lawler, G. (1997). The frontier of a Brownian path is multifractal, preprint.

[15] Lawler, G. and Puckette E., Intersection and disconnection exponents for simple random walk, in preparation.

[16] Mandelbrot, B. (1983). *The Fractal Geometry of Nature*. W. H. Freeman.

[17] Makarov, N. G. (1985). Distortion of boundary sets under conformal mappings, Proc. London Math. Soc. (3) **51**, 369-384.

[18] Olsen, L. (1995). A multifractal formalism, Adv. Math. **116**, 92-196.

[19] Puckette, E. and Werner, W. (1996). Simulations and conjectures for disconnection exponents, Electronic Communications in Probability **1**, paper no. 6, 49-64.

[20] Stanley, H. G. (1991). Fractals and multifractals: the interplay of physics and geometry, in *Fractals and Disordered Systems* (A. Bunde and S. Havlin, ed.), Springer-Verlag, Berlin, 1-50.

[21] Werner, W. (1995). An upper bound to the disconnection exponent for two-dimensional Brownian motion. Bernoulli **1**, 371-380.

[22] Werner, W. (1996). Bounds for disconnection exponents. Electronic Communications in Probability **1**, paper no. 4, 19-28.

Gregory F. Lawler
Department of Mathematics
Box 90320
Duke University
Durham, NC 27708-0320

Energy and cutsets in infinite percolation clusters

David Levin and Yuval Peres

Abstract. Grimmett, Kesten and Zhang (1993) showed that for $d \geqslant 3$, simple random walk on the infinite cluster $\mathcal{C}_\infty(\mathbf{Z}^d, p)$ of supercritical percolation on $\mathbf{Z}^d$ is a.s. transient. Their result is equivalent to the existence of a nonzero flow f on the infinite cluster such that the 2–energy $\sum_e f(e)^2$ is finite. Here we sharpen this result, and show that if $d \geqslant 3$ and $p > p_c(\mathbf{Z}^d)$, then $\mathcal{C}_\infty(\mathbf{Z}^d, p)$ supports a nonzero flow f such that the q–energy $\sum_e |f(e)|^q$ is finite for all $q > d/(d-1)$. As a corollary, we obtain that any sequence $\{\Pi_n\}$ of disjoint cutsets in $\mathcal{C}_\infty(\mathbf{Z}^d, p)$ that separate a fixed vertex from infinity, must satisfy $\sum_n |\Pi_n|^{-\beta} < \infty$ for all $\beta > 1/(d-1)$. Our proofs are based on the method of "unpredictable paths", developed by Benjamini, Pemantle and Peres (1998) and refined by Häggström and Mossel (1998).

1 Introduction

Bernoulli (bond) percolation with parameter p on an infinite graph $G = (V_G, E_G)$ is the probability measure $\mathbf{P}_p$ on $\{0,1\}^{E_G}$ where each edge in $e \in E_G$ satisfies $\mathbf{P}_p[\omega(e) = 1] = p$ and $\mathbf{P}_p[\omega(e) = 0] = 1 - p$, and the coordinate random variables $\{\omega(e)\}_{e \in E_G}$ are independent. The edge e is called *open* in ω if $\omega(e) = 1$ and *closed* if $\omega(e) = 0$. The connected components of open edges are called *clusters*. The infimum over p such that Bernoulli percolation with parameter p has an infinite cluster a.s. is called the *critical probability*, and denoted by $p_c(G)$. For $d \geqslant 2$, the cubical lattice $\mathbf{Z}^d$ satisfies $0 < p_c(\mathbf{Z}^d) < 1$, and for all $p > p_c(\mathbf{Z}^d)$ there is a.s. a unique infinite cluster, denoted $\mathcal{C}_\infty(\mathbf{Z}^d, p)$. For background on percolation, see Grimmett [5].

Grimmett, Kesten and Zhang [6] proved that if $d \geqslant 3$ and $p > p_c(\mathbf{Z}^d)$, then simple random walk on the infinite cluster $\mathcal{C}_\infty(\mathbf{Z}^d, p)$ is a.s. transient. As shown, for instance, in Doyle and Snell [4], transience of (simple random walk on) a graph G is equivalent to the existence of a nonzero *flow* f of finite 2–*energy* $\sum_{e \in E_G} f(e)^2$. (See Section 2 for formal definitions.) Thus the main result of [6] is equivalent to the existence of nonzero flows on $\mathcal{C}_\infty(\mathbf{Z}^d, p)$ with finite 2–energy a.s.

Benjamini, Pemantle and Peres [2] gave an alternative proof of this result, and extended it to high-density oriented percolation, using certain "unpredictable" random paths that have *exponential intersection tails* to construct

random flows of finite 2–energy on $\mathcal{C}_\infty(\mathbf{Z}^d, p)$. Here we adapt this approach to show that these flows have finite q–energy a.s. for $q > d/(d-1)$.
Definition. The q**–energy** of a flow f on a graph $G = (V_G, E_G)$ is

$$\mathcal{E}_q(f) := \sum_{e \in E_G} |f(e)|^q.$$

Theorem 1.1 *Let $\mathcal{C}_\infty(\mathbf{Z}^d, p)$ be the infinite cluster of independent (bond) percolation with parameter p on $\mathbf{Z}^d$. Then for $d \geqslant 3$ and $p > p_c(\mathbf{Z}^d)$, a.s.,*

$$\inf\{q : \exists \text{ a flow } f \neq 0 \text{ on } \mathcal{C}_\infty(\mathbf{Z}^d, p) \text{ with } \mathcal{E}_q(f) < \infty\} = \frac{d}{d-1}.$$

Remarks.

1. Contained in Maeda [11] is the same result for $\mathbf{Z}^d$ itself: for all $d \geqslant 2$, the infimum of q for which there is a nonzero flow of finite q–energy on $\mathbf{Z}^d$, is $d/(d-1)$. Our arguments do not determine whether Theorem 1.1 extends to $d = 2$. The assumption $d \geqslant 3$ implies that it suffices to find flows of finite q–energy when $d/(d-1) < q < 2$, whence the function $x \mapsto x^{q-1}$ is concave; this concavity is used to pass from (4) to (5) in the proof of theorem 2.2.

2. Theorem 1.1 also holds for site percolation, with the appropriate (site) p_c and with an identical proof. The proof of Theorem 1.1 also yields the same result for oriented percolation provided $p < 1$ is sufficiently large; the renormalization arguments of Hiemer [9] allow one to extend this to all p greater than the critical probability for oriented percolation.

3. The proof of transience in [6] actually yields a flow on $\mathcal{C}_\infty(p, \mathbf{Z}^3)$ with finite q–energy if $q > 1 + \log_4 3$. It might be possible to modify the construction in [6] to give an alternative proof of Theorem 1.1; however, the refinements described in Section 4 seem much harder to obtain in this manner.

A collection of edges Π is a **cutset separating v_0 from** ∞, if any infinite path emanating from v_0 must intersect Π. Nash-Williams [14] proved that if $\{\Pi_n\}_{n=1}^\infty$ is a sequence of disjoint cutsets separating v_0 from infinity in a connected transient graph, then $\sum_n |\Pi_n|^{-1} < \infty$. Theorem 1.1 provides finer information about the permissible growth rates of cutsets on supercritical infinite percolation clusters.

Corollary 1.2 *Let $d \geqslant 3$ and $p > p_c(\mathbf{Z}^d)$. With probability one, if $\{\Pi_n\}$ is a sequence of disjoint cutsets in the infinite cluster $\mathcal{C}_\infty(\mathbf{Z}^d, p)$ that separate a fixed vertex v_0 from ∞, then $\sum_n |\Pi_n|^{-\beta} < \infty$ for all $\beta > \frac{1}{d-1}$.*

This corollary captures in an interesting way the similarity of the infinite cluster to all of $\mathbf{Z}^d$; a delicate issue, that we do not address, is how the "optimal" flows and cutsets in $\mathcal{C}_\infty(\mathbf{Z}^d, p)$ behave as $p \downarrow p_c$.

Proof. Let $\beta > \frac{1}{d-1}$. By Theorem 1.1, there a.s. exists a flow $f \neq 0$ of finite $(1+\beta)$–energy on $\mathcal{C}_\infty(\mathbf{Z}^d, p)$. Suppose that v_0 is in $\mathcal{C}_\infty(\mathbf{Z}^d, p)$, and $\{\Pi_n\}$ is a sequence of disjoint cutsets in $\mathcal{C}_\infty(\mathbf{Z}^d, p)$ that separate v_0 from ∞. We may assume that f is a unit flow from v_0 to ∞, so that $\sum_{e \in \Pi_n} |f(e)| \geqslant 1$ for every n. By Jensen's inequality,

$$\forall n \quad \frac{1}{|\Pi_n|} \sum_{e \in \Pi_n} |f(e)|^{1+\beta} \geqslant \Big(\frac{1}{|\Pi_n|} \sum_{e \in \Pi_n} |f(e)| \Big)^{1+\beta} \geqslant |\Pi_n|^{-1-\beta}.$$

Multiplying by $|\Pi_n|$ and summing over n establishes the corollary. □

The rest of the paper is organized as follows. In Section 2, after recalling some terminology, we state and prove a general sufficient condition for percolation clusters to support flows of finite q–energy (Theorem 2.2). The condition involves the moment generation function for the number of common edges in a certain random path and a fixed path, that share a given edge. The proof is based on a combination of ideas from [2] and [13]. The latter paper proves that certain self-similar measures have densities in L^q for almost all parameters; although the setting is quite different, the method of passing from L^2 to L^q bounds is similar. In Section 3 we prove Theorem 1.1, by verifying that the "unpredictable" random paths constructed in [2] satisfy the condition in Theorem 2.2. Finally, Section 4 contains refinements of Theorem 1.1 involving energy gauges more general than powers; these refinements are based on the paths with optimal predictability profiles, constructed by Häggström and Mossel [8]. At this level, a difference appears (in the power of the logarithm) between the energies that can presently be bounded on supercritical percolation clusters and on all of $\mathbf{Z}^d$; it is an interesting open problem (stated precisely at the end of the paper), to determine whether this difference is an artifact of the proofs, or a real property of percolation clusters.

2 Paths With Exponential Intersection Tails

Definitions.

1. Let $G = (V_G, E_G)$ be an infinite graph with all vertices of finite degree and let $v_0 \in V_G$. Denote by $\Upsilon = \Upsilon(G, v_0)$ the collection of infinite oriented paths in G which emanate from v_0. Let $\Upsilon_1 = \Upsilon_1(G, v_0) \subset \Upsilon$ be the set of **paths with unit speed**, those paths for which the nth vertex is at distance n from v_0. A Υ-valued random element Φ may be identified with a G-valued process $\{\Phi_n\}_{n=0}^\infty$, where Φ_n is the nth vertex in Φ.

2. Let $0 < \theta < 1$. A Borel probability measure μ on $\Upsilon(G, v_0)$ has **Exponential intersection tails** with parameter θ (in short, EIT(θ)) if there exists C such that

$$\mu \times \mu\Big\{(\varphi, \psi) : |\varphi \cap \psi| \geqslant n\Big\} \leqslant C\theta^n$$

for all n, where $|\varphi \cap \psi|$ is the number of edges in the intersection of φ and ψ.

3. If such a measure μ exists for some basepoint v_0 and some $\theta < 1$, then we say that G *admits random paths with* EIT(θ).

4. The percolation cluster containing a vertex v will be denoted $\mathcal{C}(v)$.

5. To define a flow on an undirected graph G, it is convenient to consider each undirected edge as two directed edges, one in each direction. Let vw be the directed edge from v to w. A **flow** f on G with source v_0 is an antisymmetric edge function ($f(vw) = -f(wv)$) such that the net flow out of any vertex $v \neq v_0$ is zero: $\sum_w f(vw) = 0$.

Cox and Durrett [3] obtained upper bounds for the critical probability of oriented percolation using the fact that, for $d \geqslant 4$, oriented paths chosen uniformly in $\mathbf{Z}^d$ have EIT. In [2] the EIT property is exploited to prove transience of oriented supercritical clusters.

Proposition 2.1 ([2]) *Consider percolation with parameter p on G and let v_0 be a vertex in G. Suppose that μ is a probability measure on $\Upsilon_1 = \Upsilon_1(G, v_0)$ that satisfies*

$$\int_{\Upsilon_1}\int_{\Upsilon_1} p^{-|\varphi \cap \psi|}\, d\mu(\varphi)\, d\mu(\psi) < \infty\,. \tag{1}$$

Denote by φ_N the first N edges of a path φ. Then the random variables

$$Z_N := \mu\{\varphi \in \Upsilon_1 \,:\, \varphi_N \text{ is open }\} p^{-N}$$

form a nonnegative Martingale bounded in L^2, and therefore

$$\mathbf{P}_p[\mathcal{C}(v_0) \text{ is infinite }] \geqslant \mathbf{P}_p[\lim_N Z_N > 0] > 0\,.$$

Moreover, if μ satisfies EIT(θ) for some $\theta < p$, then there is a.s. a vertex v in G such that the cluster $\mathcal{C}(v)$ is transient.

The following general theorem is used to establish Theorem 1.1 in Section 3.

Theorem 2.2 *Let μ be a probability measure on the set $\Upsilon_1(G, v_0)$ of paths with unit speed from v_0. Suppose that there exists $p \in (0,1)$, $\gamma > 1$ and $C < \infty$, so that for any fixed path ψ containing edge e_l at distance l from v_0,*

$$\int_{\Upsilon_1} p^{-|\varphi \cap \psi|} \mathbf{1}_{\{\varphi \ni e_l\}} d\mu(\varphi) \leq C l^{-\gamma}. \tag{2}$$

Then the event $|\mathcal{C}(v_0)| = \infty$ has positive probability, and on this event $\mathcal{C}(v_0)$ supports a nonzero flow f with $\mathcal{E}_{1+\beta}(f) < \infty$ for all $\beta > \gamma^{-1}$.

Proof of Theorem 2.2 It suffices to consider $\beta \in (\gamma^{-1}, 1)$. If $\Gamma \subset E_G$, let $I(\Gamma)$ be the indicator of the event that all the edges in Γ are open in the percolation and let $J_e(\Gamma)$ be the indicator of the event $\{e \in \Gamma\}$. For each $N \geqslant 1$ we define an edge function f_N on the ball $B(v_0, N)$ as follows. For every directed edge $e = vw$ where w is farther from v_0 than v, let

$$f_N(e) = \int_{\Upsilon} p^{-N} I(\varphi_N) J_e(\varphi_N)\, d\mu(\varphi)\,,$$

and define $f(wv) = -f(vw)$. If v and w are at the same distance from v_0, set $f(vw) = f(wv) = 0$. Then f_N is a flow on $\mathcal{C}(v_0) \cap B(v_0, N)$ from v_0 to the complement of $B(v_0, N-1)$, i.e., for any vertex $v \in B(v_0, N-1)$ except v_0, the incoming flow to v equals the outgoing flow from v.

The expected $(1+\beta)$–energy of f_N is

$$\mathbf{E}_p \sum_{e \in E_G} \int_{\Upsilon_1} p^{-N} I(\psi_N) J_e(\psi_N)\, d\mu(\psi_N) \left\{ \int_{\Upsilon_1} p^{-N} I(\varphi_N) J_e(\varphi_N) d\mu(\varphi) \right\}^{\beta}.$$

By Fubini's Theorem this equals

$$\sum_{e \in E_G} \int_{\Upsilon_1} J_e(\psi_N) p^{-N} \mathbf{E}_p \left[I(\psi_N) \left\{ \int_{\Upsilon_1} p^{-N} I(\varphi_N \setminus \psi_N) J_e(\varphi_N) d\mu(\varphi) \right\}^{\beta} \right] d\mu(\psi) \tag{3}$$

The two factors appearing in the expectation above depend on disjoint edges, hence they are independent and the expectation of the product can be replaced by a product of expectations. Consequently (3) is equal to

$$\sum_{e \in E_G} \int_{\Upsilon_1} J_e(\psi_N) p^{-N} p^N \mathbf{E}_p \left[\left\{ \int_{\Upsilon_1} p^{-N} I(\varphi_N \setminus \psi_N) J_e(\varphi_N) d\mu(\varphi) \right\}^{\beta} \right] d\mu(\psi)\,. \tag{4}$$

An application of Jensen's inequality to the (concave) function $x \mapsto x^\beta, \beta < 1$ then yields that (4) is bounded by

$$\sum_{e \in E_G} \int_{\Upsilon_1} J_e(\psi_N) \left\{ \mathbf{E}_p \left[\int_{\Upsilon_1} p^{-N} I(\varphi_N \setminus \psi_N) J_e(\varphi_N) d\mu(\varphi) \right] \right\}^\beta \tag{5}$$

$$= \sum_{l=1}^{\infty} \int_{\Upsilon_1} \sum_{|e|=l} J_e(\psi_N) \left\{ \int_{\Upsilon_1} p^{-|\varphi_N \cap \psi_N|} J_e(\varphi_N) d\mu(\varphi) \right\}^\beta d\mu(\psi)$$

$$\leqslant \sum_{l=1}^{\infty} \int_{\Upsilon_1} \left\{ \int_{\Upsilon_1} p^{-|\varphi \cap \psi|} J_{e(l)}(\varphi) d\mu(\varphi) \right\}^\beta d\mu(\psi) \leqslant C \sum_{l=1}^{\infty} l^{-\beta\gamma}, \tag{6}$$

where $e(l) = e(l, \psi)$ is the unique edge in ψ at distance l from v_0, and we have used the hypothesis (2).

For each directed edge e, the sequence $\{f_N(e)\}_N$ is a nonnegative martingale, so it converges a.s. to a limit denoted $f(e)$. Clearly f is a flow from v_0 to infinity. The strength of f_N is precisely the random variable Z_N that appears in Proposition 2.1. The assumption (2) implies the condition (1) of Proposition 2.1, and hence $C(v_0)$ is infinite and the strength of f, $\lim_N Z_N$, is positive with positive probability. Finally, $\mathbf{E}_p[\mathcal{E}_{1+\beta}(f)] \leqslant \sup_N \mathbf{E}_p[\mathcal{E}_{1+\beta}(f_N)] < \infty$, since the right-hand side of (6) is finite for any $\beta > 1/\gamma$.

□

3 Unpredictable paths and percolation in $\mathbf{Z}^d$

Definition. For a sequence of random variables $S = \{S_n\}_{n \geqslant 0}$ taking values in a countable set V, we define its **predictability profile** $\{\mathsf{PRE}_S(k)\}_{k \geqslant 1}$ by

$$\mathsf{PRE}_S(k) = \sup \mathbf{P}[S_{n+k} = x \mid S_0, \ldots, S_n], \tag{7}$$

where the supremum is over all $x \in V$, all $n \geqslant 0$ and all histories $S_0, \ldots, S_n$.

The following was used in [2] along with Proposition 2.1 to prove the theorem of Grimmett, Kesten and Zhang. We use it in the proof of Theorem 1.1 below.

Lemma 3.1 ([2]) *Let $\{\Gamma_n\}$ be a sequence of random variables taking values in a countable set V. If the predictability profile of Γ satisfies $\sum_{k=1}^{\infty} \mathsf{PRE}_\Gamma(k) < \infty$, then there exist $C < \infty$ and $0 < \theta < 1$, such that for any sequence $\{v_n\}_{n \geqslant 0}$ in V and all $m \geqslant 1$,*

$$\mathbf{P}[\#\{n \geqslant 0 : \Gamma_n = v_n\} \geqslant m] \leqslant C\theta^m.$$

We now specialize to the case where $G = \mathbf{Z}^d$ for $d \geqslant 3$. We shall need paths whose predictability profiles are controlled. The basic building block for such paths in $\mathbf{Z}^d$ is an integer–valued nearest neighbor process:

Theorem 3.2 (Benjamini, Pemantle, Peres [2]) *For any $\alpha < 1$ there exists an integer-valued stochastic process $\{S_n\}_{n\geqslant 0}$ such that $|S_n - S_{n-1}| = 1$ a.s. for all $n \geqslant 1$ and*

$$\mathsf{PRE}_S(k) \leqslant C_\alpha k^{-\alpha} \quad \text{for some } C_\alpha < \infty, \text{ for all } k \geqslant 1 .$$

Remark. Let $\mathbf{T}_b(M)$ be the tree of depth M having b descendants of every vertex not on the boundary. The construction of S in Theorem 3.2 uses a random element σ with values in $\{-1,1\}^{\mathbf{T}_b(M)}$, which can be obtained from a variant of the Ising model at low temperature. Order the vertices on the boundary from left to right as $w_1, \ldots, w_{b^M}$. Processes $S^M, M > 1$, are defined for $n \leqslant b^M$ by $S_n^M = \sum_{k=1}^n \sigma(w_k)$, and S is then defined for all $n \geqslant 0$ using the consistency of the laws of the S^M.

Given a $\mathbf{Z}^d$-valued process Y up to time T, we define the *time-reversal* $\overleftarrow{Y}$ of Y up to time T, started at $z \in \mathbf{Z}^d$, by

$$\overleftarrow{Y}_k := z + Y_{T-k} - Y_T \text{ for } k \in [0,T] .$$

Since the processes S^M used in Theorem 3.2 to construct S are defined by summing the spins $\sigma(v)$ over v in the boundary of $\mathbf{T}_b$, the process S has the property that

$$\mathsf{PRE}_{\overleftarrow{S}}(k) \leqslant C_\alpha k^{-\alpha}. \tag{8}$$

Corollary 3.3 *For each $\frac{1}{2} < \alpha < 1$, there is a $\mathbf{Z}^d$-valued process $\Phi = \Phi^{\alpha,d}$ so that*

$$\mathsf{PRE}_\Phi(k) \leqslant C(\alpha,d) k^{-(d-1)\alpha}, \tag{9}$$

and so the random edge sequence $\{\Phi_{n-1}\Phi_n\}_{n\geqslant 1}$ is supported on Υ_1. Moreover, its time-reversal $\overleftarrow{\Phi}$, started at $z \in \mathbf{Z}^d$ and defined for times $k \leqslant M$, also satisfies for $k \leqslant M$

$$\mathsf{PRE}_{\overleftarrow{\Phi}}(k) \leqslant C(\alpha,d) k^{-(d-1)\alpha}. \tag{10}$$

Proof. Let $W_k^r = (S_k^{(r)} + k)/2$ for $r = 1, \ldots, d-1$, where $S^{(r)}$ are independent copies of the process described in Theorem 3.2. For $r = 1, \ldots, d-1$, define clocks

$$t_r(n) := \lfloor \frac{n+d-1-r}{d-1} \rfloor ,$$

and let

$$D(n) := n - \sum_{r=1}^{d-1} W^r_{t_r(n)} .$$

Write $\Phi_n = (W^1_{t_1(n)}, \ldots, W^{d-1}_{t_{d-1}(n)}, D(n))$. It is then easy to see that

$$\mathsf{PRE}_\Phi(k) \leqslant \left[\mathsf{PRE}_S(\lfloor\frac{k}{d-1}\rfloor)\right]^{d-1} \leqslant \left(\frac{C_\alpha k}{d-1}\right)^{-\alpha(d-1)} \leqslant C(\alpha,d)k^{-\alpha(d-1)}.$$

The same bound for $\mathsf{PRE}_{\Phi}(k)$ is obtained similarly, using (8). □

Proof of Theorem 1.1. Since a flow on $\mathcal{C}_\infty(\mathbf{Z}^d,p)$ is also a flow on $\mathbf{Z}^d$, there can be no flows of finite q–energy on $\mathcal{C}_\infty(\mathbf{Z}^d,p)$ for $q \leqslant d/(d-1)$.

For the remainder of the proof let $q > d/(d-1)$ and denote $\beta = q-1$. We want to show that for $p > p_c$, a.s. a flow f on $\mathcal{C}_\infty(\mathbf{Z}^d,p)$ with finite $(1+\beta)$–energy exists. Since $\beta > 1/(d-1)$, we may choose $\alpha \in (1/2,1)$ so that $\beta\alpha > 1/(d-1)$.

We first verify the hypotheses of Theorem 2.2 for $\gamma = \alpha(d-1)$. Fix a path $\psi \in \Upsilon_1$, and let $(e_0, e_1, e_2, \ldots)$ be its constituent edges. If $e = vw$, write $\underline{e}$ for v and $\overline{e}$ for w.

For any path φ, thought of as a sequence of edges, denote by φ_l the first l edges of φ and write $U(\varphi,\psi,l) := |\varphi \cap \psi| - |\varphi_l \cap \psi|$. Let Φ be the process constructed in Corollary 3.3 and let μ denote the distribution of the random edge sequence $\{\Phi_{n-1}\Phi_n\}_{n\geqslant 1}$. By Lemma 3.1, the process Φ constructed in Corollary 3.3 has the property that, given the history of the first l steps, the number of subsequent intersections with a fixed trajectory has an exponential tail:

$$\mu[\varphi : U(\varphi,\psi,l) > n \mid \mathcal{F}_l] \leqslant C_1\theta^n. \tag{11}$$

Our next goal is to verify

$$\int_{\Upsilon_1} p^{-|\varphi\cap\psi|}\mathbf{1}_{\{\varphi\in e_l\}}d\mu(\varphi) \leqslant Cl^{-\gamma}, \tag{12}$$

for p sufficiently close to 1. Denote by $\mathcal{F}_l$ the sigma-field generated by the random variables $\{\mathbf{1}_{\{e\in\varphi\}} : |e| \leqslant l\}$. The left hand side of (12) equals

$$\mu\left[p^{-|\varphi_l\cap\psi|}\mathbf{1}_{\{\varphi\ni e_l\}}\mu[p^{-U(\varphi,\psi,l)} \mid \mathcal{F}_l]\right]. \tag{13}$$

By (11), this is bounded by

$$\frac{C_1}{1-p^{-1}\theta}\sum_{m=1}^{\infty} p^{-m}\mu[\varphi \ni e_l \text{ and } |\varphi_l \cap \psi| = m]. \tag{14}$$

Let $A := \{|\varphi_{l/2}\cap\psi| \geqslant m/2\}$ and $B := \{|(\varphi_l \setminus \varphi_{l/2})\cap\psi| \geqslant m/2\}$. We have

$$\mu[|\varphi_l\cap\psi| = m \text{ and } \varphi\ni e_l] \leqslant \mu[A\cap\{\varphi\ni e_l\}] + \mu[B\cap\{\varphi\ni e_l\}]. \tag{15}$$

By (9), $\mu[\varphi \ni e_l \mid A] \leqslant C_2 l^{-\gamma}$, and by Lemma 3.1, $\mu[A] \leqslant C_1\theta^{m/2}$. Thus

$$\mu\,[A \cap \{\varphi \ni e_l\}] = \mu[\varphi \ni e_l \mid A] \cdot \mu[A] \leqslant C_3\theta^{m/2} l^{-\gamma}\,. \tag{16}$$

Let $\overleftarrow{\Phi}$ be the time–reversal of Φ started at $\overline{e_l}$, and let $\overleftarrow{B}$ be the event that $\{\overleftarrow{\Phi}_n\}_{n \leqslant l/2}$ intersects the vertices determined by ψ at least $m/2$ times. Then

$$\mu[B \cap \{\varphi \ni e_l\}] \leqslant \mathbf{P}[\overleftarrow{B} \cap \{\overleftarrow{\Phi} \ni 0\}],$$

because the number of edge intersections of two paths is bounded by the number of vertex intersections. By Lemma 3.1, $\mathbf{P}[\overleftarrow{B}] \leqslant C_1\theta^{m/2}$, and (9) implies that $\mathbf{P}[\overleftarrow{\Phi} \ni 0 \mid \overleftarrow{B}] < C_2 l^{-\gamma}$. Thus

$$\mu[B \cap \{\varphi \ni e_l\}] \leqslant \mathbf{P}[\overleftarrow{B} \cap \{\overleftarrow{\Phi} \ni 0\}] = \mathbf{P}[\overleftarrow{B}] \cdot \mathbf{P}[\overleftarrow{\Phi} \ni 0 \mid \overleftarrow{B}] \leqslant C_3\theta^{m/2} l^{-\gamma}.$$

We conclude that the right-hand side of (15) is bounded by $2C_3\theta^{m/2}l^{-\gamma}$. Thus for $p > \sqrt{\theta}$, the sum (14) is bounded by $Cl^{-\gamma}$, and (12) follows. Since $\beta > [\alpha(d-1)]^{-1}$, by Theorem 2.2, $\mathbf{P}[I(0)] > 0$, where $I(0)$ is the event that $\mathcal{C}(0)$ is infinite and supports a flow of finite $(1+\beta)$–energy. The event $\bigcup_{v \in \mathbf{Z}^d} I(v)$ does not depend on the status of any finite collection of edges, and hence by Kolmogorov's zero–one law, has probability one.

This concludes the proof for p near 1; The general case $p > p_c$ is reduced to this by the renormalization argument used in Corollary 2.1 of [2], which relies on techniques of [7],[1] and [15]; a result of Soardi and Yamasaki [17], that the existence of a flow of finite q–energy is invariant under rough isometries, is also needed. □

4 A Refinement

The concept of energy can be further generalized by defining the H–energy of a flow f as $\mathcal{E}_H(f) := \sum_e H(|f(e)|)$, where $H : [0,\infty) \to [0,\infty)$ is nondecreasing. For the rest of this section, we fix $d \geqslant 3$ and compare H–energy of flows on $\mathbf{Z}^d$ and on $\mathcal{C}_\infty(\mathbf{Z}^d, p)$. As we shall see, in both cases the "critical" gauges are obtained by logarithmic corrections to the power law $u \mapsto u^{d/(d-1)}$.

Notation. For any positive decreasing function h, let $H_h(u) := u^{d/(d-1)}/h(u)$ for $u > 0$ and $H_h(0) = 0$. If $h(u) = [\log(1+u^{-1})]^\alpha$, then we abbreviate H_h by H_α. We let C, C_i denote positive finite constants whose value is unimportant.

First we consider the case of $\mathbf{Z}^d$ itself. Let $D(l)$ be the collection of edges at distance l from the origin. T. Lyons [12] constructed a nonzero flow f_* on $\mathbf{Z}^d$ that satisfies $|f_*(e)| \leqslant Cl^{1-d}$ for any edge $e \in D(l)$. Thus

$$\sum_e H_h(|f_*(e)|) \leqslant C_1 \sum_{l=1}^{\infty} l^{d-1} \frac{(Cl^{1-d})^{d/(d-1)}}{h(Cl^{1-d})} \leqslant C_2 \sum_l \frac{1}{lh(l^{1-d})} \asymp \int_1^\infty \frac{dx}{xh(x^{-1})},$$

where $y \asymp z$ means that the ratio y/z is bounded above and below by positive constants.

Let f be a unit flow from 0. If H_h is convex, then

$$|D(l)|^{-1} \sum_{e \in D(l)} H_h(|f(e)|) \geqslant H_h \left(|D(l)|^{-1} \sum_{e \in D(l)} |f(e)| \right),$$

by Jensen's inequality. Since f is a unit flow, $\sum_{e \in D(l)} |f(e)| \geqslant 1$, so

$$\sum_{e \in D(l)} H_h(|f(e)|) \geqslant |D(l)|\, H_h(|D(l)|^{-1}) \geqslant \frac{C_3}{lh(C_4 l^{1-d})}.$$

Thus the H_h-energy of any unit flow f is at least

$$\sum_l \sum_{e \in D(l)} H_h(|f(e)|) \geqslant \sum_l \frac{C_3}{lh(C_4 l^{1-d})} \asymp \int_1^\infty \frac{dx}{xh(x^{-1})}.$$

In particular, $\mathbf{Z}^d$ supports a flow of finite H_α-energy iff $\alpha > 1$.

Proposition 4.1 *Let h be a decreasing function satisfying*

$$\sum_j^\infty \frac{1}{jh(j^{-1})} < \infty \tag{17}$$

and $h(x^2) \leqslant \kappa h(x)$ for all $x > 0$. Define $G_h(u) := H_{h^2}(u)/u$, and assume that G_h is concave. Then for $p > p_c$, there is a.s. a flow $f \neq 0$ on $\mathcal{C}_\infty(\mathbf{Z}^d, p)$ with finite H_{h^2}-energy, i.e.,

$$\sum_{e \in E(\mathbf{Z}^d)} \frac{|f(e)|^{d/(d-1)}}{h(|f(e)|)^2} < \infty.$$

In particular, $\mathcal{C}_\infty(\mathbf{Z}^d, p)$ supports a flow of finite H_α-energy for $\alpha > 2$.

We proceed as in the proof of Theorems 1.1 and 2.2, which may be consulted for notation. Let $g(y) = h(y^{1-d})$. Convergence of the sum in (17) implies that $\sum_j (jg(j))^{-1} < \infty$, and hence by Theorem 1.4 in Häggström and Mossel [8], there is an integer-valued process S with $\mathsf{PRE}_S(k) \leq Cg(k)/k$; their constructions also yield the same bound for the time-reversal of S. As in Corollary 3.3, we can define a process Φ supported on $\Upsilon_1(0, \mathbf{Z}^d)$ such that $\mathsf{PRE}_\Phi(k) \leqslant C(g(k)/k)^{d-1}$ and $\mathsf{PRE}_{\overleftarrow{\Phi}}(k) \leqslant C(g(k)/k)^{d-1}$. Let μ be the distribution of the edge sequence φ determined by Φ, and write

$$f_N(e) = \int_{\Upsilon_1} p^{-N} I(\varphi_N) J_e(\varphi_N) d\mu(\varphi)$$

for edges directed away from 0. As before it is enough to show that

$$\mathbf{E}_p \sum_{e \in E(\mathbf{Z}^d)} H_{h^2}(|f_N(e)|) = \mathbf{E}_p \sum_{e \in E(\mathbf{Z}^d)} G_h(|f_N(e)|)|f_N(e)| \tag{18}$$

is bounded uniformly in N. By Fubini's Theorem, (18) equals

$$\sum_e \int_{\Upsilon_1} \mathbf{E}_p \left[p^{-N} I(\psi_N) J_e(\psi_N) G_h \left(\int_{\Upsilon_1} p^{-N} I(\varphi_n \setminus \psi_N) J_e(\varphi_N) d\mu(\varphi) \right) \right] d\mu(\psi).$$

By independence of the status of different edges, this can be rewritten as

$$\int_{\Upsilon_1} \sum_e \mathbf{E}_p \left[p^{-N} I(\psi_N) J_e(\psi_N) \right] \mathbf{E}_p \left[G_h \left(\int_{\Upsilon_1} p^{-N} I(\varphi_n \setminus \psi_N) J_e(\varphi_N) d\mu(\varphi) \right) \right] d\mu(\psi).$$

Applying Jensen's inequality to the second expectation bounds the preceding formula by

$$\int_{\Upsilon_1} \sum_e J_e(\psi) G_h \left(\mathbf{E}_p \left[\int_{\Upsilon_1} p^{-N} I(\varphi_N \setminus \psi_N) J_e(\varphi_N) \right] \right) d\mu(\psi). \tag{19}$$

Since ψ contains one edge $e(l)$ in $D(l)$, (19) equals

$$\int_{\Upsilon_1} \sum_l G_h \left(\int_{\Upsilon_1} p^{-|\varphi_N \cap \psi_N|} J_{e(l)}(\varphi_N) d\mu(\varphi) \right) d\mu(\psi). \tag{20}$$

Arguing as in Theorem 1.1, we obtain that

$$\int_{\Upsilon_1} p^{\;|\varphi_N \cap \psi_N|} J_{e(l)}(\varphi_N) d\mu(\varphi) \leqslant \left(C_1 \frac{g(l)}{l} \right)^{d-1}.$$

Thus (20) is bounded by

$$\sum_l G_h \left(C_1 \left\{ \frac{g(l)}{l} \right\}^{d-1} \right) = \sum_l \frac{(C_2 g(l)/l)}{h\left((C_2 g(l)/l)^{d-1} \right)^2}. \tag{21}$$

Since $h((l/C_2 g(l))^{1-d}) = g(C_3 l / g(l))$, (21) is bounded by

$$\sum_l C_2 \frac{g(l)}{l} \frac{1}{g(C_3 l/g(l))^2}. \tag{22}$$

The assumption that $h(x^2) \leqslant \kappa h(x)$ implies that

$$\forall\, y > 0, \;\; g(y^2) \leqslant \kappa g(y). \tag{23}$$

Therefore $g(y)^2 \leqslant C_4 y$ for all y. Consequently,

$$\frac{1}{g(C_3 l/g(l))} \leqslant \frac{\kappa}{g(C_5 l^2/g(l)^2)} \leqslant \frac{\kappa}{g(C_6 l)},$$

where the last inequality follows since g is increasing. Thus (22) is bounded by

$$C_7 \sum_l \frac{g(l)}{l g(C_6 l)^2} \asymp \sum_l \frac{1}{l g(l)},$$

because (23) implies that $g(C_6 l) \asymp g(l)$. Hence convergence of (22) follows from convergence of $\sum_l (g(l) l)^{-1}$. □

The preceding proposition has implications for the permissible growth rate of cutsets on $\mathcal{C}_\infty(\mathbf{Z}^d, p)$. Let $\{\Pi_n\}$ be a sequence of disjoint cutsets in the percolation cluster. Assume that h satisfies the hypothesis of that proposition, and also that H_{h^2} is convex. Let f be a unit flow of finite H_{h^2}–energy on $\mathcal{C}_\infty(\mathbf{Z}^d, p)$. Then

$$\infty > \sum_e H_{h^2}(|f(e)|) \geqslant \sum_n \sum_{e \in \Pi_n} H_{h^2}(|f(e)|) \geq \sum_n \frac{|\Pi_n|^{-1/(d-1)}}{h^2(|\Pi_n|^{-1})}.$$

In particular, taking $h(u) = (\log(1 + u^{-1}))^{2+\epsilon}$ shows that

$$\sum_n |\Pi_n|^{-1/(d-1)} (\log(|\Pi_n|))^{-2-\epsilon} < \infty.$$

While we know that $\mathbf{Z}^d$ itself will support flows of finite H_h energy iff h satisfies the summability condition (17), Proposition 4.1 only gives a *sufficient* condition for finiteness of energy on $\mathcal{C}_\infty(\mathbf{Z}^d, p)$. The proof above used Theorem 1.4 of Häggström and Mossel [8], which states that for any increasing function g satisfying $\sum_j (j g(j))^{-1} < \infty$, there is a $\mathbf{Z}$–valued nearest–neighbor process with predictability profile at k bounded by $Cg(k)/k$. Hoffman [10] proved that if g does not satisfy this summability condition, then such a predictability profile cannot be attained for any nearest–neighbor process on $\mathbf{Z}$. This motivates the following.

Conjecture: For $d \geqslant 3$ and $p > p_c(\mathbf{Z}^d)$, a.s. any nonzero flow f on the infinite cluster $\mathcal{C}_\infty(\mathbf{Z}^d, p)$ must satisfy

$$\mathcal{E}_{H_2}(f) = \sum_{e: f(e) \neq 0} \frac{|f(e)|^{d/(d-1)}}{\log^2(1 + |f(e)|^{-1})} = \infty. \tag{24}$$

Added in Proof: This conjecture motivated E. Mossel and C. Hoffman to find a different construction of low energy flows on percolation clusters.

By combining their new ideas with the methods of the present paper, they showed in a recent preprint (entitled "Energy of flows on percolation clusters", to appear in *Potential Analysis*) that $\mathcal{C}_\infty(\mathbf{Z}^d, p)$ can support flows that do not satisfy (24). Moreover, under a mild regularity hypothesis on h, they proved the remarkable result that for all $d \geqslant 3$ and $p > p_c(\mathbf{Z}^d)$, the infinite cluster $\mathcal{C}_\infty(\mathbf{Z}^d, p)$ a.s. supports a nonzero flow of finite H_h energy iff $\mathbf{Z}^d$ supports such a flow.

Acknowledgements. We are grateful to the organizers of the 1997 workshop on random walks and discrete potential theory at Cortona (V. Kaimanovich, M. Picardello, L. Saloff-Coste and W. Woess) for making it a fruitful and memorable meeting. We thank E. Mossel for helpful comments on a draft of this paper.

References.

[1] P. Antal and A. Pisztora (1996). On the chemical distance in supercritical Bernoulli percolation. *Ann. Probab.* **24** 1036–1048.

[2] I. Benjamini, R. Pemantle and Y. Peres (1998). Unpredictable paths and percolation. *Ann. Probab.* **26**, 1198–1211.

[3] T. Cox and R. Durrett (1983). Oriented percolation in dimensions $d \geqslant 4$: bounds and asymptotic formulas. *Math. Proc. Camb. Phil. Soc.* **93**, 151–162.

[4] P. G. Doyle and E. J. Snell (1984). *Random walks and electrical networks.* Carus Math. Monographs **22**, Math. Assoc. Amer., Washington, D. C.

[5] G.R. Grimmett (1989). *Percolation.* Springer, New York.

[6] G. R. Grimmett, H. Kesten and Y. Zhang (1993). Random walk on the infinite cluster of the percolation model. *Probab. Th. Rel. Fields* **96**, 33–44.

[7] G. R. Grimmett and J. M. Marstrand (1990). The supercritical phase of percolation is well behaved. *Proc. Royal Soc. London Ser. A* **430**, 439–457.

[8] O. Häggström and E. Mossel (1998). Nearest-neighbor walks with low predictability profile and percolation in $2+\epsilon$ dimensions. *Ann. Probab.* **26**, 1212–1231.

[9] P. Hiemer (1998). Dynamical Renormalisation in Oriented Percolation. *Preprint.*

[10] C. Hoffman (1998). Unpredictable nearest neighbor processes. *Preprint.*

[11] F. Y. Maeda (1977). A remark on the parabolic index of infinite networks. *Hiroshima J. Math.* **7**, 147–152.

[12] Lyons, T. (1983) A simple criterion for transience of a reversible Markov chain, *Ann. Probab.* **11**, 393–402.

[13] Y. Peres and B. Solomyak (1998). Self-similar measures and intersections of Cantor sets. *Trans. Amer. Math. Soc.*, to appear.

[14] C. St. J.A. Nash-Williams (1959). Random walks and electric currents in networks. *Proc. Cambridge Philso. Soc.* **55**, 181–194.

[15] A. Pisztora (1996). Surface order large deviations for Ising, Potts and percolation models. *Probab. Th. Rel. Fields* **104**, 427–466.

[16] P. M. Soardi (1994). *Potential Theory on Infinite Networks.* Springer LNM, Berlin.

[17] P.M. Soardi and M. Yamasaki (1993). Parabolic indices and rough isometries. *Hiroshima J. Math* **23**, 333–342.

David Levin
Department of Statistics, University of California, Berkeley, CA
Yuval Peres
Mathematics Institute, The Hebrew University, Jerusalem, Israel

On discrete groups and pointwise ergodic theory

Amos Nevo

1. Introduction.

Our purpose is to present some recent developments and some open problems regarding measure preserving actions of discrete groups. The most basic example of such an action is of course the action of the group on itself by translations, and we begin by establishing some notation and recalling some definitions pertaining to this case.

1.1 The action of Γ by translations. Let Γ be a countable group, and let $\ell^1(\Gamma) = \{\mu = \sum_{\gamma\in\Gamma} \mu(\gamma)\gamma : \sum_{\gamma\in\Gamma} |\mu(\gamma)| < \infty\}$. The operation of convolution, given by $\mu * \nu(w) = \sum_{uv=w} \mu(u)\nu(v)$, together with the involution $\mu^*(w) = \overline{\mu(w^{-1})}$, define on $\ell^1(\Gamma)$ the structure of a $*$-algebra, the group algebra of Γ. Recall that μ is called symmetric if $\mu = \mu^*$.

The group Γ acts on itself by left translations, and this action gives rise to the left regular representation $\lambda = \lambda_\Gamma$ of Γ, by isometries of the Hilbert space $\ell^2(\Gamma)$, via: $\lambda(\gamma)f(w) = f(\gamma^{-1}w)$. To each element $\mu \in \ell^1(\Gamma)$ there corresponds an operator $\lambda(\mu) : \ell^2(\Gamma) \to \ell^2(\Gamma)$, given by $\lambda(\mu) = \sum_{\gamma\in\Gamma} \mu(\gamma)\lambda(\gamma)$, and satisfying $\|\lambda(\mu)\| \leqslant \|\mu\|_{\ell^1(\Gamma)}$. $\lambda(\mu)$ is the operator of (left) convolution by μ, and its norm as an operator on $\ell^2(\Gamma)$ is called the spectral norm of μ. The map $\mu \mapsto \lambda(\mu)$ is a continuous $*$-algebra homomorphism of $\ell^1(\Gamma)$ into $\mathrm{End}\,\ell^2(\Gamma)$.

A natural family of elements in $\ell^1(\Gamma)$ to consider are those which are given in explicit geometric form. To that end, assume Γ is finitely generated, and let S be a finite generating set which is symmetric, namely $S = S^{-1}$. S induces a length function on Γ, given by $|\gamma| = |\gamma|_S = \min\{n | \gamma = s_1 \cdots s_n, s_i \in S\}$, $|e| \stackrel{d}{=} 0$. The length function defines a metric on Γ, given by $d(u,v) = |uv^{-1}|$, which is invariant under right translation by Γ. The metric d determines the structure of an undirected graph, the Cayley graph $X(\Gamma, S)$, where two elements u and v are adjacent iff $d(u,v) = 1$.

Now consider the following sequences:

Definition 1.

(1) $\sigma_n = \frac{1}{\# S_n} \sum_{w\in S_n} w$, where $S_n = \{w : |w| = n\}$ is the sphere of radius n, with center e. Define also: $\sigma'_n = \frac{1}{2}(\sigma_n + \sigma_{n+1})$.

(2) $\mu_n = \frac{1}{n+1} \sum_{k=0}^{n} \sigma_k$, the average of the first $n+1$ sphere averages.

(3) $\beta_n = \frac{1}{\#B_n}\sum_{w\in B_n} w$, where $B_n = \{w : |w| \leqslant n\}$ denotes the ball of radius n with center e.

Note that the operator $\lambda(\sigma_n)$ of left convolution by σ_n is the operator of averaging a function on Γ on circles of radius n, w.r.t the metric d:

$$\lambda(\sigma_n)f(w) = \sigma_n * f(w) = \frac{1}{\#S_n}\sum_{|x|=n} f(x^{-1}w) = \frac{1}{\#S_n}\sum_{d(u,w)=n} f(u) .$$

Remark also that the definition of β_n makes sense for every length function $|\cdot|$ on Γ, not only those asociated with word length. For example, we can consider length functions of the form $|\gamma| = d(x_0, \gamma x_0)$, where (X, d) is a locally compact metric space on which Γ acts freely, properly, isometrically and co-compactly. The definitions of σ_n and μ_n make sense for every integer valued length function.

A basic goal of analysis on the group Γ is the elucidation of the properties of convolution operators, and more generally, asymptotic properties of sequences of convolution operators. Several different points of view lead to questions relating to their behaviour, and we confine ourselves here to the following two fundamental questions:

(1) Let $\mu \in \ell^1(\Gamma)$ be a probability measure. What is the spectrum of the convolution operator $\lambda(\mu) : \ell^2(\Gamma) \to \ell^2(\Gamma)$?

To illustrate the significance of this question, recall that any probability measure μ on Γ determines a random walk on Γ, with transition probabilities invariant under right translation by elements of Γ. Taking $\mu = \sigma_1$ for example, we obtain the nearest neighbour random walk on the Cayley graph $X(\Gamma, S)$. The spectral characteristics of the convolution operator $\lambda(\mu)$ are closely connected to the probabilistic properties of the random walk, as is well known (see e.g. the surveys [M-W][Wo]). For example, for a symmetric μ, $\|\lambda(\mu)\| = \lim_{n\to\infty} p_{2n}(e,e)^{1/2n}$, where $p_k(e,e)$ is the return probability to e at the k-step, a result due to H. Kesten [K].

(2) Let $\nu_n \in \ell^1(\Gamma)$ be a sequence of probability measures. Does the sequence of averaging operators $\lambda(\nu_n)f = \nu_n * f$ satisfy the strong maximal inequality in $\ell^2(\Gamma)$, namely: $\|\sup_n |\lambda(\nu_n)f|\|_{\ell^2(\Gamma)} \leqslant C\|f\|_{\ell^2(\Gamma)}$?

Maximal inequalities for convolutions in group algebras arise naturally in many different problems in harmonic analysis. The groups $\mathbb{Z}^d$ constitute a very important classical example, and for several different perspectives on the connections between maximal inequalities, harmonic analysis and pointwise ergodic theory for $\mathbb{Z}^d$ we refer e.g. to [BJR 1] [BJR 2] [Bo] [R-W] and [S-W].

1.2 General measure preserving actions. An interesting and substantial generalization of the previous set-up is obtained by considering a general measure-preserving action of Γ. We recall the basic definitions: Let $(X, \mathcal{B}, m)$ be a standard Lebesgue measure space, and assume Γ acts on X by measurable bijections preserving the σ-finite measure m. The action $(\gamma, x) \mapsto \gamma x$ induces a representation π of Γ by unitary operators on the Hilbert space $L^2(X)$, via $\pi(\gamma)f(x) = f(\gamma^{-1}x)$. This representation can be extended to the group algebra by: $\pi(\mu) = \sum_{\gamma\in\Gamma} \mu(\gamma)\pi(\gamma)$. The map $\mu \mapsto \pi(\mu)$, from $\ell^1(\Gamma)$ to $\text{End}L^2(X)$ is again a $*$-algebra homomorphism, bounded in norm by 1. We let $L^2_I(X)$ denote the closed subspace of Γ-invariant vectors in $L^2(X)$. $L^2_0(X)$ will denote the orthogonal complement to $L^2_I(X)$, and π_0 the representation of Γ on $L^2_0(X)$. We let E denote the orthogonal projection $E : L^2(X) \to L^2_I(X)$.

Both questions posed in §1.1 appear in this more general framework. Furthermore, when the measure of the space is finite (in this case we always assume $m(X) = 1$), another natural question arises. First recall that the action of Γ on X is called ergodic if $L^2_I(X)$ consists of constant functions, or equivalently, if every Γ-invariant set in X has measure zero or one. Consider the orbit $\Gamma \cdot x = \{\gamma x \mid \gamma \in \Gamma\}$ of a given point $x \in X$. Given a function f on X we can sample the values of the function at the orbit points and average them using the probability measures ν_n, namely consider $\pi(\nu_n)f(x) = \sum_{\gamma\in\Gamma} \nu_n(\gamma)f(\gamma^{-1}x)$. It is natural to ask whether these "time averages" converge as $n \to \infty$ to the "space average" of the function f on the space X, namely to the expression $\int_X f dm$.

To formulate the last problem more precisely, introduce the following:

Definition 2. Consider a measure-preserving action of Γ on a probability space $(X, \mathcal{B}, m)$. The sequence $\nu_n \in \ell^1(\Gamma)$ is called

(1) *mean ergodic sequence in* $L^2(X)$ if $\|\pi(\nu_n)f - Ef\|_{L^2(X)} \to 0$.
(2) *pointwise ergodic sequence in* $L^2(X)$ if, for every $f \in L^2(X)$, $\pi(\nu_n)f(x) \to Ef(x)$ for almost all $x \in X$, and in the L^2-norm.
(3) Let m be any σ-finite measure. The maximal function associated with the sequence ν_n, namely $f^*_\nu(x) = \sup_{n\geqslant 1} |\pi(\nu_n)f(x)|$, is said to satisfy the strong maximal inequality in $L^2(X)$ if $\|f^*_\nu\|_2 \leqslant C\|f\|_2$.

We will concentrate below on the following

1.3 Basic problems. Given a sequence of probability measures $\nu_n \in \ell^1(\Gamma)$ and a measure-preserving action of Γ on $(X, \mathcal{B}, m)$, we can now formulate the following basic problems. (in the first two, we assume $m(X) = 1$).

(1) Is ν_n a mean ergodic sequence in $L^2(X)$?
(2) Is ν_n a pointwise ergodic sequence in $L^2(X)$?
(3) Does f^*_ν satisfy the strong maximal inequality in $L^2(X)$?

(4) Does $\|\pi_0(\nu_n)\| \to 0$? If so, what is the rate of convergence ?

We will consider mostly the case of symmetric measures, which in any unitary representation are mapped to *self adjoint* operators. This allows the use of spectral methods to obtain ergodic theorems in $L^2(X)$. Spectral theory will be our main focus, and only certain brief comments will be made below on results regarding (weak or strong) maximal inequalities, as well as mean and pointwise convergence, in the spaces $L^p(X)$, $1 \leqslant p < \infty$, for $p \neq 2$.

When discussing word-hyperbolic groups, we will concentrate particularly on the case of the radial averages σ_n, β_n and μ_n defined in §1.1, which are most closely connected to the geometry of the group Γ, or more precisely, the graph $X(\Gamma, S)$.

We note that the basic problems have received considerable attention in ergodic theory, where an important goal is to generalize von-Neumann's mean ergodic theorem and Birkhoff's pointwise ergodic theorem for actions of the group $\mathbb{Z}$, to the case of actions of more general groups Γ. We refer to [O-W 1,2,3] for some ergodic theorems for amenable groups, and also to [T1] and the survey in Ch. 4 of [Si].

2. Amenable groups

2.1 Følner averages. We recall that a sequence of sets in a group Γ is called a (right) Følner sequence if it is asymptotically invariant under (right) translation, namely if $\#(F_n\gamma \Delta F_n)/\#F_n \to 0$ for every $\gamma \in \Gamma$. A group possesing a Følner sequence is called amenable, and in that case the averages most commonly considered are those arising by taking F_n to be a Følner sequence of sets and ν_n normalized counting measure on F_n.

For such averages the solution to the first problem is straightforward, as is well known. Indeed, for functions of the form $u = \pi(g)f - f$, f bounded, the Følner property of F_n easily implies that the limit of $\pi(\nu_n)(\pi(g)f - f)(x)$ as $n \to \infty$ exists and equals 0, for m-almost all $x \in X$. The mean ergodic theorem then follows by observing that the functions $\{\pi(g)f - f | f \in L^\infty(X), g \in \Gamma\}$ span a dense subspace in $L_0^2(X)$.

As to the second problem, again note that $\pi(\nu_n)u(x) \to 0$ for every $u = \pi(g)f - f$, $f \in L^\infty(X)$, by the Følner property of F_n. Hence pointwise convergence of $\pi(\nu_n)f(x)$ to $Ef(x)$ holds for a dense subspace of functions in $L_0^2(X)$, and therefore also for a dense subspace in $L^2(X)$. As is well known (see e.g. [Pe][N1]), if the maximal inequality for f_ν^* is shown to hold in $L^2(X)$, it then follows that $\pi(\nu_n)$ is pointwise ergodic sequence in $L^2(X)$ (given the mean ergodic theorem and pointwise convergence on a dense subspace). Therefore a solution to the second problem follows from a solution to the third, which we consider in the next section.

We note the following well known facts:

(1) For any probability measure $\nu \in \ell^1(\Gamma)$ and for ergodic actions preserving a (non-atomic) probability measure, we always have $\|\pi_0(\nu)\| = 1$. Indeed, when Γ is amenable, there always exists a sequence of unit vectors $f_n \in L_0^2(X)$, asymptotically invariant under Γ, namely satisfying $\|\pi_0(\gamma)f_n - f_n\| \to 0$, for every $\gamma \in \Gamma$ (see [Ro][Sc]). It follows that there does not exist an estimate of the form $\left\|\pi(\nu_n)f - \int_X f dm\right\| \leqslant \epsilon_n\|f\| \to 0$, valid for all $f \in L^2(X)$, and this provides an answer to the fourth basic problem.

(2) Conversely, assume $\|\pi_0(\nu)\| = 1$ for *some* probability measure ν, and *every* ergodic action of Γ preserving a (non-atomic) probability measure, where the support of $\mu = \nu^* * \nu$ generates Γ. Then $\|\pi_0(\mu)\| = 1$ and by the spectral theorem for the self-adjoint non-negative operator $\pi_0(\mu)$, there exists a sequence of unit vectors asymptotically invariant under $\pi_0(\mu)$, namely a sequence satisfying $\sum_{\gamma\in\Gamma} \mu(\gamma)\langle\pi_0(\gamma)f_n, f_n\rangle \to 1$. It follows that this sequence is asymptotically invariant under Γ, i.e. $\langle\pi_0(\gamma)f_n, f_n\rangle \to 1$ for each $\gamma \in \operatorname{supp} \mu$, hence for each $\gamma \in \Gamma$. It follows that Γ is amenable (see [Ro] [Sc]), and therefore $\|\pi_0(\nu)\| = 1$ for *every* probability measure ν.

(3) Finally, note that a Følner sequence defines an asymptotically invariant sequence under $\lambda(\nu)$, and hence $1 = \|\pi_0(\nu)\| = \|\lambda(\nu)\|$, for every $\nu \in \ell^1(\Gamma)$. In particular, the norm $\pi_0(\nu)$ for an *arbitrary* action is dominated by the norm of $\lambda(\nu)$, where λ is the *regular* action by left translations. In fact, this property is equivalent to the trivial representation of Γ being weakly contained in the regular representation, and so characterizes discrete amenable groups (see [Ey][Ro][Sc]).

2.2 The transfer principle for amenable groups.

As to the third problem, namely the strong maximal inequality in L^2, it has not been generally solved when ν_n are Følner averages on an amenable group. An important method - the transfer principle - which solves the problem for certain amenable groups, was introduced by A. P. Calderón [Ca], and later generalized by H. Coifman and G. Weiss [C-W], W. R. Emerson [E], C. Herz [Herz], and M. Cowling [Co2] [Co3] (see also [BPW] and [T1]). To illustrate the point, we formulate here a very simple special case of the transfer principle.

Assume ν_n is any sequence of probability measure on Γ, each with finite support. Define the operator $F_\nu^*(g) = \sup_{n\geqslant 0} |\rho(\nu_n)F(g)|$, where $\rho : \Gamma \to U(\ell^2(\Gamma))$ is the right regular representation. Recall also that $f_\nu^*(x) = \sup_{n\geqslant 0} |\pi(\nu_n)f(x)|$. The connection between these operators is given by the following:

Transfer principle for amenable groups (see [Ca, C-W, E, Herz , Co2, Co3]).

Let the amenable group Γ act measurably by measure preserving transformation on $(X, \mathcal{B}, m)$ where m is σ-finite, and $\nu_n \in \ell^1(\Gamma)$ be finitely-supported probability measures. If $\|F_\nu^*\|_{\ell^2(\Gamma)} \leqslant C\|F\|_{\ell^2(\Gamma)}$ for every $F \in \ell^2(\Gamma)$, then in any measure preserving action of Γ, for every $f \in L^2(X)$:

$$\|f_\nu^*\|_{L^2(X)} \leqslant C\|f\|_{L^2(X)} .$$

In other words, the transfer principle reduces the proof of the maximal inequality for ν_n in an *arbitrary* action, to the special case of the action of Γ on itself by right translations, namely to the case of the *regular* representation.

We will give a short proof of the tranfer principle formulated above, following the original argument in [Ca](see also [Bo]) and its generalization in [C-W][E]. Suppose for notational simplicity that ν_k is supported in the ball B_k of radius k (w.r.t. a word metric) with center e. Assume that the maximal inequality has been established for the action of the sequence $\rho(\nu_n)$ in $\ell^2(\Gamma)$. Then, given an action of Γ on a probability space X, fix $x \in X$ and consider the restriction of a function $f \in L^2(X)$ to a subset of a Γ-orbit in X of the form $B_N x = \{gx : |g| \leqslant N\}$. Regard the function obtained as a function on Γ, by defining $F_N(g) = |\pi(g)f(x)|$ if $|g| \leqslant N$ and $F_N(g) = 0$ otherwise. Now use the assumption that (for any $n \geqslant 0$) $\left\|\sup_{0\leqslant k\leqslant n} |\rho(\nu_k)(F_N)|\right\|_{\ell^2(\Gamma)} \leqslant \left\|(F_N)_\nu^*\right\|_{\ell^2(\Gamma)} \leqslant C\|F_N\|_{\ell^2(\Gamma)}$. Since $|gh| \leqslant N$ if $|g| \leq N-n$ and $|h| \leqslant n$, it follows from the definition of F_N that

$$\sum_{|g|\leqslant N-n} \sup_{0\leqslant k\leqslant n} |\pi(g)\pi(\nu_k)f(x)|^2 \leqslant \sum_{g\in\Gamma} \sup_{0\leqslant k\leqslant n} |\rho(\nu_k)F_N(g)|^2 \leqslant$$

$$\leqslant C^2\|F_N\|^2_{\ell^2(\Gamma)} = C^2 \sum_{|g|\leqslant N} |\pi(g)f(x)|^2 .$$

Since the action is measure preserving, integrating over X we obtain the inequality: $\#B_{N-n}\left\|\sup_{0\leqslant k\leqslant n} |\pi(\nu_k)f(x)|\right\|^2_{L^2(X)} \leqslant C^2\#B_N\|f\|^2_{L^2(X)}$. Now, assuming that $\lim_{N\to\infty} \frac{\#B_{N-n}}{\#B_N} = 1$ [Ca][Ca1] (e.g. when Γ has polynomial growth) the maximal inequality follows.

Clearly, this argument applies equaly well to the case of strong L^p-maximal inequality, $p > 1$. It can also be adapted to prove weak-type maximal inequalities for a large class of local and even non-local operators ([Ca][C-W][E] [Herz][Co2][Co3]). Finally, we note that to prove the transfer principle for a general discrete amenable group, it suffices to replace the balls B_n by a Følner sequence F_n, and use the fact that $\#(F_n \cdot K\Delta F_n)/\#F_n \to 0$ for every finite set $K \subset \Gamma$ [C-W].

Under additional assumptions on the Følner sets F_n it is possible to prove the maximal inequality for convolutions on the group, and complete the proof of the pointwise ergodic theorem for the sequence of Følner averages $\nu_n = (1/\#F_n)\chi_{F_n}$. Three sufficient conditions for a weak-type $(1,1)$ maximal inequality in $\ell^1(\Gamma)$ (and hence also a strong maximal inequality in every $\ell^p(\Gamma)$, $1 < p \leq \infty$) for an increasing sequence of Følner sets F_n are :

(1) $F_n = B_n$ is a ball of radius n in Γ w.r.t. an invariant metric, and $\#B_n$ has polynomial growth.
(2) $F_n = F_n^{-1}$ satisfy $F_k \cdot F_m \subset F_{k+m}$, and the doubling volume property, namely $\#F_{2n} \leqslant C\,\#F_n$ (see [Ca 1]).
(3) $\#(F_n^{-1} \cdot F_n) \leqslant C\,\#F_n$ (see [Be][Ch][E][T]).

We note that when $F_n = B_n$ are balls defined by a word metric associated with a finite symmetric set of generators of Γ, all three conditions above are equivalent.

The proofs of the maximal inequality for the convolution operators ν_n in $\ell^1(\Gamma)$ (under any one of the conditions above) generalize a method due to N. Wiener [W], which was originally used to prove the maximal inequality for the sequence β_n on $\mathbb{R}^d$ and $\mathbb{Z}^d$ (see e.g. [S3, Ch I]). The proof proceeds by establishing geometric covering properties of the sets $F_n \subset \Gamma$, in order to prove a "disjointification Lemma". The lemma asserts that given a cover of a finite set U in the group using a finite family of the sets F_n and their translates, there exists a subfamily of disjoint sets which covers at least $\epsilon(\Gamma)\#U$ of the points of U, where $\epsilon(\Gamma)$ is a fixed positive constant. For a proof of the disjointification Lemma for sets F_n satisfying condition (3) and the resulting L^1 pointwise ergodic theorem we refer to [O-W1][E][T1].

2.3 Further remarks and some open problems for amenable groups. As pointed out in [O-W1], the only groups known to contain sets that satisfy any of the three conditions listed in §2.2 are groups with polynomial growth. These conditions certainly fail for the sequence of balls B_n when the group Γ has exponential growth.

For some solvable groups of exponential growth it is still possible to choose a sequence of Følner sets for which the corresponding sequence of averages ν_n is a pointwise ergodic sequence and satisfies the strong maximal inequality. In particular, solvable groups Γ which can be represented as a finite sequence of semi-direct extensions of groups isomorphic to $\mathbb{Z}^d$, namely $\Gamma = \mathbb{Z}^{d_1} \ltimes (Z^{d_2} \ltimes (\cdots \ltimes Z^{d_k})\cdots)$ have this property. This result can be established using the arguments of [E-G], which were originally used to prove pointwise ergodic theorems for certain Følner sequences on *connected* amenable groups. For more on this result and some generalizations of it we refer to [N6].

For amenable groups not satisfying any of the conditions listed above there does not seem to exist any ergodic theorem other than the mean ergodic theorem for the Følner averages ν_n. A particularly interesting case is that of groups Γ that have subexponential growth. The sequence β_n has then a subsequence which is Følner, but it is unknown whether β_n is a pointwise ergodic sequence, and whether it satisfies the strong maximal inequality for measure preserving actions, or even for the convolution action. We note that the same questions can be asked for the ball averages β_n on any amenable group of exponential growth, as well as for the averages σ_n and μ_n. In fact, even the mean ergodic theorem has not been established in general for these sequences. We remark that the ball averages on semi-direct products usually do not form a Følner sequence, and so are not comparable to the Følner averages that appear in [E-G]. Ergodic theorems for radial averages have not been established for a general semi-direct product.

3. General groups

3.1 The subadditive maximal inequality.

For an arbitrary group Γ, it is possible to formulate a natural subadditive inequality for averages in the group algebra $\ell^1(\Gamma)$. This inequality holds for several important classes of discrete groups, and serves a purpose similar to that of the transfer principle for amenable groups, namely it implies a strong L^2 maximal inequality (for the corresponding averages) in every measure preserving action of the group. We begin with some necessary preliminaries.

Definition 3. A sequence T_n of operators on $L^2(X)$ will be called a *subadditive sequence of self adjoint Markov operators* if it satisfies the following:

(1) $T_n = T_n^*$, $\|T_n\| \leqslant 1$.
(2) $T_n f \geqslant 0$ if $f \geqslant 0$, $T_n 1 = 1$.
(3) There exist a constant $C_0 > 0$, a positive integer k, and a fixed non-negative bounded operator B on $L^2(X)$ such that:

$$T_n T_m f(x) \leqslant C_0(T_{kn} f(x) + T_{km} f(x)) + Bf(x)$$

for all bounded and nonnegative $f \in L^2(X)$.

We can now state the following subadditive maximal inequality, due independently to J. Barrionuevo [B] and [N1].

Theorem 1: Subadditive maximal inequality [B][N1]. *Let T_n be a subadditive sequence of self adjoint Markov operators. Define $f^*(x) = \sup_{n\geqslant 0} |T_n f(x)|$. Then $\|f^*\|_2 \leqslant C\|f\|_2$ for all $f \in L^2(X)$. We can take $C = 2C_0 + \|B\|$.*

Let now $\phi, \psi : \mathbb{N} \to \mathbb{N}$ be two fixed (but arbitrary) functions, and $S_n, R_n : L^2(X) \to L^2(X)$ be two sequences of linear operators satisfying

$$\left\| \sup_{n\geqslant 0} |S_n f| \right\|_2 \leqslant C\|f\|_2 \ , \left\| \sup_{n\geqslant 0} |R_n f| \right\|_2 \leqslant C\|f\|_2 \ .$$

A more general version of the subadditive maximal inequality is then:

$$T_n T_m f(x) \leqslant S_n T_{\phi(n)} f(x) + R_m T_{\psi(m)} f(x) + Bf(x) \ .$$

This condition can be shown to be sufficient for the same conclusion as in Theorem 1, using a combination of the arguments in [B] and [N1].

We note that the subadditive maximal inequality of Theorem 1 generalizes similar results due to E. M. Stein [S1, S2] and B. Weiss [We]. In particular, it was applied in [S1, S2] to prove a pointwise ergodic theorem for the even powers of positivity-preserving self-adjoint contractions on L^2. Also, it is noted in [S1, S2, We] that it implies the pointwise convergence of martingales in L^2, as well as Birkhoff's pointwise ergodic theorem in L^2. Other applications have been considered in [B]. The origin of this maximal inequality is attributed in [S1,W] to A. Kolmogoroff and G. Seliverstoff [K-S], and to R. E. A. C. Paley [P].

It is reasonable to expect that for a large class of discrete groups Γ, the sequence of radial averages $\mu_n \in \ell^1(\Gamma)$ (defined in §1.1) will satisfy the subadditive inequality given by $\mu_n * \mu_m \leqslant C(\mu_{kn} + \mu_{km}) + b$. When this inequality holds in $\ell^1(\Gamma)$, then $T_n = \pi(\mu_n)$ is a subadditive sequence of self adjoint Markov operators on $L^2(X)$ in any finite-measure-preserving action, and therefore will satisfy the strong maximal inequality in L^2.

The method of using the subadditive inequality in the group algebra to prove a maximal ergodic theorem in L^2 was developed in [N1] and applied to actions of the free group $\mathbb{F}_k$ and some related groups. More generally, for the class of word-hyperbolic groups (see [Gr]), we have:

Theorem 2. [F-N]. *Let Γ be a non-elementary word-hyperbolic group, S a finite symmetric generating set. Then there exist constants $1 < q < \infty$ and $0 < C < \infty$, depending only on (Γ, S), such that the following inequalities hold:*

a) $\sigma_t * \sigma_s \leqslant C \sum_{j=0}^{2s} q^{-(s-\frac{1}{2}j)} \sigma_{t-s+j}$ *if* $t \geqslant s$.

b) $\mu_n * \mu_m \leqslant C(\mu_{2n} + \mu_{2m})$.

Combining Theorem 1 and Theorem 2, we have:

Theorem 3. [F-N]. *Let Γ be a word-hyperbolic group, S a finite symmetric generating set. Then the sequence μ_n satisfies the strong maximal inequality in $L^2(X)$, i.e. $\|f^*_\mu\|_2 \leqslant C(\Gamma, S)\|f\|_2$ for every $f \in L^2(X)$.*

We remark that an elementary word-hyperbolic group is a finite extension of $\mathbb{Z}$, and the subadditivity of μ_n in this case is easily verified (see[S2] for the case of $\mathbb{N}$). The first inequality in Theorem 2 also holds, with $q = 1$ for elementary hyperbolic groups. Theorem 3 for elementary word-hyperbolic groups is of course a simple consequence of Wiener's maximal inequality for $\mathbb{Z}^d$-actions [W].

3.2 The pointwise ergodic theorem.

We note that given the maximal inequality for μ_n in L^2, to complete the proof of the pointwise ergodic theorem it suffices (see e.g. [Pe][N1]) to find a dense set of functions $f \in L^2$ where $\pi(\mu_n)f(x) \to \int_X f dm$, almost everywhere and in the L^2-norm. For hyperbolic groups, a sufficient condition is the existence of a dense set of functions $f \in L^2$ satisfying the following exponential mixing condition: $|\langle \pi(\gamma)f, f\rangle| \leqslant C_f \exp(-c_f|\gamma|)$, for some $c_f > 0$ (see [F-N]). If Γ is a co-compact lattice of simple Lie group of real-rank one, then the action of Γ in any ergodic G-space with finite invariant measure is exponentially mixing, hence the pointwise ergodic theorem holds for the averages $\mu_n \in \ell^1(\Gamma)$ [F-N, Thm 4], acting on a G-space.

Another possible method to complete the proof of the pointwise ergodic theorem in L^2 for μ_n for general groups, is given by the following observation. I am indebted to B. Weiss for a conversation regarding it.

Proposition 1. *Suppose μ_n satisfies the strong maximal inequality in $L^2(X)$. If in addition $\|\mu_n * \sigma_1 - \mu_n\|_{\ell^1(\Gamma)} \to 0$, then μ_n is a pointwise ergodic sequence in $L^2(X)$.*

As an example, note that Proposition 1 applies to free groups with free generators. Indeed, for $\mathbb{F}_r$ (with S a free set of generators) it is easily seen that $\sigma_n * \sigma_1 = \frac{1}{2r}\sigma_{n-1} + (1 - \frac{1}{2r})\sigma_{n+1}$ for all $n \geqslant 1$. Therefore $\mu_n * \sigma_1 = \frac{1}{n+1}(\frac{1}{2r}\sigma_0 + \sigma_1 + \cdots + \sigma_n + (1-\frac{1}{2r})\sigma_{n+1})$. Hence $\mu_n * \sigma_1 - \mu_n = \frac{1}{n+1}((\frac{1}{2r}-1)\sigma_0 + (1 - \frac{1}{2r})\sigma_{n+1}) \to 0$ as $n \to \infty$. Combining this observation with Theorem 2 (which is much simpler in the case of free groups with free generators, see [N1]), Theorem 3 and Proposition 1 imply that the sequence $\mu_n \in \ell^1(\mathbb{F}_r)$ is a pointwise ergodic sequence in L^2. We remark that an extension of this result to free group actions on von-Neumann algebras has recently been established in [Wa]. Furthermore, the foregoing argument can be used to show the following : Let π be a representation of $\mathbb{F}_r$ by isometries of a reflexive separable Banach space V, and P_1 the projection onto the space of $\pi(\sigma_1)$-invariant vectors (see [D-S] vol. 1, Ch. VIII, §5, cor. 4). Then $\pi(\mu_n)w$ converges in norm to $P_1 w$, for every $w \in V$. If V is strictly convex, the space of invariants for $\pi(\sigma_1)$ coincide with the space of invariants for $\pi(\Gamma)$, and so a mean ergodic theorem is obtained.

3.3 Some remarks and open problems for general groups.

As already noted in §3.2, the subadditive inequality for μ_n in the group algebra, namely

$$\mu_n * \mu_m \leqslant C(\mu_{kn} + \mu_{km}) + b$$

seems likely to hold for a large class of discrete groups, and there are yet no examples where it has been shown to fail. The inequality also makes sense for locally compact groups, and it does in fact hold for the groups $G = SO^0(n,1)$, $n > 2$. By Theorem 2, it also holds for co-compact lattices in these groups. However, it is unknown whether it holds for higher-real-rank simple Lie groups or their lattices.

Another possible approach to prove that the sequence μ_n satisfies the strong maximal inequality in L^2 was developed in [N-S1], for free groups with free generators. Indeed, in this case there are constants $C, k > 0$ such that for all $n \geqslant 0$:

$$\mu_n \leqslant \frac{C}{kn+1} \sum_{j=0}^{kn} \sigma_1^{*j}$$

and hence the strong maximal inequality for every $1 < p \leqslant \infty$, as well as the weak type $(1,1)$ maximal inequality follow from the Hopf-Dunford-Schwartz theorem for the Markov operator σ_1 (see e.g. [D-S][Pe]). The proof of this convolution estimate is based on the following fact: the distribution of the probability measure σ_1^{*j} on $\mathbb{F}_r$ depends only on the word length, and is very similar to the binomial distribution $(\frac{1}{2r}\delta_{-1} + \frac{2r-1}{2r}\delta_1)^{*j}$ on $\mathbb{Z}$ (see [N-S1]). It is an interesting problem to obtain estimates on the distribution of convolution powers σ_1^{*j} that will allow the conclusion that the average μ_n (namely the uniform average of the spherical averages σ_i, $0 \leqslant i \leqslant n$) is dominated by the uniform average of the convolution powers σ_1^{*j}, $0 \leqslant j \leqslant kn$, for various classes of groups.

Finally, we remark that the problem of establishing a mean ergodic theorem for μ_n for general groups Γ (and of course also of establishing a pointwise ergodic theorem) remains unresolved, and no counter-example has been shown to exist.

4. Free groups with free generators

4.1 Pointwise ergodic theorems for non-Abelian free groups.

Ergodic theorems for actions of the free groups occupy a special role in the general theory, for the following reasons:

(1) Ergodic theorems for free groups constitute the most direct generalization of von-Neumann's and Birkhoff's theorems. Recall that theses claim that given an arbitrary invertible measure preserving transformation T on a probability space X, the averages $(T^{-n}f + \cdots + T^n f)/(2n+1)$ converge (in L^2 norm and almost everywhere) to the limit $\tilde{f}$, where $\tilde{f}$ is the conditional expectation of f w.r.t. the σ-algebra of T-invariant sets. The ergodic

theorems for $\mathbb{F}_2$ (see [G] and [N1] for the mean and pointwise version, respectively) assert that given *two* arbitrary invertible measure preserving transformations T and S, there is a natural way to average the values of a function f along the orbits of the group generated by T and S, so as to obtain the same conclusion. (2) An ergodic theorem for the free group $\mathbb{F}_r$ obviously implies an ergodic theorem for any group Γ that can be generated by r elements, since any of its actions can be regarded as an action of $\mathbb{F}_r$. The weights with which one must average on the orbits of Γ are those determined by the factor map of a given presentation $\varphi : \mathbb{F}_r \to \Gamma$. These weights are not usually comparable to radial averages with respect to a finite symmetric set of generators on Γ, and we remark that it is an interesting problem to understand what weights do occur in specific situations. (3) The free group exhibits a great deal more symmetry than most other groups. A quantitative manifestation of this fact is the existence of a very large group of symmetries of the Cayley graph $X(\mathbb{F}_r, S)$ (w.r.t. free generators), which is a $2r$-regular tree T_{2r}. These symmetries, or graph automorphisms, are isometries of the Cayley graph, and commute with the radial operators $\lambda(\sigma_n)$ of averaging a function on the graph on spheres of a fixed radius. Using these symmetries, for radial averages a rather complete picture can be developed, which shed some light on the possibilities (and limitations) of various generalizations.

To ilustrate the point, consider the following free products, which are all non-elementary word-hyperbolic groups:

(1) $\mathbb{F}_k$ = the free group on k generators, $1 < k < \infty$, with $S = \{x_i, x_i^{-1}\}_{i=1}^{k}$, where $x_1, \ldots, x_k$ are free generators.

(2) $\Gamma(r,h) = G_1 * G_2 \cdots * G_r$ = the free product of r finite groups each of order h, where $r \geqslant 2$, $h \geqslant 2$, $r + h > 4$, with $S = \bigcup_{i=1}^{r} G_i \setminus \{e\}$.

Denote by E' the orthogonal projection on the subspace of $L^2(X)$ consisting of functions satisfying: $\pi(\sigma_n)f = (-1)^n f$, and let E be the projection on the space of invariants as in §1.2. Recall that $\sigma'_n = \frac{1}{2}(\sigma_n + \sigma_{n+1})$, and let (X, m) be a Γ-space where $E' \neq 0$.

Theorem 4: Ergodic theorems for radial averages on free groups [N1]. *For Γ and (X, m) as above*

(1) *σ_n and β_n satisfy the maximal inequality in $L^2(X)$, but are not mean (and hence not pointwise) ergodic sequences in $L^2(X)$.*

(2) *The sequences σ'_n and μ_n are pointwise ergodic sequences in $L^2(X)$.*

(3) *σ_{2n} converges to $E + c(\Gamma)E'$, which is a conditional expectation operator w.r.t. a Γ-invariant sub-σ-algebra iff $\Gamma = \Gamma(r,2)$ or $\Gamma = F_k$.*

(4) *β_{2n} converges to $E + c(\Gamma)\frac{q(\Gamma)-1}{q(\Gamma)+1}E'$, which is not a conditional expectation operator on a Γ-invariant sub-σ-algebra.*

The convergence is for each function $f \in L^2(X)$, pointwise almost every-

where and in the norm of $L^2(X)$.

We note the following regarding Theorem 4 and its proof for $\mathbb{F}_r$.

(1) Mean convergence of the averages above was established by Y. Guivarc'h [G], using a method which generalizes von-Neumann's original proof of the mean ergodic theorem for $\mathbb{Z}$-actions. The essential point is the use of the spectral theory of the *commutative* convolution algebra $A(\mathbb{F}_r)$ generated by σ_n, $n \geq 0$ in $\ell^1(\mathbb{F}_r)$. This algebra was introduced into ergodic theory by V. I. Arnol'd and A.L. Krylov [A-K], who used its spectral theory to prove that given a dense free subgroup Γ of the orthogonal group $SO^0(m, \mathbb{R})$, $m \geqslant 2$ and a unit vector x in $\mathbb{R}^m$, the points $\{w(x) \,:\, w \in \Gamma\,,\ |w| = n\}$ become equidistributed in the unit sphere.

(2) The commutativity of the convolution subalgebra $A(\mathbb{F}_r)$ follows, by induction, from the elementary identity: $\sigma_1 * \sigma_n = \frac{1}{2r}\sigma_{n-1} + (1 - \frac{1}{2r})\sigma_{n+1}$, that holds in $\ell^1(\mathbb{F}_r)$. It is also a consequence of the commutativity of the algebra of radial averages on a regular tree, in keeping with observation (3) above, since these algebras are isomorphic.

(3) The convolution identity implies that every complex character of the algebra $A(\mathbb{F}_r)$ can be viewed as a function on $\mathbb{N}_+$, satisfying a second order linear recurrence relation with constant coefficients. Solving the recurrence (see e.g. [C][G][FT-P]) it is seen that actually the characters decay *exponentially* with n, except for the trivial character and the special character $sgn(\sigma_n) = (-1)^n$. The existence of the special character is responsible for the interesting fact that σ_n and β_n are *not* mean ergodic sequences.

(4) More explicitly, the spectrum of $A(\mathbb{F}_r)$ is given as follows ($q = 2r-1$) :

$$\varphi_z(\sigma_n) = \mathbf{c}(z)q^{-nz} + \mathbf{c}(1-z)q^{-n(1-z)}, \quad z \neq \frac{1}{2} + \frac{ij\pi}{\log q},$$

$$\mathbf{c}(z) = \frac{q^{1-z} - q^{z-1}}{(q+1)(q^{-z} - q^{z-1})}$$

and:

$$\varphi_z(\sigma_n) = \left(1 + n\frac{q-1}{q+1}\right)(-1)^{jn}q^{-\frac{n}{2}}, \quad z = \frac{1}{2} + \frac{ij\pi}{\log q}$$

A necessary and sufficient condition for φ_z to be continuous is that it be bounded, and this condition is equivalent to $0 \leqslant \operatorname{Re} z \leqslant 1$. The unitary representation of $\mathbb{F}_r$ in $L^2(X)$, extended to $\ell^1(\mathbb{F}_r)$, assignes to σ_1 a self-adjoint operator. Consequently, the values $\varphi_z(\sigma_1) = \gamma(z)$ are real, for those φ_z that occur in the spectrum of σ_1 in $L^2(X)$. It is easily verified that $\gamma(z)$ is real iff $\operatorname{Re} z = \frac{1}{2}$, or $\operatorname{Im} z = \frac{ij\pi}{\log q}$. The image of this set under γ is the real spectrum of $A(\mathbb{F}_r)$. Note that for z and $1 - z$ the same character obtains, so we can assume that $0 \leqslant \operatorname{Re} z \leqslant \frac{1}{2}$. Note also that the characters corresponding to $z = s$ and to $z = s + \frac{ij\pi}{\log q}$ differ by sign only: $\varphi_{s+\frac{ij\pi}{\log q}}(\sigma_n) = (-1)^{jn}\varphi_s(\sigma_n)$. In particular, the sign character ε, given by $\varepsilon(\sigma_n) = (-1)^n$, is obtained at the points $z = \frac{i(2j+1)\pi}{\log q}$. The real spectrum is naturally divided to the principal series, where $\operatorname{Re} z = \frac{1}{2}$, and the complementary, where $\operatorname{Re} z \neq \frac{1}{2}$.

(5) The mean ergodic theorem for σ_n' is proved in [G] as follows: by the spectral theorem for the operator $\pi(\sigma_1)$ acting on $L^2(X)$

$$\langle \pi(\sigma_1)f, f\rangle = \int_{\mathrm{Sp}A(\mathbb{F}_r)} \varphi_z(\sigma_1)d\nu_f(z)$$

Where $\varphi_z : A(\mathbb{F}_r) \to \mathbb{C}$ are the continuous real characters of the algebra, and ν_f the spectral measure determined by f. It now suffices to check that $\varphi_z(\sigma_n') \to 0$, for *every* non-trivial character φ_z.

(6) The proof of the pointwise ergodic theorems for σ_n' [N1] proceeds by applying the method of Littlewood-Paley-Stein square functions to the spectral theory of $A(\mathbb{F}_r)$, in order to establish a strong maximal inequality in L^2. This approach generalizes the method of proof used to establish the maximal inequality for powers of a single self-adjoint operator in [S1].

(7) The method of square functions was applied together with Stein's complex interpolation theorem, to prove the strong maximal inequality in L^p, $1 < p \leqslant \infty$ for the averages σ_n in [N-S1]. A weak-type maximal inequality for μ_n in L^1 was also established in [N-S1].

(8) Theorem 4 shows that the behaviour of the sequences $\pi(\sigma_n)$ and $\pi(\beta_n)$ is rather delicate. $\pi(\beta_n)$ and $\pi(\sigma_n)$ do not converge, but the subsequences $\pi(\sigma_{2n})$ and $\pi(\beta_{2n})$ do converge, however the limit is not the ergodic mean. Analyzing σ_n and β_n for general groups, and even for general hyperbolic groups, therefore seems an interesting challenge. The sequence μ_n seems to have more stable behaviour, and as noted in §3, other methods may be available for analyzing it.

Finally, we note the basic observation [N1] that the spectral method outlined above applies whenever the group algebra of a group G contains a commutative subalgebra admitting an explicit spectral theory. A large class of interesting examples arises from groups G containing a compact subgroup K such that the convolution algebra of bi-K-invariant measures on G is commutative, namely from Gelfand pairs (G, K). This class includes semi-simple algebraic groups over locally compact non-discrete fields, as well as many motion groups. We refer to [N1-3][N-S1,2] [M-N-S] for proofs of ergodic theorems for radial averages using spectral methods and harmonic analysis on semi-simple Lie (and algebraic) groups, and to [N-T] for the ergodic theory of some radial averages on the Heisenberg group.

5. Discrete subgroups of Lie groups

5.1 Spectral transfer.

In general, the convolution algebra generated by the radial averages $\sigma_k \in \ell^1(\Gamma)$ is not commutative, and the spectral methods described in §4 are not applicable. However, for a discrete subgroup Γ of a semi-simple algebraic group G, it is still possible to use (non-commutative) harmonic analysis and the spectral theory of G, to prove ergodic theorems for the actions of Γ on G-spaces. Before describing the method, we first recall the definition of property T due to D. A. Kazhdan [Ka].

Definition 4. [Ka] Let G be a locally compact second countable (lcsc) group. G has property T of Kazhdan if there exist a compact subset $Q \subseteq G$ and $\epsilon > 0$ satisfying the following: Given a strongly continuous unitary representation $(\pi, \mathcal{H})$ which does not contain G-invariant unit vectors, for every unit vector $v \in \mathcal{H}$ there exists $g \in Q$ with $\| \pi(g)v - v \| \geqslant \epsilon$. A set Q as above is called a Kazhdan set and ϵ is called a Kazhdan constant for Q.

As is well known (see e.g. [Ka, HV]), the group of k-points of a simple algebraic group defined over a locally compact non-discrete field k has Kazhdan's property T, provided its k-rank is at least 2, for example the groups $SL_n(k)$, $n \geqslant 3$. When $k = \mathbb{R}$, the simple $\mathbb{R}$-rank one groups $Sp(n, 1, \mathbb{R})$ also have property T. Furthermore, any lattice subgroup of a group possessing property T has this property as well (see e.g. [HV, Ch. 3][Z]).

Let now ν be a probability measure on a discrete group Γ, and suppose the group generated by the support of $\mu = \nu^* * \nu$ equals Γ. We note the following well known facts.

(1) If Γ has property T, then there exists $\alpha(\nu) < 1$, such that in every ergodic probability-measure-preserving action of Γ, $\|\pi_0(\nu)\| \leqslant \alpha(\nu)$. Indeed, otherwise there would exist a unitary representation τ of Γ on a space $\mathcal{H}_\tau$, without invariant unit vectors and with $\|\tau(\nu)\| = 1$. Then $\|\tau(\mu)\| = 1$ and by the spectral theorem, the self-adjoint non-negative operator $\tau(\mu)$ will have a sequence of asymptotically invariant unit vectors in $\mathcal{H}_\tau$. It follows (see §2.1, remark 2) that Γ will have an asymptotically invariant sequence and this contradicts property T.

(2) Conversely, if there exists at least one probability measure $\nu \in \ell^1(\Gamma)$ whose support generates Γ, and satisfying the estimate $\|\pi_0(\nu)\| \leqslant \alpha(\nu) < 1$ for every ergodic action, then Γ has Kazhdan's property T (and hence *every* measure as in (1) satisfy a similar estimate). Indeed the operator $\pi(\nu)$ cannot have an asymptotically invariant sequence of unit vectors in $L_0^2(X)$, and hence Γ does not have an asymptotically invariant sequence in $L_0^2(X)$. But by [Co-We], property T is characterized by the absence of asymptotically invariant sequence of sets in *all* ergodic actions. We refer to [Ro],[Sc] and [H-R-V 1, 2] for more details on these matters.

(3) Note that given a unitary representation π and a probability measure $\mu \in \ell^1(\Gamma)$, if $\|\pi(\mu)\| \leqslant 1-\epsilon$, then for any unit vector v:

$$\epsilon \leqslant 1 - \|\pi(\mu)v\| \leqslant \|\pi(\mu)v - v\| = \|\int_G (\pi(g)v - v)d\mu(g)\|$$
$$\leqslant \int_G \|\pi(g)v - v\| d\mu(g).$$

It follows that for at least one $g \in \operatorname{supp}\mu = Q$, $\|\pi(g)v - v\| \geqslant \epsilon$. Hence, by definition, (Q, ϵ) are Kazhdan constants for the representation π. In other words, every estimate of the form $\|\pi(\mu)\| \leqslant 1-\epsilon$ yields also Kazhdan constants, when μ has finite support.

(4) Clearly the estimate $\|\pi_0(\mu)\| \leqslant \alpha(\mu) < 1$ for a probability measure $\mu \in \ell^1(\Gamma)$, implies that the sequence of convolution powers μ^{*n} is a pointwise ergodic sequence, and in fact the L^2-norm convergence to the ergodic mean is at an exponential rate: $\left\|\pi(\mu)^n f - \int_X f dm\right\|_2 \leqslant \alpha^n \|f\|_2$. The same applies also to the rate of pointwise convergence (see Theorem 6 below). However, it is usually difficult to compute or even estimate the weights arising as convolution powers on Γ and therefore it is of interest to establish ergodic theorems for averages defined geometrically, such as spheres and balls w.r.t. a word metric.

We now recall a result from [N4], which is based on a spectral transfer principle due to M. Cowling [Co2] and also R. Howe [H] and C. C. Moore [Mo] [H-M]) (see also [H-T][Oh]). The spectral transfer principle asserts that any unitary representation π without invariant unit vectors of a simple Lie group of real rank at least 2 satisfies: $\pi^{\otimes n} \subseteq \infty \cdot \lambda_G$, for some $n = n(G)$ depending only on G. Using Jensen's inequality, the following norm inequality can be established:

Theorem 5. [N4] *Let $G = \prod_{i=1}^N G_i$ be a connected semi-simple Kazhdan Lie group with finite center and no non-trivial compact factors. Then there exists an even $n = n(G) = 2k$, with the following property. Let $\mu \in M(G)$ be an arbitrary Borel probability measure on G. Then:*

(i) *Given any strongly continuous unitary representation π of G with no G_i-invariant unit vector (for all $1 \leqslant i \leqslant N$), $\|\pi(\mu)\| \leqslant \|\lambda_G(\mu)\|^{\frac{1}{n(G)}}$.*

(ii) *In particular, if G acts on (X, B, m), each G_i acts ergodically ($1 \leqslant i \leqslant N$), and π_0 denotes the representation of G in $L_0^2(X)$, then $\|\pi_0(\mu)\| \leq \|\lambda_G(\mu)\|^{\frac{1}{n(G)}}$.*

(iii) *If μ is supported on a closed unimodular subgroup $H \subset G$, then $\|\pi_0(\mu)\| \leq \|\lambda_H(\mu)\|^{1/n(G)}$.*

As noted in [N4], the last result bears an obvious similarity to the transfer principle for amenable groups of §2.2, since it bounds the norms of the operators $\pi_0(\mu)$ in an *arbitrary* representation π_0, in terms of the spectral norm of $\lambda_G(\mu)$ acting as a convolution operator in the *regular* representation.

Typically, for a sequence μ_k of probability measures on G whose supports satisfy appropriate growth conditions, it is possible to obtain an exponential decay estimate, namely $\|\lambda_G(\mu_k)\| \leqslant C\ \exp(-\delta k)$, for some $\delta > 0$ depending on the sequence. Therefore, in any measure preserving ergodic action: $\|\pi_0(\mu_k)\| \leqslant C\ \exp(-\delta k/n(G))$. Note that in contrast, in the case of amenable groups, $\|\pi_0(\mu_k)\| = 1$ for all $k \geqslant 0$ (if the measure is non-atomic, see §2.1 and [Ro][Sc])

As will be noted further below, such decay estimates for averages on a group imply maximal inequalities for the maximal function $f_\mu^*(x) = \sup_{k\geqslant 0} |\pi(\mu_k)f(x)|$, and even for the exponential-maximal function defined by : $\sup_{k\geqslant 0} \exp(k\delta/2)|\pi(\mu_k)f(x)|$.

Many natural probability preserving actions of general semi-simple algebraic groups (not necessarily possesing property T), satisfy $\pi_0^{\otimes n(X)} \subset \infty \cdot \lambda_G$. Then by Theorem 5 $\|\pi_0(\mu)\| \leqslant \|\lambda_G(\mu)\|^{1/n(X)}$ for *every* Borel probability measure $\mu \in M(G)$. We therefore formulate the following:

Theorem 6. *(see* [N4]*) Let G be a locally compact second countable group, and X a measure-preserving action of G. Assume $\pi_0^{\otimes n(X)} \subset \infty \cdot \lambda_G$, and that the probability measures ν_k satisfy $\|\lambda_G(\nu_k)\| \leqslant Ce^{-\theta k}$. Then*

(1) *ν_k satisfies an exponential maximal inequality in $L_0^2(X)$, with parameter $\frac{\theta}{2n(X)}$, namely for $f \in L_0^2(X)$:*

$$\left\| \sup_{k\geqslant 0} e^{\frac{\theta}{2n}k} |\pi_0(\nu_k)f| \right\|_{L_0^2(X)} \leqslant C\|f\|_{L_0^2(X)} \cdot$$

(2) *In particular, if $m(X) - 1$, the convergence of $\pi(\nu_k)$ to the ergodic mean is exponential with rate $\frac{\theta}{2n(X)}$ for L^2-functions, namely for $f \in L^2(X)$*

$$|\pi(\nu_k)f(x) - \int_X fdm| \leqslant C(x,f)e^{-\frac{\theta}{2n}k} .$$

We remark that the exponential rate of convergence to the ergodic mean, as well as the exponential maximal inequality, constitute a new phenomenon that does not arise in the ergodic theory of amenable groups (see remark (1) §2.1). This phenomenon was first established for *radial* averages on semi-simple Kazhdan Lie groups in [M-N-S]. For examples of actions of G and of absolutely continuous (or singular) averages ν_k satisfying the conditions of the foregoing theorem we refer to [M-N-S] and [N4]. We note also that exponential decay estimates as above imply an exponential maximal inequality in L^p, $1 < p < \infty$, by Riesz-Thorin interpolation [N4], and therefore also an exponential rate of pointwise convergence to the ergodic mean for L^p-function. The problem of establishing ergodic theorems in L^1, however, is largely unresolved.

5.2 Ergodic theorems for discrete hyperbolic subgroups.

We now restrict our attention to averages supported on discrete word-hyperbolic (see [Gr]) subgroups $\Gamma \subset G$. Recall the following convolution theorem due to P. Jolissaint [J] (see also [J-V]):

Theorem [J]. *Let Γ be a non-elementary word-hyperbolic group, $S = S^{-1}$ a finite symmetric set of generators, and $L(g) = d(e,g)$ the associated length function. Then there exist two positive constants $s = s(\Gamma, S)$ and $C = C(\Gamma, S)$, such that for any finitely supported function $f \in \ell^1(\Gamma)$:*

$$\|\lambda(f)\| \leqslant C\|f \cdot (1+L)^s\|_{\ell^2(\Gamma)} \, .$$

The foregoing result can be used to prove the following

Proposition 2. [N4] *Let Γ denote a non-elementary word-hyperbolic group. Let $S = S^{-1}$ be a finite symmetric set of generators, and let $q_\sigma = q_\sigma(\Gamma, S) = \lim_{n\to\infty} \frac{1}{n} \log |S_n|$ (see [Coo] for the existence of the limit). Then:*

(1) $\|\lambda_\Gamma(\sigma_k)\| \leqslant Bq^{-k}$ *and* $\|\lambda_\Gamma(\beta_k)\| \leqslant Bq^{-k}$ *for any* $1 < q < \sqrt{q_\sigma}$ *and some* $B = B_q > 0$.

(2) σ_k *and* β_k *satisfy the strong exponential maximal inequality in* $\ell^2(\Gamma)$, *and in particular*

$$\|f^*_\sigma\|_{\ell^2(\Gamma)} \leqslant C\|f\|_{\ell^2(\Gamma)} \quad \text{and} \quad \|f^*_\beta\|_{\ell^2(\Gamma)} \leqslant C\|f\|_{\ell^2(\Gamma)} \, .$$

The previous result computes the spectral norm of operators in $\ell^1(\Gamma)$ acting by convolutions in the regular representation. In order to estimate the norms of these operators acting in $L^2_0(X)$, we now add the spectral assumption $\pi_0^{\otimes n} \subset \infty \cdot \lambda_\Gamma$. This assumption will hold for example when Γ is a discrete subgroup of a simple non-compact Kazhdan Lie group G and X is an ergodic G-space. Indeed then $\pi_0^{\otimes n} \subset \infty \cdot \lambda_G$, and since Γ is a discrete subgroup, we have $(\pi_0 \big| \Gamma)^{\otimes n} \subset \infty \cdot \lambda_G \big| \Gamma = \infty \cdot \lambda_\Gamma$, and Theorem 5 gives $\|\pi_0(\mu)\| \leqslant \|\lambda_\Gamma(\mu)\|^{1/n}$ for any probability measure $\mu \in \ell^1(\Gamma)$. Proposition 2 together with the foregoing remarks can be combined to give the following conclusion.

Theorem 7. *(see* [N4]*) Let G be an lcsc group, and let $\Gamma \subset G$ be a non-elementary word-hyperbolic discrete subgroup of G with a given finite symmetric set of generators S. Assume G acts on a probability space X*

and $\pi_0^{\otimes n} \subset \infty \cdot \lambda_G$. Then the sequence $\nu_k = \sigma_k$ of sphere averages and the sequence $\nu_k = \beta_k$ of ball averages in $\ell^1(\Gamma)$ satisfy the conclusions of Theorem 6. In particular, these conclusions hold when G is a semi-simple Kazhdan Lie group with finite center and no compact factors, and Γ any discrete non-elementary word-hyperbolic subgroup.

We remark that Theorem 7 improves [F-N, Thm 4] (see also §3.2), but only under the additional assumption that the action satisfies $\pi_0^{\otimes n} \subset \infty \cdot \lambda_G$, which does not always hold, e.g. for the groups $G = SO^0(n, 1)$.

Examples. [N4]

(1) Theorem 7 applies to co-compact lattices Γ in the simple Kazhdan groups $G = Sp(n, 1)$, $n \geqslant 2$, which are word-hyperbolic groups. In every ergodic action of G, the sphere averages on Γ converge at an exponential rate to the ergodic mean.

(2) Let $\Gamma = \mathbb{F}_r$, the free non-abelian group on r generators, and assume $S = S^{-1}$ is a free set of generators. Then (see §4.1):

$$\|\lambda(\sigma_k)\| = \varphi_{1/2}(\sigma_k) \leqslant (1+k)(2r-1)^{-k/2} \leqslant Bq^{-k} \quad \text{if} \quad q < \sqrt{2r-1}\,.$$

When $\mathbb{F}_r$ is emebedded as a discrete subgroup of a group G, for every representation π_0 of G satisfying $\pi_0^{\otimes n} \subset \infty \cdot \lambda_G$, Theorem 7 implies $\|\pi_0(\sigma_k)\| + \|\pi_0(\beta_k)\| \leqslant Bq^{-k/n}$ if $q < \sqrt{2r-1}$.

5.3 The regular representation of lattice subgroups.

In this section, we let G be a locally compact second countable (lcsc) group, and $\Gamma \subset G$ a discrete lattice subgroup. Given a sequence of compact sets $U_n \subset G$ satisfying $m_G(U_n) \to \infty$, we can consider the sets $A_n = U_n \cap \Gamma$ of lattice points, (which we assume are non-empty) and define the averages $\alpha_n \in \ell^1(\Gamma)$ by:

$$\alpha_n = \frac{1}{|A_n|} \sum_{\gamma \in A_n} \gamma\,.$$

For example, when G is a simple non-compact Lie group we can take U_n as the bi-K-invariant balls $B_n = \{g \mid d(gK, K) \leqslant n\}$, where $K \subset G$ is a maximal compact subgroup, and d the canonical G-invariant metric on G/K.

It is a natural question whether the norm of the discrete convolution operator $\lambda_G(\alpha_n)$ on $L^2(G)$ can be estimated by the convolution norm of the corresponding absolutely continuous averaging operator $u_n = \frac{1}{m_G(U_n)} \chi_{U_n}$. A partial answer to this question is provided by a norm estimate given as follows.

Let ψ be continuous compactly supported functions on G, so that $\tilde{\psi}(x) = \tilde{\psi}(x\Gamma) = \sum_{\gamma \in \Gamma} \psi(x\gamma)$ satisfies $\tilde{\psi} \in L^2(G/\Gamma)$. Let $F_k = \chi_{U_k}/m(U_k)$ be the normalized characteristic function of the sets U_k. The functions $\psi^* * F_k * \psi$ are continuous on G, and therefore $f_k = (\psi^* * F_k * \psi)\Big| \Gamma$ are well defined functions on Γ. Using a norm estimate due to [C-S] and stated in [Bi-St], it is possible to prove:

Proposition 3. [N4] *Let Γ be a lattice in an lcsc group G. Then for the sequences F_k and f_k defined above:*
(1) $\|\lambda_\Gamma(f_k)\| \leqslant C\|\lambda_G(F_k)\|$.
(2) If $\|\lambda_G(F_k)\| \leqslant Ce^{-\theta k}$ for the sequence of functions $F_k \in L^1(G)$ then $f_k \in \ell^1(\Gamma)$ satisfy, in every unitary representation π_0 of Γ such that $\pi_0^{\otimes n} \subset \infty \cdot \lambda_\Gamma$, the estimate: $\|\pi_0(f_k)\| \leqslant C\exp(-\theta k/n)$.

Consider now two arbitrary probability measures μ_1 and μ_2 on an lcsc group G and note the following simple norm estimate:

Proposition 4. [N4] *If $\mu_1 \leqslant C\mu_2$, and $\pi_0^{\otimes n} \subset \infty \cdot \lambda_G$, then $\|\pi_0(\mu_1)\| \leqslant \|C\lambda_G(\mu_2)\|^{1/n}$.*

The foregoing propositions imply the following. First, let us say that μ_1 is a bounded perturbation of μ_2 with constant C if $\mu_1 \leqslant C\mu_2$. We then have:

Proposition 5. [N4] *Let G be an lcsc group, and let ν_k be any sequence of probability measures on G that satisfies the estimate $\|\lambda_G(\nu_k)\| \leqslant B\exp(-\delta k)$, for some $\delta > 0$. Let $\tilde{\nu}_k$ be probability measures which are bounded perturbations of ν_k with fixed constant C. Then $\|\pi_0(\tilde{\nu}_k)\| \leqslant BC^{1/n}\exp(-\delta/nk)$, provided $\pi_0^{\otimes n} \subset \infty \cdot \lambda_G$.*

In particular, consider the sequence of measures f_k of Proposition 3. If the expression $\liminf_{k\to\infty} \sum_{\gamma\in\Gamma} f_k(\gamma) > 0$ we can normalize f_k, and the resulting probability measures will satisfy the same exponential decay estimate given by Proposition 3 (with a different constant C). Then by Proposition 4, any sequence of measures on $\Gamma \subset G$ which is a bounded perturbation of the normalized version of f_k satisfies an exponential decay estimate in norm, in any unitary representation π_0 of Γ which satisfies $\pi_0^{\otimes n} \subset \infty \cdot \lambda_\Gamma$.

This fact can be applied to prove the following convolution estimate in the regular representation of a lattice subgroup of a connected semi-simple Lie group G with finite center and no compact factors. Let K be a maximal compact subgroup, and consider $\beta_k^G = \frac{1}{vol B_k}\chi_{B_k}(g)$, where B_k is the bi-K-invariant ball of radius k in G. Let $B_k(\Gamma) = B_k \cap \Gamma$, and let $b_k = \frac{1}{|B_k(\Gamma)|}\sum_{\gamma\in B_k(\Gamma)} \gamma$.

Theorem 8. [N4] *Let $\Gamma \subset G$ be a lattice in a connected semi-simple Lie group G, with finite center and no compact factors. Then the sequence*

$b_k \in \ell^1(\Gamma)$ satisfies $\|\lambda_\Gamma(b_k)\| \leqslant C\exp -\theta k$. Here $\theta > 0$ satisfies $\|\lambda_G(\beta_t^G)\| \leqslant C\exp(-\theta t)$. In particular the sequence b_k satisfies the strong exponential maximal inequality in $\ell^2(\Gamma)$.

5.4 Unitary induction and general representations of lattice subgroups.

Let now Γ be a lattice in an lcsc group G, and consider the following well known facts about unitary induction of representations from Γ to G (see e.g.[HV, Ch. 3]). Let $D \subset G$ be a left fundamental domain, namely $G = D\Gamma$. Then D maps bijectively to G/Γ under the canonical map, and $s : d\Gamma \mapsto d \in D$ is a measurable section. We normalize the G-invariant measure on G/Γ to have total mass one. Writing $gd = d'\gamma$, the map $\alpha(g,d) = \gamma$ is a cocycle $\alpha : G \times G/\Gamma \to \Gamma$. Given any probability measure μ on G, we can use α to define a measure on $\tilde{\mu}$ on Γ, whose total mass is 1. The weight at a given element γ is given by the $\mu \times m_{G/\Gamma}$-measure of the set $\{(g, y\Gamma) \in G \times G/\Gamma \mid \alpha(g^{-1}, y\Gamma)^{-1} = \gamma\}$. In other words, $\tilde{\mu}$ is the distribution function of $\alpha(g^{-1}, y\Gamma)^{-1}$ on the measure space $(G \times G/\Gamma, \mu \times m_{G/\Gamma})$. Let now τ be a unitary representation of Γ, and consider the representation π_τ induced to G. The representation space is given by $\{f : G/\Gamma \to \mathcal{H}_\tau \mid \int_{G/\Gamma} \|f(y\Gamma)\|^2 dm_{G/\Gamma} < \infty\}$, and the G-action by $\pi_\tau(g)f(y\Gamma) = \tau(\alpha(g^{-1}, y\Gamma)^{-1})f(g^{-1}y\Gamma)$. It follows that if $v \in \mathcal{H}_\tau$, and $F_v : G/\Gamma \to \mathcal{H}_\tau$ is the constant function $F_v(y\Gamma) = v$, then $\pi_\tau(\mu)F_v = \tau(\tilde{\mu})v$. Therefore, we conclude:

Proposition 6. [N4] *Let G be a semi-simple Kazhdan Lie group with finite center and no compact factors, and let Γ be a lattice in G. Let β_k^G denote the ball averages on G, and $\tilde{\beta}_k$ the corresponding probability measures on Γ obtained as above by some fixed choice of a fundamental domain. If τ is any unitary representation of Γ without Γ-invariant unit vectors, then*

$$\|\tau(\tilde{\beta}_k)\| \leqslant \|\pi_\tau(\beta_k^G)\| \leqslant \|\lambda_G(\beta_k^G)\|^{1/n(G)} \leqslant \exp(-k\theta/n(G))$$

and $\tilde{\beta}_k$ satisfy the conclusions of Theorem 6.

It is natural to ask whether the averages b_k defined before Theorem 8 also satisfy the estimate: $\|\tau(b_k)\| \leqslant C\exp -k\theta$, where τ is *any* unitary representation of Γ without Γ-invariant unit vectors and $\theta > 0$ depends only on Γ. This estimate is valid for $\tau = \lambda_\Gamma$ by Theorem 8, but the general problem remains unresolved.

In this context, we recall also the following well known fact, due to P. de la Harpe and A. Valette [HV]:

Proposition. [HV, Ch. 3, Proof of Lemme 3] *Let G be an lcsc group, Γ a lattice subgroup, normalize $m(G/\Gamma) = 1$, and let (ϵ, Q) be a set of Kazhdan* *constants for G. Let $Y \subset G/\Gamma$ be any compact subset satisfying for every $k \in Q$: $m(Y \Delta kY) < \frac{\epsilon^2}{2} m(Y)$. Let $s : G/\Gamma \to G$ be any measurable section which is bounded on compact sets, and $\alpha : G \times G/\Gamma \to \Gamma$ be the corresponding cocycle defined above. Then if $L = \alpha(Q^{-1} \times Y) \subset \Gamma$, L^{-1} and $\epsilon/2$ are Kazhdan constants for Γ.*

Originally explicit Kazhdan constants for connected semi-simple Lie groups were obtained using the exponential decay estimates of matrix coefficients of K-finite vectors in irreducible representation of G, estimates due mainly to M. Cowling [Co1] (see also [H]). For an elegant elementary derivation of Kazhdan constants using this method we refer to [H-T, ch. V, §4].

We note also that to make the estimate $\|\pi_0(\mu)\| \leqslant \|\lambda(\mu)\|^{1/n}$ of operator norms explicit, one needs an estimate of the integer $n = n(G, X)$ such that $\pi_0^{\otimes n} \subset \infty \cdot \lambda_G$. For semi-simple Kazhdan groups $n = n(G)$ depends on G only, and was calculated in [H][H-T] and [Li]. For a concise recent discussion of $n(G)$ we refer to [Oh].

6. Some further remarks and open problems for Kazhdan groups

6.1 Best possible estimate of operator norms.

The spectral transfer principle ([N4], see §5.1) asserts that for a simple non-compact Kazhdan Lie group G, $\pi_0^{\otimes n} \subset \infty \cdot \lambda_G$, and by Theorem 5 (§5.1) $\|\pi_0(\mu)\|_{L_0^2(X)} \leqslant \|\lambda_G(\mu)\|^{1/n}$ for any probability measure μ. If μ is supported on a closed unimodular subgroup H, then $\|\pi_0(\mu)\|_{L_0^2(X)} \leqslant \|\lambda_H(\mu)\|^{1/n}$. The best possible estimate of this kind is of course when $n = 1$. A basic fact in the representation theory of semi-simple groups, due to D. A. Kazhdan [Ka] (see also [Z], [H-T], and [Co1]) is that certain subgroups H of G have the following property: For every unitary representation π of G without invariant unit vectors, and every (absolutely continuous) Borel probability measure on H, we have $\|\pi(\mu)\| \leqslant \|\lambda_H(\mu)\|$. Equivalently, the restriction of π to H is weakly contained in $\infty \cdot \lambda_H$ (for the equivalence see [Ey]).

For example, subgroups $H \subset G$ of the form $H = SL_2(\mathbb{R}) \ltimes \mathbb{R}^2$, where $SL_2(\mathbb{R})$ acts on $\mathbb{R}^2$ via the defining representation have this restriction property, provided only that any $\mathbb{R}^2$-invariant unit vector is also G-invariant. A particularly elegant, elementary, and quantitative argument establishing this fact is due to R. Howe and E. Tan [H-T, Ch 4]. It gives rise to the following best possible estimates for operator norms of (singular) measures supported on $SL_2(\mathbb{R}) \subset H$. Consider $a_t = \text{diag}\,(e^t, e^{-t}) \in SL_2(\mathbb{R})$ and the spherical averages $s_t = m_K * \delta_{a_t} * m_K$ on $SL_2(\mathbb{R})$, where m_K is normalized Lebesgue measure on the circle group K. We then have $\|\pi_0(s_t)\| \leqslant \|\lambda_{SL_2(\mathbb{R})}(s_t)\| = \Xi_2(a_t)$, where $\Xi_2(g)$ is the Harish Chandra

Ξ-function for $SL_2(\mathbb{R})$. Therefore $\|\pi_0(s_t)\| \leqslant C_2(1+t)e^{-t}$ (see e.g. [H-T], Ch. V). We note that similar conclusions apply to many other subgroups $H \subset G$, and we refer to [N4], Prop. 4.1 for more details.

Restricting the representation further to subgroup of $SL_2(\mathbb{R}) \subset H \subset G$, we can estimate operator norms of averages supported on discrete subgroups by the operator norm of the averages acting in the regular representation. As an example, we cite the following recent result regarding norm estimates in the regular representation :

Theorem. [CVJ] *Let $\Gamma = \langle T, r\rangle$ be a one relator group, T a set of generators with $|T| \geqslant 4$, $T \cap T^{-1} = \emptyset$, and let $S = T \cup T^{-1}$. Then $\|\lambda_\Gamma(\sigma_1\| \leqslant \frac{2\sqrt{|S|-2}+1}{|S|} < 1$. Here $\sigma_1 = \sigma_1(\Gamma, S) = \frac{1}{|S|}\sum_{s\in S} s \in \ell^1(\Gamma)$.*

For example, when $\Gamma = \Lambda_g$ is a fundamental group of a closed surface of genus $g \geqslant 2$, and $S = S^{-1}$ is the set of standard generators, the estimate obtained in [CVJ] is $\|\lambda(\sigma_1)\| \leqslant g^{-1}\sqrt{2g-1}$. We note that for certain sequences ν_n in a one-relator group it is possible to obtain exponential decay of $\lambda_\Gamma(\nu_n)$, using the Haagerup-type inequality contained in [CVJ] (see also §5.2). Then Theorem 6 applies and gives ergodic theorems for $\pi(\nu_n)$, acting in a G-space.

Another example is obtained when Γ is a discrete non-elementary word-hyperbolic subgroup of $SL_2(\mathbb{R}) \subset H$, and S a symmetric set of generators for Γ. Then $\|\pi_0(\sigma_n)\| = \|\lambda_\Gamma(\sigma_n)\| \leqslant Bq^{-n}$ for any $q < \sqrt{q_\sigma}$, as noted in §5.2.

6.2 Some open problems for group actions with a spectral gap.
It was seen in Section 5 that for certain discrete subgroups of semi-simple algebraic groups, the spectral transfer principle yields estimates of the form $\|\lambda_\Gamma(\nu_n)\| \leqslant C\exp(-\theta n)$, $\theta > 0$ and then by Theorem 6 the sequence $\pi(\nu_n)$ is a mean and pointwise ergodic sequence and satisfies the strong exponential maximal inequality, in every action satisfying $\pi_0^{\otimes n} \subset \infty \cdot \lambda_\Gamma$.

Another family of groups for which similar conclusions hold is the family of triangle Kazhdan groups constructed in [CMSZ]. Indeed, exponential decay estimates were obtained in [CMS] for convolution norms of a natural family of radial averages.

Discrete groups with property T were constructed recently by [Zu] and [B-S]. For these, however, estimates of convolution norms have not yet been established.

Before noting some of the natural problems that arise in this context, let us first define a unitary representation π of Γ to be mixing if all its matrix coefficients belong to $C_0(\Gamma)$, namely vanish at infinity.

Let now Γ be an arbitrary Kazhdan group, S a finite symmetric set of generators. Does σ_n satisfy $\|\tau(\sigma_n)\| \leqslant C\exp(-\theta n)$ for every mixing unitary representation τ of Γ, and some $\theta = \theta(\Gamma) > 0$? We note that it is not known whether, in general, $\|\tau(\sigma_n)\| \to 0$ for every such τ, or whether this estimate holds even for the special case $\tau = \lambda_\Gamma$ of the regular representation. As noted in §5.4, these problems are not settled also for the averages b_k on Kazhdan lattices introduced there, except the case $\tau = \lambda_\Gamma$, considered in Theorem 8.

More generally, let X be a mixing probability-measure-preserving action of a group Γ, namely the representation in $L_0^2(X)$ is mixing. Let S be a finite symmetric generating set, and assume that $\|\pi_0(\sigma_1)\| < 1$. Does it follow that $\|\pi_0(\sigma_n)\| \to 0$? Does it follow that $\|\pi_0(\sigma_n)\| \leqslant C\exp(-\theta n)$ for some $\theta = \theta(\Gamma, S, X) > 0$? The same questions are not resolved even for the regular representation of a non-amenable group. Finally, note that a (weak) mixing assumption is necessary, otherwise actions with discrete spectrum of $\mathbb{F}_r$ provide counterexamples to $\|\pi_0(\sigma_n)\| \to 0$, e.g. when the sign character is realized, and the generating set S is taken as a ball of radius 2 w.r.t free generators.

References.

[A-K] V.I. Arnold and A.L. Krylov, *Uniform distribution of points on a sphere and some ergodic properties of solutions of linear ordinary differential equations in the complex plane*, Sov. Math. Dokl. vol. 4, no. 1 (1962), 1-5.

[B-S] W. Ballman and J. Światkowsky, *On L^2-cohomology and property T for automorphism groups of polyhedral cell complexes*, Geometric and Functional Analysis vol. 7 (1997), 615-645.

[B] J. Barrionuevo, *Estimates for some Kakeya-type maximal operators*, Trans. A.M.S. vol. 335, no. 2 (1993), 667-682.

[BJR 1] A. Bellow, R. Jones, and J. Rosenblatt, *Convergence for moving averages*, Erg. Th. and Dyn. Sys. vol. 10 (1990), 43-62.

[BJR 2] A. Bellow, R. Jones, and J. Rosenblatt, *Almost everywhere convergence of convolution powers*, Erg. Th. and Dyn. Sys. vol. 14 (1994), 415-432.

[BPW] E. Berkson, M. Paluszynski, and G. Weiss, *Transference couples and their applications to convolution operators and maximal operators*, Lecture Notes in Pure and applied Math., Dekker, New York vol. 175, 1996, pp. 69-84.

[Be] T. Bewley, *Extensions of the Birkhoff and von-Neumann ergodic theorems to semigroup actions*, Ann. Inst. H. Poincaré vol. VII, no 4 (1971), 283-291.

[Bi-St] C. Bishop and T. Steger, *Representation theoretic rigidity in $PSL(2,\mathbb{R})$*, Acta. Math. vol. 170 (1993), 121-149.

[Bo] J. Bourgain, *On the maximal ergodic theorem for certain subsets of integers*, Israel Journal of Math. vol. 61 (1988), 39-72 ; 73-83.

[Ca] A. P. Calderón, *Ergodic theory and translation-invariant operators*, Proc. Nat. Acad. Sci. U.S.A. vol. 59 (1968), 349-353.

[Ca 1] A. P. Calderón, *A general ergodic theorem*, Annals of Math. vol. 58, no. 1 (1953), 182-191.

[C] P. Cartier, *Harmonic analysis on trees*, Proc. Symp. Pure Math. A. M. S. vol. 26 (1972), 419-424.

[Ch] J. Chatard, *Application des propriétés de moyenne d'un groupe localement compact á la théory ergodique*, Ann. Inst. H. Poincaré vol. VI no 4 (1970), 307-326.

[C-M-S-Z] D. I. Cartwright, A. M. Mantero, T. Steger and A. Zappa, *Groups acting simply transitively on the vertices of a building of type $\tilde{A}_2$*, Geometriae Dedicata vol. 47 (1993), 143-166.

[C-M-S] D. I. Cartwright, W. Mlotkowski and T. Steger, *Property (T) and $\tilde{A}_2$ groups*, Ann. Inst. Fourier (Grenoble) vol. 44 (1993), 213-248.

[C-W] R. R. Coifman and G. Weiss, *Transference Methods in Analysis*, CBMS Regional conference series in Math. vol. 31 (1976).

[Co-We] A. Connes and B. Weiss, *Property T and asymptotically invariant sequences*, Israel Jnl. of Math. vol. 37 (1980), 209-210.

[Coo] M. Coornaert, *Measures de Patterson-Sullivan sur le bord d'un space hyperbolique au sens de Gromov*, Pacific J. Math. vol. 159 no. 2 (1993), 241-270.

[Co1] M. Cowling, *Sur les coefficients des representations unitaires des groupes de Lie simples*, Springer Lecture Notes in Mathematics vol. 739 (1979), 132-178.

[Co2] M. Cowling, *Harmonic analysis on semigroups*, Annals of Math. vol. 11 (1983), 267-283.

[Co3] M. Cowling, *Herz's "principe de majoration" and the Kunze-Stein phenomenon*, UNSW preprint (1996).

[C-S] M. Cowling and T. Steger, *Irreducibility of restrictions of unitary representations to lattices*, J. Reine. Angew. Math. vol. 420 (1991), 85-98.

[CVJ] P.A. Cherix, A. Valette and P. Jolissaint, *On spectra of simple random walks on one-relator groups*, Pacific J. Math. vol. 175 (1996), 417-438.

[D-S] N. Dunford and J. T. Schwartz, *Linear Operators* vol. *I, II*, Interscience, New York, 1963.

[E] W. R. Emerson, *The pointwise ergodic theorem for amenable groups*, Amer. J. Math. vol. 96 (1974), 472-487.

[E-G] W. R. Emerson and F. P. Greenleaf, *Group structure and the pointwise ergodic theorem for connected amenable groups*, Adv. in Math. vol. 14 (1974), 153-172.

[Ey] P. Eymard, *L'algébre de Fourier d'un groupe localement compact.*, Bull. Soc. Math. de France vol. 92 (1964), 181-236.

[FT-P] A. Figà-Talamanca and M. A. Picardello, *Harmonic Analysis on Free Groups*, Lecture Notes in Pure and Applied Math. vol. 87, Dekker, New York, 1983.

[FT-N] A. Figà-Talamanca and C. Nebbia, *Harmonic analysis and representation theory for groups acting on homogeneous trees*, London Mathematical Society Lecture Notes Series vol. 162, Cambridge Univ. Press, 1992.

[F-N] K. Fujiwara and A. Nevo, *Maximal and pointwise ergodic theorems for word hyperbolic groups*, Erg. Th. and Dyn. Sys. vol. 18 (1998), 843-858.

[Gr] M. Gromov, *Hyperbolic groups*, Essays in group theory, Ed. S.M. Gersten, MSRI Publ. vol. 8, Springer 1987, pp. 75-263.

[G] Y. Guivarc'h, *Généralisation d'un théorème de von Neumann*, C.R. Acad. Sc. Paris vol t. 268, (1969), 1020-1023.

[HRV 1] P. de la Harpe, A. G. Robertson and A. Valette,, *On the spectrum of the sum of generators of a finitely generated group I*, Israel J. Math vol. 81 (1993), 65-96.

[HRV 2] P. de la Harpe, A. G. Robertson and A. Valette, *On the spectrum of the sum of generators of a finitely generated group II*, Colloq. Math. vol. LXV (1993), 87-102.

[HV] P. de la Harpe and A. Valette, *Propriété (T) de Kazhdan pour les groupes localement compacts*, Astérisque vol. 175 (1989), Soc. Math. de France.

[Herz] C. S. Herz, *The theory of p-spaces with an application to convolution operators*, Trans. A. M. S vol. 154 (1971), 69-82.

[H] R. Howe, *On a notion of rank for unitary representations of the classical groups*, Harmonic Analysis and Group Representations, C.I.M.E., 1980, pp. 223-232.

[H-M] R. Howe and C. C. Moore, *Asymptotic properties of unitary representations*, J. Func. Anal. vol. 32 (1979), 72-96.

[H-T] R. Howe and E.C. Tan, *Non-Abelian Harmonic Analysis*, Springer Verlag, 1992.

[J] P. Jolissaint, *Rapidly decreasing functions in reduced C^*-algebras of groups*, Trans. AMS vol. 317 (1990), 167-196.

[J-V] P. Jolissaint and A. Valette, *Normes de Sobolev et convoluteures bornes sur $L^2(G)$*, Ann. Inst. Fourier, Grenoble, vol. 41 (1991), 797-822.

[Ka] D.A. Kazhdan, *On a connection between the dual space of a group and the structure of its closed subgroups*, Func. Anal. Appl. vol. 1 (1967), 63-65.

[K] H. Kesten, *Symmetric random walks on groups*, Trans. A. M. S. vol. 92 (1959), 336-354.

[K-S] A. Kolmogoroff and G. Seliverstoff, *Sur la convergence des sèries de Fourier*, Comptes Rendous vol. 178 (1924), 303-306.

[Li] J-S. Li, *The minimal decay of matrix coefficients for classical groups*, In: Harmonic analysis and its applications in China, 1995.

[M-N-S] G. A. Margulis, A. Nevo and E. M. Stein, *Analogs of Wiener's ergodic theorems for semi-simple Lie groups II*, Submitted.

[M-W] B. Mohar and W. Woess, *A survey on spectra of infinite graphs*, Bull. London. Math. Soc. vol. 21 (1989), 209-234.

[Mo] C.C. Moore, *Exponential decay of correlation coefficients for geodesic flows*, Group Representation, Ergodic Theory, Operator Algebras and Mathematical Physics, Math. Sci. Res. Inst. Publ. vol. 6, Springer Verlag, 1987, pp. 163-181.

[N1] A. Nevo, *Harmonic analysis and pointwise ergodic theorems for non-commuting transformations*, Journal of the A.M.S. vol. 7 (1994), 875-902.

[N2] A. Nevo, *Pointwise ergodic theorems for radial averages on simple Lie groups I*, Duke Math. Jnl. vol. 76 (1994), 113-140.

[N3] A. Nevo, *Pointwise ergodic theorems for radial averages on simple Lie groups II*, Duke Math. Jnl. vol. 86 (1997), 239-259.

[N4] A. Nevo, *Spectral transfer and pointwise ergodic theorems for Kazhdan groups*, Math. Research Letters vol. 5 (1998), 305-325.

[N5] A. Nevo, *Spectral transfer and spectral multipliers for semi-simple groups*, Preprint.

[N6] A. Nevo, *Lie groups and pointwise ergodic theory*, In preparation.

[N-S1] A. Nevo and E. M. Stein, *A generalization of Birkhoff's pointwise ergodic theorem*, Acta Mathematica vol. 173 (1994), 135-154.

[N-S2] A. Nevo and E. M. Stein, *Analogs of Wiener's ergodic theorems for semi-simple Lie groups I*, Annals of Mathematics vol. 145 (1997), 565-595.

[N-T] A. Nevo and S. Thangavelu, *Pointwise ergodic theorems for radial averages on the Heisenberg group*, Advances in Mathematics vol. 127 (1997), 307-334.

[Oh] H. Oh, *Representations with minimal decay of matrix coefficients and tempered subgroups*, To appear in Bull. Math. Soc. France vol. 126 (1998).

[O-W1] D. Ornstein and B. Weiss, *The Shannon-McMillan-Breiman theorem for a class of amenable groups*, Israel Jnl. of Mathematics vol. 44 (1983), 53-60.

[O-W2] D. Ornstein and B. Weiss, *Subsequence ergodic theorems for amenable groups*, Israel Journal of Mathematics vol. 79 (1992), 113-127.

[O-W3] D. Ornstein and B. Weiss, *Entropy and isomorphism theorems for actions of amenable groups*, J.d'Analyse Mathématique vol. 48, No. 1 (1987), 1-141.

[P] R. E. A. C. Paley, *A proof of a theorem of averages*, Proc. London Math. Soc. vol. 31 (1930), 289-300.

[Pe] K. Petersen, *Ergodic Theory*, Cambridge University Press, 1983.

[Ro] J. Rosenblatt, *Uniqueness of invariant means for measure-preserving transformations*, Trans. A. M. S. vol. 265 (1981), 623-636.

[R-W] J. Rosenblatt and M. Wierdl, *Pointwise ergodic theory via harmonic analysis*, London Math. Soc. Lecture Notes Series, Cambridge University Press vol. 205, Proc. Conf. on Ergodic Theory held at Alexandria, Egypt 1993 (1993), 3-151.

[R] J. C. Rota, *An "Alternierende Verfahren" for general positive operators*, Bull. A. M. S. vol. 68 (1962), 95-102.

[Sc] K. Schmidt, *Amenability, Kazhdan's property T, strong ergodicity and invariant means for ergodic group-actions*, Ergod. Th. and Dyn. Sys. vol. 1 (1981), 233-236.

[Si] Ya. G. Sinai (Ed.), *Dynamical Systems II*, Encylopedia of Mathematical Sciences, Springer Verlag vol2, 1989.

[S1] E.M. Stein, *On the maximal ergodic theorem*, Proc. Nat. Acad. Sci. U.S.A. vol. 47 (1961), 1894-1897.

[S2] E.M. Stein, *Topics in harmonic analysis related to the Littlewood–Paley theory*, Annals of Math. Studies, vol. 63 (1970), Princeton Univ. Press.

[S3] E. M. Stein, *Harmonic Analysis*, Princeton Univ. Press, 1993.

[S-W] E.M. Stein and S. Wainger, *Discrete analogs of singular Radon transforms*, Bull. (New Series) A.M.S. vol. 23 no. 2 (1990), 537-544.

[T] A. Templeman, *Ergodic theorems for general dynamic systems*, Soviet. Math. Dokl. vol. 8 no. 5 (1967), 1213-1216.

[T1] A. Templeman, *Ergodic Theorems for Group Actions*, Kluwer Academic Publisher vol. 78, 1992.

[Wa] T. E. Walker, *Ergodic theorems for free group actions on von-Neumann algebras*, J. Funt. Anal. vol. 150 (1997), 26-47.

[We] B. Weiss, *Positive cones in Hilbert space and a maximal inequality*, Inequalities III, Ed. O. Shisha, Academic Press,, 1972.

[W] N. Wiener, *The ergodic theorem*, Duke Math. Jnl. vol. 5 (1939), 1-18.

[Wo] W. Woess, *Random walks on infinite graphs and groups - a survey of selected topics*, Bull. London Math. Soc. vol. 26 (1994), 1-60.

[Z] R.J. Zimmer, *Ergodic Theory and Semisimple Groups*, Birkhauser, Boston, 1984.

[Zu] A. Zuk, *La propriété (T) de Kazhdan pour les groupes agissant sur les polyedres*, C. R. Acad. Sci. Paris vol. 323, *Serie I* (1996), 453-458.

Amos Nevo
Technion, Haifa, Israel
email: anevo@tx.technion.ac.il

Random walk and isoperimetry on discrete subgroups of Lie groups

Christophe Pittet and Laurent Saloff-Coste

Abstract. This paper reviews certain aspects of the proof of the following known result. Let Γ be an amenable discrete subgroup of a Lie group having finitely many connected components. Then Γ is finitely generated and, if Γ has exponential growth, the isoperimetric profile $I^{\uparrow}(n) = \inf_{\#A \geqslant n} \#\partial A$ satisfies $cn(\log n)^{-1} \leqslant I^{\uparrow}(n) \leqslant Cn(\log n)^{-1}$ and the probability of return of simple random walk to it starting state satisfies $c_1 \exp(-C_1 n^{1/3}) \leqslant \mathbf{P}^x(X_{2n} = x) \leqslant C_2 \exp(-c_2 n^{1/3})$.

1 Introduction

Let G be a finitely generated group. For any finite generating set S satisfying $S = S^{-1}$, consider the Cayley graph (G, S) with vertex set G and an edge from x to y if and only if $y = xs$ for some $s \in S$. Thus, edges are oriented but this is merely a convention since (x, y) is an edge if and only if (y, x) is an edge. We allow the identity element id to be in S in which case our graph has a loop at each vertex. Clearly the graph (G, S) is invariant under the left action of G. Denote by $|x|_S$ the distance from the neutral element id to x in the Cayley graph (G, S), that is, $|x|_S$ is the minimal number k of elements of S needed to write x as $x = s_1 s_2 \dots s_k$, $s_i \in S$. We also set $d_S(x, y) = |x^{-1}y|_S$. This is a left invariant distance on G. The volume growth function of (G, S) is defined by

$$V_S(n) = \#\{x \in G : |x|_S \leqslant n\}.$$

For any finite set $A \subset G$, let

$$\partial_S A = \{(x, s) \in G \times S : x \in A,\ xs \notin A\}$$

be the boundary of A. The isoperimetric profile $I_S^{\uparrow}$ is defined by

$$I_S^{\uparrow}(n) = \inf_{\#A \geqslant n} \#\partial_S A.$$

This is an increasing function of n.

The simple random walk on (G, S) is the Markov process $(X_i)_0^\infty$ with values in G which evolves as follows. If the current state is x, the next state is a neighbor of x chosen uniformly at random. This implicitely defines a probability measure $\mathbf{P}_S$ on $G^{\mathbb{N}}$ such that

$$\mathbf{P}_S(X_n = y / X_0 = x) = \mu_S^{(n)}(x^{-1}y)$$

where

$$\mu_S(g) = \frac{1}{\#S}\mathbf{1}_S(g)$$

and $\mu^{(n)}$ is the n-fold convolution power of μ. Following usual notation we will also write $\mathbf{P}_S^x(\cdot) = \mathbf{P}_S(\cdot / X_0 = x)$ for the law of the walk based on S and started at $x \in G$. To avoid parity problems, we consider only the probability of return at even times and set

$$\phi_S(n) = \mathbf{P}_S^{\mathrm{id}}(X_{2n} = \mathrm{id}) = \mu_S^{(2n)}(\mathrm{id}).$$

It is well known and easy to see that $\phi_S(n) = \max_g \mu_S^{(2n)}(g)$ and it follows that ϕ_S is a non-increasing function.

Given two positive non-increasing functions u, v defined on the positive real axis, write $u \preceq v$ if there exist $C \geqslant 1$ such that $\forall\, t > 0, \quad u(t) \leqslant Cv(t/C)$. Write $u \simeq v$ if $u \preceq v$ and $v \preceq u$. When a function is defined only on the integers, we extend it to the positive real axis by linear interpolation. We will use the same name for the original function and its extention. In particular, we view $V_S, I_S^{\dagger}, \phi_S$ as defined on the positive real axis. In this paper, we are interested in the behavior of $I_S^{\dagger}(n)$, $\phi_S(n)$ for large n up to the equivalence relation $\simeq$.

For completness we also consider the alternative isoperimetric profile J_S defined by

$$J_S(n) = \sup_{\#A \leqslant n} \frac{\#A}{\#\partial_S A}.$$

This is an increasing function of n. Group invariance shows that, in fact,

$$J_S(n) = \sup_{n/2 \leqslant \#A \leqslant n} \frac{\#A}{\#\partial_S A}.$$

Thus $n/J_S(n)$ gives a lower bound on the size of the boundary of sets of cardinality of order n. Note that whether or not $I_S^{\dagger}(n) \simeq n/J_S(n)$ for all Cayley graphs is not known.

We now recall the important notion of quasi-isometry. Let (G, S) and (H, T) be two Cayley graphs as above. A map $\psi : G \to H$ is called a quasi-isometry if there exists a constant $C > 0$ with the following properties:

1) $C^{-1}d_S(x, y) - C \leqslant d_T(\psi(x), \psi(y)) \leqslant C d_S(x, y)$ for all $x, y \in G$.

2) Any $h \in H$ is at distance atmost C of $\psi(G)$.

This notion clearly induces an equivalence relation between Cayley graphs; (G,S) and (H,T) are quasi-isometric if there exists a quasi-isometry ψ from G to H. Examples of quasi-isometric Cayley graphs are as follows. Two Cayley graphs (G,S), (G,T) of the same group G with different generating sets S,T are quasi-isometric. The Cayley graphs (G,S) and (H,T) where $H = G/F$ with F a finite normal subgroup of G are quasi-isometric. The Cayley graphs (G,S) and (H,T) where $H \subset G$ is a subgroup of finite index are quasi-isometric.

If (G,S) and (H,T) are quasi-isometric, it is well known and easy to see that $V_S \simeq V_T$, $I_S^\dagger \simeq I_T^\dagger$ and $J_S \simeq J_T$. Moreover, the authors recently proved that also $\phi_S \simeq \phi_T$. See [16]. In particular it follows that the behavior of $V_S, I_S^\dagger, J_S, \phi_S$ up to the equivalence relation $\simeq$ depends only of the group G and not of the choice of the finite generating set S.

The aim of this paper is to discuss in some detail the proof of the following theorem.

Theorem 1.1 *Assume that Γ is a discrete subgroup of a Lie group having finitely many connected components. Then*

either Γ *is non-amenable in which case, if Γ is finitely generated, then* $V(n) \simeq \exp(n)$ *and*

$$\phi(n) \simeq \exp(-n), \quad J(n) \simeq 1, \quad I^\dagger(n) \simeq (1+n);$$

or Γ *is amenable, finitely generated, and*

(a) *either $V(n) \simeq \exp(n)$, in which case*

$$\phi(n) \simeq e^{-n^{1/3}}, \quad J(n) \simeq \log_*(n), \quad I^\dagger(n) \simeq \frac{1+n}{\log_*(n)};$$

(b) *or $V(n) \not\simeq \exp(n)$, in which case there exists an integer d such that* $V(n) \simeq (1+n)^d$ *and*

$$\phi(n) \simeq (1+n)^{-d/2}, \quad J(n) \simeq (1+n)^{1/d}, \quad I^\dagger(n) \simeq (1+n)^{1-1/d}.$$

The proof of this theorem is divided in two parts of very different flavor. The first part of the proof consists in showing that classical structure theorems imply that only two possibilities arise: either Γ is non-amenable or it contains a polycyclic group of finite index. The second part consists in proving the desired estimates for polycyclic groups.

We know recall the ingredients of the first part of the proof.

Theorem 1.2 *Let G be a Lie group with a finite number of connected components and let Γ be a discrete subgroup of G. The following properties are equivalent.*

1. *The group Γ contains a finite index subgroup which is polycyclic*

2. *The group Γ contains no free subgroup on two generators.*

The proof of this theorem is as follows. By Tits alternative [18], any subgroup Γ of a Lie group G with finitely many connected components either contains a free group on two generators or a finite index subgroup which is solvable. In the second alternative, if we assume in addition that Γ is discrete in G, then it follows from a theorem of Mostow [12] that Γ contains a polycyclic subgroup of finite index. See [25], Proposition 4.1 and also [14].

To explain how this result relates to Theorem 1.1, recall the celebrated theorems of Følner and Kesten [4, 11] asserting respectively that

(i) A finitely generated group is non-amenable if and only if $I^{\uparrow}(n) \simeq (1+n)$.

(ii) A finitely generated group is non-amenable if and only if $\phi(n) \simeq \exp(-n)$.

Recall also that any group containing a free group on two generators is non-amenable and that passing to a subgroup of finite index does not alter the behavior of $I^{\uparrow}, J$ or ϕ. Recall finally that any polycyclic group is finitely generated.

All these facts reduce the proof of Theorem 1.1 to the following set of results.

Theorem 1.3 *Assume that Γ is polycyclic. Then either there exists an integer d such that $V(n) \simeq (1+n)^d$ or $V(n) \simeq \exp(n)$.*

This theorem follows from the results of J. Wolf [25] and H. Bass [2]. Indeed, Wolf showed that any polycyclic group either contains a finite index nilpotent group or has exponential volume growth. Bass computed the exact degree of polynomial growth of any nilpotent group N, showing that $V(n) \simeq (1+n)^d$ where d is given by

$$d = \sum_1^c \ell \ \mathbf{rk}(N_\ell/N_{\ell+1})$$

where $N_1 = N$, $N_2 = [N, N]$, $N_3 = [N, N_2], \dots$ is the lower central series of N and c the class of nilpotency of N.

Theorem 1.4 *Let Γ be a finitely generated group. If $V(n) \simeq (1+n)^d$ then $I^{\uparrow}(n)^{1-1/d}$, $J(n) \simeq (1+n)^{1/d}$, $\phi(n) \simeq (1+n)^{-d/2}$.*

This result was first proved, in a somewhat different form, by N. Varopoulos [21]. More precisely, Varopoulos proved the estimates of $I^{\uparrow}, J$ and ϕ given in Theorem 1.4 for nilpotent groups. By Gromov theorem [5] and Bass' result quoted above, nilpotent groups are exactly those for which V has a polynomial behavior. Note that Theorem 1.4 can be proved without using Gromov's result. Note also that each of the conclusions in Theorem 1.4, that is $I^{\uparrow}(n)^{1-1/d}$, $J(n) \simeq (1+n)^{1/d}$ and $\phi(n) \simeq (1+n)^{-d/2}$ implies $V(n) \simeq (1+n)^d$. See [14] for details.

Theorem 1.5 *Assume that* Γ *is polycyclic and has exponential growth. Then*

1. *The isoperimetric profile satisfies* $J(n) \simeq \log_* n$, $I^{\uparrow}(n) \simeq n/\log_*(n)$.

2. *The probability of return after* $2n$ *steps satisfies* $\phi(n) \simeq \exp(-n^{1/3})$.

This theorem gathers results of Alexopoulos [1], Coulhon and Saloff-Coste [3], Pittet [13] and Varopoulos [21]. Some of the details of the proofs are discussed below.

Examples described in [14, 15] show that many other behaviors appear for $I^{\uparrow}$, J and ϕ in the classe of solvable groups having exponential growth.

2 Polycyclic groups having exponential growth.

The aim of this section is to discuss the proof of Theorem 1.5. We will distinguish two parts. The first part can be stated as follows.

Theorem 2.1 *Fix* $\alpha \in [0,1]$. *Let* Γ *be a finitely generated group satisfying* $V(n) \succeq \exp(n^\alpha)$. *Then*

$$\phi(n) \preceq \exp\left(-n^{\alpha/(\alpha+2)}\right) \tag{2.1}$$

and

$$I^{\uparrow}(n) \succeq \frac{1+n}{[\log(2+n)]^{1/\alpha}}, \quad J(n) \preceq [\log(2+n)]^{1/\alpha}. \tag{2.2}$$

Inequality (2.1) was first proved by N. Varopoulos in [21, 22]. Alternative proofs are given in [9] and [23].

Inequalities (2.2) were first proved by Coulhon and Saloff-Coste [3], improving previous results of [21]. We will not go into the detail of these proofs but observe that no deep algebra is involved.

The missing part of Theorem 1.5 is contained in the next statement.

Theorem 2.2 *Let* Γ *be a polycyclic group having exponential volume growth. Then*

$$\phi(n) \succeq \exp\left(-n^{1/3}\right) \tag{2.3}$$

and

$$I^{\uparrow}(n) \preceq \frac{1+n}{\log(2+n)}, \quad J(n) \succeq \log(2+n). \tag{2.4}$$

The lower bound (2.3) is due to G. Alexopoulos [1] whereas (2.4) was first proved by Pittet [13]. The proofs of these inequalities are discussed below in some detail. The proof of (2.4) in [13] depends on an unpublished result of Grunewald. We will indicate below how one can modify the proof in [13] and avoid using Grunewald's result.

3 Proof of Theorem 2.2: the algebraic part

The proof of Theorem 2.2 must take advantage of the special structure of polycyclic groups. Let G be a polycyclic group. Then there exists an integer m such that G^m is torsion free and G/G^m is finite [8]. Hence we can assume from now on that G is torsion free. According to a result of Malcev ([17], 15.1.4), any finitely generated polycyclic group G contains a (finitely generated) subgroup of finite index whose derived group is nilpotent. Thus, passing to a subgroup of finite index, we can assume that we have an exact sequence

$$0 \to N \to G \to \mathbb{Z}^r$$

where N is a finitely generated nilpotent group.

For the time being, assume that G is a finitely generated group which admits an exact sequence

$$0 \to N \to G \to \mathbb{Z}^r$$

with N finitely generated. Let B be a finite symmetric generating set of N. Let a_i, $i = 1, \ldots, r$, be elements of G which project onto the canonical basis of $\mathbb{Z}^r$ and set

$$A = \{a_i^\epsilon; i = 1, \ldots, r,\ \epsilon = \pm 1\} \tag{3.5}$$

and $S = A \cup B$. Clearly, S generates G. For any $k = (k_1, \ldots, k_r) \in \mathbb{Z}^r$, set

$$\mathbf{k} = a_1^{k_1} \cdots a_r^{k_r}.$$

By hypothesis, the set

$$K = \{\mathbf{k} : k \in \mathbb{Z}^r\} \subset G$$

is a section of $\mathbb{Z}^r$ in G. Given $\mathbf{k} = a_1^{k_1} \cdots a_r^{k_r} \in K$, we set $|\mathbf{k}| = \sum_1^r |k_i|$.

Clearly, any $x \in G$ can be written uniquely as $x = y\,\mathbf{k}$ with $y \in N$ and $\mathbf{k} \in K$. The next two lemmas address the following questions. Consider an element $x = s_1 \ldots s_\ell \in G$ written as a word of length ℓ in the alphabet S. Write $x = y\,\mathbf{k}$ with $y \in N$, $\mathbf{k} \in K$ as above. Can one controle the minimal length $|y|_B$ of $y \in N$ with respect to the generating set B? Can one controle $|\mathbf{k}|$?

Lemma 3.1 *Referring to the notation introduced above, there exists an integer q such that, for any $x \in G$ and any $s \in S$, if $x = y\,\mathbf{k}$ with $y \in N$, $\mathbf{k} \in K$, then*

$$xs = z\,\mathbf{k}' \text{ with } z \in N,\ \mathbf{k}' \in K \text{ and } |z|_B \leqslant |y|_B + q^{|\mathbf{k}|},\ |\mathbf{k}'| \leqslant |\mathbf{k}| + 1.$$

This lemma is proved in [13] and also in [1] although it is not explicitely stated there. See also [14] for details. The proof is elementary. The result described in this lemma will be explicitely used in the proof of (2.4). Here we note that Lemma 2.3 easily implies by induction that any $x = s_1 \cdots s_\ell \in G$ can be written $x = y\,\mathbf{k}$ with $y \in N$, $\mathbf{k} \in K$ and $|y|_B \leqslant q^\ell$, $|\mathbf{k}| \leqslant \ell$. The next lemma is similar to this result but somewhat more subtle. It will be needed in the proof of (2.3).

Lemma 3.2 *Referring to the setting introduced above, let ψ be the canonical projection $G \mapsto G/N = \mathbb{Z}^r$. Given a sequence $\sigma = (s_i)_1^\ell \in S^\ell$, let*

$$k(i) = \psi(s_1 \cdots s_i) = (k_1(i), \ldots, k_r(i)),\ \ \mathbf{k}(i) = a_1^{k_1(i)} \cdots a_r^{k_r(i)}$$

and set

$$t(\sigma) = \max_{1 \leqslant i \leqslant \ell} |\mathbf{k}(i)|.$$

Then there exists an integer q such that, for any integer ℓ and any sequence $\sigma = (s_i)_1^\ell \in S^\ell$, we have $s_1 \ldots s_\ell = y\,\mathbf{k}$ with $y \in N$, $\mathbf{k} \in K$ and

$$|y|_B \leqslant \ell q^{t(\sigma)},\ \ |\mathbf{k}| \leqslant t(\sigma).$$

Proof: We proceed by induction on ℓ. If $\ell = 0$ there is nothing to prove. Assuming that the desired result is true for any sequence of length ℓ of elements of S, we prove it for $\sigma = (s_1, \ldots, s_{\ell+1})$, $s_i \in S$. To this end, write

$$x = s_1 \cdots s_{\ell+1} = x' s_{\ell+1}$$

with $x' = s_1 \cdots s_\ell$. By the induction hypothesis, $x' = y'\,\mathbf{k}'$ with $|y'|_B \leqslant \ell\, q^{t(\sigma')}$, $|\mathbf{k}'| \leqslant t(\sigma')$, $\sigma' = (s_1, \ldots, s_\ell)$. By Lemma 2.3, we have

$$x = x' s_{\ell+1} = y'\,\mathbf{k}'\, s_{\ell+1} = y\ \mathbf{k}$$

with

$$|y|_B \leqslant |y'|_B + q^{|\mathbf{k}'|} \leqslant \ell\, q^{t(\sigma')} + q^{|\mathbf{k}'|} \leqslant (\ell + 1) q^{t(\sigma')}$$

and $\mathbf{k} = \psi(s_1 \cdots s_{\ell+1})$ thus $|\mathbf{k}| \leqslant t(\sigma)$. The proof is complete since $t(\sigma') \leqslant t(\sigma)$.

4 Følner sets

Assume that G admits an exact sequence

$$0 \to N \to G \to \mathbb{Z}^r$$

where N is a finitely generated nilpotent group. We want to construct finite sets such that

$$\#\partial A \leqslant \frac{C\#A}{\log(2+\#A)}.$$

When G has exponential volume growth we know from Theorem 2.1 that this inequality is optimal: such sets are Følner sets, up to a multiplicative constant.

In order to produce finite sets having small boundary in G we need first to produce such sets in N. This is done in the next lemma.

Lemma 4.1 *Let (N, B) be the Cayley graph of a finitely generated group N. Assume that $V_B(n) \simeq (1+n)^d$ for some $d > 0$. Then, for any integer $R \geqslant 2$, there exists an increasing sequence of finite sets $D(i)$ such that*

1. *For each i, $D(i) = B(\mathrm{id}, \rho(i))$ for some $\rho(i)$ satisfying*

$$R^{2i} \leqslant \rho(i) \leqslant R^{2(i+1)} - (R^2-1)R^i.$$

In particular there exist c_0, C_0 such that $c_0 R^{2di} \leqslant \#D(i) \leqslant C_0 R^{2di}$.

2. *There exists C_1 such that, for each i and $\rho(i)$ as above, the set $C(i) = B(\mathrm{id}, \rho(i) + (R^2-1)R^i) \setminus B(\mathrm{id}, \rho(i))$ satisfies $\#C(i) \leqslant C_1 R^{(2d-1)i}$.*

Proof: Consider the annulus

$$A(i) = B(\mathrm{id}, R^{2(i+1)}) \setminus B(\mathrm{id}, R^{2i}).$$

Subdivise this annulus into the union of the annulii

$$A(i,\ell) = B(\mathrm{id}, R^{2i} + (\ell+1)R^i(R^2-1)) \setminus B(\mathrm{id}, R^{2i} + \ell R^i(R^2-1))$$

where ℓ varies from 0 to $R^i - 1$. Then

$$\begin{aligned} \#A(i) &= \sum_{\ell=0}^{R^i-1} \#A(i,\ell) \\ &\geqslant R^i \min_{\ell\in\{0,\dots,R^i-1\}} \{\#A(i,\ell)\}. \end{aligned}$$

It follows that there exist a constant C_1 and an integer $\ell_0 \in \{0, \dots, R^i-1\}$ such that

$$\#A(i,\ell_0) \leqslant C_1 R^{(2d-1)i}.$$

Setting

$$D(i) = B(\mathrm{id}, \rho(i)), \quad \rho(i) = R^{2i} + \ell_0(R^2 - 1)R^i,$$

we find that $C(i) = A(i, \ell_0)$ and thus $\#C(i) \leqslant C_1 R^{(2d-1)i}$ as desired. This ends the proof of Lemma 2.5.

Proof of (2.4): We will prove the following more precise result.

Lemma 4.2 *Let G be a finitely generated group. Assume that there is an exact sequence*

$$0 \to N \to G \to \mathbb{Z}^r$$

with N a nilpotent finitely generated group. Let S be a finite symmetric generating set of G. Then there exist a constant C, an integer M, and an increasing sequence of finite sets $F(i)$ with the following properties.

1. *For each positive integer j there is at least one i such that*

$$M^j \leqslant \#F(i) \leqslant M^{j+1}.$$

2. *For each integer i,*

$$\#\partial F(i) \leqslant \frac{C\#F(i)}{\log_*[\#F(i)]}.$$

Before embarking with the proof, observe that this yields

$$J(n) \succeq \log_*(n), \quad I^{\uparrow}(n) \preceq \frac{n}{\log_* n}$$

for any group G as in the lemma. By passing to a subgroup of finite index as explained at the beginning of Section 2.1, this proves (2.4).

Returning to the proof of Lemma 2.6, let B a finite generating set of N, and let A be as in (2.5). Without loss of generality we can assume that $S = A \cup B$. The group N being nilpotent there exists a $d \geqslant 0$ such that the volume growth function of (N, B) satisfies $V_B(n) \simeq (1+n)^d$. Thus Lemma 2.5 applies to the Cayley graph (N, B). Let q be the integer given by Lemma 2.3. Fix $R = q$ and let $D(i) \subset N$ be the sets given by Lemma 2.5. In G, consider the finite subsets

$$F(i) = \{x = y\mathbf{k} \in G : y \in D(i) \subset N, \ \ \mathbf{k} \in K, |\mathbf{k}| \leqslant i\}.$$

These sets clearly form an increasing sequence and there exists two positive constants c_2, C_2 such that

$$c_2 i^r R^{2di} \leqslant \#F(i) \leqslant C_2 i^r R^{2di}.$$

Hence

$$\sup_i \frac{\#F(i+1)}{\#F(i)} < \infty$$

which shows that one can pick M large enough so that conclusion 1 of Lemma 2.6 is satisfied.

The main part of the proof is to estimate the size of $\partial F(i)$. Fix i. Let $\rho(i)$, $R^{2i} \leqslant \rho(i) < R^{2(i+1)}$ be the integer given by Lemma 2.5 so that $D(i) = B_N(\mathrm{id}, \rho(i)) \subset N$. By Lemma 2.3, if $y \in F(i)$ and $s \in S$, $ys = z\,\mathbf{k}'$ with $|z|_B \leqslant \rho(i) + R^i$ and $|\mathbf{k}'| \leqslant i+1$. It follows that $\partial F(i)$ is contained in the union of the sets

$$\{x = y\,\mathbf{k} \in G : y \in N,\ \mathbf{k} \in K,\ |y|_B \leqslant \rho(i),\ |\mathbf{k}| = i\} \times S$$

and

$$\{x = y\,\mathbf{k} \in G : y \in C(i),\ \mathbf{k} \in K,\ |\mathbf{k}| \leqslant i+1\} \times S$$

with $C(i) \subset N$ as in Lemma 2.5. Of course,

$$\#\{\mathbf{k} \in K,\ |\mathbf{k}| = i\} \simeq (1+i)^{r-1}.$$

Hence there exist constants C_3, C_4 such that

$$\#\partial F(i) \leqslant C_3\left((1+i)^{r-1}R^{2di} + (1+i)^r R^{(2d-1)i}\right) \leqslant C_4\left(\frac{1}{i} + R^{-i}\right)\#F(i).$$

Since $\log_*[\#F(i)] \simeq i$, it follows that there exists a constant C such that

$$\#\partial F(i) \leqslant \frac{C\#F(i)}{\log_*[\#F(i)]}.$$

This ends the proof of Lemma 2.6.

5 Lower bound on ϕ

In this section we prove the lower bound (2.3) for the probability of return to the starting state. We follow the proof of [1]. By invariance under quasi-isometry and as explained at the beginning of Section 2.1, it suffices to prove this result for groups G admitting an exact sequence

$$0 \to N \to G \to \mathbb{Z}^r$$

where N is a finitely generated nilpotent group. Let ψ denote the canonical projection $\psi : G \mapsto G/N = \mathbb{Z}^r$. Let us recall the notation introduced in Section 2.1. Fix a symmetric generating set B in N and $a_1, \ldots, a_r \in G$ such that $\psi(a_1) = e_1, \ldots, \psi(a_r) = e_r$ where $(e_1, \ldots, e_r)$ is the canonical basis of $\mathbb{Z}^r$.

Set $S = A \cup B$ where $A = \{a_i^{\pm 1}, i = 1, \ldots, r\}$. Recall that $K = \{\mathbf{k} : k \in \mathbb{Z}^r\}$ where

$$\mathbf{k} = a_1^{k_1} \cdots a_r^{k_r}$$

and that any $x \in G$ can be written uniquely as $x = y\mathbf{k}$ with $y \in N$ and $\mathbf{k} \in K$. For $(k_1, \ldots, k_r) \in \mathbb{Z}^r$ set $|k| = |\mathbf{k}| = \sum_1^r |k_i|$.

Set $\mu(g) = \frac{1}{\#S}\mathbf{1}_S(g)$ and consider the left-invariant random walk $(X_n)_{n \geqslant 0}$ on G with independent increments $X_n^{-1}X_{n+1}$ of law μ. In other words, $X_n = \xi_0\xi_1 \ldots \xi_n$ where each ξ_i, $i \geqslant 1$, is picked independently and uniformly (i.e., according to μ) in S. This induces a family of probability measure $\mathbf{P}_x$, indexed by the starting point $x \in G$, on the path space $G^{\mathbb{N}}$ so that $\mathbf{P}_x(X_n = y) = \mu^{(n)}(x^{-1}y)$.

Let $\alpha = \#A$, $\beta = \#B$ so that $\#S = \alpha + \beta$, and consider the random variable $Z_n = \psi(X_n) \in \mathbb{Z}^r$. Then

$$Z_n = Z_0 + \sum_1^n \zeta_i$$

where $\zeta_i = \psi(X_{i-1}^{-1}X_i) = \psi(\xi_i)$ with ξ_i is distributed according to μ. Thus $(Z_n)_{n \geqslant 0}$ can be interpreted as a random walk on $\mathbb{Z}^r$ with independent identically distributed increments ζ_i of law ν given by

$$\nu(\zeta = 0) = \frac{\beta}{\alpha + \beta}, \quad \nu(\zeta = \pm e_j) = \frac{\alpha}{2r(\alpha + \beta)}, \; j = 1, \ldots, r. \tag{5.6}$$

The law ν induces a family of probability measures $\mathbb{P}_z$, indexed by the starting point $z \in \mathbf{Z}^r$, on the path space $[\mathbb{Z}^r]^{\mathbb{N}}$ so that $\mathbb{P}_z(Z_n = z') = \nu^{(n)}(z' - z)$.

Let

$$M_n = \max_{0 \leqslant j \leqslant n} |Z_j|. \tag{5.7}$$

Let q be the integer given by Lemma 2.4 and consider the sets

$$D_m^\ell = \{x = y\,\mathbf{k} \in G : |y|_B \leqslant \ell\, q^m, \;\; |\urcorner| \leqslant m\}.$$

Lemma 2.4 implies the following.

Lemma 5.1 *With the notation introduced above, for all integers ℓ and m,*

$$\mathbf{P}_{\mathrm{id}}(X_\ell \in D_m^\ell) \geqslant \mathbb{P}_0(M_\ell \leqslant m).$$

Proof: For any sequence $\sigma = (s_1, \ldots, s_\ell)$, let

$$k(i) = \psi(s_1 \cdots s_i), \quad t(\sigma) = \max_{1 \leqslant i \leqslant \ell} |k(i)|.$$

Then Lemma 2.4 shows that

$$\{x = s_1 \cdots s_\ell : \sigma = (s_i)_1^\ell \in S^\ell,\ \ t(\sigma) \leqslant m\}$$
$$\subset \{x = y\,\mathbf{k} : y \in N,\ \mathbf{k} \in K,\ |y|_B \leqslant \ell q^m,\ |k| \leqslant m\}.$$

Thus

$$\mathbb{P}_0(M_\ell \leqslant m) = \mathbf{P}_{\mathrm{id}}\left(\sup_{1\leqslant i\leqslant \ell} |Z_i| \leqslant m\right) \leqslant \mathbf{P}_{\mathrm{id}}(X_\ell \in D^\ell_m).$$

To finish the proof of (2.3), we need the following result concerning random walk on $\mathbb{Z}^r$.

Lemma 5.2 *Let ν be the probability measure on $\mathbb{Z}^r$ given by (2.6) with $\alpha, \beta >$ 0. Consider the random walk $(Z_n)_{n\geqslant 0}$ generated by ν and let $(M_n)_{n\geqslant 0}$ be as in (2.7). Then there exists a constant C such that*

$$\mathbb{P}_0(M_\ell \leqslant m) \geqslant e^{-C\ell/m^2}.$$

This lemma is essentially well known although it seems hard to find a reference in text books. A proof is outlined in [1, 19] and details can be found in [14].

End of the proof of (2.3): Recall that

$$\mu^{(2\ell)}(\mathrm{id}) = \sup_{g\in G} \mu^{(2\ell)}(g).$$

Thus, for any finite set $D \subset G$,

$$\mathbf{P}_{\mathrm{id}}(X_{2\ell} \in D) = \sum_{g\in D} \mu^{(2\ell)}(g) \leqslant [\#D]\, \mu^{(2\ell)}(\mathrm{id}).$$

Applying this to $D = D^{2\ell}_m$, and using Lemma 2.7, we get

$$\phi_S(\ell) = \mu^{(2\ell)}(\mathrm{id}) \geqslant \left[\#D^{2\ell}_m\right]^{-1} \mathbb{P}_0(M_{2\ell} \leqslant m). \tag{5.8}$$

Since the group N is nilpotent, $V_B(n) \simeq (1+n)^d$ for some integer $d \geqslant 0$. Thus there exists a constant C_0 such that

$$\#D^{2\ell}_m \leqslant C_0(1+m)^r(1+\ell)^d q^{dm} \leqslant C_0(1+\ell+m)^{r+d} q^{dm}. \tag{5.9}$$

By Lemma 2.8, there exists a constant C_1 such that $\mathbb{P}_0(M_{2\ell} \leqslant m) \geqslant e^{-C_1\ell/m^2}$. Then (2.8), (2.9) yield

$$\phi_S(\ell) \geqslant \exp\left(-C_1\frac{\ell}{m^2} - C_2\, m - C_3 \log \ell\right)$$

for all $\ell, m \geqslant 2$ with $C_2 = d\log q$, $C_3 = \log(r+d)$. Taking $m = \ell^{1/3}$ yields the desired result (2.3).

References.

[1] Alexopoulos G. (1992) *A lower estimate for central probabilities on polycyclic groups.* Canadian Math. J., 44, 897-910.

[2] Bass H. (1972) *The degree of polynomial growth of finitely generated nilpotent groups.* Proc. London Math. Soc., 25, 603-614.

[3] Coulhon T. and Saloff-Coste L., (1993) *Isopérimétrie sur les groupes et les variétés.* Rev. Math. Iberoamericana, 9, 293-314.

[4] Følner E., (1955) *On groups with full Banach mean value.* Math. Scand. 3, 243-254.

[5] Grigorchuk R. (1991) *On growth in group theory.* In Proceeding of the International Congress of Mathematicians, Kyoto, 1990.

[6] Gromov M. (1981) *Groups of polynomial growth and expanding maps.* Publ. Math. IHES, 53, 53-73.

[7] Gromov M. (1993) *Asymptotic invariants of infinite groups.* Geometric group Theory, Vol. II, (G.A Niblo & M. A. Roller Eds.) London Mathematical Society Lecture Note Series 182, Cambridge University Press.

[8] Hall Ph. (1957) *Nilpotent groups.* In Collected Works of Philip Hall, Oxford Univesity Press.

[9] Hebisch W. and Saloff-Coste L., (1993) *Gaussian estimates for Markov chains and random walks on groups.* Ann. Prob. 21, 673-709.

[10] Kanai M. (1985) *Rough isometries, and combinatorial approximations of geometries of non-compact riemannian manifolds.* J. Math. Soc. Japan, 37, 391-413.

[11] Kesten H., (1959) *Symmetric random walks on groups.* Trans. Amer. Math. Soc. 92, 336-354.

[12] Mostov G.D. (1954) *Fundamental groups of homogeneous spaces.* Ann. of Math. 60, 1-27.

[13] Pittet Ch. (1995) *Følner sequences on polycyclic groups.* Rev. Math. Iberoamericana, 11, 675-685.

[14] Pittet Ch. and Saloff-Coste L. (1997) *A survey on the relationships between volume growth, isoperimetry, and the behavior of simple random walk on Cayley graphs, with examples.* Preprint.

[15] Pittet Ch. and Saloff-Coste L. (1997) *Amenable groups, isopreimetric profiles and random walks.* In Proceedings of the 1996 Canberra Geometric group Theory Conference.

[16] Pittet Ch. and Saloff-Coste L. (1997) *On the stability of the behavior of random walks on groups.*

[17] Robinson D.J.S. (1993) *A Course in the Theory of Groups.* Graduate texts in Mathematics, Springer.

[18] Tits J. (1972) *Free subgroups in linear groups.* J. Algebra, 20, 250-270.

[19] Varopoulos N. (1983) *A potential theoretic property of solvable groups.* Bull. Sci. Math. 108, 263-273.

[20] Varopoulos N. (1986) *Théorie du potentiel sur des groupes et des variétés.* CRAS 302, série I, 203-205.

[21] Varopoulos N. (1991) *Analysis and geometry on groups* In Proceeding of the International Congress of Mathematicians, Kyoto, 1990.

[22] Varopoulos N. (1991) *Groups of superpolynomial growth.* In Proceeding of the Kyoto ICM satellite conference on Harmonic Analysis, Springer.

[23] Varopoulos N., Saloff-Coste L. and Coulhon T. *Analysis and geometry on groups.* Cambridge University Press.

[24] Woess W. (1994) *Random walks on infinite graphs and groups — A survey on selected topics.* Bull. London Math. Soc. 26, 1-60.

[25] Wolf J. (1968) *Growth of finitely generated solvable groups and curvature of Riemannian manifolds.* J. Diff. Geom., 2, 424-446.

Ch. Pittet

CNRS, Laboratoire Emile Picard

Université Paul Sabatier, Toulouse

L. Saloff-Coste

CNRS, Statistique et Probabilités

Université Paul Sabatier, Toulouse

Distance distortion on Lie groups

Nicholas Th. Varopoulos

0. Introduction

Let G be a locally compact group that is generated by some symmetric compact neighborhood $\Omega \subset G$ of the identity e. For every $g \in G$ we define the length, or the distance from e, of an element $g \in G$

$$|g| = \inf\{n;\ \ \Omega^n = \Omega \cdots \Omega \ni g\}\,. \tag{0.1}$$

where we set $\Omega^\circ = \{e\}$.

The above definition depends, of course, but obviously in an inessential way, on Ω. From this a left invariant distance can be defined on G by $d(x,y) = |x^{-1}y|$.

Let $H \subset G$ be a closed subgroup that is also assumed to be compactly generated. For every $h \in H$ we can then define two different lengths $|h|_G$ and $|h|_H$ depending on whether we define $|h|$ as in (0.1) considering h as an element of G or H. It is clear that $|h|_G \leqslant C|h|_H$ for some $C > 0$ that does not depend on h. It is known that no estimate in the opposite direction can be given (cf. [1]).

Let $\phi : \mathbb{R}_+ \to \mathbb{R}_+$ be some positive function. We shall say that H has ϕ distance distortion in G if:

$$|h|_H \leqslant C\phi(C|h|_G) + C \quad \forall\ \ h \in H$$

for some $C > 0$ (that is independent of h). In this paper ϕ will only admit "three different values":

(i) $\phi(t) \equiv t$ for $t > 0$ and then we shall say that we have 0-distance distortion between H and G (0-distortion in short).

(ii) $\phi(t) = t^c$ for some $c > 0$ and then we shall say that we have polynomial distortion.

(iii) $\phi(t) = e^t$ and then we shall say that we have exponential distortion.

In this paper I shall prove

Theorem G. *Let G be some connected real Lie group and let $H \subset G$ be some solvable closed subgroup. Then H is compactly generated and has exponential distortion.*

The above theorem admits a number of important, but almost immediate, corollaries.

Corollary 1 *The conclusion of Theorem G holds if we assume that H is connected but not necessarily solvable.*

Corollary 2 *The conclusion of Theorem G holds if G is a discrete polycyclic group and H any subgroup (cf. [3]).*

Corollary 3 *The conclusion of Theorem G holds if we assume that H is a closed amenable subgroup (cf. Proposition in §2).*

We have in particular:

Theorem A. *Let G be some amenable connected Lie group and let $H \subset G$ be some closed subgroup. Then H is compactly generated and has exponential distance distortion in G.*

The definition of amenability is standard (cf. [4]). What is less standard is the definition of an R-group. These are the connected Lie groups for which $adX : \mathfrak{r} \to \mathfrak{r}$, $X \in \mathfrak{r}$ (= the Lie algebra of R) has only pure imaginary eigenvalues. We shall extend that definition, in the obvious way, for Lie groups that have only a finite number of components. It is easy to see (cf. [5]) that R-groups are amenable. We have then:

Theorem R. *Let G be an R-group and let $H \subset G$ be some closed subgroup. Then H is compactly generated and has polynomial distance distortion in G.*

The above theorem has an obvious converse and gives therefore a characterization of connected R-groups.

More generally, if G is a locally compact group that is generated by some compact symmetric neighborhood Ω of e, we say that G has polynomial volume growth if

$$\gamma(n) = \text{Haar measure of } \Omega^n = \mathcal{O}\left(n^c\right) \quad \forall n = 1, 2, \cdots$$

for some $c > 0$. It is then well known that if G as above is a Lie group (with possibly infinitely many connected components) and if G_0 is the connected component of the identity then, if G has polynomial volume growth, then G_0 is an R-group and G/G_0 is nilpotent by finite (cf. [5], [23]). From these facts it is very easy to deduce from Theorem R the following:

Corollary *Let G be some locally compact compactly generated group of polynomial volume growth and let $H \subset G$ be some closed subgroup. Then*

H is compactly generated and the distance distorsion between G and H is polynomial.

A number of side results will be proved in the course of the proofs (e.g. if G is a compact extension of $\mathbf{R}^n$ then every subgroup has 0-distortion). To keep the presentation reasonably light I shall not state these results here.

For counterexamples, background material, and the proof of some special cases of the above Theorems the reader could consult [1], [6], [7], [8].

Three parts can be distinguished in this paper:

1) The Appendix A.1-A.4: where we collect together standard facts most of which are known to the experts but I was unable to find appropriate references.
2) §1 - §3: where we use the results from the Appendix to reduce the proof of Theorem G to Theorem R.
3) §4 - §7: where we develop an "ad hoc" geometric machinery to give the proof of Theorem R.

§4 - §7 can be read independently. These sections very nearly give a classification of the closed subgroups of a special but important class of R-groups.

As a closing remark I should point out that Corollary 2 on polycyclic groups can be proved independently by methods involving *discrete* groups. If one imitates the proof that we have given to reduce Theorem G to Theorem R one can reduce the problem to nilpotent groups. The problem is then much simpler and most of the machinery developed in this paper becomes redundant. Seeing things in this light, one realizes that the main geometric difficulty of our problem lies in controlling, in a quantitative way, the winding of $\mathbf{R}^m$ as an analytic subgroup of $\mathbf{T}^n$.

1. The Exponential Coordinates of the Second Kind

Let $\mathfrak{r}$ be a solvable real Lie algebra and let us choose some basis $(X_1, \cdots, X_n) \subset \mathfrak{r}$ such that each $I_j = \text{span}\,(X_1, \cdots, X_j)$ is an ideal in I_{j+1}. The exponential coordinates of the second kind

$$t = t(\gamma) = (t_1, \cdots, t_n) \in \mathbf{R}^n \quad \forall \quad \gamma \in R$$

can then be defined on the simply connected Lie group R that corresponds to $\mathfrak{r}$ (cf. [9]). We have then

$$\gamma = \exp(t_1 X_1) \cdots \exp(t_n X_n) \tag{1.1}$$

that gives a bijective diffeomorphism between R and $\mathbf{R}^n$.

It is of course evident that:

$$|\gamma|_R \leqslant |t(\gamma)|_{\mathbf{R}^n} \quad \forall \quad \gamma \in R. \tag{1.2}$$

It is also easy to see that:

(i) If $(X_1, \cdots, X_m)$ is a basis of the nilradical of $\mathfrak{r}$, for some $1 \leqslant m \leqslant n$, then
(cf. [6], [7], [8])

$$|t(\gamma)|_{\mathbf{R}^n} \leqslant C \exp(C|\gamma|_R) + C \quad \forall \quad \gamma \in R \tag{1.3}$$

for some $C > 0$ independent of γ.

(ii) If $\mathfrak{r}$ is an R-algebra then

$$|t(\gamma)|_{\mathbf{R}^n} \leqslant C|\gamma|_R^C + C \quad \forall \quad \gamma \in R; \tag{1.4}$$

one simply uses induction on the dimension to see this (cf. [6], [7], [8]).

Using these facts one can very easily prove Theorems G and R for simply connected solvable Lie groups (cf. 3.18.12 of [9]).

Let now R be some general connected solvable (real) Lie group and let N be the closed nilradical. We have then $R/N \cong \mathbf{R}^a \times \mathbf{T}^b$. It follows therefore that we can find $R_1 \subset R$ a co-compact (normal) closed subgroup such that $R_1/N \cong \mathbf{R}^a$. We shall concentrate on these groups and assume for the rest of this section that $R/N \cong \mathbf{R}^a$. Let $T \subset N$ be the maximal torus (cf. A4) and let $(X_1, \cdots, X_n)$ the basis of $\mathfrak{r}$ be chosen so that $(X_1, \cdots, X_p)$ and $(X_1, \cdots, X_m)$ (for some $1 \leqslant p \leqslant m \leqslant n$) are bases of the Lie algebras of T and N respectively. If we chose appropriate exponential coordinates of the second kind for $\widetilde{R}$ (the universal cover of R) we easily see that we obtain exponential coordinates of the second kind for R itself of the form:

$$\begin{gathered}\gamma = \exp(t_1 X_1) \cdots \exp(t_n X_n) \quad \forall \quad \gamma \in R \\ (0 \leqslant t_j < 2\pi \quad \forall \quad 1 \leqslant j \leqslant p \quad t_k \in \mathbf{R} \quad , \quad \forall p+1 \leqslant k \leqslant n).\end{gathered} \tag{1.5}$$

The estimates (1.2) (1.3) (1.4) clearly still hold and arguing as in [9] §3.18 we obtain that any closed subgroup $T \subset R_1 \subset R$ is exponentially distorted and also that R_1 is polynomially distorted if R is an R-group. These facts can also be seen by simply considering $R_1/T \subset R/T$ that are simply connected groups.

It is obviously vital that the exponential coordinates (1.1), (1.5) are written down in the correct order (i.e. $I_1 \subset I_2 \subset \cdots \subset \mathfrak{t} \subset I_{p+1} \cdots$ where $\mathfrak{t}$ is the Lie algebra of T). But some variations on that order are possible. By considering γ^{-1} we can, for instance, consider the product of the one parameter subgroups the "other way around":

$$\gamma = \exp(t'_n X_n) \cdots \exp(t'_1 X_1)$$

and clearly $t'(\gamma) = -t(\gamma^{-1}) \in \mathbf{R}^n$. But in (1.1) or (1.5) we can also clearly write:

$$\gamma = \exp(t''_m X_m) \cdots \exp(t''_1 X_1) \exp(t_{m+1} X_{m+1}) \cdots \exp(t_n X_n);$$

$$(t''_1, \cdots, t''_m, t_{m+1} \cdots) \in \mathbf{R}^n, \quad \gamma \in R.$$

These remarks will be used in §6.

We shall also need later on a slight refinement of the above easy distance distortion results. We shall assume that G is a connected real Lie group that contains some closed normal connected and simply connected solvable R-group $R_0 \subset G$ such that G/R_0 is compact. Let $R_1 \subset R_0$ be another closed and simply connected subgroup that is normal in G and let

$$\pi : G \to G/R_1 = G^*$$

be the canonical projection. Let further $G_1^* \subset G^*$ be some compact subgroup and $G_1 = \pi^{-1}(G_1^*) \subset G$. Then we can assert that the distance distortion between G and G_1 is polynomial. Indeed there exist some compact subset $K \subset\subset G_1$ such that $G_1 = R_1 K$. Let $g_1 = r_1 k \in G_1$ $(r_1 \in R_1 \ , \ k \in K)$. We have then:

$$|g_1|_{G_1} \approx |r_1|_{G_1} \approx |r_1|_{R_1} \underset{p}{<} |r_1|_{R_0} \approx |r_1|_G \approx |r_1 k|_G = |g_1|_G$$

[Here the notation $\underset{p}{<}$ indicates "polynomial distortion". This notation is formally defined and systematically used in §6. The sign $\approx$ indicates, as usual, that the two quantities are within multiplicative constants.] This gives the required polynomial distorsion between G and G_1.

2. Technical Lemmas

The reader will have to refer constantly to the Appendix for the definitions and for the background material that is used in this section.

Lemma 1. *Let G be some compactly generated locally compact group and let H be some closed subgroup. Then:*

(i) If H is co-compact in G then H is compactly generated and of 0-distortion.

(ii) If there exist finitely many central elements $z_1, \cdots, z_n \in G$ such that $G = \overline{Gp(H; z_1, \cdots, z_n)}$ then H is compactly generated and has 0-distortion $z_1, \cdots, z_n \in G$. Furthermore if we assume that H is amenable then G is amenable too.

(iii) If $G \cong H \propto L$ for some locally compact group L and if we assume that H is compactly generated then the distance distortion between G and H is at most exponential. If we assume in addition that G is a connected R-group then that distance distortion is polynomial.

(iv) If $H \subset H_1 \subset G$ is another closed subgroup that contains H co-compactly and if we assume that H is compactly generated and that has

exponential (resp. polynomial; 0) distance distortion with G then H_1 is also compactly generated and has exponential (resp. polynomial; 0) distance distortion with G.

The above Lemma (part (ii)) has the following:

Corollary. *Let $\pi : G_1 \to G_2$ be an almost embedding [cf. A.1] between connected Lie groups and assume that $H \subset G_1$ is some closed subgroup such that $\overline{\pi(H)}$, the closure of $\pi(H)$ in G_2, is compactly generated. Then H is compactly generated. If we assume furthermore that the distance distortion between $\overline{\pi(H)}$ and G_2 is exponential (resp. polynomial; 0) then the same holds for the distance distorsion between H and G_1.*

Proof (i) Let $h \in H$ and $e = g_0, g_1, \cdots, g_n \in G$ such that $g_n = h$, $n \leqslant C|h|_G$; let further $h_j \in H$ be such that $g_j \in h_j K$, $g_j^{-1} g_{j+1} \in K$ where $C > 0$ and $K \subset\subset G$ is some fixed compact subset independent of h. We clearly have $h_j^{-1} h_{j+1} \in \left(K^2 K^{-1}\right) \cap H$. This completes the proof.

(ii) This is proved by induction on n. We may therefore assume that $n = 1$. In that case G_0/H can only be a compact group or $\mathbb{Z}$ (since is generated as a locally compact group by a single element (cf. A.3)). In the second case $G \cong H \oplus \mathbb{Z}$ in the first H is co-compact in G. In both cases our assertion follows.

(iii) Let $A \subset H$(resp. $B \subset L$) [so that $A^n = A \cdot A \cdots A$, $A^B = \{bab^{-1};\ a \in A,\ b \in B\} \subset H$] be a generating compact neighborhood of e in H (resp. L). Then clearly $A^B \subset A^n$ for some $n \geqslant 1$ and therefore $A^{B^p} \subset A^{n^p}$. For arbitrary $l_1, l_2 \cdots \in B$ we have

$$l_1 A l_2 A \cdots l_n A = A^{l_1} A^{l_1 l_2} \cdots A^{l_1 l_2 \cdots l_n} l_1 l_2 \cdots l_n$$

From these remarks the exponential distortion follows at once. The assertion about R-groups will not be needed (cf. [6], [8] for the details).

The proof of (iv) is evident.

Proof of the Corollary. Let $H_1 = \pi^{-1}\left(\overline{\pi(H)}\right)$. Then $H_1 = \overline{Gp\{H;\ z_1, \cdots, z_n\}}$ where $z_1, \cdots, z_n \in G_1$ are central elements and H_1 is clearly compactly generated (observe that the center Z of a connected Lie group G is compactly generated — indeed $Z \cap$ Rad G and Im (Z) in $G \to G/\text{Rad}\,(G)$ are compactly generated, cf. [12], [18]). It follows that H is compactly generated by Lemma (ii). We shall now prove the distance distortion result for the exponential distortion (the polynomial and 0-distortion can be treated similarly) and for that (again by Lemma (ii)) we may assume that $H = H_1 = \pi^{-1}\left(\overline{\pi(H)}\right)$ and therefore that $\pi(H) = H_2 \subset G_2$ is closed.

Let $h \in H$. Then $|\pi(h)|_{G_2} \leqslant |h|_{G_1}$ and therefore $|\pi(h)|_{H_2} \leqslant C\exp(C|h|_{G_1})$. We can thus find $z \in \operatorname{Ker}\pi$ such that $|hz|_H \leqslant C\exp(C|h|_{G_1})$. But this implies that $|z|_{G_1} \leqslant C'\exp(C|h|_{G_1})$. By the definition of an almost embedding this gives: $|z|_H \leqslant |z|_{\operatorname{Ker}\pi} \leqslant C\exp(C|h|_{G_1})$. Our assertion follows.

To see the significance of part (iv) in the above Lemma observe that in every connected real Lie group G we can find some solvable, compactly generated, co-compact closed subgroup. [Indeed we can clearly assume that $G = Q \propto NAK$ is simply connected where Q is its radical and NAK is the Iwasawa decomposition of the Levi factor. Then there exists a discrete subgroup $Z \subset K$ that is central (in G) and such that K/Z is compact (cf. [18] [13]). The group $QNAZ \subset G$ is then solvable]. From this we see in particular that Corollary 1 follows immediately from Theorem G.

Lemma 2. *Let G be some connected real Lie group and let $H \subset G$ be some closed amenable subgroup. Then H is compactly generated.*

By considering $\widetilde{G}$ the universal cover of G and by Lemma 1 (ii) and the appendix it is clear that we may assume that $G = GL(n;\mathbb{C})$. When H is assumed to be solvable the result is well known. Indeed if $\overline{H}$ is the Zariski closure of H (say the complex Zariski closure) in GL then $\overline{H}$ is solvable and has only a finite number of components; then the result is contained in ([12] 3.8).

To prove the Lemma for a general amenable subgroup H we shall need to use the following deep theorem of J. Tits [10].

Theorem (Tits). *Let G be some real Lie group with only finitely many components and let $\Gamma \subset G$ be an amenable subgroup (as a discrete group) but otherwise arbitrary. Then there exists a (normal) solvable subgroup of finite index $\Gamma_1 \subset \Gamma$.*

The above result is normally stated for $G \cong GL(n;\mathbb{C})$. But by passing to the universal cover of the component of the identity of G and by the results in A.1 we can easily reduce the above to the classical Tits theorem. [The fact that $\Gamma_1 \subset \Gamma$ can be assumed normal follows for free, so to speak, by considering $\bigcap_{\gamma\in\Gamma} \gamma^{-1}\Gamma_1\gamma = \bigcap_j \gamma_j^{-1}\Gamma_1\gamma_j$ where γ_j $(j = 1, \cdots, k)$ is a finite set of coset representatives].

To complete the proof of the Lemma in the general case let $H_0 \subset H \subset \overline{H} \subset GL$ be the component of the identity and the Zariski closure of H. Then $\overline{H}$ normalizes H_0 and the image of H by $\overline{H} \to \overline{H}/H_0$ is a (closed) discrete amenable group. It contains therefore a solvable subgroup of finite index, which, being closed, is finitely generated (cf. [12]). It follows that H/H_0 is finitely generated and this proves our assertion.

A proof of the following proposition can be found in [5]. I will however give below an independent proof.

Proposition *Let G be some Lie group with finitely many components and let $H \subset G$ be some closed amenable subgroup. Then H contains some co-compact normal solvable subgroup $H_1 \subset G$.*

When H is connected we can of course take H_1 to be the radical of H. The above proposition can therefore be thought as some kind of generalization of the Levi decomposition (cf. 3.18.13 of [9]) in this more general context.

Before we embark in the proof we shall make a reduction and show that for the proof we may assume that $G = GL(n; \mathbb{C})$ (for some $n \geqslant 1$).

Indeed by considering the universal cover of the component of the identity we see that we may assume that G is connected and simply connected, so that we have an almost embedding $\theta : G \to GL(n; \mathbb{R})$. The real Zariski closure $\overline{\theta(H)} \subset GL$ is then amenable [cf. A.2]. We let $X \subset \overline{\theta(H)}$ be the component of e in this group and, by considering a subgroup $H_1 \subset H$ of finite index, we may clearly assume that $H \subseteq \theta^{-1}(X)$. Now $\theta^{-1}(X) = \overline{Gp\left(\widetilde{X};\ \operatorname{Ker}\theta\right)}$ where $\widetilde{X} \subset G$ is some analytic subgroup that "lifts" X. It follows that $L = \left(\theta^{-1}(X)\right)_0$, the component of the identity, satisfies $L \supset \widetilde{X}$ and therefore that $\theta^{-1}(X)/L$ is a compactly generated, totally disconnected, abelian Lie group and is therefore isomorphic to $\mathbb{Z}^n\times$ (finite group) (cf. A.3). It is furthermore clear that the image of $(\operatorname{Ker}\theta)$ on $\theta^{-1}(X)/L$ is dense and therefore is the whole group. Therefore, by further reducing H to $H_2 \subset H_1 \subset H$, another subgroup of finite index, we may assume that H can be identified with some closed subgroup of $L \oplus Z$ where L is an amenable connected, which (by passing to the universal cover) may assumed to be simply connected, Lie group and Z is a finitely generated abelian discrete group. It follows therefore that we can find $\varphi : L \oplus Z \to GL(n, \mathbb{R})$ such that $\operatorname{Ker}\varphi$ is finite [to see this we use A.1 to almost embed L in GL with a finite kernel (or even zero kernel). We then use some obvious embedding of Z in GL and take the "direct sum embedding"]. The group $\varphi(H)$ is then a closed amenable subgroup of $GL(n; \mathbb{R})$ and since $\operatorname{Ker}\varphi$ is finite it is clear that it suffices to prove the proposition for $\varphi(H) \subset GL(n; \mathbb{R})$. This completes our reduction.

We shall now prove the proposition with $G = GL(n; \mathbb{C})$. Towards that we shall need

Lemma (a). *Let $H_0 \subset H \subset GL(n; \mathbb{C})$ be two closed subgroups such that H_0 is the component of the identity in H and such that the group H/H_0 is amenable. Then we can find some normal closed subgroup $H_0 \subset L \subset H$*

and $m \geqslant 1$ such that

$$[H : L] < +\infty \quad ; \quad L^m \subset H_0$$

(where we denote as usual $L^{m+1} = [L^m, L^m]$ the commutator subgroup and $L^0 = L$).

Proof. Consider $\overline{H}$ the Zariski closure of H in $GL(n; \mathbb{C})$ which by our hypothesis normalizes H_0. The group $\overline{H}/H_0$ has finitely many components and, if $\pi : \overline{H} \to \overline{H}/H_0$ is the canonical projection, we can find a solvable subgroup $\Gamma \subset \pi(H)$ of finite index (by Tits). Then the group $L = \pi^{-1}(\Gamma)$ satisfies all the conditions of the Lemma.

Lemma (b). *Let K be some locally compact group and let $K_0 \subset K$ be some closed normal subgroup that is isomorphic with some semisimple (real) compact Lie group with trivial center, let $C(K_0) \subset K$ be the centralizer of K_0 in K [i.e. $x \in C(K_0)$ if and only if $x \in K$ and $xk = kx \ \forall k \in K_0$]. Then $C(K_0)$ is normal and closed and satisfies*

$$C(K_0) \cap K_0 = \{e\} \ , \ [K : K_0 C(K_0)] < +\infty.$$

The above says that the product $C(K_0) \oplus K_0$ is direct and has finite index in K.

Proof. The first assertion is immediate. The second follows from the fact that if we quotient the group Aut K_0 of automorphisms of K_0 by the subgroup of Inner automorphisms of K_0 we obtain a finite group. This fact is classical and is usually formulated in terms of Lie algebras, namely that the group Aut $(\mathfrak{k}_0)$/ Int $(\mathfrak{k}_0)$ is finite, where now $\mathfrak{k}_0$ is the Lie algebra of K_0, (cf. Ch IX §5 of [18]). Observe that for every $\alpha \in$ Aut (K_0) the mapping $\alpha \mapsto d\alpha$ induces an identification Aut $(K_0) \subset$ Aut$(\mathfrak{k}_0)$ and that for that identification Inner Aut $(K_0) \subset$ Int $(\mathfrak{k}_0)$. Therefore this gives a mapping

$$\frac{\text{Aut}(K_0)}{\text{Inner Aut }(K_0)} \to \frac{\text{Aut }(\mathfrak{k}_0)}{\text{Int }(\mathfrak{k}_0)}$$

which is (1-1). Indeed if $\alpha \in$ Aut (K_0) is such that $d\alpha \in$ Int $(\mathfrak{k}_0)$ then by the definition of Int $(\mathfrak{k}_0)$ there exist $k_1, \cdots, k_n \in K_0$ (in some small enough neighborhood of e) such that $d\alpha =$ Ad $(k_1) \cdots$ Ad $(k_n) =$ Ad $(k_1 \cdots k_n)$ and therefore α is an inner automorphism. The fact that $C(K_0)$ is closed and normal is of course evident by the definition.

We are finally in a position to give the proof of the proposition (with $G = GL$). Let H_0 be the component of the identity of H let $Q \subset H_0$ be the radical and let

$$H' = \alpha^{-1}(Z(H_0/Q)) \subset H_0 \subset H$$

where $\alpha : H_0 \to H_0/Q$ is the canonical mapping. [i.e. the inverse image of the (finite) center of H_0/Q]. Let also $\pi : H \to H/H' = K$ be the canonical mapping and let $K_0 = \pi(H_0)$ be the component of the identity of K. It is then clear that $H_0 = \pi^{-1}(K_0)$ and that the conditions of Lemma (a) are verified. Therefore we can find a subgroup L such that:

$$H_0 \subset L \subset H \quad ; \quad [H : L] < +\infty \quad ; \quad L^m \subset H_0$$

(for some $m \geqslant 1$). Let $C(K_0) \subset K$ be the centralizer of K_0 in K and clearly, by the definition of H', K_0 and $C(K_0)$ satisfy the conditions and the conclusion of Lemma (b).
Wc have therefore:

$$\pi(L) \supset K_0 \quad ; \quad [K : \pi(L)] < +\infty \quad ; \quad [K : C(K_0) \oplus K_0] < +\infty.$$

We see therefore that the group

$$\pi(L) \cap (C(K_0) \oplus K_0) = (\pi(L) \cap C(K_0)) \oplus K_0$$

is of finite index in K:

$$[K : (\pi(L) \cap C(K_0)) \oplus K_0] < +\infty. \tag{2.1}$$

Let $H' \subset A = \pi^{-1}[C(K_0)] \cap L \triangleleft H$ which is clearly a normal closed subgroup of H such that

$$\pi(A) = \pi(L) \cap C(K_0).$$

It follows that, if $m \geqslant 1$ is large enough,

$$(\pi(A))^m = \pi(A^m) \subset \pi(L^m) \cap C(K_0) \subset K_0 \cap C(K_0) = \{e\}.$$

This means that $A^m \subset H'$ and since, by the definition, H' is solvable it follows that the group A is solvable too.

On the other hand, since $\operatorname{Ker} \pi = H' \subset A$, we have

$$H/A \cong \pi(H)/\pi(A) = K/\pi(L) \cap C(K_0).$$

This implies by (2.1) that H/A is a compact group. If we set $H_1 = A$ all the conditions of our proposition are verified.

3. Reduction of the Theorems.

Reductions of Theorems A and R.

We shall first show that in both theorems A and R we can reduce the proof to the special case where G is a solvable group.

First of all the subgroup H can certainly be assumed solvable because of the Proposition in §2 (or, if H is assumed to be connected, just by the Levi decomposition). We then consider $\pi : G \to G/Q$ where Q is the radical of G and $G^* = \pi^{-1}\left[\left(\overline{\pi(H)}\right)_0\right]$, where $\left(\overline{\pi(H)}\right)_0$ is the component of the identity of $\overline{\pi(H)}$, which, being a closed subgroup of the compact group G/Q, has only a finite number of components. The group G^* is co-compact in G and is connected and solvable (since $\left(\overline{\pi(H)}\right)_0$ is) and $[H : H \cap G^*] < +\infty$. Our reduction follows. (Observe incidentally that if we assume that H is connected then $H = H \cap G^*$).

Now let G be an arbitrary solvable connected real Lie group and let $N \subset G$ be its nilradical, let further $T \subset N$ be the maximal torus in N (cf. A.4) which is fully invariant in N and thus normal in G. We can then consider $\theta = G \to G/T = G^*$ which is also a connected solvable group.

Let now $H \subset G$ be an arbitrary closed subgroup. Then $\theta(H) \subset G^*$ is closed and is connected if H is. The fact that T is compact implies very easily that both Theorems A and R for $H \subset G$ will hold as soon as they are known to hold for $\theta(H) \subset G^*$.

If $N^* = N/T$ is the nilradical of G^* and if we denote by $\pi : G^* \to G^*/N^* \cong \mathbf{R}^a \times \mathbf{T}^b$ the canonical projection then the subgroup $R = \pi^{-1}(\mathbf{R}^a) \subset G^*$ is closed, co-compact, simply connected, and normal.

The conclusion is that we can reduce both Theorems A and R to G being solvable and of the above special form (i.e. $\exists R \triangleleft G,\ \ G/R$ compact and R simply connected solvable). [In this reduction the "connectedness of H" is preserved].

The second, much subtler set of reductions consists in assuming in both Theorems A and R that G is solvable and H closed, but not necessarily connected. We shall reduce this situation to the analogous problem where now H is assumed to be connected. First of all, by considering the universal cover $\pi : \widetilde{G} \to G$ and $\widetilde{H} = \pi^{-1}(H) \subset \widetilde{G}$ we see that we can assume that G is simply connected (even if H is connected the new $\widetilde{H}$ is not necessarily connected) and therefore almost embeddable: $\theta : G \to GL(n;\mathbf{R})$ in GL with Ker θ finite (cf. A.1), since G is amenable. In fact we can even assume that Ker $\theta = \{e\}$ (but finite kernel is just as good for our purposes). From this it follows that we can assume $H \subset G \subset GL(n;\mathbf{R})$ [since the finite Ker θ does not affect either Theorem A or R]. We can now consider $\overline{G}$, the real Zariski closure of G in $GL(n;\mathbf{R})$. $\overline{G}$ will be solvable if G is, and (less trivially) amenable if G is (cf. A.2). Furthermore, if G is an R-group, $\overline{G}$ will be an R-group. If $\left(\overline{G}\right)_0$ is the component of the identity of $\overline{G}$ and $H_1 = H \cap \left(\overline{G}\right)_0$ then $[H : H_1] < +\infty$ and we can apply the results of $A.3$ to $\left(\overline{G}\right)_0 \supset H_1$ to find $H_1 \subset \widehat{H} \subset \left(\overline{G}\right)_0$ closed subgroup with finitely many topological components that contains H_1 co-compactly.

If we combine the above with our previous considerations we finally see the following:

Reduction. *In both Theorems A and R we may assume that H is connected and that G is solvable and contains* $R \triangleleft G$ *a closed simply connected normal analytic subgroup such that* G/R *is compact.*

Observe that then we can even assume that $G = R \propto T$ where $T \cong \mathbf{T}^a$ (by Ch. XV 3.7 of [11] and the fact that the only compact subgroup of R is $\{e\}$). If we use the theory of the semisimple splitting (cf. [19]) we easily see that in the above reduction we can even assume R to be nilpotent. These additional further reductions do not seem to simplify the proofs of the Theorems. The details will therefore be omitted.

Reduction of Theorem G.

We shall make essential use of the following basic theorem of A. Borel (cf. [14], [20]).

Theorem (Borel). *Let G be some connected complex Lie group and let* $H \subset G$ *be some solvable subgroup (otherwise arbitrary). We can find* $B \subset G$ *some closed connected solvable co-compact subgroup such that* $[H : H \cap B] < +\infty$.

Indeed let Q be the radical of G and let $\pi : G \to G/Q$ be the canonical projection. G/Q is a complex affine algebraic group (cf. [14], [20]). Let $\overline{\pi(H)} \subset G/Q$ be the Zariski closure of $\pi(H)$ in G/Q and $H_0 = \left(\overline{\pi(H)}\right)_0$ the connected component of the identity. If $H_0 \subset B_0 \subset G/Q$ is some Borel subgroup (cf. [14]) then $B = \pi^{-1}(B_0)$ satisfies all the conditions of the above theorem. [We use here the fact that B_0 is connected, but we could also restrict attention to the component of the identity of B_0 and avoid having to use this (non trivial) fact].

Let now G be arbitrary and let $\theta : G \to G_c$ an almost embedding of G in some complex group as in A.1. By the corollary of Lemma 1 §2 it follows that we can reduce Theorem G for $H \subset G$ (where H is some closed subgroup) to $\overline{\theta(H)} \subset G_c$, which by A.1 can be identified to closed subgroups of $GL(n; \mathbf{R})$. I.e. we can reduce Theorem G to

$$H \subset G \subset GL(n; \mathbf{R}).$$

But then we can consider the real algebraic closure $\overline{G}$ of G and $H_1 = H \cap (\overline{G})_0$ as before and also some subgroup $H_1 \subset \widehat{H}_1 \subset (\overline{G})_0$ with finitely many components that contains H_1 co-compactly. Therefore we have the following:

Reduction. *In Theorem G we may assume that G is a connected solvable group and H is a connected subgroup.*

In this above reduction (we cannot, in general, assume that G is simply connected but:) we may assume that G is a semidirect product

$$G = U \cdot \mathbf{C}_*^a \subset GL(n; \mathbf{C})$$

where U is unipotent and $\mathbf{C}_*^a$ *is a complex torus.*

The last point follows by standard structure theorems for algebraic groups (over algebraically closed or 0-characteristic fields, cf. [17] Tome III Ch. V §3.5; [14], [20]).

Reduction of Theorem G to Theorem R.

Reduction. *The proof of Theorem G can be reduced to the proof of Theorem R.*

As we already saw, we may assume in Theorem G that: H is connected and

$$H \subset U \cdot \mathbf{C}_*^a \cong (U \propto \mathbf{T}^a) \propto \mathbf{R}^a = G_0 \propto \mathbf{R}^a \quad (\subseteq GL(n; \mathbf{C}) \subset GL(2n; \mathbf{R}))$$

where U is a simply connected nilpotent group and G_0 is an R-group (cf. A.2). [Observe that here G_0 is already embedded as a *real* algebraic group in $GL(2n; \mathbf{R})$ (cf. A.2) and therefore at this point we can avoid having to embed an arbitrary R-group in $GL(n; \mathbf{R})$ and use the fact that the Zariski closure is an R-group].

Let $H_0 = H \cap G_0$. We shall show that if we can assume the polynomial distortion between H_0 and G_0 to hold we can deduce the exponential distortion between H and G. This will complete the proof of the above reduction.

Let $\pi : G \to G/G_0 \cong \mathbf{R}^a$ and let $X_1, \cdots, X_n \in \mathfrak{h}$ (= the Lie algebra of H) be such that $d\pi(X_j)$ $(j = 1, \cdots, n)$ is a basis of the Lie algebra of $\pi(H)$. Let:

$$\sigma(t) = \exp(t_1 X_1) \cdots \exp(t_n X_n) \in H \qquad t = (t_1, \cdots, t_n) \in \mathbf{R}^n$$

Let $x \in H$. Then $|\pi(x)|_{\mathbf{R}^a} \leqslant |x|_G$. Therefore there exists $t \in \mathbf{R}^n$ such that

$$|t| \leqslant C|x|_G \quad ; \quad \pi(\sigma(t)) = \pi(x).$$

It follows that $\widetilde{x} = x\sigma(-t) \in H_0$ satisfies

$$|\widetilde{x}|_G \leqslant C|x|_G \ ; \ |x|_H \leqslant C(|\widetilde{x}|_H + |x|_G). \tag{3.1}$$

But by the exponential distortion between G and G_0 (cf. Lemma 1 §2) it follows that

$$|\widetilde{x}|_{G_0} \leqslant (C \exp C|x|_G).$$

If we assume that Theorem R holds for the pair $H_0 \subset G$, this in turn implies

$$|\widetilde{x}|_H \leqslant |\widetilde{x}|_{H_0} \leqslant C \exp(C|x|_G).$$

This, together with (3.1), proves the required exponential distortion $|x|_H \leqslant C \exp(C|x|_G)$ and completes the proof of the reduction.

A direct approach to Corollary 2.

Assume for simplicity, and without loss of generality (cf. [3], [12]), that Γ is a polycyclic discrete group and that $\Gamma_1 = [\Gamma, \Gamma]$ is nilpotent and such that $\Gamma/\Gamma_1 \cong \mathbb{Z}^n$. Then the argument we have just given to reduce Theorem G to Theorem R adapts very easily and reduces the proof of Corollary 2 to the case when the polycyclic groups are in fact nilpotent. This is because the distance distortion between Γ and Γ_1 can be directly seen to be exponential . This last point is not trivial but it can be seen by adapting the methods of [6] to the discrete situation. I shall not give the details but simply point out that the critical role that Cartan subalgebras played in [6] is now taken over by nilpotent almost complements in polycyclic groups (cf. [3]).

4. Euclidean by compact groups and the tangent at infinity.

Let G be some connected real Lie group and let $H \subset G$ be some closed normal subgroup that is isomorphic to $\mathbf{R}^n$. We shall tacitly use the exponential mapping to identify H with its Lie algebra $\mathfrak{h} \cong \mathbf{R}^n$. The Ad-action of G on $\mathfrak{h}$ induces clearly an action of G/H on $\mathfrak{h}$ and gives an identification of G/H with a subgroup of $GL(\mathfrak{h})$. We shall assume in this section that $\overline{(G/H)}$, the closure of G/H in $GL(\mathfrak{h})$, is compact.

Let $R \subset G$ be some solvable subgroup of G (but otherwise arbitrary). It is then well known (cf. 3.6 of [21], 3.7 of [9]) that we can find a subgroup $R_1 \subset R$ of finite index that has the following properties:

(i) H is completely reducible under the action of $\mathrm{Ad}(R_1)$ into 1 or 2-dimensional subspaces, i.e., we have the decomposition

$$H = \widetilde{H}_1 \oplus \cdots \oplus \widetilde{H}_q \oplus H_1 \oplus \cdots \oplus H_p$$

that is stable by the action of R_1 and $\dim \widetilde{H}_j = 1$, $\dim H_i = 2$.

(ii) If we assume in addition that R is connected, then we can take $R_1 = R$ and

$$\widetilde{H} = \widetilde{H}_1 \oplus \cdots \oplus \widetilde{H}_q \ ; \quad H = \widetilde{H} \oplus H_1 \oplus \cdots \oplus H_p \tag{4.1}$$

The action of R on $\widetilde{H}$ is trivial and the action of R on H_j can be identified with a "complex rotation":

$$\mathrm{Ad}(r)z = e^{i\theta_j(r)}z \quad \forall \quad z \in \mathbb{C}$$

where we identify H_j with the complex numbers $\mathbb{C}$ and $r \mapsto \theta_j(r)$ $(j = 1, \cdots, p)$ is a non trivial group homomorphism $R \to \mathbb{R}$ (mod 2π).

The proof of the above facts follows well known standard lines that use the solvability, e.g. Lie's theorem for (ii), and the compactness that guarantees that every invariant subspace of H has a direct complement. For simplicity, I shall assume in what follows that $R = R_1$. We shall now impose two additional conditions on R.

(a) If $\pi : G \to G/H$ is the natural projection then $\overline{\pi(R)} \subset G/H$ is compact (i.e. $\pi(R)$ is precompact in G/H).

(b) The commutator subgroup $[R, R] = Gp(xyx^{-1}y^{-1};\ x, y \in R)$ is precompact in G (i.e. lies in some compact subset of G).

Property (a) implies that there exists two precompact subsets for $K, K^* \subset G$ such that has all $\gamma \in R$ we have:

$$\gamma = h(\gamma)k(\gamma) = k^*(\gamma)h^*(\gamma) \quad \forall \quad h, h^* \in H, \quad k \in K, \quad k^* \in K^*. \tag{4.2}$$

In our immediate applications, in this paper, G/H will itself be compact. Condition (a) is then verified automatically and furthermore $G \cong H \cdot K \cong H \propto (G/H)$ where $K \subset G$ is some compact subgroup (cf. 3.7 of [11]). The decomposition (4.2) is then none other than the semidirect decompostion of the elements of G.

Property (b) actually says that there exists a compact subset $L \subset\subset G$ such that

$$\gamma_1\gamma_2 \in \gamma_2\gamma_1 L \quad \forall \quad \gamma_1, \gamma_2 \in R \tag{4.3}$$

If we use (4.2), (4.3) we deduce that:

$$H \ni (\mathrm{Ad}(\gamma_1) - I)h(\gamma_2) - (\mathrm{Ad}(\gamma_2) - I)h(\gamma_1) = \mathcal{O}(1) \tag{4.4}$$

where I denotes the identity mapping on $\mathfrak{h}$.

Let us denote by $h(\gamma) = (\widetilde{h}(\gamma), h_1(\gamma), \cdots, h_p(\gamma))$ the decomposition of $h(\gamma)$ in (4.1). Let us assume that $1 \leqslant j \leqslant p$ and $\gamma_0 \in R$ are such that $\mathrm{Ad}(\gamma_0)\,|_{H_j} \neq I$. Then (4.4) implies that

$$|(\mathrm{Ad}(\gamma_0) - I)h_j(\gamma)| = \left|\sin \frac{\theta_j(\gamma_0)}{2}\right| |h_j(\gamma)| = \mathcal{O}(1) \forall \quad \gamma \in R$$

and therefore $h_j(\gamma) = \mathcal{O}(1)$. It follows in particular that for every $j = 1, \cdots, p$ we have just two possibilities:
(i) $h_j(\gamma) = 0(1) \quad ; \quad \gamma \in R$.
(ii) $h_j(\gamma) \neq 0(1)$ and $\mathrm{Ad}(R)\mid_{H_j} = I$.

In case (ii) we also have

$$h_j(\gamma_1\gamma_2)k(\gamma_1\gamma_2) = h_j(\gamma_1)k(\gamma_1)h_j(\gamma_2)k(\gamma_2)$$

$$= \left[h_j(\gamma_1) + h_j^{\gamma_1}(\gamma_2)\right] k(\gamma_1)k(\gamma_2) = \left[h_j(\gamma_1) + h_j(\gamma_2)\right] k(\gamma_1)k(\gamma_2) \ \forall \ \gamma_1, \gamma_2 \in R$$

where we denote as usual $x^y = y^{-1}xy \quad (x, y \in G)$.

From this it clearly follows that:

$$|h_j(\gamma_1) + h_j(\gamma_2) - h_j(\gamma_1\gamma_2)|_{G^*} = \mathcal{O}(1) \tag{4.5}$$

where $G^* = \pi^{-1}\left(\overline{\pi(R)}\right) \subset G$. Since $H \subset G^*$ is co-compact this implies that

$$h_j(\gamma_1) + h_j(\gamma_2) - h_j(\gamma_1\gamma_2) = \mathcal{O}(1)$$

in $H = \mathbf{R}^n$. We shall now need the following

Lemma. *Let $h : \mathbf{R} \to \mathbf{R}$ be such that*

$$|h(x) + h(y) - h(x+y)| \leqslant C$$

for every $x, y \in \mathbf{R}$ and for some $C > 0$. Then there exists $\alpha \in \mathbf{R}$ such that

$$h(t) = \alpha t + o(|t|) \quad \text{for} \quad |t| \to \infty$$

We can clearly assume that there exists $x \in \mathbf{R}$ such that $h(x) > 10C$. Indeed if $h(t) = \mathcal{O}(1)$ there is nothing to prove, and since $h(z) - h(-z) = \mathcal{O}(1) \quad (z \in \mathbf{R})$, we can assume that $\overline{\lim_{z\to\infty}}\, h(z) = +\infty$. We can also clearly assume that $h(0) = 0$ and therefore that $h(t) = \varphi(t)t$ for some $\varphi : \mathbf{R} \to \mathbf{R}$. Since, by the above, $h(nx) \geqslant 9nC \quad (n \geqslant 1)$ and since, by our hypothesis,

$$|h(x_1) - h(x_2)| = 0(|x_1 - x_2| + 1) \quad \forall \quad x_1, x_2 \in \mathbf{R} \tag{4.6}$$

it follows that

$$0 \leqslant \alpha = \underline{\lim_{t\to\infty}}\, \varphi(t) \leqslant \beta = \overline{\lim_{t\to\infty}}\, \varphi(t).$$

We shall prove that $\alpha = \beta$ and this will complete the proof of the Lemma. Assume by contradiction that $\alpha < \alpha' < \beta' < \beta$. We can find then arbitrary large $m_0 \geqslant 1$ such that $h(m_0) < \alpha' m_0$ and therefore

$$h(pm_0) \leqslant \alpha' m_0 p + Cp = \left(\alpha' + \frac{C}{m_0}\right) m_0 p \ \forall \ p \geqslant 1.$$

Similarly there exists an arbitrarly large $n_0 \geqslant 1$ such that

$$h(pn_0) \geqslant \left(\beta' - \frac{C}{n_0}\right) n_0 p \ \ \forall \ \ p \geqslant 1$$

Thus, if m_0, n_0 are large enough, we obtain thus a contradiction with (4.6).

We shall put together all the above considerations and, to avoid unnecessary complications, we shall assume that $R/[R,R] \cong \mathbb{R}^a \times \mathbb{Z}^b$. We have then

Proposition. *There exists a uniquely defined linear functional* $L : R/[R,R] \to H \cong \mathbb{R}^n$ *and relatively compact subsets* $K_1, K_2 \subset G$ *[that are* **not** *uniquely defined] such that* $\gamma = k_1(\gamma)[L(\gamma)+o(|\gamma|_*)] = [L(\gamma)+o(|\gamma|_*)]k_2(\gamma)$; $k_i(\gamma) \in K_i$, $\gamma \in R$, $i = 1,2)$, *where* $|\gamma|_*$ *indicates the norm of* $\pi(\gamma)$ *in* $R/[R,R] \cong \mathbb{R}^a \times \mathbb{Z}^b$ $(\pi : R \to R/[R,R]$ *being the canonical projection). Furthermore we have:*

$$L(R) \subset \widetilde{H} = \{h \in H; \ \ (\mathrm{Ad}(r)h = h \ \ \forall r \in R\} \tag{4.7}$$

The uniqueness is automatic from the very definition of L.

Remark (inessential for what follows). The above mapping could clearly be singular. But if for instance $R \cong \mathbb{R}^a$ and $R \subset G$ is a closed subgroup, then L is not singular. Indeed

$$\left|\{\gamma \in \mathbb{Z}^a \subset R \ ; \ \ |L(\gamma)| \leqslant N\}\right| \geqslant CN^a$$

for some $C > 0$ and $N = 1, 2, \cdots$ large enough. By an easy counting argument we see therefore that if L was singular, we would be able to find two sequences $\gamma_j, \gamma'_j \in \mathbb{Z}^a \subset R$, $\gamma_j \neq \gamma'_j$ $(j = 1, 2, \cdots)$ such that $\gamma_j^{-1}\gamma'_j \to 0$ $(j \to \infty)$. This clearly gives us a contradiction. We can make an analogous remark when $R \cong \mathbb{R}^a \times \mathbb{Z}^b$.

One final observation will be essential in what follows: If we consider the decomposition:

$$\gamma = h(\gamma)k(\gamma) = k^*(\gamma)h^*(\gamma)$$

(which depends on the choice of K) it is not a priori true that:

$$h(\gamma) \in \widetilde{H}_R = \{h \in H; \ \ \mathrm{Ad}(\gamma)h = h, \ \ \gamma \in R\}. \tag{4.8}$$

But by the above considerations (when R satisfies (a) and (b) of course), by modifying h boundedly, i.e. by replacing, if necessary, $h(\gamma)$ by an appropriate $\widetilde{h}(\gamma) \in H$ such that $|h(\gamma) - \widetilde{h}(\gamma)|_H = \mathcal{O}(1)$, we can make sure that (4.8) holds. We clearly have then

$$\gamma = h(\gamma)k(\gamma) = k(\gamma)h(\gamma).$$

We shall now denote by $\mathfrak{g}$ the Lie algebra of G and define $\mathcal{C} \subset \mathfrak{g}$ by the property that $x \in \mathcal{C}$ if and only if the canonical image $\dot{x}$ of x in $\mathfrak{g}/\mathfrak{h}$, generates a precompact subgroup $(e^{t\dot{x}};\ t \in \mathbf{R}) \subset G/H$. [i.e. $Alg(\dot{x}) \subset \mathfrak{g}/\mathfrak{h}$ is compact in the sense of A.4]. It is clear that $\mathcal{C}$ is a cone (i.e. $x \in \mathcal{C},\ \lambda \in R$ implies that $\lambda x \in \mathcal{C}$). We can then uniquely define a homogeneous mapping $L : \mathcal{C} \to \mathfrak{h} \cong H$ (i.e. $L(\lambda x) = \lambda Lx$) such that

$$d_{G^*}\left(e^{tX},\ (LX)t\right) = o(|t|)\ \ \forall\ X \in \mathcal{C}\ ,\ \ |t| \to \infty.$$

We have identified in the above $(LX)t \in H \subset G$ and $G^* = \pi^{-1}\left(\overline{\pi(R)}\right)$ as in (4.5) (where now $R \subset G$ is the one parameter subgroup generated by X). An alternative way to define LX is to write

$$e^{tX} = h(t)k(t) = k^*(t)h^*(t)$$

as in (4.2) and then:

$$|h(t) - (LX)t|_H\ ,\ \ |h^*(t) - (LX)t|_H = o(|t|),\ \ |t| \to \infty$$

We clearly have:
$LX = \mathcal{O} \Leftrightarrow \left|e^{tX}\right|_G = O(1) \Leftrightarrow Alg\,(X) \subset \mathfrak{g}$ is compact in G (in the sense of A.4).

The important case for us is when G/H is compact. In that case we have $\mathcal{C} = \mathfrak{g}$ and also

$$LX \neq 0 \Leftrightarrow \left|e^{tX}\right|_G \geqslant C|t|;\ \ t \in \mathbf{R},\ \ |t| \geqslant 1;$$

$$LX_1 = LX_2 \Leftrightarrow d_G\left(e^{tX_1}, e^{tX_2}\right) = o(|t|);\ \ |t| \to \infty$$

for some $C > 0$. Observe that, in a sense, the "ray" $(LX)t$ $(t \in \mathbf{R})$ can be thought as the asymptotic direction (i.e. the tangent) of e^{tX} at infinity.

Let now $\theta = G \to G$ be a group automorphism that is such that $\theta(H) \subset H$ and let us also denote by $\theta = \mathfrak{g} \to \mathfrak{g}$ the corresponding algebra automorphism. For every $X \in \mathcal{C}$ we clearly have

$$e^{tX} = h\left(e^{tX}\right)k_1(t)\ \ ;\ \ \theta\left(e^{tX}\right) = \theta h\left(e^{tX}\right)\theta k_1(t)\ \ ;\ \ e^{t\theta X} = h\left(e^{t\theta X}\right)k_2(t)$$

because it is clear that $\theta\mathcal{C} = \mathcal{C} \subset \mathfrak{g}$, where $k_1, k_2, \theta k_1$ all lie in some fixed compact subset of G (for $t \in \mathbf{R}$). It follows that

$$\left|h\left(e^{t\theta X}\right) - \theta h\left(e^{tX}\right)\right|_H = \mathcal{O}(1)\ \ \forall\ \ t \in \mathbf{R}$$

which by the definition of L implies that

$$\theta\mathcal{C} = \mathcal{C}\ \forall\ \ L(\theta X) = \theta LX\ \ \forall\ \ X \in \mathcal{C}.$$

The important special case of the above that will be used is when $\theta : x \to g^{-1}xg$ $(x, g \in G)$. Then $\theta =$ Ad$(g) : \mathfrak{g} \to \mathfrak{g}$ so that

$$L((\mathrm{Ad}(g))X) = \mathrm{Ad}g(LX) \ \forall \ g \in G, \ \ X \in \mathcal{C}. \tag{4.9}$$

By (4.7) we also have for the Lie algebra bracket operation

$$[X, LX] = 0 \quad \forall X \in \mathcal{C}. \tag{4.10}$$

From (4.9) we **cannot** "a priori" deduce in general that

$$L([Y, X]) = [Y, LX] \quad \forall X, Y \in \mathfrak{g}. \tag{4.11}$$

But as we shall see in §5 there are some important special cases for which (4.11) holds.

5. Further properties of the L-mapping.

A Lemma on nilpotent groups.

Lemma. *Let N be some nilpotent connected real Lie group and let $\mathfrak{n}$ be its Lie algebra. Then for all $X, Y \in \mathfrak{n}$ we have:*

$$\lim_{t \to \infty} \frac{1}{t} |\exp(t[X, Y])|_N = 0$$

Let $X_1 = X$, $X_2 = Y, X_3, \cdots, X_n$ be some basis of $\mathfrak{n}$ (we can clearly assume that X, Y are linearly independent) and let $\tilde{\mathfrak{n}}$ be the free nilpotent algebra of the same rank as $\mathfrak{n}$ constructed from the free generators $\widetilde{X}_1, \widetilde{X}_2 \cdots \widetilde{X}_n$. Let $\widetilde{N}$ be the corresponding group and let $\tilde{\mathfrak{n}} \to \mathfrak{n}$; $\widetilde{N} \to N$ be the canonical maps and $\varphi_\lambda = \tilde{\mathfrak{n}} \to \tilde{\mathfrak{n}}$ $\varphi_\lambda = \widetilde{N} \to \widetilde{N}$ be the one parameter family of homomorphisms defined by $\varphi_\lambda\left(\widetilde{X}_j\right) = \lambda \widetilde{X}_j$ $(j = 1, \cdots, n)$ (i.e. the dilation structure on $\widetilde{N}$ cf. [22]). For the canonical distance constructed by the above basis on $\widetilde{N}$ (cf. [2]) we clearly have

$$|\varphi_\lambda(g)|_{\widetilde{N}} = \lambda |g|_{\widetilde{N}} \ \forall \ \lambda > 0, \ g \in \widetilde{N}.$$

and since $\varphi_\lambda\left(\left[\widetilde{X}, \widetilde{Y}\right]\right) = \lambda^2 \left[\widetilde{X}, \widetilde{Y}\right]$ $(\lambda > 0)$ we have:

$$\left|\exp\left(\lambda^2 \left[\widetilde{X}, \widetilde{Y}\right]\right)\right|_{\widetilde{N}} = \left|\varphi_\lambda \exp\left(\left[\widetilde{X}, \widetilde{Y}\right]\right)\right|_{\widetilde{N}} = \lambda \left|\exp\left[\widetilde{X}, \widetilde{Y}\right]\right|_{\widetilde{N}}$$

Our assertion follows since

$$\frac{1}{t}\left|\exp\left(t\,[X,Y]\right)\right|_N \leqslant \frac{1}{t}\left|\exp\left(t\left[\widetilde{X},\widetilde{Y}\right]\right)\right|_{\widetilde{N}} \underset{t\to\infty}{\longrightarrow} 0$$

for the appropriate distance $|\ |_N$ on N.

The L-mapping on a nilpotent subalgebra.

Let $G, \mathfrak{g}$ and $\mathbf{R}^n \cong H \subset G$ be as in §4 and to avoid unnecessary complications let us assume, once and for all from here onwards, that G/H is compact [as already pointed out this makes G isomorphic with the semidirect product of $\mathbf{R}^n$ with some compact group. But this additional information will not be essential for our purposes]. Let now $\mathfrak{n} \subset \mathfrak{g}$ be some nilpotent subalgebra and let $\mathcal{I} \subset \mathfrak{n}$ be the maximal compact subalgebra (with respect to the group G) that it contains (cf. A.4).

An immediate consequence of the above Lemma and of the properties of the L-mapping (that now is defined everywhere on $\mathfrak{n}$,see the end of §4) is the following basic inclusion

$$[\mathfrak{n},\mathfrak{n}] \subset \mathcal{I} \subset \mathfrak{n}.$$

This implies in particular, once more, that $\mathcal{I}$ is an ideal in $\mathfrak{n}$.

Let $N(\mathcal{I}) \subset N \subset G$ be the analytic subgroups generated in G by $\mathfrak{n}$ and $\mathcal{I}$. $N(\mathcal{I})$ is precompact and is normalized by N and, by the above, the commutator of N satisfies $[N,N] \subset N(\mathcal{I})$. The conditions (a) and (b) are therefore satisfied for the subgroup $N \subset G$ (i.e. $R = N$). On the other hand, by a well known formula of Zassenhaus, (cf. [27], [15], [2]: this formula is, in some sense, an "inversion" of the B.C.H. formula 2.15 of [9]) we have

$$\exp(X+Y) \in e^X e^Y N(\mathcal{I}) \forall X, Y \in \mathfrak{n}.$$

Together with (4.8) this implies

$$\left|h(\exp(X+Y)) - h\left(e^X\right) - h\left(e^Y\right)\right|_H = 0(1) \quad ;\forall\, X, Y \in \mathfrak{n}$$

and therefore:

$$L(X+Y) = L(X) + L(Y)\ \forall\ \ X, Y \in \mathfrak{n}.$$

This and what was said in §4 imply that the restriction of the mapping L on $\mathfrak{n}$ is linear and that $\mathcal{I} = \operatorname{Ker} L$.

From the above and §4 it follows that

$$g \in G \quad \operatorname{Ad}(g)\mathfrak{n} \subset \mathfrak{n} \Rightarrow \ \operatorname{Ad}(g)\mathcal{I} \subset \mathcal{I}$$

and therefore

$$x \in \mathfrak{g}, \quad [x, \mathfrak{n}] \subset \mathfrak{n} \Rightarrow [x, \mathcal{I}] \subset \mathcal{I}.$$

The other important property of L that we have to recall [cf. (4.7)] is:

$$\mathrm{Im}(L|\mathfrak{n}) = L(\mathfrak{n}) \subset \widetilde{H} = \{x \in \mathfrak{h}; \ [\mathfrak{n}, x] = 0\}.$$

Observe that now (4.11) holds with $X \in \mathfrak{n}$, $Y \in \mathfrak{g}$ such that $[Y, \mathfrak{n}] \subset \mathfrak{n}$, because of (4.9) with $g = e^{tY}$ and the linearity of L on $\mathfrak{n}$.

Consequences of the B.C.H. formula.

We preserve here all the notations introduced in this (and the previous) section and let $\mathfrak{n} \subset \mathfrak{g}$ be some nilpotent subalgebra as before. Let $X \in \mathfrak{n}$, $Y \in \mathfrak{g}$ be such that:

$$[Y, \mathfrak{n}] \subset \mathfrak{n} \quad ; \quad [LX, Y] = 0 \tag{5.1}$$

by formula (4.11), that as we saw now holds, this implies that

$$L[X, Y] = [LX, Y] = 0$$

and therefore that

$$[X, Y] \in \mathcal{I}. \tag{5.2}$$

Let as before $N(\mathcal{I}) \subset G$ be the analytic subgroup generated by $\mathcal{I}$. Then the inclusions $\mathrm{Ad}(e^{\tau Y})\mathcal{I} \subset \mathcal{I}$, $\mathrm{Ad}(e^{tX})\mathcal{I} \subset \mathcal{I}$, $\mathrm{Ad}(\exp(tX + \tau Y))\mathcal{I} \subset \mathcal{I}$ imply that

$$\begin{aligned} &e^{\tau Y} N(\mathcal{I}) e^{-\tau Y} \subset N(\mathcal{I}); \quad e^{tX} N(\mathcal{I}) e^{-tX} \subset N(\mathcal{I}); \\ &\exp(tX + \tau Y) N(\mathcal{I}) \exp(-tX - \tau Y) \subset N(\mathcal{I}) \end{aligned} \tag{5.3}$$

and by the B.C.H. formula, for $|t|, |\tau| \leqslant \varepsilon_0$ small enough and X, Y as in (5.1), we have:

$$[\exp(tX), \exp(\tau Y)] = \exp(tX)\exp(\tau Y)\exp(-tX)\exp(-\tau Y) \in N(\mathcal{I}), \tag{5.4}$$

$$\exp(tX)\exp(\tau Y) \in \exp(tX + \tau Y) N(\mathcal{I}). \tag{5.5}$$

(5.4) says that for small enough t, τ the two elements $e^{tX}, e^{\tau Y} \in G$ commute modulo $N(\mathcal{I})$. On the other hand the relations (5.3) allow us to "break the commutation" of e^{tX} and $e^{\tau Y}$ in small enough steps and to deduce therefore that (5.4) holds for arbitrary $t, \tau \in \mathbf{R}$. Similarly for arbitrary $t, \tau \in \mathbf{R}$ and $n \geqslant 1$ large enough, by raising the relation

$$\exp\left(\frac{t}{n}X\right) \exp\left(\frac{\tau}{n}Y\right) \in \exp\left(\frac{1}{n}(tX + \tau Y)\right) N(\mathcal{I})$$

to the nth power and using (5.3), (5.4) we deduce that (5.5) holds for arbitrary $t, \tau \in \mathbf{R}$.

If we recall that $N(\mathcal{I}) \subset G$ is a precompact group we deduce from (5.5) that

$$|h(\exp(-tY)\exp(tX)) - h(\exp(t(X-Y)))|_H = \mathcal{O}(1) \tag{5.6}$$

But by the definition of L we also have that

$$\begin{aligned}\exp(-tY)\exp(tX) &= (-tLY + o(t))k_Y(t)(tLX + o(t))k_X(t) \\ &= [t(LX - LY) + o(t)]k_Y(t)k_X(t) \end{aligned} \tag{5.7}$$

because, by our hypothesis (5.1), $LX \in H \subset G$ and $\exp(-tY)$ (and therefore also LX and $k_Y(t)$) commute in G. Here of course $k_X(t), k_Y(t)$ lie in some fixed compact subset of G ($t \in \mathbf{R}$). We have:

Lemma 5.1 *Let $X \in \mathfrak{n}$, $Y, \in \mathfrak{g}$ be such that*

$$LX = LY \quad , \quad [Y, \mathfrak{n}] \subset \mathfrak{n}. \tag{5.8}$$

We then have

$$L(X - Y) = 0. \tag{5.9}$$

Indeed, condition (5.8) and (4.10) imply that (5.1) is verified and therefore that (5.6) and (5.7) hold. These and (5.8) imply the required result (5.9).

We shall finish this section by formalizing property (5.2) of the mapping L in the way that it will be used in the next section. For this let $\mathfrak{n}_1 \subset \mathfrak{g}$ be **another** nilpotent subalgebra of $\mathfrak{g}$ such that

$$[\mathfrak{n}, \mathfrak{n}_1] \subset \mathfrak{n}$$

and let us denote as before

$$\widetilde{H}_1 = \{h \in \mathfrak{h} \ ; \ [h, \mathfrak{n}_1] = 0\}.$$

Let further $L_0 = L|\mathfrak{n}$; $L_1 = L|_{\mathfrak{n}_1}$. We clearly have $L_1(\mathfrak{n}_1) \subset \widetilde{H}_1$ and by (5.2)

$$\left[L_0^{-1}\left(\widetilde{H}_1\right), \mathfrak{n}_1\right] \subset \mathcal{I}.$$

We can restate this in the following

Lemma 5.2 *Let $\mathfrak{n}, \mathfrak{n}_1 \subset \mathfrak{g}$ be two nilpotent subalgebras such that $[\mathfrak{n}, \mathfrak{n}_1] \subset \mathfrak{n}$ and let $\mathcal{I}, L_0, L_1$ be as above. Then we have:*

$$\left[L_0^{-1}\left(L_1(\mathfrak{n}_1)\right), \mathfrak{n}_1 + \mathfrak{n}\right] \subset \mathcal{I}.$$

Solvable algebras.

Let $\mathfrak{g}$, $G \supset H$ and the mapping L be here as before. We shall consider some solvable Lie algebra $\mathfrak{r}$ and an algebra homomorphism $\alpha : \mathfrak{r} \to \mathfrak{g}$. We shall also assume that G/H is compact and therefore that the mapping L is defined everywhere on the solvable subalgebra $\alpha(\mathfrak{r}) \subset \mathfrak{g}$. We shall denote again by $L : \mathfrak{r} \to \mathfrak{h}(\cong H)$ the composition mapping $L \circ \alpha$. We shall argue directly with this new mapping rather than $L : \alpha(\mathfrak{r}) \to \mathfrak{h}$. Let $\mathfrak{n} \subset \mathfrak{r}$ be the nilradical of $\mathfrak{r}$ and let $\mathfrak{n}_1 \subset \mathfrak{r}$ be some Cartan subalgebra. This algebra is nilpotent (cf. Tome III Ch. VI §4 of [17]) and satisfies

$$\mathfrak{r} = \mathfrak{n} + \mathfrak{n}_1 \quad ; \quad [\mathfrak{n}, \mathfrak{n}_1] \subset \mathfrak{n}.$$

I shall also denote;

$$L_0 = L|\mathfrak{n} \ ; \ L_1 = L|\mathfrak{n}_1 \ ; \ \mathcal{I} = \mathcal{I}_0 = \text{Ker } L_0 \subset \mathfrak{n} \ ; \ \mathcal{I}_1 = \text{Ker } L_1 \subset \mathfrak{n}_1.$$

Clearly $\alpha(\mathfrak{n} \cap \mathcal{I}_1) \subset \alpha(\mathfrak{n}) \subset \mathfrak{g}$ is a compact subalgebra of $\mathfrak{g}$. Therefore

$$\mathfrak{n} \cap \mathcal{I}_1 \subset \mathcal{I}.$$

It is important to observe also that one has

$$\begin{aligned} &[\mathfrak{r}, \mathfrak{r}] \subset \mathfrak{n} \quad ; \quad [\mathfrak{r}, \mathcal{I}] \subset \mathcal{I} \ ; \\ &[\mathfrak{n}, \mathfrak{n}] \subset \mathcal{I} \quad ; \quad [\mathcal{I}_1, \mathcal{I}_1] \subset \mathcal{I}_1 \cap \mathfrak{n} \subset \mathcal{I} \,. \end{aligned} \tag{5.10}$$

We clearly have: $\mathcal{I} \subset \mathfrak{n} \subset \mathfrak{n} + \mathcal{I}_1 \subset \mathfrak{n} + \mathfrak{n}_1 = \mathfrak{r}$ and this allows us to construct a **basis** of

$$X_1, \cdots, X_\mu, X_{\mu+1}, \cdots, X_\nu \ ; \ Y_1, \cdots, Y_\lambda, Y_{\lambda+1}, \cdots, Y_\rho \tag{5.11}$$

such that $(X_j \ ; \ 1 \leqslant j \leqslant \mu)$ is a basis of $\mathcal{I}$, $(X_j \ ; \ 1 \leqslant j \leqslant \nu)$ is a basis of $\mathfrak{n}$, $Y_1, \cdots, Y_\lambda \in \mathcal{I}_1$, the X's together with $Y_1, \cdots, Y_\lambda$ form a basis of $\mathfrak{n} + \mathcal{I}_1$ and finally all the Y's lie in $\mathfrak{n}_1$. Because of the inclusions (5.10) it is also clear that we can choose the first part $(X_1, \cdots, X_\mu)$ of this basis so that the basis (5.11) "gives coordinates of the second kind" in the sense of §1 (i.e. the subspace generated by the first k-elements is normalized by the $(k+1)^{\text{th}}$ element of the basis).

The above basis will be now modified in an appropriate way. First of all we shall, as we may, choose the $Y_{\lambda+1}, \cdots, Y_\rho$ (i.e. the last part of the basis) in such a way that for some $\lambda < \sigma \leqslant \rho$ the subspace $\mathcal{L} = \text{Span}\,(L_1(Y_{\sigma+1}), \cdots, L_1(Y_\rho)) \subset H$ satisfies

$$L_0(\mathfrak{n}) \cap \mathcal{L} = \{0\} \tag{5.12}$$

and $\{L_1(Y_{\lambda+1}), \cdots, L_1(Y_\sigma)\}$ is a basis of the space:

$$L_0(\mathfrak{n}) \cap L_1(\text{Span}\,(Y_1, \cdots, Y_\rho)).$$

Let us then choose $\xi_{\lambda+1}, \cdots, \xi_\sigma \in \mathfrak{n}$ so that

$$L_0(\xi_j) = L_1(Y_j) \quad \forall \quad \lambda < j \leqslant \sigma$$

and let us set $Y'_j = Y_j - \xi_j \quad (\lambda < j \leqslant \sigma)$. Thus we obtain a new basis:

$$X_1, \cdots, X_\nu, \;\; Y_1, \cdots, Y_\lambda, \;\; Y'_{\lambda+1}, \cdots, Y'_\sigma, \;\; Y_{\sigma+1}, \cdots, Y_\rho \tag{5.13}$$

and this new basis, ordered as in (5.13), can give coordinates of the second kind as before. We shall denote by

$$\mathfrak{f} = \text{ Span }(X_1, \cdots, X_\mu, \; Y_1, \cdots, Y_\lambda, \; Y'_{\lambda+1}, \cdots, Y'_\sigma)$$

the subspace generated and we shall show that $\mathfrak{f} \subset \mathfrak{r}$ is a subalgebra of $\mathfrak{r}$ and also that it is a compact subalgebra (i.e. $\alpha(\mathfrak{f}) \subset \mathfrak{g}$ is a compact subalgebra with respect to the group G).

The fact that $\mathfrak{f}$ is an algebra follows from (5.10) and Lemma 5.1 together with the fact that $\xi_j \in L_0^{-1} L_1(\mathfrak{n}_1) \;\; (\lambda < j \leqslant \sigma)$. We even have the stronger assertion

$$[\mathfrak{f}, \mathfrak{f}] \subset \mathcal{I}. \tag{5.14}$$

The fact that $\mathfrak{f}$ is compact is a consequence of the fact that both $\mathcal{I}$ and $\mathcal{I}_1$ are clearly compact and so is each 1-dimensional subalgebra generated by each $Y'_j \;\; (\lambda < j \leqslant \sigma)$ by Lemma 5.1. This together with (5.14) allows us to prove step by step that each subalgebra $\mathfrak{f}_j = \text{ Span }(\mathcal{I}, \mathcal{I}_1, Y'_{\lambda+1} \cdots Y'_{\lambda+j})$ is compact (cf. A.4.; in fact because of (5.14) the proof here is more direct than in A.4). We shall now show that the algebra $\mathfrak{f}$ is "maximal compact" and that it has some additional properties.

For this goal let R be some connected real Lie group whose algebra is $\mathfrak{r}$ and let $\alpha : R \to G$ be some continuous homomorphism that induces the homomorphism $\alpha : \mathfrak{r} \to \mathfrak{g}$ as above. Thus we can identify $\text{Ad}\,(\alpha(R))|_{\mathfrak{h}} = \mathcal{R}$ with a subgroup of $GL(\mathfrak{h})$ that has compact closure $\overline{\mathcal{R}} \subset GL(\mathfrak{h})$. We shall abusively denote by $\text{Ad}\,(\gamma) = \text{Ad}\,(\alpha(\gamma)) \;\; (\gamma \in R)$ the action of $\gamma \in R$ on $\mathfrak{h}$. Then we have

$$\text{Ad}\,(\gamma) L_0(\mathfrak{n}) \subset L_0(\mathfrak{n}) \quad \forall \gamma \in R \tag{5.15}$$

because $adY(L_0(\mathfrak{n})) = L_0(adY\mathfrak{n}) \subset L_0(\mathfrak{n}) \;\; (\forall Y \in \mathfrak{r})$ by (4.11) (or simply by (4.9)). We also have

$$C^{-1}(|h| + |l_0|) \leqslant |h + \text{ Ad}\,(\gamma) l_0|_H \leqslant C(|h| + |l_0|) \quad \forall \; h \in \mathcal{L} \;, \;\; l_0 \in L_0(\mathfrak{n}) \text{ and}$$

$\gamma \in R$
for some $C > 0$ independent of l_1, l_0 and γ. This holds because of (5.12) and because of (5.15) which implies

$$\overline{\gamma} L_0(\mathfrak{n}) \subset L_0(\mathfrak{n}) \quad ; \quad \overline{\gamma} \in \overline{\mathcal{R}} \subset GL(\mathfrak{h})$$

and finally because of the compactness of $\overline{\mathcal{R}}$ which guarantees that

$$|\overline{\gamma} l|_H \approx |l|_H \quad \forall \ \overline{\gamma} \in \overline{\mathcal{R}} \ , \quad l \in H.$$

From the above remarks it is easy to deduce that $\mathfrak{f}$ is maximal compact in $\mathfrak{r}$ in the sense that if $\alpha(\mathfrak{f}) \subset \mathfrak{p} \subset \alpha(\mathfrak{r})$ is some compact subalgebra of $\mathfrak{g}$ (with respect to G) then $\mathfrak{p} = \alpha(\mathfrak{f})$. This point however will not be needed and the details will be left to the interested reader.

6. The Geometry of the embedding of a solvable subgroup.

In this section G will denote some connected real Lie group that contains a closed normal connected and simply connected solvable subgroup $R_0 \subset G$ such that G/R_0 is compact. It is clear that $\mathfrak{r}_0 \supset \mathfrak{n}_0 \supset [\mathfrak{n}_0, \mathfrak{n}_0], \cdots$ (where $\mathfrak{r}_0$ is the Lie algebra of R_0 and $\mathfrak{n}_0$ is the nilradical of $\mathfrak{r}_0$) generate subgroups $R_0 \supset R_1 \supset \cdots$ that are connected simply connected closed and normal in G and such that $R_j/R_{j+1} \cong \mathbb{R}^a$ is a Euclidean space. We shall in particular consider

$$\pi : G \to G^* = G/R_1 \tag{6.1}$$

(observe that R_1 could reduce to $\{0\}$) and now G^* contains a closed normal subgroup $H \subset G^*$ such that $H \cong \mathbb{R}^a$ where $a \geqslant 1$ (for otherwise G^* is compact).

Let now $R \subset G$ be some solvable closed connected subgroup of G let $N \subset R$ be the nilradical and assume that $R/N \cong \mathbb{R}^p$ is a Euclidean space. We shall also denote by T the maximal torus of N (as in §1) and by $\mathfrak{t} \subset \mathfrak{n} \subset \mathfrak{r}$ the Lie algebra of T and the nilradical of the Lie algebra $\mathfrak{r}$ of R. The composed mapping $R \to G^*$ induces a mapping $\alpha : \mathfrak{r} \to \mathfrak{g}^*$ (= the Lie algebra of G^*) and we can use α (as in the last section of §5) to define $\mathcal{I}, \mathcal{I}_1$ etc. ($\mathfrak{n}_1 \subset \mathfrak{r}$ will be here as in §5 some fixed Cartan subalgebra of $\mathfrak{r}$). It is of course clear that $\mathfrak{t} \subset \mathcal{I}$ and, in the choice of the bases (5.11), (5.13), we can clearly ensure that $X_1, \cdots, X_\tau$ is a basis of $\mathfrak{t}$ (for some $1 \leqslant \tau \leqslant \mu$).

Let us denote by $\widetilde{R} = R(\mathfrak{f}) \subset R$ the analytic subgroup of R that corresponds to $\mathfrak{f} \subset \mathfrak{r}$. The fact that $\mathfrak{t} \subset \mathfrak{f}$ implies then (c.f §1) that $\widetilde{R} \subset R$ is a closed subgroup and that when R is assumed to be an R-group then the distance distortion between R and $\widetilde{R}$ is polynomial. We shall now use exponential coordinates with respect to the bases (5.13) to express the elements $\gamma \in R$ as follows (cf. §1):

$$\begin{aligned}\gamma = &\exp(t_\nu X_\nu)\cdots\exp(t_\mu X_\mu)\cdots\exp(t_1 X_1)\\ &\exp(\tau_1 Y_1)\cdots\exp(\tau_{\lambda+1} Y'_{\lambda+1})\cdots\exp(\tau_{\sigma+1} Y_{\sigma+1})\cdots\end{aligned} \tag{6.2}$$

where $0 \leqslant t_j < 2\pi$ for $j = 1, \cdots, \tau$ and all the other coordinates run through $\mathbf{R}$.

Using the above decomposition we can then write

$$\gamma = \exp(t_\nu X_\nu)\cdots\exp(t_{\mu+1} X_{\mu+1})\gamma_{\mathfrak{f}}\exp(\tau_{\sigma+1} Y_{\sigma+1})\cdots\exp(\tau_\rho Y_\rho) \tag{6.3}$$

or, bearing in mind that $[\mathfrak{n}, \mathfrak{n}] \subset \mathcal{I}$, $[\mathfrak{n}_1, \mathfrak{n}_1] \subset \mathcal{I}_1$, we can also write:

$$\gamma = \exp(t_\nu X_\nu + \cdots + t_{\mu+1} X_{\mu+1})\gamma'_{\mathfrak{f}}\exp(\tau_{\sigma+1} Y_{\sigma+1} + \cdots + \tau_\rho Y_\rho) \tag{6.4}$$

where $\gamma_{\mathfrak{f}}, \gamma'_{\mathfrak{f}} \in \widetilde{R}$. It will be convenient in what follows to introduce a special notation. For variable quantities $A, B \geqslant 0$ we shall say that $A \underset{p}{<} B$ if there exist a fixed $C > 0$, independent of A and B, such that $A \leqslant CB^C + C$. We shall also say that $A \underset{p}{=} B$ if $A \underset{p}{<} B$ **and** $B \underset{p}{<} A$.

In (6.3), (6.4) we write:

$$T(\gamma) = |t_\nu| + \cdots |t_{\mu+1}| + |\tau_{\sigma+1}| + \cdots + |\tau_\rho|, \quad \gamma \in R.$$

Then, with the above notation we have for $\gamma \in R \subset G$:

$$|\gamma_{\mathfrak{f}}|_R + T(\gamma) \underset{p}{=} |\gamma|_R + T(\gamma) \underset{p}{=} |\gamma'_{\mathfrak{f}}|_R + T(\gamma) \tag{6.5}$$

$$|\gamma_{\mathfrak{f}}|_G + T(\gamma) \underset{p}{=} |\gamma|_G + T(\gamma) \underset{p}{=} |\gamma'_{\mathfrak{f}}|_G + T(\gamma) \tag{6.7}$$

$$|\gamma_{\mathfrak{f}}|_R \underset{p}{=} |\gamma|_{\widetilde{R}} \quad ; \quad |\gamma'_{\mathfrak{f}}|_R \underset{p}{=} |\gamma'_{\mathfrak{f}}|_{\widetilde{R}}. \tag{6.8}$$

In what follows we shall not need **both** (6.3) and (6.4), just one of the two will suffice. I have maintained both (6.3) and (6.4), here for the convenience of the reader with the hope that it will make the picture clearer. Observe that the nilpotency of $\mathfrak{n}$ and $\mathfrak{n}_1$ is used in an essential way to prove the right hand estimates of (6.5), (6.7).

The other vital estimate that we shall need is that there exists $C > 0$ such that:

$$C^{-1}T(\gamma) \leqslant |\pi(\gamma)|_{G^\bullet} \leqslant CT(\gamma) \quad \forall \quad \gamma \in R. \tag{6.9}$$

Indeed by the exponential coordinates (6.2), (6.3), (6.4) and the definition of L we see that

$$\pi(\gamma) = (L_0(t_\nu X_\nu + \cdots + t_{\mu+1}X_{\mu+1}) + o(|t|))k(L_1(\tau_{\sigma+1}Y_{\sigma+1} + \cdots + \tau_\rho Y_\rho) + o(|t|))$$

(6.10)

$$= (l_0(t_\nu, \cdots) + o(|t|))\; k(l_1(\tau_{\sigma+1}, \cdots) + o(|\tau|))$$

where $k \in \pi(R) \cap K$ (since $k = \pi(\gamma_{\mathfrak{f}})$ or $\pi(\gamma'_{\mathfrak{f}})$ and $\gamma_{\mathfrak{f}}, \gamma_{\mathfrak{f}'} \in \widetilde{R}$) and $K \subset G^*$ is some fixed compact subset. From (6.10) the right hand estimate follows at once since each of the three factors in (6.10) satisfies this estimate.

To prove the left hand estimate of (6.9) we rewrite (6.10) in the form:

$$\pi(\gamma) = k(\mathrm{Ad}(k)l_0 + l_1 + o(T(\gamma)))$$

and by using the same argument as in the end of §5 we see that:

$$|\mathrm{Ad}(k)l_0 + l_1|_H \approx T(\gamma).$$

This proves our assertion. From (6.9) it is clear in particular that $\pi(R) \subset G^*$ is precompact if and only if $T(\gamma) = \mathcal{O}(1)$ and that this holds if and only if $T(\gamma) = 0$ $(\forall \gamma \in R)$ and if and only if $\widetilde{R} = R$.

7. The proof of Theorem R.

We are now going to place ourselves in the setup of the previous section with $G \supset R_0, R$ and to preserve all the notations of that section. However we shall further assume that G is an R-group. We shall not assume "a priori" here that the closed connected solvable subgroup R satisfies $R/N \cong \mathbf{R}^a$ and allow the possibility that $R/N \cong \mathbf{R}^a \times \mathbf{T}^b$. We shall then use a double induction on $\dim R_0$ and $\dim R$ (alternatively, induction on $\dim R + \dim R_0$) to prove the polynomial distance distortion between R and G. By §3 this will complete the proof of Theorem R.

First of all by the inductive hypothesis we may assume that $b = 0$ and that $R/N \cong \mathbf{R}^a$ (i.e. that we are **exactly** in the situation of §6). Indeed if $p : R \to R/N$ is the canonical mapping and if $b \neq 0$ we consider $p^{-1}(\mathbf{R}^a) \subset R$ which is co-compact in R and of lower dimension and we use Lemma 1 (iv) of §2 to reduce our problem to a lower dimension for R.

We shall then consider $\pi : G \to G^*$ as in (6.1) and distinguish two possibilities

(i) $\pi(R) \subset G^*$ is precompact: (i.e. $T(\gamma) = 0$ $\forall \gamma \in R$)

(ii) $\pi(R) \subset G^*$ is not precompact.

In case (i) we have $R \subset G^\# = \pi^{-1}(\overline{\pi(R)}) \subset G$ and the new group $G^\#$ has the same properties as the group G, but the dimension of $G^\# \supset R_0^\# =$

$R_1 = \text{Ker}\,\pi$ (i.e. the simply connected solvable co-compact subgroup) is lower. Now, the polynomial distance distortion between R and $G^{\#}$, and therefore also between R and G (see the end of §1), follows by the inductive hypothesis. We can therefore assume that we are in case (ii) and therefore that the subgroup $\widetilde{R} = R(\mathfrak{f}) \subset R$ constructed in §6 satisfies (see the end of §6):

$$\dim \widetilde{R} < \dim R. \tag{7.1}$$

In case (ii) (preserving the notation of §6) we shall fix some $\gamma \in R \subset G$ and distinguish two possibilities:

case (a): $T(\gamma) \geqslant c|\gamma|_R^\theta$

case (b): $T(\gamma) \leqslant c|\gamma|_R^\theta$

where $0 < c,\ \theta \ll 1$ are fixed and will be chosen later.

In case (a) (cf. (6.9)) we clearly have

$$|\gamma|_G \geqslant |\pi(\gamma)|_{G^*} \approx T(\gamma) \geqslant c|\gamma|_R^\theta$$

and the required polynomial distortion follows.

Let us assume therefore that we are in case (b). Because of the estimates (6.5), (6.6), (6.7) we have:

$$|\gamma|_R \underset{p}{<} |\gamma_{\mathfrak{f}}|_R \underset{p}{<} |\gamma_{\mathfrak{f}}|_{\widetilde{R}} \underset{p}{<} |\gamma_{\mathfrak{f}}|_G \quad (\text{mod } T(\gamma)) \tag{7.2}$$

[$x \underset{p}{<} y \quad (\text{mod } T(\gamma))$ stands for $x \underset{p}{<} y + T(\gamma)$]. The last estimate in (7.2) is a consequence of the inductive hypothesis and (7.1). By using (7.2) we deduce that for some C (independent of γ) we have

$$|\gamma|_R \leqslant C\left(|\gamma|_G^C + (T(\gamma))^C + C\right)$$

and, if c, θ in case (b) have been chosen small enough, this implies that

$$|\gamma|_R \leqslant C|\gamma|_G^C + C$$

thereby proving the required polynomial distortion. The proof of the inductive step is thus complete.

The above reduction of $\dim R_0 + \dim R$ by at least one unit at a time can clearly go on as long as we have not reached a point where R is compact (i.e. $R \cong \mathbb{T}^a$). At that point of course the induction starts trivially. The proof of Theorem R is therefore complete.

I shall finish this section with a few words of how one deduces the corollary of Theorem R. Assume for simplicity that G is a Lie group and let G_0 be the component of the identity. Let $\pi : G \to G/G_0 = N$. We can then assume (cf. [23]) that N is nilpotent. Let $H \subset G$ be a subgroup. Let

$N \supset N_1 \supset \cdots N_p = \{0\}$ be some central series of N. By an argument that is very similar to the reduction of Theorem G to Theorem R in §3 one can use induction on p. More explicitly we can reduce the problem successively to $H \cap \pi^{-1}(N_1)$, $H \cap \pi^{-1}(N_2), \cdots \subset G$ until at the end we just have to consider $H \cap G_0$. Since G_0 is a connected R-Lie group the result follows from Theorem R. The interested reader can fill in the details.

If G is not a Lie group then there exists some compact subgroup $K \subset G$ such that G/K is a Lie group (cf. [24]). The proof of the corollary is therefore easily reduced to the Lie group case.

Appendix

A.1 The Complexification and the Ado-Iwasawa Theorem.

Let G be some simply connected Lie group let $\mathfrak{g}$ be its Lie algebra. We can then complexify $\mathfrak{g}_c = \mathfrak{g} \otimes \mathbb{C}$ and construct G_c the corresponding complex simply connected group. We have then $\theta : G \to G_c$ a canonical local analytic isomorphism and $J : G_c \to G_c$ an involution that corresponds to the canonical complex conjugation on $\mathfrak{g}$. It is very easy to see that if $\Omega \subset G_c$ is a small enough neighborhood of e in G_c that is stable by J then $\Omega \cap \theta(G) = \{x \in \Omega;\ Jx = x\}$. From this it follows that $\theta(G)$ is "locally" and hence "globally" closed. The mapping θ has in general a discrete non trivial kernel, e.g. if G is a semisimple real Lie group with infinite center then Ker θ is *always* infinite. If G is solvable then θ is bijective. Indeed in that case $\theta(G)$, being connected, is simply connected (cf. [9] 3.18.12) and therefore Ker $\theta \cong \pi_1(\theta(G)) = \{e\}$.

If G is a general simply connected Lie group we can write it as a semidirect product $G \cong Q \propto NAK$ where Q is the radical and NAK is the Iwasawa decomposition of a semisimple Levi factor S. Let $Z(S)$ be the center of S. Then the following facts are either standard or well known and easy to prove:

(i) $Z(S)$ is of zero distortion in S, and hence in G, and is a finitely generated discrete abelian group (cf. [13]).

(ii) There exists $Z \subset Z(S)$ a subgroup of finite index that is central in G (cf. [11], [7], [13]).

(iii) When G is amenable then S is compact and $Z(S)$ is finite.

In view of the above it is convenient to introduce the following terminology:

"We say that $\pi : G_1 \to G_2$ is an almost embedding if $\pi(G_1)$ is closed in G_2 and if Ker π is a discrete 0-distorted central subgroup of G_1".

It is clear that we can consider the corresponding simply connected complex groups Q_c and S_c and form the semidirect product $G_c = Q_c \propto S_c$ and the injection $\theta : G = Q \propto S \to G_c$. By the above remarks it follows that θ is an almost embedding and if we assume in addition that G is amenable then Ker θ is finite.

Let now G be an arbitrary simply connected complex group and let N be the nilradical, so that $G/N \cong \mathbb{C}^a \times H$ where H is complex semisimple (cf. [9] 3.16.3). It follows that $G/N \underset{\to}{\subset} GL(n;\mathbb{R})$ can be linearized (i.e. identified to a closed subgroup of GL (cf. [14], [20]). It also is a classical fact (the Ado-Iwasawa theorem: cf. 3.18.16 and Exercise 41 Ch. 3 of [9]) that we can find a continuous homomorphism $\varphi : G \to GL(m;\mathbb{R})$ so that φ is bijective on N and $\varphi(N)$ is a closed subgroup. Let $\xi : G \to G/N \underset{\to}{\subset} GL(n;\mathbb{R})$ be the composed mapping and let $\xi \oplus \varphi : G \to GL(n+m; \mathbf{R})$ be the direct sum embedding. It is then easy to verify that $\xi \oplus \varphi$ is bijective and identifies G with a closed subgroup of $GL(n+m; \mathbf{R})$.

Remark. (not essential in this paper): The above argument can be adapted to show that when G is real simply connected and amenable then G can be embedded as a closed subgroup of $GL(n; \mathbf{C})$ for some $n \geqslant 1$. Indeed $G/N = \mathbf{R}^a \times H$ as above but now H is compact and therefore embeddable in GL.

For all the above the reader could also refer to Exercise 42 p. 284 of [28].

A.2 The real Zariski Topology.

There are two equivalent ways of defining the *real* Zariski topology on $GL(n; \mathbf{R})$. This topology is the Zariski topology of the *real* affine algebraic variety $GL(n; \mathbf{R})$ (cf. [17] Tome II, [20]). Equivalently we can embed $GL(n; \mathbf{R}) \subset GL(n; \mathbf{C})$ so that $GL(n; \mathbf{R}) = GL(n; \mathbf{C}) \cap M_{n\times n}(\mathbf{R})$. For any $A \subset GL(n; \mathbf{R})$ we can then consider $\overline{A}_c$ the complex Zariski closure of A in $GL(n; \mathbf{C})$ and then define the real algebraic closure $\overline{A} = \overline{A}_c \cap GL(n; \mathbf{R})$ (cf. [12], [4]).

One should observe that we can in turn embed canonically $GL(n; \mathbf{C}) \subseteq GL(2n; \mathbf{R})$ as a closed subgroup (for the real Zariski topology). In this embedding an element $\sigma \in GL(n; \mathbf{C})$ is unipotent in $GL(n; \mathbf{C})$ if and only if it is unipotent in $GL(2n; \mathbf{R})$ [i.e. $(\sigma - I)^n = 0$ for some $n \geqslant 1$]. We also have:

$$\left\{ |\det \sigma|^2, \quad \sigma \in GL(n; \mathbf{C}) \right\} = \left\{ \det \sigma, \quad \sigma \in GL(2n; \mathbf{R}) \right\}. \qquad (A.2.1)$$

For the above real Zariski topology on $GL(n; \mathbf{R})$ the closed subsets have only a finite number of topological components (cf. [16], [12]).

It follows that we can consider two different Zariski topologies on $GL(n; \mathbf{C})$: its intrinsic complex Zariski topology and the strictly stronger real Zariski topology induced by $GL(2n; \mathbf{R})$. To illustrate the difference let $G = U \cdot \mathbf{C}_*^a \subset GL(n; \mathbf{C})$ be some algebraic solvable complex subgroup that is the semidirect product of its unipotent radical U and of a maximal torus $\mathbf{C}_*^a$ (cf. Ch.V §3.5 of [17] Tome III and [14],[20]). Let z_i $(i = 1, \cdots, a)$ be the canonical coordinates on $\mathbf{C}_*^a$, or can be identified with regular functions (i.e. polynomials on the matrix coefficients and on $(\det)^{-1}$) on G. From (A.2.1) it then follows that $|z_i|^2$ $(i = 1, \cdots, a)$ are regular functions on G for the real algebraic structure induced by $G \subset GL(2n; \mathbf{R})$. We conclude therefore that the maximal compact (for the Hausdorff topology) subgroup $\mathbf{T}^a = (|z_i|^2 = 1 \;\; i = 1, \cdots, a) \subset \mathbf{C}_*^a$ gives rise to a subgroup $G_0 = U \cdot \mathbf{T}^a$

that is closed for the real Zariski topology (induced by $GL(2n;\mathbb{R})$) and we also have a semidirect product $G \cong G_0 \propto \mathbb{R}^a$ in the Lie group sense.

Let now $Q \subset GL(n;\mathbb{R})$ be an analytic solvable subgroup such that its Lie algebra $\mathfrak{q} \subset \mathfrak{gl}(n;\mathbb{R})$ is an R-algebra. Let $\overline{Q}$ be the real Zariski closure of Q and let $\mathfrak{q}'$ be the Lie algebra of $\overline{Q}$. We shall show that $\mathfrak{q}'$ is an R-algebra. First of all by (Th. 13 of§14 of [17] Tome II) we have $[\mathfrak{q}', \mathfrak{q}'] \subset \mathfrak{q}$. To show that $\mathfrak{q}'$ is an R-algebra it suffices therefore to prove that the Ad-action of $\overline{Q}$ on $\mathfrak{q}$ has unimodular eigenvalues.

Since $Q \subset GL(n;\mathbb{R})$ we can identify the complexified Lie algebra $\mathfrak{q}_c = \mathfrak{q} \otimes \mathbb{C}$ with $\mathfrak{q} \subset \mathfrak{gl}(n;\mathbb{C}) = M_{n\times n}(\mathbb{C})$. For each $T \in GL(n;\mathbb{R})$ let us denote $\mathrm{Ad}T : M_{n\times n}(\mathbb{C}) \to M_{n\times n}(\mathbb{C})$ $\quad (\mathrm{Ad}T)\xi = T^{-1}\xi T \quad (\xi \in M_{n\times n})$. Then the Ad action of Q on $\mathfrak{q}_c$ extends to the above Ad action of $GL(n;\mathbb{C})$ on $M_{n\times n}(\mathbb{C})$. Since Q is solvable we can find a filtration

$$M_{n\times n}(\mathbb{C}) \supset \mathfrak{q}_c = V_0 \supseteq V_1 \supseteq \cdots \supseteq V_k = \{0\} \tag{A.2.2}$$

such that $\dim_{\mathbb{C}} V_j/V_{j+1} = 1 \quad (j = 0, 1, \cdots, k-1)$ that is stable by the $\mathrm{Ad}(Q)$ action. Explicitly this means that, if we fix $e_j \in V_j \backslash V_{j-1} \quad (j = 0, 1, \cdots, k-1)$, then for all $T \in Q$ we have:

$$(\mathrm{Ad}T)e_j - \lambda_j(T)e_j \in V_{j+1} \quad j = 0, 1, \cdots, k-1 \tag{A.2.3}$$

where $\lambda_j(T) \in \mathbb{C}$ is determined uniquely by $(A.2.3)$ and $|\lambda_j(T)| = 1$ since Q is an R-group. Furthermore, since $(e_0, \cdots, e_{k-1})$ is a basis of V_0, we can extend the definition of $\lambda_j(T)$ to all T in the stabilizer H_0 of V_0 (i.e. $\{H_0 = [\xi \in GL(n;\mathbb{R}) \;\; ; \;\; \mathrm{Ad}\xi V_0 \subset V_0\}$). For such a T we simply define $\lambda_j(T)$ to be the e_j-coefficient of $\mathrm{Ad}(T)e_j \in V_0$. We have then

$$|\lambda_j(T)|^2 = (\mathrm{Re}\,\lambda_j(T))^2 + (\mathrm{Im}\,\lambda_j(T))^2 = P(T_{i,j};\;\; i,j = 1, \cdots, n;\;\; (\det T)^{-1}) \tag{A.2.4}$$

where $T = (T_{i,j};\;\; i,j = 1, \cdots, n) \in GL(n;\mathbb{R}) \subset M_{n\times n}(\mathbb{R})$ and P is a polynomial function (since $(\mathrm{Ad}T)\, e_j = T^{-1}e_jT$).

Let us now define $H \subset GL(n;\mathbb{R})$ as follows: $h \in H$ if and only if

$$\mathrm{Ad}hV_j \subset V_j, \quad \mathrm{Ad}he_j - \lambda_j(h)e_j \in V_{j+1}, \quad |\lambda_j(h)| = 1 \forall \;\; j = 0, 1, \cdots, k-1.$$

It is clear that H is a subgroup and that $\lambda_0(h), \cdots, \lambda_{k-1}(h) \quad (h \in H)$ are the eigenvalues of the action of h on V_0. $(A.2.4)$ implies that H is a real affine algebraic group.

We obviously have $Q \subset H$ and therefore also $\overline{Q} \subset H$, (here $\overline{Q}$ denotes the real algebraic closure), $\overline{Q}$ (and also H) has the additional property that $\mathrm{Ad}(\overline{Q})\mathfrak{q} \subset \mathfrak{q}$. Therefore the eigenvalues of the action of $\mathrm{Ad}(\overline{Q})$ on $\mathfrak{q}$, which are the same as the eigenvalues of the action on $\mathfrak{q}_c$, are unimodular. This proves our assertion.

This is all that will be needed in this paper, but for the sake of completeness we shall also consider more general R-algebras. Indeed the above considerations extend to a more general analytic subgroup $G \subset GL(n;\mathbb{R})$. Let $\mathfrak{g} = \mathfrak{q} \propto \mathfrak{s} \subset \mathfrak{gl}(n;\mathbb{R})$ be the Lie algebra of G and $\mathfrak{q}$ be its radical together with a Levi decomposition. Let $\mathfrak{q}'$ be the smallest algebraic subalgebra that contains $\mathfrak{q}$. Then $\mathfrak{s}$ nomalizes $\mathfrak{q}'$ so that $\mathfrak{g}' = \mathfrak{q}' + \mathfrak{s} \subset \mathfrak{gl}(n;\mathbb{R})$ is an algebra. Since $\mathfrak{g}'/\mathfrak{q}'$ (being a quotient of $\mathfrak{s}$) is semisimple it follows that the radical of $\mathfrak{g}'$ is $\mathfrak{q}'$. Therefore $\mathfrak{g}'$ is algebraic (cf. §3.5 of Ch.V of [17] Tome III) and clearly $\mathfrak{g}'/\mathfrak{q}'$ is of compact type if $\mathfrak{s}$ is of compact type. Furthermore by the above considerations, if $\mathfrak{g}$ is an R-algebra so are $\mathfrak{q}$ and $\mathfrak{q}'$ and therefore so is $\mathfrak{g}'$.

Proposition. *Let G be some amenable locally compact group. Let $\varphi : G \to GL(n;\mathbb{R})$ be some continuous mapping and let $\overline{G} = \overline{\varphi(G)} \subset GL(n;\mathbb{R})$ be the real Zariski closure of $\varphi(G)$ in $GL(n;\mathbb{R})$. Then*
(i) $\overline{G}$ is amenable.
(ii) If G is a connected Lie group of R-type then $\overline{G}$ is an R-group.

The proof of part (i) of this proposition when G is a connected Lie group was given just above. The general case is a Theorem of C.C. Moore (cf.4.1.15 of [4]). The proof of part (ii) was given above.

Remark. (not used in this paper). There is a part (iii) to the above proposition:
(iii) If G is a locally compact group of polynomial volume growth $\overline{G}$ is an R-group.

Proof. use [24] and assume that G is a Lie group. Then argue separately on the component of the identity G_0 that is an R-group and on G/G_o that is almost nilpotent [23].

A.3 The co-compact connected hull.

Let G be a real linear algebraic group that is solvable and can be put in the form

$$G = N \cdot A \subset GL(n;\mathbb{R})$$

where N is unipotent connected and simply connected group and A is abelian (cf. §3.5 of Ch.V of [17] Tome III). Let $R \subset G$ be some closed subgroup (for the Hausdorff topology). Then we have:

Proposition. *There exists some (Hausdorff) closed subgroup $R \subset \widehat{R} \subset G$ that contains R co-compactly and has only a finite number of topological components.*

Let us recall here that if $H_1 \subset H_2 \subset G$ are two arbitrary subgroups of some topological group we say that H_2 contains H_1 co-compactly if there exists some compact subset $K \subset\subset G$ such that $H_2 = H_1 K = K^{-1} H_1$. It is clear that if $H_1 \subset H_2 \subset H_3 \subset G$ and if H_1 is co-compact in H_2 and H_2 is co-compact in H_3 then H_1 is co-compact in H_3. What is also very easy to see is that an arbitrary subgroup $X \subset \mathbb{R}^n$ is contained co-compactly in the vector subspace $\hat{X} \subset \mathbb{R}^n$ that it spans. From this we can deduce that if $X \subset A$ is an arbitrary subgroup of some abelian Lie group ($\cong \mathbb{T}^a \times \mathbb{R}^b \times$ Finite see tge end of $A.3$) with a finite number of components then there exists a closed subgroup $X \subset \hat{X} \subset A$ with finitely many components such that X is co-compact in $\widehat{X}$ consider the universal cover $\pi^{-1}(X) \subset \mathbb{R}^{a+b} \xrightarrow[\pi]{} \mathbb{R}^a \times \mathbb{T}^b$ and let $\hat{X} = \pi\left[(\pi^{-1}(X))^\wedge + \pi^{-1}(\mathbb{T}^b)\right]$.

Let now $R_0 = R_N = N \cap R \subset N$ and let $R_0^* = \overline{R}_0 = \overline{R}_N \subset N$ be the real Zariski closure of R_0 in $GL(n;\mathbb{R})$. It is then well known that $\overline{R}_0$ is a real connected and simply connected subgroup of N that contains R_0 co-compactly (cf. 2-5 of [12]). Let us also consider R/R_N which is an abelian compactly generated Lie group (because R is compactly generated cf. 3.8 of [12]). It follows that $R/R_N \cong \mathbb{R}^a \times \mathbb{T}^b \times \mathbb{Z}^c \times \prod_j \mathbb{Z}(m_j)$. By this we can find a finite family of (continuous or discrete parameter) one parameter subgroups $Z_j \subset R$ $(j = 1, \cdots, n)$ such that

$$[Z_j, R_0] \subset R_0 \quad ; \quad [Z_j, Z_k] \subset R_0 \quad \forall \quad j,k = 1, \cdots n$$

(we denote here $[A, B] = \{aba^{-1}b^{-1};\ a \in A, b \in B\}$) and such that $R_n = R$ and the subgroups

$$R_j = R_0 Z_1 \cdots Z_j \quad ; \quad (j = 0, 1, \cdots, n)$$

are all closed. One should observe that we do not assume that the Z_j's themselves are closed subgroups. This, a priori, can only be achieved if $\mathbb{T}^b \times \prod_j \mathbb{Z}(m_j) = \{0\}$ and therefore only if we "reduce" R to some co-compact subgroup $R_1 \subset R$. We shall however not need to do this here.

We shall now denote by $\overline{Z}_j$ the (real) Zariski closure of Z_j in $GL(n;\mathbb{R})$ and by Z_j^* some Hausdorff closed subgroup that, as before, contains Z_j co-compactly and has only a finite number of components. We clearly have (by standard properties of Zariski closures):

$$[Z_j^*, \overline{R}_0] \subset [\overline{Z}_j, \overline{R}_0] \subset \overline{R}_0 \quad ; \quad [Z_j^*, Z_k^*] \subset [\overline{Z}_j, \overline{Z}_k] \subset \overline{R}_0 \quad \forall \quad j,k = 1, \cdots, n. \tag{A.3.1}$$

It follows in particular that the sets

$$R_j^* = R_0^* Z_1^* \cdots Z_j^* \quad ; \quad j = 0, 1, \cdots, n$$

are subgroups with only finitely many topological components. It also follows that R_{j+1}^* normalizes R_j^* $(j = 0, 1, \cdots, n-1.)$ We shall prove by induction on j that each group R_j^* is closed and that it contains R_j co-compactly. For $j = n$ this will give a proof of our proposition.

Assume that this fact holds for some $j = 0, 1, \cdots$ (it clearly holds for $j = 0$). We have

$$R_{j+1}^* = R_j^* Z_{j+1}^* = R_j^* Z_{j+1} K_1$$

$$R_j^* Z_{j+1} = K_2 R_j Z_{j+1} = K_2 R_{j+1} = R_{j+1} K_2^{-1}$$

where $K_1, K_2 \subset GL$ are compact sets, because $R_j^* Z_{j+1}$ is a group (cf. (A.3.1)) and $(R_j^* Z_{j+1})^{-1} = R_j^* Z_{j+1}$. It follows that $R_{j+1}^* = R_{j+1} K_2^{-1} K_1$ and this proves both assertions of the inductive step (in one stroke!) and completes the proof of our proposition.

One should observe that the real Zariski closure of the group H

$$H = \left\{ \begin{pmatrix} e^n & 0 \\ 0 & e^{\alpha n} \end{pmatrix} \ , \ n \in \mathbb{Z} \right\} \subset GL(2; \mathbb{R}),$$

where α is irrational, is the group

$$\overline{H} = \left\{ \begin{pmatrix} x & 0 \\ 0 & y \end{pmatrix} \ ; \ x, y \in \mathbb{R}_* = \mathbb{R} \backslash \{0\} \right\}$$

that does not contain H co-compactly. In the above proposition therefore we cannot in general just take $\widehat{R} =$ the real Zariski closure of R. Indeed the component of the identity of $(\overline{H})_0$ being a Lie group, must contain the one parameter subgroup $K = \{y = x^\alpha; \ x > 0\}$. The group K is clearly not algebraic and neither is any other Lie subgroup of the same dimension (with finitely many components) that contains K. It follows that the algebraic closure $\overline{K}$ of K has dimension 2].

Note: The fact that every compactly generated abelian Lie group can be written as $G = G_0 \times D$ where $G_0 \cong \mathbb{R}^a \times \mathbb{T}^b$ (cf. Th. 44 of [26]) and $D \cong \mathbb{Z}^c \times \prod_j \mathbb{Z}(m_j)$ can be seen by lifting the obvious generators of $G/G_0 = D$ in G in such a way that the order of these generators is preserved — this is possible since G_0 is divisible. One consequence of this is that if G contains a dense subgroup that is isomorphic to $\mathbb{Z}$ (such groups are called monothetic, see [25]) then G is either compact or $G \cong \mathbb{Z}$.

A.4 Compact embeddings.

Let G be some Lie group let $\mathfrak{g}$ be its Lie algebra. For every subalgebra $\mathfrak{g}_1 \subset \mathfrak{g}$ we consider the analytic subgroup $G_1 \subset G$ that corresponds to $\mathfrak{g}_1$. Note that the topological closure $\overline{G_1}$ of G_1 and the corresponding subalgebra $\mathfrak{g}_1$ satisfy $G_1 \subset \overline{G_1} \subset G$ and $\mathfrak{g}_1 \subset \overline{\mathfrak{g}}_1 \subset \mathfrak{g}$. If $\mathfrak{h} \subset \mathfrak{g}$ is some subspace that is normalized by $\mathfrak{g}_1$ (i.e. $[\mathfrak{h}, \mathfrak{g}_1] \subset \mathfrak{h}$) then $\text{Ad}(x)\mathfrak{h} \subset \mathfrak{h}$ for all $x \in G_1$ and therefore for all $x \in \overline{G_1}$. It follows that $[\mathfrak{h}, \overline{\mathfrak{g}}_1] \subset \mathfrak{h}$. (cf. Ch. XVI of [11]).

Let $\mathfrak{g}_1, \mathfrak{g}_2 \subset \mathfrak{g}$ be subalgebras such that $\mathfrak{g}_3 = \mathfrak{g}_1 + \mathfrak{g}_2$ is a subalgebra. Now $\mathfrak{g}_1$ normalizes $\mathfrak{g}_3$ and therefore $[\overline{\mathfrak{g}}_1, \mathfrak{g}_2] \subset \mathfrak{g}_3$ and $\mathfrak{g}_3 \subset \overline{\mathfrak{g}}_1 + \mathfrak{g}_2 = \mathfrak{h} \subset \mathfrak{g}$ is a subalgebra that is normalized by $\mathfrak{g}_2$. It follows that $\overline{\mathfrak{g}}$ normalizes $\mathfrak{h}$ and that therefore $\overline{\mathfrak{g}}_1 + \overline{\mathfrak{g}}_2$ is a subalgebra. We shall say that a subalgebra $\mathfrak{g}_1$ is compact if the group $\overline{G_1}$ is compact. Then we have:

Proposition. *Let $\mathfrak{g}_1, \mathfrak{g}_2 \subset \mathfrak{g}$ be two subalgebras such that $\mathfrak{g}_1 + \mathfrak{g}_2 = \mathfrak{g}_3$ is a subalgebra. Then if $\mathfrak{g}_1, \mathfrak{g}_2$ are compact so is $\mathfrak{g}_3$ and $\overline{\mathfrak{g}}_3 = \overline{\mathfrak{g}}_1 + \overline{\mathfrak{g}}_2$.*

Proof. Let H be the analytic group that corresponds to $\overline{\mathfrak{g}}_1 + \overline{\mathfrak{g}}_2$ and $\alpha = \overline{G_1} \times \overline{G_2} \to H \subset G$ be the analytic mapping $\alpha(x_1, x_2) = x_1 x_2 \in H$. By dimensional considerations it is clear that α is an open mapping. Clearly, the image of α is compact. Therefore it follows that $\alpha(\overline{G_1} \times \overline{G_2})$ is compact and is the whole of H. This proves that $\overline{\mathfrak{g}}_1 + \overline{\mathfrak{g}}_2$ is a compact algebra and is therefore equal to $\overline{\mathfrak{g}}_3$.

It is clear that every compact subalgebra $\mathfrak{g}_1$ is contained in some maximal compact subalgebra. There could exist in general several maximal compact subalgebras in $\mathfrak{g}$. This maximal compact subalgebra is unique when $\mathfrak{g}$ is nilpotent. Indeed it is well known (cf. 3.6 of [9]) that every connected nilpotent Lie group N contains a uniquely defined (fully invariant) central maximal torus T and that N/T contains no nontrivial compact subgroup.

More generally, let G be a general Lie group and let $\mathfrak{g}$ be its Lie algebra. Let $\mathfrak{n} \subset \mathfrak{g}$ be some nilpotent subalgebra. Then there exists a unique maximal compact subalgebra $\mathcal{I} \subset \mathfrak{n}$ that contains all the compact subalgebras of $\mathfrak{g}$ that lie in $\mathfrak{n}$. Indeed $\overline{\mathfrak{n}} \subset \mathfrak{g}$ is also nilpotent and $\mathcal{I}$ is the intersection with $\mathfrak{n}$ of the maximal compact subalgebra of $\overline{\mathfrak{n}}$ which is defined with respect to the Lie group $\overline{N} \subset G$ (i.e. the closure in G of the analytic subgroup N generated by $\mathfrak{n}$). It is also evident by the above that $\mathcal{I}$ is an ideal in $\mathfrak{n}$ (or even in $\overline{\mathfrak{n}}$). Furthermore if $\mathfrak{z}(\overline{\mathfrak{n}})$ is the center of $\overline{\mathfrak{n}}$ it follows that:

$$\mathcal{I} \subset \mathfrak{z}(\overline{\mathfrak{n}}) \cap \mathfrak{n} \subset \mathfrak{z}(\mathfrak{n}).$$

References.

[1] M. Gromov: Asymptotic Invariants of infinite Groups. Prepublication I.H.E.S., Bures s. Yvette, France.

[2] N. Th.Varopoulos, L.Saloff-Coste, T. Coulhon: Analysis and Geometry on Groups, Cambridge Tracts on Mathematics No. 102.

[3] D. Segal: Polycyclic groups, Cambridge Tracts in Mathematics No. 82.

[4] R.J. Zimmer: Ergodic Theory and Semisimple Groups, Birkhäuser, 1984.

[5] Y. Guivarc'h: Croissance polynomiale et periodes des fonctions harmoniques, Bull. Soc. Math. France 101 (1973) pp. 333-379.

[6] N. Th. Varopoulos: Diffusion on Lie groups II, Can. J. Math. Vol 46 (5) (1994) pp. 1073-1093.

[7] N. Th. Varopoulos: Analysis on Lie groups (to appear)

[8] N. Th. Varopoulos & S. Mustapha: Lecture notes, Univ. Paris VI.

[9] V.S. Varadarajan: Lie groups, Lie algebras and their Representations, Prentice-Hall, 1984.

[10] J. Tits: Free subgroups in Linear groups. J. Algebra 20 250-270 (1972).

[11] G. Hochschild: The Structure of Lie Groups, Holden-Day.

[12] M.S. Raghunathan: Discrete Subgroups of Lie Groups, Ergebrisse No. 68, Springer-Verlag.

[13] N. Th. Varopoulos: Hardy-Littlewood theory on unimodular groups. Ann. Inst. Henri Poincare 31 (4) (1995) pp. 669-688.

[14] J.E. Humphreys: Linear Algebraic Groups, Graduate Texts in Math., Springer Verlag.

[15] L. Hörmander: Hypoelliptic second order differential equations, Acta Math. 119 (1967) 147-171.

[16] H. Whitney: Elementary structure of real algebraic varieties. Annals of Math. 66 (3) (1957) pp.545-556.

[17] C. Chevalley: Theorie des groupes de Lie, Tomes II & III, Hermann 1955.

[18] S. Helgason: Differential Geometry, Lie Groups and Symmetric Spaces, Acad. Press Vol. 80, 2nd ed., 1978.

[19] L. Auslander L.W. Green: G-induced Flows, Amer. J. Math. 88 (1966) pp.43-60.

[20] A.L. Onischik, E.B. Vinberg: Lie groups and Algebraic groups, Springer Verlag, 1988.

[21] B.A.F. Wehrfritz: Infinite Linear Groups. Ergebrisse No 76, Springer-Verlag.

[22] G. Folland, E. Stein: Hardy spaces on Homogeneous Groups, Princeton University Press 1982.

[23] M. Gromov: Groups of polynomial growth and expanding maps, Publ. Math, I.H.E.S. 53 (1981) pp. 53-78.

[24] V. Losert: On the structure of groups with polynomial growth, Math. Zeitschr. 195 109-117 (4).

[25] W. Rudin: Fourier Analysis on Groups, Wiley.

[26] L. Pontrjagin: Topological Groups, Princeton University Press.

[27] W. Magnus, A. Karrass, D. Solitar: Combinational Group Theory, Interscience, 1966.

[28] N. Bourbaki: Groupes et algebres de Lie, Chapitre 2 & 3.

N. Th. Varopoulos
UFR Mathématiques
Université Paris VI
4 Place Jussieu, 75025 Paris, France

List of Participants

(The affiliations are those at the time of the meeting, some have changed.)

Georges Alexopoulos (Université de Paris-Sud, France)
Laura Atanasi (Università degli Studi di Roma, Italy)
Alano Ancona (Université de Paris-Sud, France)
Martine Babillot (Université Paris 6, France)
Paolo Baldi (Università di Roma - Tor Vergata, Italy)
Martin T. Barlow (University of Britisch Columbia, Vancouver, Canada)
Richard F. Bass (University of Washington, Seattle, USA)
Itai Benjamini (The Weizmann Institute of Science, Rehovot, Israel)
Daniela Bertacchi (Università di Milano, Italy)
Walter Betori (Università di Perugia, Italy)
Philippe Bougerol (Université Paris 6, France)
Marek Bozejko (Uniwersytet Wroclawski, Wroclaw, Poland)
Sara Brofferio (Université Paris 6, France)
Robert Brooks (Technion, Haifa, Israel)
Donald I. Cartwright (University of Sydney, Australia)
Enrico Casadio Tarabusi (Università degli Studi di Roma, Italy)
Tullio Ceccherini-Silberstein (Università degli Studi dell'Aquila, Italy)
Yves Colin de Verdiere (Institut Fourier, Saint Martin d'Heres, France)
Thierry Coulhon (Université de Cergy-Pontoise, France)
Thierry Delmotte (Université de Cergy-Pontoise, France)
C. De Santis (Università di Roma - Tor Vergata, Italy)
Persi Diaconis (Cornell University, Ithaca, NY, USA)
Fausto Di Biase (Università di Roma - Tor Vergata, Italy)
Jozef Dodziuk (CUNY Graduate School, New York, USA)
Tomasz Dubejko (Strombek-Bever, Belgium)
Luigi Fontana (Università di Milano, Italy)
Hillel Furstenberg (The Hebrew University, Jerusalem, Israel)
Alberto Gandolfi (Università di Roma - Tor Vergata, Italy)
Peter Grabner (Technische Universität Graz, Austria)
Rostislav I. Grigorchuk (Steklov Mathematical Institute, Moscow, Russia)
Geoffrey R. Grimmett (University of Cambridge, England)
Olle Häggström (Chalmers University of Technology, Göteborg, Sweden)
W. Hansen (Universitt Bielefeld, Germany)
Pierre de la Harpe (Université de Genève, Switzerland)
Tony Hinrichs (Washington University, St. Louis, Missouri, USA)

Tony Hinrichs (Washington University, St. Louis, Missouri, USA)
Alessandra Iozzi (University of Maryland, College Park, Maryland, USA)
Tommaso Isola (Università di Roma - Tor Vergata, Italy)
Marco Isopi (Università degli Studi di Roma, Italy)
Vadim A. Kaimanovich (Université de Rennes 1, France)
Harry Kesten (Cornell University, Ithaca, NY, USA)
Yuri Kifer (The Hebrew University, Jerusalem, Israel)
Adam Koranyi (Lehman College, CUNY, New York, USA)
Gabriella Kuhn (Università di Milano, Italy)
K. F. Lai (University of Sydney, Australia)
Gregory Lawler (Duke University, Durham, NC, USA)
Torbjörn Lundh (University of Cambridge, England)
Russell Lyons (Indiana University, Bloomington, USA)
M. Marias (Aristotle University, Thessaloniki, Greece)
Howard Masur (University of Illinois, Chicago, USA)
Frederic Mathéus (Ecole Normale Superieure de Lyon, France)
Camillo Melzi (Université Paris 6, France)
Wojciech Mlotkowski (Uniwersytet Wroclawski, Wroclaw, Poland)
Stefano Mortola (Politecnico di Milano, Italy)
Martin Moskowitz (CUNY Graduate School, New York, USA)
Frédéric Mouton (Institut Fourier, Saint Martin d'Heres, France)
Tatjana Nagnibeda-Smirnova (ETH Zürich, Switzerland)
Amos Nevo (Technion, Haifa, Israel)
Sam Northshield (SUNY College at Plattsburgh, New York, USA)
Stefano Ortobelli (Università di Bergamo, Italy)
Marc Peigné (Universit de Rennes 1, France)
Yuval Peres (The Hebrew University, Jerusalem, Israel)
Massimo A. Picardello (Università di Roma - Tor Vergata, Italy)
Giovanna Regoli (Università di Perugia, Italy)
Sandrine Roussel (Université Paul Sabatier, Toulouse, France)
Emmanuel Russ (Université de Cergy-Pontoise, France)
Laurent Saloff-Coste (Université Paul Sabatier, Toulouse, France)
Maura Salvatori (Università di Milano, Italy)
Oded Schramm (The Weizmann Institute of Science, Rehovot, Israel)
Paolo M. Soardi (Università di Milano, Italy)
Ken Stephenson (University of Tennessee, Knoxville, USA)
Olof Svensson (Lulea, Sweden)
Christiane Takacs (Universität Linz, Austria)
John Taylor (McGill University, Montreal, Canada)

Fabio Tolli (Université Paris 6, France)
Alain Valette (Université de Neuchatel, Switzerland)
Nicholas Th. Varopoulos (Université Paris 6, France)
George Willis (The University of Newcastle, Australia)
Wolfgang Woess (Università di Milano, Italy)
Armen H. Zemanian (SUNY at Stony Brook, New York, USA)
Fabio Zucca (Università di Milano, Italy)
Andrzej Zuk (Ecole Normale Superieure de Lyon, France)